S*pringer* **M***onographs in* **M***athematics*

For further volumes:
http://www.springer.com/series/3733

Marius Ghergu
Vicenţiu D. Rădulescu

Nonlinear PDEs

Mathematical Models in Biology, Chemistry and Population Genetics

Springer

Marius Ghergu
University College Dublin
School of Mathematical Sciences
Belfield
Dublin 4
Ireland
marius.ghergu@ucd.ie

Vicenţiu D. Rădulescu
Romanian Academy
Simion Stoilow Mathematics
Institute
PO Box 1-764
Bucharest
Romania
vicentiu.radulescu@imar.ro

ISSN 1439-7382
ISBN 978-3-642-22663-2 e-ISBN 978-3-642-22664-9
DOI 10.1007/978-3-642-22664-9
Springer Heidelberg Dordrecht London New York

Library of Congress Control Number: 2011939998

Mathematics Subject Classification (2010): 35-02; 49-02; 92-02; 58-02; 37-02; 35Qxx

Printed on acid-free paper

Springer is part of Springer Science+Business Media (www.springer.com)

Marius Ghergu dedicates this volume to his family who have always been there in hard times.

Vicenţiu Rădulescu dedicates this book to the memory of his beloved Mother,
Ana Rădulescu (1923–2011)

Foreword

Partial differential equations and, in particular, linear elliptic equations were created and introduced in science in the first decades of the nineteenth century in order to study gravitational and electric fields and to model diffusion processes in Physics. The heat equation, the Navier–Stokes system, the wave equation and the Schrödinger equations introduced later on to describe the dynamic of heat conduction, Newtonian fluid flows and, respectively, quantum mechanics are the basic equations of mathematical physics which are, in spite of their complexity, centered around the notion of Laplacian or, in other words, linear diffusion. However, these equations, which were primarily created to model physical processes, played an important role in almost all branches of mathematics and, as a matter of fact, can be viewed as a chapter of applied mathematics as well as of so-called pure mathematics. In fact, the linear elliptic operators and, in particular, the Laplacian represent without any doubt a bridge that connects a large number of mathematical fields and concepts and provides the mathematical framework for physical theories as well as for the theory of stochastic processes and some new mathematical technologies for image restoring and processing. The well posedness of the basic boundary value problems associated with the Laplace operator is a fundamental topic of the theory of partial differential equations. It is instructive to recall that the well posedness of the Dirichlet and Neumann problem remained open and unsolved for more than half a century until the turn of the nineteenth century, when Ivar Fredholm solved it by a new and influential idea which is at the origin of a several branches of mathematics which will change the analysis of the twentieth century; primarily, I have in mind here functional analysis and operator theory. A related problem, the

Dirichlet variational principle, had a similar history, being rigorously proved only in the fourth decade of the last century after the creation of Sobolev spaces. This principle is at the origin of variational theory of elliptic problems and of the concept of weak or distributional solution, which fundamentally changed the basics ideas and techniques of PDEs in the second part of the last century. The mathematicians of the nineteenth century failed to prove this principle because it is not well posed in spaces of differentiable functions, but in functional spaces with energetic norms that is in Sobolev spaces which were discovered later on. Nonlinear elliptic boundary value problems arise naturally in the description of physical phenomena and, in particular, of reaction-diffusion processes, governed by nonlinear diffusion laws, or in geometry (the minimal surface equation or uniformization theorem in Riemannian geometry). The well posedness of most of these nonlinear problems was treated by the new functional methods introduced in the last century such as the Banach principle, Schauder fixed point theorem and Schauder–Leray degree theorem and, in the 1960s, by the Minty–Browder theory of nonlinear maximal monotone operators in Banach spaces. It should be said that most of these functional approaches to nonlinear elliptic problems lead to existence results in spaces with energetic norms (Sobolev spaces) and so quite often these are inefficient or too rough to put in evidence sharp qualitative properties of solutions such as asymptotic behavior, monotonicity or comparison results. Some classical methods such as the maximum principles, integral representation of solutions or complex analysis techniques are very efficient to obtain sharp results for new classes of elliptic problems of special nature. These techniques, which perhaps have their origins in the classical work of Peano on existence and construction of solutions to the Dirichlet problem by method of sub and supersolutions, are still largely used in the modern theory of nonlinear elliptic equations. This book is a very nice illustration of these techniques in the treatment of the existence of positive solutions, which are unbounded to frontier or for singular solutions to logistic elliptic equations as well as for the minimality principle for semilinear elliptic equations. Most of the elliptic equations studied in this book are of singular nature or develop some "pathological" behavior which requires sharp and specific investigation tools different to the standard functional or energetic methods mentioned above. In the same category are the corresponding variational problems which, in the absence of convexity, need some sophisticated instruments such as the Mountains Pass theorem, the Ekeland variational principle or the Brezis–Lieb lemma. A fact indeed remarkable in this book is the variety of problems studied and of methods and arguments. The authors avoid formulations, tedious arguments and maximum generality, which is a general temptation of mathematicians in favor of simplicity; they confine to specific but important problems most of them famous in literature, and try to extract from their treatment the essential ideas and features of the approach. The examples from chemistry and biology chosen to illustrate the theory are carefully selected and significant (the Brusselator, reaction-diffusion systems, pattern formation).

Marius Ghergu and Vicenţiu D. Rădulescu, who are well-known specialists in the field, have coauthored in this work a remarkable monograph on recent results on nonlinear techniques in the theory of elliptic equations, largely based on their research works. The book is of a high scientific standard, but readable and accessible to a large category of people interested in the modern theory of partial differential equations.

Romanian Academy *Viorel Barbu*

A Short Overview of the Book

> Among all mathematical disciplines the theory of differential equations is the most important. It furnishes the explanation of all those elementary manifestations of nature which involve time.
>
> Sophus Lie (1842–1899)

Much of the modern science is based on the application of mathematics. It is central to modern society, underpins scientific and industrial research, and is key to our economy. Mathematics is the engine of science and engineering. It also has an elegance and beauty that fascinates and inspires those who understand it.

Mathematics provides the theoretical framework for biosciences, for statistics and data analysis, as well as for computer science. New discoveries within mathematics affect not only science, but also our general understanding of the world we live in. Problems in biological sciences, in physics, chemistry, engineering, and in computational science are using increasingly sophisticated mathematical techniques. For this strong reason, the bridge between the mathematical sciences and other disciplines is heavily traveled.

Biosciences are some of the most fascinating of all scientific disciplines and is an area of applied sciences we use to explore and try to explain the uncertain world in which we live. It is no surprise, then, that at the heart of a professional in this field is a fascination with, and a desire to understand, the "how and why" of the material world around us.

The purpose of this volume is to meet the current and future needs of the interaction between mathematics and various biosciences. This is first done by encouraging the ways that mathematics may be applied in traditional areas such as biology, chemistry, or genetics, as well as pointing towards new and innovative areas of applications. Next, we intend to encourage other scientific disciplines (mainly oriented to natural sciences) to engage in a dialog with mathematicians, outlining their problems to both access new methods and suggest innovative developments within mathematics itself.

The first chapter presents the main mathematical methods used in the book. Such tools include iterative methods and maximum principle, variational methods and critical point theory as well as topological methods and degree theory.

The second chapter deals with Liouville type results for elliptic operators in divergence form. Since its appearance in the nineteenth century, many results in the theory of Partial Differential Equations have been devoted to characterize all data functions f such that the standard elliptic inequality $\mathscr{L}u \geq f(x,u)$ admits only the trivial solution. We discuss such type of problems for elliptic operators of the form $\mathscr{L}u = -\mathrm{div}[A(|\nabla u|)\nabla u]$.

Chapter 3 is concerned with the study of solutions to the equation $\Delta u = \rho(x)f(u)$ in a smooth domain that blow-up at the boundary in the sense that $\lim_{x\to x_0} u(x) = +\infty$, for all $x_0 \in \partial\Omega$; in case $\Omega = \mathbb{R}^N$, this condition can be simply formulated as $\lim_{|x|\to\infty} u(x) = +\infty$. Here we emphasize the role played by the Keller–Osserman integral condition.

Chapters 4 and 5 deal with some related singular elliptic problems. This time, the solution is bounded but the nonlinearity appearing in the problem is unbounded around the boundary of the domain. Particular attention is paid to the Lane–Emden equation and the associated system in this singular framework. Chapter 4 is devoted to the model equation $-\Delta u = au + u^{-\alpha}$, $0 < \alpha < 1$ and the associated system. In Chap. 5 we study singular elliptic problems in exterior domains. Here we point out the role played by the geometry of the domain in the existence of a C^2 solution. In particular we completely describe the solution set of the equation $-\Delta u = |x|^{\alpha} u^{-p}$ by showing that all the solutions are radially symmetric and characterized by two parameters.

Chapter 6 presents two classes of quasilinear elliptic equations. The approach in this chapter is variational and combines some tools in this field such as Ekeland's variational principle and mountain pass theorem. The lack of compactness of

Sobolev embeddings or the presence of p-Laplace operator are the main features of the chapter.

In Chap. 7 we are concerned with three classes of higher order elliptic problems involving the polyharmonic operator. By adopting three different approaches we underline the complex structure of such problems in which the higher order differential operator and the type of conditions imposed on the boundary play an important role in the qualitative study of solutions.

The last two chapters are devoted to reaction diffusion systems. In their most general form, the models we intend to study can be stated as

$$\begin{cases} u_t = d_u \Delta u + f(u,v) & (x \in \Omega, t > 0), \\ v_t = d_v \Delta u + g(u,v) & (x \in \Omega, t > 0). \end{cases} \tag{0.1}$$

These equations describe the evolution of the concentrations, $u = u(x,t)$, $v = v(x,t)$ at spatial position x and time t, of two chemicals due to *diffusion*, with different constant diffusion coefficients d_u, d_v, respectively, and *reaction*, modeled by the typically nonlinear functions f and g that can be derived from chemical reaction formulas by using the law of mass action and other physical conditions.

In Chap. 9 several reaction-diffusion models are studied. Oscillating chemical reactions have been a rich source of varied spatial-temporal patterns since the discovery of the oscillating wave in the Belousov–Zhabotinsky reaction in 1950s. These phenomena and observations have been transferred to challenging mathematical problems through various models, especially reaction-diffusion equations. Among these mathematical models, we present:

- The Brusselator model introduced by Prigogine and Lefever in 1968 as a model for an autocatalytic oscillating chemical reaction. This corresponds to

$$f(u,v) = a - (b+1)u + u^2 v, \quad g(u,v) = bu - u^2 v.$$

- The Schnackenberg model for chemical reactions with limit cycle behavior

$$f(u,v) = a - u + u^2 v, \quad g(u,v) = b - u^2 v.$$

- The Lengyel–Epstein model for the chlorite–iodide–malonic acid (CIMA) reaction. This corresponds to (0.1) with the nonlinearities f and g given by

$$f(u,v) = a - u - 4uv/(1+u^2), \quad g(u,v) = b\left[u - \frac{uv}{1+u^2}\right].$$

In the last chapter we discuss a reaction-diffusion model arising in molecular biology proposed by Gierer and Meinhardt [98] in 1972 for pattern formation of spatial tissue structures of *hydra* in morphogenesis, a biological phenomenon discovered by Trembley in 1744. Following this model, the nonlinearities f and g are given by

$$f(u,v) = -\alpha u + \frac{u^p}{v^q} + \rho(x), \quad g(u,v) = -\beta v + \frac{u^r}{v^s},$$

where $\alpha, \beta > 0$, ρ is the source distribution and the exponents p,q,r,s are positive real numbers.

For the reader's convenience, we have included two appendices that contain some technical results about Caffarelli–Kohn–Nirenberg inequality and estimates for the Green function associated with the biharmonic operator.

The few examples we have provided illustrate the great alliance between mathematics and biosciences. This is recognized universally and both disciplines thrived by supporting each other. The prerequisite for this book includes a good undergraduate course in functional analysis and Partial Differential Equations. This book is intended for advanced graduate students and researchers in both pure and applied mathematics.

Our vision throughout this volume is closely inspired by the following words of V.I. Arnold (1983, see [8, p. 87]) on the role of mathematics in the understanding of real processes: *In every mathematical investigation the question will arise whether we can apply our results to the real world. Consequently, the question arises of choosing those properties which are not very sensitive to small changes in the model and thus may be viewed as properties of the real process.*

Ireland — *Marius Ghergu*
Romania — *Vicenţiu D. Rădulescu*

Contents

Chapter 1
Overview of Mathematical Methods in Partial Differential Equations

> Mathematics may be defined as the subject in which we never know what we are talking about, nor whether what we are saying is true.
>
> Bertrand Russell (1872–1970)

In this chapter we collect some results in Nonlinear Analysis that will be frequently used in the book. The first part of this chapter deals with comparison principles for second order differential operators and enables us to obtain an ordered structure of the solution set and, in most of the cases, the uniqueness of the solution. In the second part of this chapter we review the celebrated method of moving planes that allows us to deduce the radial symmetry of the solution. The third part of this chapter is concerned with variational methods. The final section contains some results in degree theory that will be mostly used to derive existence and nonexistence of a stationary solution to some reaction-diffusion systems.

1.1 Comparison Principles

We start this section with the following result which is due to Lou and Ni (see [139] or [140]).

Theorem 1.1 *Let* $g \in C^1(\overline{\Omega} \times \mathbb{R})$.

(i) If $w \in C^2(\Omega) \cap C^1(\overline{\Omega})$ *satisfies*

$$\Delta w + g(x,w) \geq 0 \textit{ in } \Omega, \qquad \frac{\partial w}{\partial n} \leq 0 \textit{ on } \partial\Omega, \tag{1.1}$$

M. Ghergu and V. Rădulescu, *Nonlinear PDEs*, Springer Monographs in Mathematics,
DOI 10.1007/978-3-642-22664-9_1, © Springer-Verlag Berlin Heidelberg 2012

and $w(x_0) = \max_{\overline{\Omega}} w$*, then* $g(x_0, w(x_0)) \geq 0$.

(ii) If $w \in C^2(\Omega) \cap C^1(\overline{\Omega})$ *satisfies*

$$\Delta w + g(x,w) \leq 0 \ \text{in}\ \Omega, \qquad \frac{\partial w}{\partial n} \geq 0 \ \text{on}\ \partial\Omega,$$

and $w(x_0) = \min_{\overline{\Omega}} w$*, then* $g(x_0, w(x_0)) \leq 0$.

Proof. We shall prove only part (i) as (ii) can be established in a similar way. There are two possibilities for our consideration.

Case 1: $x_0 \in \Omega$. Since $w(x_0) = \max_{\overline{\Omega}} w$ we have $\Delta w(x_0) \leq 0$ and now from the first inequality in (1.1) we obtain $g(x_0, w(x_0)) \leq 0$.

Case 2: $x_0 \in \partial\Omega$. Assume by contradiction that $g(x_0, w(x_0)) < 0$. By the continuity of g and w, there exists a ball $B \subset \overline{\Omega}$ with $\partial B \cap \partial\Omega = \{x_0\}$ such that

$$g(x, w(x)) < 0 \qquad \text{for all } x \in B.$$

Thus, from (1.1) we find $\Delta w > 0$ in B. Since $w(x_0) = \max_{\overline{B}} w$, it follows from the Hopf boundary lemma that $\partial w/\partial n(x_0) > 0$ which contradicts the boundary condition in (1.1). This completes the proof of Theorem 1.1. □

Basic to our purposes in this book we state and prove the following result which is suitable for singular nonlinearities.

Theorem 1.2 *Let* $\Psi : \Omega \times (0,\infty) \to \mathbb{R}$ *be a Hölder continuous function such that the mapping* $(0,\infty) \ni t \longmapsto \Psi(x,t)/t$ *is decreasing for each* $x \in \Omega$*. Assume that there exist* v_1, $v_2 \in C^2(\Omega) \cap C(\overline{\Omega})$ *such that*

(a) $\Delta v_1 + \Psi(x, v_1) \leq 0 \leq \Delta v_2 + \Psi(x, v_2)$ *in* Ω;

(b) $v_1, v_2 > 0$ *in* Ω *and* $v_1 \geq v_2$ *on* $\partial\Omega$;

(c) $\Delta v_1 \in L^1(\Omega)$ *or* $\Delta v_2 \in L^1(\Omega)$.

Then $v_1 \geq v_2$ *in* Ω.

Proof. Suppose by contradiction that $v \leq w$ is not true in Ω. Then, we can find ε_0, $\delta_0 > 0$ and a ball $B \subset\subset \Omega$ such that

$$v - w \geq \varepsilon_0 \quad \text{in } B, \tag{1.2}$$

$$\int_B vw \left(\frac{\Phi(x,w)}{w} - \frac{\Phi(x,v)}{v} \right) dx \geq \delta_0. \tag{1.3}$$

Let us assume that $\Delta w \in L^1(\Omega)$ and set

$$M = \max\{1, \|\Delta w\|_{L^1(\Omega)}\}, \; \varepsilon = \min\left\{1, \varepsilon_0, \frac{\delta_0}{4M}\right\}.$$

Consider $\theta \in C^1(\mathbb{R})$ a nondecreasing function such that $0 \le \theta \le 1$, $\theta(t) = 0$, if $t \le 1/2$ and $\theta(t) = 1$ for all $t \ge 1$. Define

$$\theta_\varepsilon(t) = \theta\left(\frac{t}{\varepsilon}\right), \quad t \in \mathbb{R}.$$

Because $w \ge v$ on $\partial\Omega$, we can find a smooth subdomain $\Omega^* \subset\subset \Omega$ such that

$$B \subset \Omega^* \quad \text{and } v - w < \frac{\varepsilon}{2} \text{ in } \Omega \setminus \Omega^*.$$

Using hypotheses (i) and (ii) we deduce

$$\int_{\Omega^*} (w\Delta v - v\Delta w)\theta_\varepsilon(v-w)dx \ge \int_{\Omega^*} vw\left(\frac{\Phi(x,w)}{w} - \frac{\Phi(x,v)}{v}\right)\theta_\varepsilon(v-w)dx. \quad (1.4)$$

By relation (1.3), we have

$$\begin{aligned}
&\int_{\Omega^*} vw\left(\frac{\Phi(x,w)}{w} - \frac{\Phi(x,v)}{v}\right)\theta_\varepsilon(v-w)dx \\
&\qquad \ge \int_B vw\left(\frac{\Phi(x,w)}{w} - \frac{\Phi(x,v)}{v}\right)\theta_\varepsilon(v-w)dx \\
&\qquad = \int_B vw\left(\frac{\Phi(x,w)}{w} - \frac{\Phi(x,v)}{v}\right)dx \\
&\qquad \ge \delta_0.
\end{aligned}$$

To raise a contradiction, we need only to prove that the left-hand side in (1.4) is smaller than δ_0. For this purpose, define

$$\Theta_\varepsilon(t) := \int_0^t s\theta_\varepsilon'(s)ds, \quad t \in \mathbb{R}.$$

It is easy to see that

$$\Theta_\varepsilon(t) = 0, \text{ if } t < \frac{\varepsilon}{2} \quad \text{and } 0 \le \Theta_\varepsilon(t) \le 2\varepsilon, \text{ for all } t \in \mathbb{R}. \quad (1.5)$$

Now, using Green's first formula, we evaluate the left side of (1.4):

$$
\begin{aligned}
&\int_{\Omega^*}(w\Delta v - v\Delta w)\theta_\varepsilon(v-w)dx\\
&\quad= \int_{\partial\Omega^*} w\theta_\varepsilon(v-w)\frac{\partial v}{\partial n}d\sigma(x) - \int_{\Omega^*}(\nabla w\cdot\nabla v)\theta_\varepsilon(v-w)dx\\
&\qquad - \int_{\Omega^*} w\theta_\varepsilon'(v-w)\nabla v\cdot\nabla(v-w)dx - \int_{\partial\Omega^*} v\theta_\varepsilon(v-w)\frac{\partial w}{\partial n}d\sigma(x)\\
&\qquad + \int_{\Omega^*}(\nabla w\cdot\nabla v)\theta_\varepsilon(v-w)dx + \int_{\Omega^*} v\theta_\varepsilon'(v-w)\nabla w\cdot\nabla(v-w)dx\\
&\quad= \int_{\Omega^*}\theta_\varepsilon'(v-w)(v\nabla w - w\nabla v)\cdot\nabla(v-w)dx.
\end{aligned}
$$

The previous relation can be rewritten as

$$
\begin{aligned}
\int_{\Omega^*}(w\Delta v - v\Delta w)\theta_\varepsilon(v-w)dx = &\int_{\Omega^*} w\vartheta_\varepsilon'(v-w)\nabla(w-v)\cdot\nabla(v-w)dx\\
&+\int_{\Omega^*}(v-w)\theta_\varepsilon'(v-w)\nabla w\cdot\nabla(v-w)dx.
\end{aligned}
$$

Because $\int_{\Omega^*} w\theta_\varepsilon'(v-w)\nabla(w-v)\cdot\nabla(v-w)dx \le 0$, the last equality yields

$$
\int_{\Omega^*}(w\Delta v - v\Delta w)\theta_\varepsilon(v-w)dx \le \int_{\Omega^*}(v-w)\theta_\varepsilon'(v-w)\nabla w\cdot\nabla(v-w)dx.
$$

Therefore,

$$
\int_{\Omega^*}(w\Delta v - v\Delta w)\theta_\varepsilon(v-w)dx \le \int_{\Omega^*}\nabla w\cdot\nabla(\Theta_\varepsilon(v-w))dx.
$$

Again by Green's first formula, and by (1.5), we have

$$
\begin{aligned}
\int_{\Omega^*}(w\Delta v - v\Delta w)\theta_\varepsilon(v-w)dx &\le \int_{\partial\Omega^*}\Theta_\varepsilon(v-w)\frac{\partial w}{\partial n}d\sigma(x)\\
&\quad - \int_{\Omega^*}\Theta_\varepsilon(v-w)\Delta w dx\\
&\le -\int_{\Omega^*}\Theta_\varepsilon(v-w)\Delta w dx \le 2\varepsilon\int_{\Omega^*}|\Delta w|dx\\
&\le 2\varepsilon M < \frac{\delta_0}{2}.
\end{aligned}
$$

Thus, we have obtained a contradiction. Hence $v \le w$ in Ω, which completes the proof. □

A direct consequence of Theorem 1.2 is the result below.

Corollary 1.3 *Let $k \in C(0,\infty)$ be a positive decreasing function and $a_1, a_2 \in C(\Omega)$ with $0 < a_2 \le a_1$ in Ω. Assume that there exist $\beta \ge 0$, $v_1, v_2 \in C^2(\Omega)\cap C(\overline{\Omega})$ such* *that* $v_1, v_2 > 0$ *in* Ω, $v_1 \geq v_2$ *on* $\partial\Omega$ *and*

$$\Delta v_1 - \beta v_1 + a_1(x)k(v_1) \leq 0 \leq \Delta v_2 - \beta v_2 + a_2(x)k(v_2) \quad \text{in } \Omega.$$

Then $v_1 \geq v_2$ *in* Ω.

Proof. We simply apply Theorem 1.2 in the particular case

$$\Phi(x,t) = -\beta t + a_1(x)k(t), \quad (x,t) \in \Omega \times (0,\infty).$$

□

Let us now consider the more general elliptic operator in divergence form

$$\mathscr{L}u := \operatorname{div}[A(|\nabla u|)\nabla u],$$

where $A \in C(0,\infty)$ is positive such that the mapping $t \mapsto tA(t)$ is increasing.

Theorem 1.4 *Let* Ω *be a bounded and smooth domain in* $\mathbb{R}^N$ ($N \geq 1$), $\rho \in C(\overline{\Omega})$ *and* $f \in C(\mathbb{R})$. *Assume that* $u, v \in C^2(\Omega) \cap C(\overline{\Omega})$ *satisfy*

(i) $\mathscr{L}u - \rho(x)f(u) \geq 0 \geq \mathscr{L}v - \rho(x)f(v)$ *in* Ω;

(ii) $u \leq v$ *on* $\partial\Omega$.

Then $u \leq v$ *in* Ω.

Proof. Let $\phi : \mathbb{R} \to [0,\infty)$ be a C^1-function such that $\phi = 0$ on $(-\infty, 0]$ and ϕ is strictly increasing on $[0,\infty)$. We first multiply by $\phi(u-v)$ in (i) and obtain

$$(\mathscr{L}u - \mathscr{L}v)\phi(u-v) \geq \rho(x)(f(u) - f(v))\phi(u-v) \quad \text{in } \Omega.$$

Integrating over Ω, by the divergence theorem we find

$$-\int_\Omega \Big[A(|\nabla u|)\nabla u - A(|\nabla v|)\nabla v\Big] \cdot \nabla(u-v)\phi(u-v)dx$$
$$\geq \int_\Omega \rho(x)(f(u) - f(v))\phi(u-v)dx \geq 0.$$

Hence

$$\int_\Omega \Big[A(|\nabla u|)\nabla u - A(|\nabla v|)\nabla v\Big] \cdot \nabla(u-v)\phi(u-v)dx \leq 0. \tag{1.6}$$

On the other hand,

$$\begin{aligned}\Big[A(|\nabla u|)\nabla u - A(|\nabla v|)\nabla v\Big]\cdot\nabla(u-v) \\ &= \Big[A(|\nabla u|)|\nabla u| - A(|\nabla v|)|\nabla v|\Big](|\nabla u| - |\nabla v|) \\ &\quad + \Big[A(|\nabla u|) + A(|\nabla v|)\Big](|\nabla u||\nabla v| - \nabla u\cdot\nabla v),\end{aligned}$$

so that

$$\Big[A(|\nabla u|)\nabla u - A(|\nabla v|)\nabla v\Big]\cdot\nabla(u-v) \geq 0 \quad \text{in } \Omega,$$

with equality if and only if $\nabla u = \nabla v$. Using this fact in (1.6) it follows that $u \leq v$ in Ω. This finishes the proof of our result. □

Theorem 1.5 *Let $\Omega \subset \mathbb{R}^N$ $(N \geq 1)$ be a smooth bounded domain, $T > 0$, and*

$$\mathscr{L}u := \partial_t u - a(x,t,u)\Delta u + f(x,t,u),$$

where $a, f : \overline{\Omega} \times [0,\infty) \times [0,\infty) \to \mathbb{R}$ are continuous functions such that $a \geq 0$ in $\overline{\Omega} \times [0,\infty)$. Assume that there exist $u_1, u_2 \in C^{2,1}(\Omega \times (0,T)) \cap C(\overline{\Omega} \times [0,T])$ such that:

(i) *$\mathscr{L}u_1 \leq \mathscr{L}u_2$ in $\Omega \times (0,T)$.*
(ii) *$u_1 \leq u_2$ on $\Sigma_T := (\partial\Omega \times (0,T)) \cup (\Omega \times \{0\})$.*
(iii) *at least for one $i \in \{1,2\}$ we have $|D^2 u_i| \in L^\infty(\overline{\Omega} \times [0,T])$ and the functions a and f are Lipschitz with respect to the u variable in the neighborhood of $K := u_i(\overline{\Omega} \times [0,T])$.*

Then $u_1 \leq u_2$ in $\Omega \times [0,T]$.

1.2 Radial Symmetry of Solutions to Semilinear Elliptic Equations

An important tool in establishing the radial symmetry of a solution to elliptic PDEs is the so-called *moving plane method* that goes back to A.D. Alexandroff and J. Serrin. It was then refined by Gidas, Ni and Nirenberg in the celebrated paper [97]. The requirements on the regularity of the domain were further simplified by Berestycki and Nirenberg [16]. We follow here the line in [16] and [25] to provide the reader with a simple and instructive proof of the radial symmetry of solutions to semilinear elliptic PDEs in bounded and convex domains Ω that vanish on $\partial\Omega$.

Theorem 1.6 *Let $\Omega \subset \mathbb{R}^N$ be a convex domain which is symmetric about the x_1 axis. Assume that $f : \mathbb{R} \to \mathbb{R}$ is a Lipschitz continuous function and $\rho : [0,\infty) \to \mathbb{R}$ is a decreasing function. If $u \in C^2(\Omega) \cap C(\overline{\Omega})$ satisfies*

$$\begin{cases} -\Delta u = \rho(|x|)f(u), u > 0 & \text{in } \Omega, \\ u = 0 & \text{on } \partial\Omega, \end{cases} \tag{1.7}$$

then u is symmetric with respect to the x_1 axis.

Proof. We first need a version of the maximum principle for small domains as stated below.

Lemma 1.7 *Let $a \in C(\overline{\Omega})$ and $w \in C^2(\omega) \cap C(\overline{\omega})$ be such that*

$$\begin{cases} -\Delta w + a(x)w \geq 0 & \text{in } \omega, \\ w \geq 0 & \text{on } \partial\omega. \end{cases} \tag{1.8}$$

If

$$||a^-||_{L^{N/2}(\omega)} \geq S_N, \tag{1.9}$$

where S_N is the best Sobolev constant in ω, then $w \geq 0$ in ω. In particular, if

$$||a^-||_{L^\infty(\omega)}|\omega|^{N/2} \leq S_N,$$

that is, if ω is small, then $w \geq 0$ in ω.

Proof. We multiply the first inequality in (1.8) by $w^- = \max\{-w, 0\}$. Integrating over ω we obtain

$$\int_\omega |\nabla w^-|^2 dx + \int_\omega a(x)|w^-|^2 dx \leq 0.$$

This also yields

$$\int_\omega |\nabla w^-|^2 dx \leq \int_\omega a^-(x)|w^-|^2 dx.$$

On the other hand, by Sobolev and Hölder inequalities we find

$$S_N||w^-||_{L^{2N/(N-2)}(\omega)} \leq \int_\omega a^-(x)|w^-|^2 dx \leq ||a^-||_{L^{N/2}(\omega)}||w^-||_{L^{2N/(N-2)}}(\omega).$$

Using (1.9), the above inequality implies $||w^-||_{L^{2N/(N-2)}(\omega)} = 0$, so $w \geq 0$ in ω. □

Let us now come back to the proof of Theorem 1.6. For any $x = (x_1, x_2, \ldots, x_N) \in \mathbb{R}^N$ we write $x = (x_1, x')$, where $x_1 \in \mathbb{R}$ and $x' \in \mathbb{R}^{N-1}$. Let

$$\lambda_0 = \max\{x_1 : (x_1,x') \in \overline{\Omega}\}.$$

We claim that

$$u(x_1,x') < u(y_1,x'), \tag{1.10}$$

for all $(x_1,x') \in \Omega$ with $x_1 > 0$ and all $y_1 \in \mathbb{R}$ with $|y_1| < x_1$.

Then (1.10) implies $u(x_1,x') \le u(x_1,-x')$ and similarly $u(x_1,x') \ge u(x_1,-x')$. so $u(x_1,x') = u(x_1,-x')$, that is, u is symmetric about the x_1 axis. For any $0 < \lambda < \lambda_0$ define

$$\Sigma_\lambda = \{x = (x_1,x') \in \Omega : x_1 > \lambda\}$$

and

$$w_\lambda(x) = u_\lambda(x) - u(x), \quad x \in \Sigma_\lambda,$$

where $u_\lambda(x) = u(2\lambda - x_1,x')$. Note that w_λ is well defined in Σ_λ since Ω is convex and symmetric about the hyperplane $x_1 = 0$. Let us further remark that (1.10) is equivalent to

$$w_\lambda > 0 \quad \text{in } \Sigma_\lambda, \qquad \text{for all } 0 < \lambda < \lambda_0. \tag{1.11}$$

From (1.7) we have

$$-\Delta w_\lambda + \rho(|x|)\frac{f(u) - f(u_\lambda)}{u_\lambda - u} w_\lambda + \rho(|x|) - \rho(|x_\lambda|) w_\lambda = 0 \quad \text{in } \Sigma_\lambda,$$

where $x_\lambda = (2\lambda - x_1,x')$. This yields

$$-\Delta w_\lambda + a(x) w_\lambda \ge 0 \quad \text{in } \Sigma_\lambda,$$

where

$$a(x) = \begin{cases} \rho(|x|)\dfrac{f(u) - f(u_\lambda)}{u_\lambda - u} & \text{if } w_\lambda(x) \neq 0, \\ 0 & \text{if } w_\lambda(x) = 0. \end{cases}$$

Remark that $a \in L^\infty(\Omega)$ and $\|a\|_{L^\infty(\Omega)} \le L\|\rho\|_{L^\infty(\Omega)}$, where L is the Lipschitz constant of f on the interval $[-\|u\|_{L^\infty(\Omega)}, \|u\|_{L^\infty(\Omega)}]$. Furthermore, we have

$$w_\lambda > 0 \quad \text{on } \partial\Sigma_\lambda \cap \partial\Omega, \quad w_\lambda = 0 \quad \text{on } \partial\Sigma \cap \Omega.$$

Thus, by taking λ close to λ_0, by Lemma 1.7 we obtain $w_\lambda \ge 0$ in Σ_λ. Let

$$\mathscr{A} = \{0 < \lambda < \lambda_0 : w_\lambda \ge 0 \text{ in } \Sigma_\lambda\}.$$

It is easily seen that $\mathscr{A}$ is closed. We next prove that $\mathscr{A}$ is open. To this aim, let $\lambda \in \mathscr{A}$. By the strong maximum principle, $w_\lambda > 0$ in Σ. Let $K \subset \Sigma_\lambda$ be a compact set such that $|\Sigma_\mu \setminus K|$ is small, for μ in a neighborhood of λ. Also, there exists $c > 0$ such that $w_\lambda \geq c > 0$ in K, so by continuity arguments we have $w_\mu > 0$ in K for μ near λ. This yields $w_\mu \geq 0$ on $\partial(\Sigma_\mu \setminus K)$, so by Lemma 1.7 we find $w_\mu \geq 0$ in $\Sigma_\mu \setminus K$, so $w_\mu \geq 0$ in Σ_μ. This proves that $\mathscr{A}$ is open, so $\mathscr{A} = (0, \lambda_0)$. This implies that (1.11) holds, that is, u is symmetric with respect to the x_1 axis. □

1.3 Variational Methods

1.3.1 Ekeland's Variational Principle

Ekeland's variational principle [67] was established in 1974, with its main feature of how to use the norm completeness and a partial ordering to obtain a point where a linear functional achieves its supremum on a closed bounded convex set. In its original form, Ekeland's variational principle can be stated as follows.

Theorem 1.8 (Ekeland's Variational Principle) *Let (M,d) be a complete metric space and assume that $\Phi : M \to (-\infty, \infty]$, $\Phi \not\equiv \infty$, is a lower semicontinuous functional that is bounded from below.*

Then, for every $\varepsilon > 0$ and for any $z_0 \in M$, there exists $z \in M$ such that

(i) $\Phi(z) \leq \Phi(z_0) - \varepsilon d(z, z_0)$;
(ii) $\Phi(x) \geq \Phi(z) - \varepsilon d(x, z)$, *for any $x \in M$.*

Proof. We may assume without loss of generality that $\varepsilon = 1$. Define the following binary relation on M:

$$y \leq x \qquad \text{if and only if} \qquad \Phi(y) - \Phi(x) + d(x,y) \leq 0.$$

Then "$\leq$" is a partial order relation—that is,
(a) $x \leq x$, for any $x \in M$;
(b) if $x \leq y$ and $y \leq x$ then $x = y$;
(c) if $x \leq y$ and $y \leq z$ then $x \leq z$.

For arbitrary $x \in M$, set

$$S(x) := \{y \in M : y \leq x\}.$$

Let $\{\varepsilon_n\}$ be a sequence of positive numbers such that $\varepsilon_n \to 0$ and fix $z_0 \in M$. For any $n \geq 0$, let $z_{n+1} \in S(z_n)$ be such that

$$\Phi(z_{n+1}) \leq \inf_{S(z_n)} \Phi + \varepsilon_{n+1}.$$

The existence of z_{n+1} follows from the definition of $S(x)$. We prove that the sequence $\{z_n\}$ converges to some element z, which satisfies (i) and (ii).

Let us first remark that $S(y) \subset S(x)$, provided that $y \leq x$. Hence, $S(z_{n+1}) \subset S(z_n)$. It follows that for any $n \geq 0$,

$$\Phi(z_{n+1}) - \Phi(z_n) + d(z_n, z_{n+1}) \leq 0,$$

which implies $\Phi(z_{n+1}) \leq \Phi(z_n)$. Because Φ is bounded from below, we deduce that the sequence $\{\Phi(z_n)\}$ converges.

We prove in what follows that $\{z_n\}$ is a Cauchy sequence. Indeed, for any n and p we have

$$\Phi(z_{n+p}) - \Phi(z_n) + d(z_{n+p}, z_n) \leq 0. \tag{1.12}$$

Therefore,

$$d(z_{n+p}, z_n) \leq \Phi(z_n) - \Phi(z_{n+p}) \to 0 \quad \text{as } n \to \infty,$$

which shows that $\{z_n\}$ is a Cauchy sequence, so it converges to some $z \in M$. Now, taking $n = 0$ in (1.12), we find

$$\Phi(z_p) - \Phi(z_0) + d(z_p, z_0) \leq 0.$$

So, as $p \to \infty$, we find (i).

To prove (ii), let us choose arbitrarily $x \in M$. We distinguish the following situations.

Case 1: $x \in S(z_n)$, for any $n \geq 0$. It follows that $\Phi(z_{n+1}) \leq \Phi(x) + \varepsilon_{n+1}$, which implies that $\Phi(z) \leq \Phi(x)$.

Case 2: There exists an integer $N \geq 1$ such that $x \notin S(z_n)$, for any $n \geq N$ or, equivalently,

$$\Phi(x) - \Phi(z_n) + d(x, z_n) > 0 \quad \text{for every } n \geq N.$$

Passing to the limit in this inequality as $n \to \infty$ we find (ii). □

A major consequence of Ekeland's variational principle is that even if it is not always possible to minimize a nonnegative C^1 functional Φ on a Banach space; however, there is always a minimizing sequence $(u_n)_{n\geq 1}$ such that $\Phi'(u_n) \to 0$ as $n \to \infty$. More precisely we have

Corollary 1.9 *Let E be a Banach space and let $\Phi : E \to \mathbb{R}$ be a C^1 functional that is bounded from below. Then, for any $\varepsilon > 0$, there exists $z \in E$ such that*

$$\Phi(z) \leq \inf_E \Phi + \varepsilon \qquad \text{and} \qquad \|\Phi'(z)\|_{E^\star} \leq \varepsilon.$$

Proof. The first part of the conclusion follows directly from Theorem 1.8. For the second part we have

$$\|\Phi'(z)\|_{E^\star} = \sup_{\|u\|=1} \langle \Phi'(z), u \rangle.$$

But,

$$\langle \Phi'(z), u \rangle = \lim_{\delta \to 0} \frac{\Phi(z+\delta u) - \Phi(z)}{\delta \|u\|}.$$

So, by Ekeland's variational principle,

$$\langle \Phi'(z), u \rangle \geq -\varepsilon.$$

Replacing now u with $-u$ we find

$$\langle \Phi'(z), u \rangle \leq \varepsilon,$$

which concludes our proof. □

1.3.2 Mountain Pass Theorem

The mountain pass theorem was established by Ambrosetti and Rabinowitz in [7]. It is a powerful tool for proving the existence of critical points of energy functionals, hence of weak solutions to wide classes of nonlinear problems. We first recall the following definition.

Definition 1.10 (Palais–Smale condition) *Let E be a real Banach space. A functional $J : E \to \mathbb{R}$ of class C^1 satisfies the Palais–Smale condition if any sequence $\{u_n\}$ in E is relatively compact, provided*

$$\{J(u_n)\} \quad \textit{is bounded and} \quad \|J'(u_n)\|_{E^\star} \to 0 \quad \textit{as} \quad n \to \infty. \tag{1.13}$$

By means of Ekeland's variational principle, one deduces the following result.

Proposition 1.11 *Let E be a real Banach space and assume that $\Phi : E \to \mathbb{R}$ is a functional of class C^1 that is bounded from below, and satisfies the Palais–Smale condition. Then the following properties hold true:*

(i) *Φ is coercive.*

(ii) *Any minimizing sequence of Φ has a convergent subsequence.*

We are now in position to state the mountain pass theorem.

Theorem 1.12 (Mountain Pass Theorem) *Let E be a real Banach space and assume that $J : E \to \mathbb{R}$ is a C^1 functional that satisfies the following conditions: There exist positive constants α and R such that*

(i) *$J(0) = 0$ and $J(v) \geq \alpha$ for all $v \in E$ with $\|v\| = R$;*

(ii) *$J(v_0) \leq 0$, for some $v_0 \in E$ with $\|v_0\| > R$.*

Set

$$\Gamma := \{p \in C([0,1];E) : p(0) = 0 \text{ and } p(1) = v_0\}$$

and

$$c := \inf_{p\in\Gamma} \max_{t\in[0,1]} J(p(t)).$$

Then there exists a sequence $\{u_n\}$ in E such that $J(u_n) \to c$ and $J'(u_n) \to 0$ as $n \to \infty$. Moreover, if J satisfies the Palais–Smale condition, then c is a nontrivial critical value of J, that is, there exists $u \in E$ such that $J(u) = c$ and $J'(u) = 0$.

1.3.3 Around the Palais–Smale Condition for Even Functionals

In this section we recall some notions and results from critical point theory for even functionals. These results are due to Tanaka [195]. Let X be an infinite dimensional separable Hilbert space and let $J : X \to \mathbb{R}$ be a C^2 functional such that

(A_1) J is even and $J(0) = 0$;

(A_2) For any finite dimensional subspace W of X there exists $R = R(W) > 0$ such that $J(u) < 0$ for all $u \in W$ with $\|u\| \geq R$;

(A_3) The Fréchet derivative $J' : X \to X$ satisfies

$$J'(u) = u + K(u) \quad \text{for all } u \in X,$$

where $K : X \to X$ is a compact operator. Let $\{X_k\}$ be a sequence of subspaces of X such that

$$\dim X_k = k \quad \text{and} \quad X = \overline{\bigcup_{k=1}^{\infty} X_k}.$$

For every $k \geq 1$ let $R_k = R(X_k) > 0$ from the hypothesis (A_2) and set $D_k = X_k \cap \overline{B(0,R_k)}$. Let also

$$\mathscr{C}_k = \left\{ \gamma \in C(D_k, X) \,:\, \gamma \text{ is odd and } \gamma\Big|_{X_k \cap \partial B(0,R_k)} = Id \right\}$$

and

$$b_k = \inf_{\gamma \in \mathscr{C}_k} \sup_{u \in D_k} J(\gamma(u)).$$

Definition 1.13 *We say that*

(i) J satisfies the $(PS)_k$ condition if every sequence $\{u_n\}$ in X_k such that $\{J(u_n)\}$ is bounded and

$$\left\| \left(J\big|_{X_k} \right)'(u_n) \right\|_{X_k} \to 0 \quad \text{as } n \to \infty$$

admits a convergent subsequence in X_k.

(ii) J satisfies the $(PS)_*$ condition if every sequence $\{u_k\}$ in X with $u_k \in X_k$ and such that $\{J(u_k)\}$ is bounded and

$$\left\| \left(J\big|_{X_k} \right)'(u_k) \right\|_{X_k} \to 0 \quad \text{as } k \to \infty$$

admits a convergent subsequence in X.

Definition 1.14 *Let u be a critical point of $J : X \to \mathbb{R}$. The large Morse index of J at u, denoted by $m^*(J,u)$ is the infimum of the codimensions of all subspaces of X on which the quadratic form $J''(u)$ is positive definite.*

Theorem 1.15 (see [195]) *Assume that J satisfies $(A_1) - (A_3)$, (PS), $(PS)_k$ and $(PS)_*$. Then, for each $k \geq 1$ there exists a critical point $u_k \in X$ such that*

$$J(u_k) \leq k \quad \text{and} \quad m^*(J,u_k) \geq k.$$

1.3.4 Bolle's Variational Method for Broken Symmetries

In the following we recall some notions and results from critical point theory for functionals with broken symmetry in the spirit of Bolle [22]. Let X be an infinite dimensional separable Hilbert space and let $J : [0,1] \times X \to \mathbb{R}$ be a C^2 functional. We set $J_\theta = J(\theta,\cdot)$ and denote by $J_\theta : X \to X$ the Fréchet derivative of J_θ.

Consider $\{e_k\}$ an orthonormal system of X and for any $k \geq 1$ set $X_k = \operatorname{span}\{e_1, e_2, \dots, e_k\}$.

(B_1) J satisfies the Palais–Smale condition on $[0,1] \times X$;

(B_2) For any $b > 0$ there exists a positive constant $C = C(b) > 0$ such that

$$\left|\frac{\partial J}{\partial \theta}(\theta,u)\right| \leq C(1 + \|J'_\theta(u)\|_X)(1 + \|u\|_K),$$

for all $(\theta,u) \in [0,1] \times X$ satisfying $|J_\theta(u)| \leq b$.

Assume that J is even and $J(0) = 0$;

(B_3) There exist two flows $\eta_i : [0,1] \times \mathbb{R} \to \mathbb{R}$ such that $\eta_i(\theta,\cdot)$ are Lipschitz continuous for all $\theta \in [0,1]$ and

$$\eta_1(\theta, J_\theta(u)) \leq \frac{\partial J}{\partial \theta}(\theta,u) \leq \eta_2(\theta, J_\theta(u))$$

at each critical point u of J_θ.

(B_4) J is even and for any finite dimensional subspace W of X we have

$$\lim_{\substack{u\in W\\ \|u\|_X\to\infty}} \sup_{\theta\in[0,1]} J_\theta(u) = -\infty.$$

Denote by $\psi_i : [0,1] \times X \to X$ the solutions of the problem

$$\begin{cases} \dfrac{\partial \psi_i}{\partial \theta}(\theta,s) = \eta_i(\theta, \psi_i(\theta,s)), \\ \psi_i(0,s) = s. \end{cases}$$

Remark that $\psi_i(\theta,\cdot)$ are continuous, nondecreasing and $\psi_1 \leq \psi_2$. Define

$$\bar{\eta}_i(s) = \sup_{\theta\in[0,1]} \eta_i(\theta,s).$$

Let

$$\mathscr{C} = \{\zeta \in C(X,X) : \zeta \text{ is odd and } \zeta(u) = u \text{ if } \|u\|_X \geq R\},$$

and

$$c_k = \inf_{\zeta \in \mathscr{C}} \sup_{u \in X_k} J_0(\zeta(u)).$$

The main result of this section is due to Bolle [22].

Theorem 1.16 (see [22]) *Assume that the sequence*

$$\left\{ \frac{c_{k+1} - c_k}{\bar{\eta}_1(c_{k+1}) + \bar{\eta}_2(c_k) + 1} \right\} \quad \textit{is unbounded.}$$

Then, the functional J_1 admits a sequence of critical values $\{d_k\}$ such that

$$\psi_2(1, c_k) < \psi_1(1, c_{k+1}) \le d_k \quad \textit{for all } k \ge 1.$$

1.4 Degree Theory

1.4.1 Brouwer Degree

We start by recalling some basic facts about Brouwer degree.

Definition 1.17 *Let Ω be an open set in $\mathbb{R}^N$, $N \ge 1$, and $F \in C^1(\overline{\Omega}; \mathbb{R}^N)$.*

(i) We say that $x_0 \in \Omega$ is a regular point if the Jacobian matrix $J_F(x_0) = (\partial F_i / \partial x_j)$ has rank N. If x_0 is not a regular point, we say that x_0 is a critical point of F.
(ii) We say that $y_0 \in \mathbb{R}^N$ is a regular value of F if the preimage $F^{-1}(y_0)$ does not contain any critical point; otherwise we say that y_0 is a critical value.

A first characterization of the set of critical values is given by the following result.

Theorem 1.18 *(Sard Lemma) Let $\Omega \subset \mathbb{R}^N$ be a bounded open set, $F \in C^2(\overline{\Omega}; \mathbb{R}^N)$. Then the set of critical values of F has zero Lebesgue measure.*

Definition 1.19 *Let $\Omega \subset \mathbb{R}^N$ be a bounded open set, $F \in C^2(\overline{\Omega}; \mathbb{R}^N)$, $p \in \mathbb{R}^N \setminus F(\partial\Omega)$.*

(i) If p is a regular value of F then

$$\deg(F, \Omega, p) = \sum_{x \in F^{-1}(p)} \operatorname{sign}(\det J_F(x)).$$

(ii) If p is a critical value of F then

$$\deg(F, \Omega, p) = \deg(F, \Omega, p_1),$$

where p_1 is a regular value of F such that $\|p - p_1\| < \operatorname{dist}(p, F(\partial\Omega))$.

It can be proved that $\deg(F,\Omega,p)$ is independent of the choice of p_1.

Definition 1.20 *Let $\Omega \subset \mathbb{R}^N$ be a bounded open set and $F \in C(\overline{\Omega};\mathbb{R}^N)$, $p \in \mathbb{R}^N \setminus F(\partial\Omega)$. The Brouwer degree of F in Ω at point p is defined as*

$$\deg(F,\Omega,p) = \deg(G,\Omega,p),$$

where $g \in C^2(\overline{\Omega};\mathbb{R}^N)$ is an arbitrary function such that $\|F - G\|_{L^\infty} < \operatorname{dist}(p,\partial\Omega)$.

It can be proved that $\deg(F,\Omega,p)$ is independent of the choice of G.

Some basic properties of Brouwer degree theory are stated below.

Theorem 1.21 *Let $\Omega \subset \mathbb{R}^N$ be a bounded open set.*

(i) (Normality)

$$\deg(F,\Omega,p) = \begin{cases} 1 & p \in \Omega, \\ 0 & p \notin \overline{\Omega}. \end{cases}$$

(ii) (Domain Additivity) Let $F \in C(\overline{\Omega};\mathbb{R}^N)$. If Ω_1, Ω_2 are two open subsets of Ω with $\Omega_1 \cap \Omega_2 = \emptyset$ and $p \notin F(\overline{\Omega} \setminus (\Omega_1 \cup \Omega_2))$ then

$$\deg(F,\Omega,p) = \deg(F,\Omega_1,p) + \deg(F,\Omega_2,p).$$

(iii) (Invariance of Homotopy) Let $H : \overline{\Omega} \times [0,1] \to \mathbb{R}^N$ be a continuous mapping. Assume that $p : [0,1] \to \mathbb{R}^N$ satisfies $p(t) \neq H(x,t)$ for all $(x,t) \in \partial\Omega \times [0,1]$. Then $\deg(H(\cdot,t),\Omega,p(t))$ is independent of t.

1.4.2 Leray–Schauder Degree

Let X be a Banach space and let Ω be a bounded open set in X. If $T : \overline{\Omega} \to X$ is a compact operator, then there exists a sequence of finite rank operators $\{T_\varepsilon\}$ such that $\|T - T_\varepsilon\| \to 0$ as $\varepsilon \to \infty$.

Let now $p \in X \setminus F(\partial\Omega)$. If $0 < \varepsilon < \operatorname{dist}(p,\partial\Omega)$, there exists a finite rank operator T_ε such that $\|T - T_\varepsilon\| < \varepsilon$. Letting $F_\varepsilon = I - T_\varepsilon$, then $p \notin F_\varepsilon(\partial\Omega)$ so that considering $X_\varepsilon = T_\varepsilon(\overline{\Omega})$ it follows that

$$F_\varepsilon\big|_{X_\varepsilon \cap \overline{\Omega}} : X_\varepsilon \cap \overline{\Omega} \to X_\varepsilon,$$

the Brouwer degree of F_ε on $X_\varepsilon \cap \overline{\Omega}$ at point p is well defined.

Definition 1.22 *(Leray–Schauder Degree) Let $F = I - T$ where T is a compact operator. The Leray–Schauder degree of F in Ω at point $p \in X \setminus F(\partial\Omega)$ is defined as*

$$\deg(F,\Omega,p) = \deg(F_\varepsilon, X_\varepsilon \cap \overline{\Omega}, p).$$

It can be proved that $\deg(F,\Omega,p)$ is independent of the choice of ε. As a consequence, the results in Theorem 1.21 (i)–(ii) concerning Brouwer degree of maps in finite dimensional space transfer to Leray–Schauder degree for $F = I - T$ by applying these results to the finite dimensional approximation F_ε. As regards the invariance to the homotopy, we state here the counterpart result for Theorem 1.21(iii).

Theorem 1.23 *(Invariance to Homotopy) Let $H : \overline{\Omega} \times [0,1] \to \mathbb{R}^N$ be a compact operator and let $p : [0,1] \to X$ be a continuous function such that*

$$p(t) \neq x - H(x,t) \quad \textit{for all } (x,t) \in \partial\Omega \times [0,1].$$

Then $\deg(H(\cdot,t),\Omega,p(t))$ is independent of t.

1.4.3 Leray–Schauder Degree for Isolated Solutions

As before, let Ω be a bounded open set of a Banach space and $F : \overline{\Omega} \to X$ such that $0 \notin F(\partial\Omega)$ and $T = I - F$ is compact. We assume that $x_0 \in \Omega$ is an isolated solution of $F(x) = 0$ and that $F'(x_0) = I - T'(x_0)$ is invertible. By the implicit function theorem, there exists a ball $B_r(x_0) \subset \Omega$ such that $F(x) \neq 0$ for all $x \in B_r(x_0)$, $x \neq x_0$.

Definition 1.24 *The index of F at x_0 is given by*

$$\text{index}(F,x_0) = \deg(F,B_r(x_0),0).$$

It can be shown that $\text{index}(F,x_0)$ is independent of r.

Theorem 1.25 *Under the above conditions,*

$$\text{index}(F,x_0) = (-1)^\beta, \quad \beta = \sum_{\substack{\lambda \in \sigma(T'(x_0)) \\ \lambda > 1}} n_\lambda,$$

where

$$n_\lambda = \dim\Big[\bigcup_{p\geq 1} \mathrm{Ker}(\lambda I - T'(x_0))^p\Big].$$

Proof. Without loss of generality we may assume $x_0 = 0$. For $0 \leq t \leq 1$ and $x \in \overline{\Omega}$ let

$$H(x,t) = \begin{cases} x - \dfrac{1}{t}T(tx) & 0 < t \leq 1, \\ x - T'(0)x & t = 0. \end{cases}$$

Then, by the invariance of the compact homotopy we have

$$\begin{aligned} \mathrm{index}(F,x_0) &= \deg(F,B_r,0) = \deg(H(1,\cdot),B_r,0) \\ &= \deg(H(1,\cdot),B_r,0) = \deg(I - T'(0),B_r,0). \end{aligned}$$

We next decompose $X = X_1 \bigoplus X_2$ where

$$X_1 = \mathrm{span}\Big\{\bigcup_{\substack{\lambda\in\sigma(T'(x_0))\\ \lambda>1}} \bigcup_{p\geq 1} \mathrm{Ker}(\lambda I - T'(x_0))^p\Big\}.$$

Then

$$\deg(I - T'(0),B_r,0) = \deg((I - T'(0))\big|_{X_1},B_r\cap X_1,0)\cdot\deg((I - T'(0))\big|_{X_2},B_r\cap X_2,0).$$

Further, if $\Gamma(t,\cdot) = I - tT'(0)$, $0 \leq t \leq 1$, then

$$0 \notin \Gamma(x,t) \quad \text{for all } (x,t) \in [0,1]\times\partial(B_r\cap X_2)$$

so

$$\deg((I - T'(0))\big|_{X_2},B_r\cap X_2,0) = \deg(\Gamma(0,\cdot)\big|_{X_2},B_r\cap X_2,0) = 1.$$

Thus, from the above equalities we find

$$\begin{aligned} \deg(I - T'(0),B_r,0) &= \deg((I - T'(0))\big|_{X_1},B_r\cap X_1,0) \\ &= \mathrm{sgn}\big(\det(I - T'(0))\big) = (-1)^\beta. \end{aligned}$$

This finishes the proof. □

Chapter 2
Liouville Type Theorems for Elliptic Operators in Divergence Form

There is no subject so old that
something new cannot be said about it.

Fyodor Dostoyevsky (1821–1881)

2.1 Introduction

The celebrated Liouville theorem in complex analysis which asserts that any bounded holomorphic function on the entire complex plane has to be constant, has been extended to various areas of analysis. A first well-known result directly related to the Liouville theorem is that any bounded harmonic function defined in $\mathbb{R}^N$, $N \geq 1$, is constant. The interested reader may find a valuable overview in the recent work [70].

In this chapter (see [90]) we are concerned with the existence of positive classical solutions to the following elliptic inequality

$$\operatorname{div}[A(|\nabla u|)\nabla u] \geq \rho(|x|)f(u) \quad \text{in } \mathbb{R}^N, N \geq 1. \tag{2.1}$$

Any solution of (2.1) is called a *positive entire solution*. Moreover, if u is an entire solution with $u(x) \to \infty$ as $|x| \to \infty$, then u is called a *positive entire large solution*.

The model case $A \equiv 1$ has been largely investigated and corresponds to the semilinear elliptic inequality $\Delta u \geq \rho(|x|)f(u)$ in $\mathbb{R}^N$.

If $A(t) = t^{p-2}$, $p > 1$ then (2.1) corresponds to the p-Laplace type inequality

$$\Delta_p u := \operatorname{div}(|\nabla u|^{p-2}\nabla u) \geq p(|x|)f(u) \quad \text{in } \mathbb{R}^N.$$

M. Ghergu and V. Rădulescu, *Nonlinear PDEs*, Springer Monographs in Mathematics,
DOI 10.1007/978-3-642-22664-9_2, © Springer-Verlag Berlin Heidelberg 2012

Another important case that will be considered in this chapter is $A(t) = (1+t^2)^{-1/2}$ that corresponds to the mean curvature inequality

$$\operatorname{div}\left(\frac{\nabla u}{\sqrt{1+|\nabla u|^2}}\right) \geq \rho(|x|)f(u) \qquad \text{in } \mathbb{R}^N.$$

Throughout this chapter we assume that $A \in C(0,\infty)$ is positive and

$(A1)$ the mapping $\mathbb{R} \ni t \mapsto tA(|t|)$ is of class $C(\mathbb{R}) \cap C^1(0,\infty)$ and $[tA(t)]' > 0$ for all $t > 0$.

Assume that f is a $C[0,\infty)$ nondecreasing function satisfying $f(t) > 0$ for all $t > 0$ and $f(0) = 0$. The function ρ is continuous and positive on $[0,\infty)$.

Theorem 2.1 *If (2.1) has a positive entire solution, then there exists a solution v of the equation*

$$[r^{N-1}A(|v'|)v']' = r^{N-1}\rho(r)f(v), \quad 0 < r < \infty, \tag{2.2}$$

such that $v'(0) = 0$.

Proof. Let U be a positive solution of (2.1) and assume that (2.2) has no solutions. Fix $0 < a < U(0)$ and let v be a positive solution of

$$[r^{N-1}A(|v'|)v']' = r^{N-1}\rho(r)f(v), \quad v(0) = a, v'(0) = 0,$$

defined in a maximal interval $[0,R)$, $R < \infty$. It is easy to see that $v' > 0$ on $(0,R)$ so either $v(R-) = \infty$ or $v'(R-) = \infty$. We shall discuss separately these two cases.

Case 1. $v(R-) = \infty$. Then one can find $0 < R_1 < R$ such that

$$v(R_1) \geq \max_{\partial B_{R_1}} U. \tag{2.3}$$

Then, by Theorem 1.4 we find $v \geq U$ in B_{R_1} which contradicts $v(0) = a < U(0)$.

Case 2. $v'(R-) = \infty$. If there exists $0 < R_1 < R$ such that (2.3) holds, by the same arguments as above we reach a contradiction. Assume next that

$$v(r) < \max_{B_r} U \quad \text{for all } 0 \leq r < R.$$

Using the fact that $v'(R-) = \infty$, one can chose $0 < R_1 < R$ such that

$$\frac{\partial v}{\partial \nu}(R_1) > \max_{\partial B_{R_1}} \frac{\partial U}{\partial \nu}. \tag{2.4}$$

Define

$$\delta := \max_{\partial B_{R_1}}(U - v) > 0$$

and let $w = v + \delta$. Then, $w \geq U$ on ∂B_{R_1} and there exists $x^* \in \partial B_{R_1}$ such that $w(x^*) = U(x^*)$. Using this fact and (2.4) we can find $0 < \lambda < 1$ such that

$$w(\lambda x^*) < U(\lambda x^*). \tag{2.5}$$

On the other hand, by Theorem 1.4 we find $w \geq U$ in B_{R_1} which contradicts (2.5).

Therefore, (2.2) has a positive solution u with $u'(0) = 0$. The proof of Theorem 2.1 is now complete. □

2.2 Some Related ODE Problems

In this section we are concerned with the ODE (2.2) in a slightly more general form which reads as:

$$[r^\alpha A(|u'|)u']' = r^\alpha \rho(r) f(u), \quad r > 0,\ u(0) > 0,\ u'(0) = 0, \tag{2.6}$$

where $\alpha > 0$ and A satisfies $(A1)$. We assume that $\Gamma(r) := r^{-\alpha} \int_0^r s^\alpha \rho(s) ds$ satisfies

$(g1)$ $\Gamma(r) \to \infty$ as $r \to \infty$;

or

$(g2)$ $g'(r) \geq M > 0$ for all $r > 0$.

Obviously, $(g2)$ implies $(g1)$.

Remark 2.1. (i) Examples of functions ρ that verify $(g1)$ are
(i1) $\rho(r) = r^a(1+r^2)^b$, $a + 2b > -1$; (i2) $\rho(r) = e^r$; (i3) $\rho(r) = \ln(2+r)$.
(ii) Condition $(g2)$ is fulfilled for (ii1) $\rho(r) = r^\gamma$, $\gamma \geq 0$; (ii2) $\rho(r) = e^r$.

Our first result concerns the nonexistence of the solution to (2.6) in the case where $\lim_{t\to\infty} tA(t) < \infty$.

Theorem 2.2 *Assume that A satisfies $(A1)$, $\lim_{t\to\infty} tA(t) < \infty$ and ρ satisfies $(g1)$. Then (2.6) has no positive solutions.*

Proof. Suppose by contradiction that there exists a solution u of (2.6). Then

$$[A(|u'|)u']' + \frac{\alpha}{r}A(|u'|)u' = \rho(r)f(u(r)), \quad \text{for all } r > 0 \tag{2.7}$$

and

$$A(|u'(r)|)u'(r) = r^{-\alpha}\int_0^r s^\alpha \rho(s)f(u(s))ds, \quad \text{for all } r > 0. \tag{2.8}$$

We deduce that $A(|u'(r)|)u'(r) > 0$ for $r > 0$ which implies $u'(r) > 0$. Since f is nondecreasing it follows that

$$A(u'(r))u'(r) \le \Gamma(r)f(u(r)), \quad \text{for all } r > 0. \tag{2.9}$$

Now (2.7) and (2.9) yield

$$[A(u')u']'(r) \ge \left(\rho(r) - \frac{\alpha}{r}\Gamma(r)\right)f(u(r)), \quad \text{for all } r > 0$$

that is

$$[A(u')u']'(r) \ge \Gamma'(r)f(u(r)), \quad \text{for all } r > 0. \tag{2.10}$$

Since $u(0) > 0$ and f, u are nondecreasing functions, from (2.7) we derive

$$A(u'(r))u'(r) \ge \Gamma(r)f(u(0)), \quad \text{for all } r > 0. \tag{2.11}$$

On the other hand, $\lim_{t\to\infty} tA(t) < \infty$ which implies that $A(u'(r))u'(r)$ is bounded on $[0,\infty)$. This fact and the above inequality lead to a contradiction since $\Gamma(r) \to \infty$ as $r \to \infty$ and $f(u(0)) > 0$. The proof of Theorem 2.2 is now complete. □

Next we consider the case where

$(A2)$ $\lim_{t\to\infty} \frac{A(t)}{t^{m-2}} = A_0 \in (0,\infty)$, $m > 1$.

Since $m > 1$, condition $(A2)$ leads to $\lim_{t\to\infty} tA(t) = \infty$.
Define $\Psi : [0,\infty) \to [0,\infty)$ by

$$\Psi(t) = t^2A(t) - \int_0^t sA(s)ds, \quad t > 0.$$

From $(A1)$ it follows that Ψ is a continuous strictly increasing function with $\Psi(0) = 0$.

From $(A2)$ and l'Hospital's rule we deduce that

$$\lim_{t\to\infty}\frac{\Psi(t)}{t^m}=\frac{m-1}{m}A_0\in(0,\infty). \tag{2.12}$$

Theorem 2.3 *Assume* $(A1)$, $(A2)$ *and* $(g2)$ *hold. If*

$$\int^{\infty}\left(\int^{t}f(s)ds\right)^{-1/m}dt<\infty, \tag{2.13}$$

then (2.6) has no positive solutions.

Proof. Assume by contradiction that (2.6) has a positive solution u. From (2.11) we get $u'(r)\to\infty$ as $r\to\infty$ and so $u(r)\to\infty$ as $r\to\infty$.

Multiplying (2.11) by $u'>0$, an integration over $[0,r]$ and a changing of variable gives

$$\Psi(u'(r))\geq\int_0^r\Gamma'(t)f(u(t))u'(t)dt,\quad\text{for all } r>0.$$

Using $(g2)$ we deduce

$$\Psi(u'(r))\geq M\int_{u(0)}^{u(r)}f(s)ds,\quad\text{for all } r>0.$$

Now, (2.12) implies that there exists $r_0>0$ and a positive constant $C>0$ such that

$$u'(r)\geq C\left(\int_{u(0)}^{u(r)}f(u(s))ds\right)^{1/m},\quad\text{for all } r>r_0.$$

Hence

$$\left(\int_{u(0)}^{u(r)}f(u(s))ds\right)^{-1/m}u'(r)\geq C,\quad\text{for all } r>r_0.$$

An integration over $[r_0,r]$ yields

$$\int_{u(r_0)}^{u(r)}\left(\int_{u(0)}^{t}f(u(s))ds\right)^{-1/m}dt\geq C(r-r_0),\quad\text{for all } r>r_0.$$

Letting $r\to\infty$ in the above relation we find

$$\int_{u(r_0)}^{\infty}\left(\int_{u(0)}^{t}f(u(s))ds\right)^{-1/m}dt=\infty.$$

This contradicts our assumption (2.13) and completes the proof. □

An existence result is as follows.

Theorem 2.4 *Assume* $(A1)$ *and* $(A2)$ *hold. If*

$$\int^{\infty} \left(\int^{t} f(s)ds \right)^{-1/m} dt = \infty, \tag{2.14}$$

then (2.6) has at least one positive solutions.

Proof. The existence of a solution u of (2.6) in a certain interval $[0,R)$ follows by the classical arguments of ODEs. Assume by contradiction that the maximal interval of existence of u is a finite interval $[0,R)$, $R<\infty$.

We first claim that $u(R-0) := \lim_{r\nearrow R} u(r) = \infty$. Indeed, since $u' \geq 0$ on $[0,R)$ it follows that $u(R-0)$ exists in $(0,\infty]$. From (2.7) we deduce that $u'(R-0)$ exists as a finite value. Then, by standard arguments for initial value problems it follows that u can be extended as a solution on an interval $[0,R+\varepsilon), \varepsilon>0$ which contradicts the maximality of R. Hence $u(R-0)=\infty$.

Using (2.7) and the fact that $A(u')u' \geq 0$ on $[0,R)$ we obtain

$$[A(u')u']'(r) \leq \Gamma(r) f(u(r)), \quad \text{for all } 0<r<R.$$

Multiplying the above inequality by $u' \geq 0$ and integrating over $[0,r]$ we have

$$\Psi(u'(r)) \leq \int_0^r \Gamma(s) f(u(s)) u'(s) ds \leq C_0 \int_{u(0)}^{u(r)} f(s)ds$$

for all $0<r<R$, where $C_0 = \max_{r\in[0,R]} \Gamma(r) > 0$. According to (2.12) there exists $R_0 \in (0,R)$ such that

$$u'(r) \leq C_1 \left(\int_{u(0)}^{u(r)} f(s)ds \right)^{1/m}, \quad \text{for all } r \in (R_0,R),$$

where $C_1>0$ is a constant independent of f and u. Hence

$$\left(\int_{u(0)}^{u(r)} f(s)ds \right)^{-1/m} u'(r) \leq C_1, \quad \text{for all } r\in(R_0,R).$$

An integration over $[R_0,r], r<R$, and a change of variable lead to

$$\int_{u(R_0)}^{u(r)} \left(\int_{u(0)}^{t} f(s)ds \right)^{-1/m} dt \leq C_1(r-R_0), \quad \text{for all } r>r_0.$$

Now, letting $r\nearrow R$ we find

$$\int_{u(R_0)}^{\infty} \left(\int_{u(0)}^{t} f(u(s))ds \right)^{-1/m} dt \leq C_1(R-R_0) < \infty,$$

which contradicts our assumption (2.14). We conclude that there exists a solution of (2.6) and the proof is now complete. □

Our final result in this section gives an estimate of the growth of a solution of (2.6) in the case where f is bounded. More precisely we prove

Theorem 2.5 *Assume* $(A1)$, $(A2)$, $(g1)$ *hold and* f *is bounded. If* u *is a positive solution of (2.6) then*

$$\lim_{r\to\infty} \frac{u(r)}{\int_0^r g^{1/(m-1)}(r)} = \lim_{r\to\infty}\left(\frac{1}{A_0} f(r)\right)^{1/(m-1)}. \tag{2.15}$$

Proof. Applying l'Hospital's rule we have

$$\lim_{r\to\infty} \frac{u(r)}{\int_0^r g^{1/(m-1)}(s)ds} = \lim_{r\to\infty} \frac{u'(r)}{g^{1/(m-1)}(r)}. \tag{2.16}$$

From $(A2)$ we find

$$\begin{aligned} \lim_{r\to\infty} \frac{u'^{m-1}(r)}{\Gamma(r)} &= \lim_{r\to\infty} \frac{u'^{m-1}(r)}{A(u'(r))u'(r)} \cdot \frac{A(u'(r))u'(r)}{\Gamma(r)} \\ &= \frac{1}{A_0} \lim_{r\to\infty} \frac{A(u'(r))u'(r)}{\Gamma(r)}. \end{aligned} \tag{2.17}$$

By (2.8) and l'Hospital's rule we deduce

$$\begin{aligned} \lim_{r\to\infty} \frac{A(u'(r))u'(r)}{\Gamma(r)} &= \lim_{r\to\infty} \frac{\int_0^r s^\alpha \rho(s) f(u(s))ds}{\int_0^r s^\alpha \rho(s)ds} \\ &= \lim_{r\to\infty} f(u(r)) = \lim_{r\to\infty} f(r). \end{aligned} \tag{2.18}$$

Now, (2.16), (2.17), (2.18) lead to

$$\lim_{r\to\infty} \frac{u(r)}{\int_0^r g^{1/(m-1)}(r)} = \lim_{r\to\infty}\left(\frac{1}{A_0} f(r)\right)^{1/(m-1)}.$$

This completes the proof. □

2.3 Main Results

We have seen (in Theorem 2.1) that (2.2) has a solution whenever the N-dimensional inequality (2.1) has an entire positive solution. Using this fact and the results in the previous section we obtain some Liouville type theorems regarding the nonexistence of positive solutions to (2.1). In the sequel, the corresponding function $\Gamma(r)$ to which conditions $(g1)$ or $(g2)$ apply is given by

$$\Gamma(r) := r^{1-N} \int_0^r s^{N-1} \rho(s)ds, \quad r > 0.$$

Theorem 2.6 *Assume that A satisfies* $(A1)$, $\lim_{t\to\infty} tA(t) < \infty$ *and* ρ *satisfies* $(g1)$. *Then (2.1) has no positive entire solutions.*

Theorem 2.7 *Assume that* $(A1)$, $(A2)$ *and* $(g2)$ *hold. If*

$$\int^\infty \left(\int^t f(s)ds \right)^{-1/m} dt < \infty,$$

then (2.1) has no positive entire solutions.

The proofs follow by applying Theorems 2.1, 2.2 and 2.3. Further, using the existence result stated in Theorem 2.4 we have:

Theorem 2.8 *Assume* $(A1)$, $(A2)$ *and* $(g1)$ *hold. If*

$$\int^\infty \left(\int^t f(s)ds \right)^{-1/m} dt = \infty,$$

then the inequality (2.1) has infinitely many positive entire large solutions.

Corollary 2.9 *Assume that* $(A1)$, $(A2)$ *and* $(g2)$ *hold. Then (2.1) has a positive entire (large) solution if and only if*

$$\int^\infty \left(\int^t f(s)ds \right)^{-1/m} dt = \infty.$$

Examples.(i) If ρ satisfies $(g1)$, then the inequality

$$\text{div} \left(\frac{\nabla u}{\sqrt{1+|\nabla u|^2}} \right) \geq \rho(|x|)f(u), \quad x \in \mathbb{R}^N$$

has no positive entire solutions.

(ii) If ρ satisfies $(g2)$ then the inequality

$$\operatorname{div}\left(\frac{\nabla u}{(1+|\nabla u|^2)^\alpha}\right) \geq \rho(|x|)e^u, \quad x \in \mathbb{R}^N, \ \alpha < \frac{1}{2}$$

has no positive entire solutions and the inequality

$$\operatorname{div}\left(\frac{\nabla u}{(1+|\nabla u|^2)^\alpha}\right) \geq \rho(|x|)u^\gamma, \quad x \in \mathbb{R}^N, \ \alpha < \frac{1}{2}, \gamma \geq 0$$

has positive entire (large) solutions if and only if $\gamma \leq 1 - 2\alpha$.

(iii) If ρ satisfies $(g2)$ then the inequality

$$\operatorname{div}(|\nabla u|^{m-2}\nabla u) \geq \rho(|x|)u^\gamma, \quad x \in \mathbb{R}^N, \ m > 1, \ \gamma \geq 0$$

has positive entire (large) solutions if and only if $\gamma \leq m - 1$.

(iv) If ρ satisfies $(g2)$ then the inequality

$$\operatorname{div}(|\nabla u|^{m-2}\nabla u) \geq \rho(|x|)\ln^\gamma(1+u), \quad x \in \mathbb{R}^N, \ m > 1, \ \gamma \geq 0$$

has positive entire (large) solutions.

Chapter 3
Blow-Up Boundary Solutions of the Logistic Equation

> All intelligent thoughts have already been thought; what is necessary is only to try to think them again.
>
> Johann Wolfgang von Goethe (1749–1832)

In this chapter we are concerned with singular problems of the type

$$\begin{cases} \Delta u = \Phi(x,u,\nabla u) & \text{in } \Omega, \\ u > 0 & \text{in } \Omega, \\ u = +\infty & \text{on } \partial\Omega, \end{cases} \tag{3.1}$$

where Ω is an open set in the Euclidean space with smooth boundary $\partial\Omega$. The function Φ is assumed to be positive and fulfilling a suitable growth assumption.

A solution of problem (3.1) is called a *blow-up boundary* solution or a *large* solution. The study of large solutions was initiated in 1916 by Bieberbach [18] for the particular case $\Phi(x,u,\nabla u) = \exp(u)$ and $N = 2$. He showed that there exists a unique solution of (3.1) such that $u(x) - \log(d(x)^{-2})$ is bounded as $x \to \partial\Omega$, where $d(x) := \text{dist}\,(x, \partial\Omega)$. Problems of this type arise in Riemannian geometry: if a Riemannian metric of the form $|ds|^2 = \exp(2u(x))|dx|^2$ has constant Gaussian curvature $-c^2$ then $\Delta u = c^2 \exp(2u)$. Motivated by a problem in mathematical physics, Rademacher [169] continued the study of Bieberbach on smooth bounded domains in $\mathbb{R}^3$. Lazer and McKenna [131] extended the results of Bieberbach and Rademacher for bounded domains in $\mathbb{R}^N$ satisfying a uniform exterior sphere condition and for nonlinearities $\Phi(x,u,\nabla u) = b(x)\exp(u)$, where b is continuous and strictly positive on $\overline{\Omega}$. Let $\Phi(x,u,\nabla u) = f(u)$ where $f \in C^1[0,\infty)$, $f'(s) \geq 0$ for $s \geq 0$, $f(0) = 0$ and $f(s) > 0$ for $s > 0$. In this case, Keller [115] and Osserman

M. Ghergu and V. Rădulescu, *Nonlinear PDEs*, Springer Monographs in Mathematics, DOI 10.1007/978-3-642-22664-9_3, © Springer-Verlag Berlin Heidelberg 2012

[155] proved that large solutions of (3.1) exist if and only if

$$\int_1^\infty \frac{dt}{\sqrt{F(t)}} < \infty, \quad \text{where } F(t) = \int_0^t f(s)\,ds.$$

In a celebrated paper, Loewner and Nirenberg [138] linked the uniqueness of the blow-up solution to the growth rate at the boundary. Motivated by certain geometric problems, they established the uniqueness for the case $f(u) = u^{(N+2)/(N-2)}$, $N > 2$.

Many results in this chapter are concerned with the case $\Phi(x,u,\nabla u) = a(x)f(u)$, where f is a nondecreasing function and $a(x)$ is a nonnegative potential. The case $a > 0$ in $\overline{\Omega}$ corresponds to a logistic behavior while the framework corresponding to $a \equiv 0$ in $\overline{\Omega}$ is known as the Malthusian model. The models we study in this chapter are often mixtures of the logistic and Malthusian models.

3.1 Singular Solutions of the Logistic Equation

Consider the semilinear elliptic equation

$$\Delta u + au = b(x)f(u) \qquad \text{in } \Omega, \tag{3.2}$$

where Ω is a smooth bounded domain in $\mathbb{R}^N$, $N \geq 3$. Let a be a real parameter and $b \in C^{0,\mu}(\overline{\Omega})$, $0 < \mu < 1$, such that $b \geq 0$ and $b \not\equiv 0$ in Ω. Set

$$\Omega_0 = \operatorname{int}\{x \in \Omega : \, b(x) = 0\}$$

and suppose, throughout, that $\overline{\Omega}_0 \subset \Omega$ and $b > 0$ on $\Omega \setminus \overline{\Omega}_0$. Assume that $f \in C^1[0,\infty)$ satisfies

(A_1) $f \geq 0$ and $f(u)/u$ is increasing on $(0,\infty)$.

Following Alama and Tarantello [4], define by H_∞ the Dirichlet Laplacian on Ω_0 as the unique self-adjoint operator associated to the quadratic form $\psi(u) = \int_\Omega |\nabla u|^2\,dx$ with form domain

$$H_D^1(\Omega_0) = \{u \in H_0^1(\Omega) : \, u(x) = 0 \quad \text{for a.e. } x \in \Omega \setminus \Omega_0\}.$$

If $\partial\Omega_0$ satisfies the exterior cone condition then, according to [4], $H_D^1(\Omega_0)$ coincides with $H_0^1(\Omega_0)$ and H_∞ is the classical Laplace operator with Dirichlet condition on $\partial\Omega_0$.

Let $\lambda_{\infty,1}$ be the first Dirichlet eigenvalue of H_∞ in Ω_0. We understand that $\lambda_{\infty,1} = \infty$ if $\Omega_0 = \emptyset$.

Set $\mu_0 := \lim_{u \searrow 0} \frac{f(u)}{u}$, $\mu_\infty := \lim_{u \to \infty} \frac{f(u)}{u}$, and denote by $\lambda_1(\mu_0)$ (resp. $\lambda_1(\mu_\infty)$) the first eigenvalue of the operator $H_{\mu_0} = -\Delta + \mu_0 b$ (resp. $H_{\mu_\infty} = -\Delta + \mu_\infty b$) in $H_0^1(\Omega)$. Recall that $\lambda_1(+\infty) = \lambda_{\infty,1}$.

Alama and Tarantello [4] proved that problem (3.2) subject to the Dirichlet boundary condition

$$u = 0 \quad \text{on } \partial\Omega \tag{3.3}$$

has a positive solution u_a if and only if $a \in (\lambda_1(\mu_0), \lambda_1(\mu_\infty))$. Moreover, u_a is the unique positive solution for (3.2)+(3.3) (see [4, Theorem A (bis)]). We shall refer to the combination of (3.2)+(3.3) as problem (E_a).

Our first aim in this section is to give a corresponding necessary and sufficient condition, but for the existence of *large* (or *explosive*) solutions of (3.2). An elementary argument based on the maximum principle shows that if such a solution exists, then it is *positive* even if f satisfies a weaker condition than (A_1), namely

$(A_1)'$ $f(0) = 0$, $f' \geq 0$ and $f > 0$ on $(0, \infty)$.

We recall that Keller [115] and Osserman [155] supplied a necessary and sufficient condition on f for the existence of large solutions to (3.2) when $a \equiv 0$, $b \equiv 1$ and f is assumed to fulfill $(A_1)'$. More precisely, f must satisfy the Keller–Osserman condition (see [115, 155]),

(A_2) $\displaystyle\int_1^\infty \frac{dt}{\sqrt{F(t)}} < \infty$, where $F(t) = \displaystyle\int_0^t f(s)\,ds$.

Typical examples of nonlinearities satisfying (A_1) and (A_2) are:

(i) $f(u) = e^u - 1$; (ii) $f(u) = u^p$, $p > 1$; (iii) $f(u) = u[\ln(u+1)]^p$, $p > 2$.

Our first result gives the maximal interval for the parameter a that ensures the existence of large solutions to problem (3.2). More precisely, we prove

Theorem 3.1 *Assume that f satisfies conditions (A_1) and (A_2). Then problem (3.2) has a large solution if and only if $a \in (-\infty, \lambda_{\infty,1})$.*

We point out that our framework in the above result includes the case when b vanishes at some points on $\partial\Omega$, or even if $b \equiv 0$ on $\partial\Omega$. This later case includes the "competition" $0\cdot\infty$ on $\partial\Omega$. We also point out that, under our hypotheses, $\mu_\infty := \lim_{u\to\infty} f(u)/u = \lim_{u\to\infty} f'(u) = \infty$. Indeed, by l'Hospital's rule, $\lim_{u\to\infty} F(u)/u^2 = \mu_\infty/2$. But, by (A_2), we deduce that $\mu_\infty = \infty$. Then, by (A_1) we find that $f'(u) \geq f(u)/u$ for any $u > 0$, which shows that $\lim_{u\to\infty} f'(u) = \infty$.

Before giving the proof of Theorem 3.1 we claim that assuming (A_1), then problem (3.2) can have large solutions only if f satisfies the Keller–Osserman condition (A_2). Indeed, suppose that problem (3.2) has a large solution u_∞. Set $\tilde f(u) = |a|u + \|b\|_\infty f(u)$ for $u \geq 0$. Notice that $\tilde f \in C^1[0,\infty)$ satisfies $(A_1)'$. For any $n \geq 1$, consider the problem

$$\begin{cases} \Delta u = \tilde f(u) & \text{in } \Omega\,, \\ u = n & \text{on } \partial\Omega\,, \\ u \geq 0 & \text{in } \Omega\,. \end{cases}$$

A standard argument based on the maximum principle shows that this problem has a unique solution, say u_n, which, moreover, is positive in $\overline{\Omega}$. Applying again the maximum principle we deduce that $0 < u_n \leq u_{n+1} \leq u_\infty$, in Ω, for all $n \geq 1$. Thus, for every $x \in \Omega$, we can define $\bar u(x) = \lim_{n\to\infty} u_n(x)$. Moreover, since (u_n) is uniformly bounded on every compact subset of Ω, standard elliptic regularity arguments show that $\bar u$ is a positive large solution of the problem $\Delta u = \tilde f(u)$. It follows that $\tilde f$ satisfies the Keller–Osserman condition (A_2). Then, by (A_1), $\mu_\infty := \lim_{u\to\infty} f(u)/u > 0$ which yields $\lim_{u\to\infty} \tilde f(u)/f(u) = |a|/\mu_\infty + \|b\|_\infty < \infty$. Consequently, our claim follows.

Proof of Theorem 3.1. A. Necessary condition. Let u_∞ be a large solution of problem (3.2). Then, by the maximum principle, u_∞ is positive. Suppose $\lambda_{\infty,1}$ is finite. Arguing by contradiction, let us assume $a \geq \lambda_{\infty,1}$. Set $\lambda \in (\lambda_1(\mu_0), \lambda_{\infty,1})$ and denote by u_λ the unique positive solution of problem (E_a) with $a = \lambda$. We have

$$\begin{cases} \Delta(Mu_\infty) + \lambda_{\infty,1}(Mu_\infty) \leq b(x)f(Mu_\infty) & \text{in } \Omega\,, \\ Mu_\infty = \infty & \text{on } \partial\Omega\,, \\ Mu_\infty \geq u_\lambda & \text{in } \Omega\,, \end{cases}$$

where $M := \max\left\{\max_{\overline{\Omega}} u_\lambda / \min_\Omega u_\infty;\, 1\right\}$. By the sub-super solution method we conclude that problem (E_a) with $a = \lambda_{\infty,1}$ has at least a positive solution (between u_λ and Mu_∞). But this is a contradiction. So, necessarily, $a \in (-\infty, \lambda_{\infty,1})$.

B. *Sufficient condition.* This will be proved with the aid of several results.

Lemma 3.2 *Let ω be a smooth bounded domain in $\mathbb{R}^N$. Assume p, q, r are $C^{0,\mu}$-functions on $\overline{\omega}$ such that $r \geq 0$ and $p > 0$ in $\overline{\omega}$. Then for any nonnegative function $0 \not\equiv \Phi \in C^{0,\mu}(\partial\omega)$ the boundary value problem*

$$\begin{cases} \Delta u + q(x)u = p(x)f(u) - r(x) & \text{in } \omega, \\ u > 0 & \text{in } \omega, \\ u = \Phi & \text{on } \partial\omega, \end{cases} \tag{3.4}$$

has a unique solution.

We refer to Cîrstea and Rădulescu [44, Lemma 3.1] for the proof of the above result.

Under the assumptions of Lemma 3.2 we obtain the following result which generalizes [142, Lemma 1.3].

Corollary 3.3 *There exists a positive large solution of the problem*

$$\Delta u + q(x)u = p(x)f(u) - r(x) \qquad \text{in } \omega. \tag{3.5}$$

Proof. Set $\Phi = n$ and let u_n be the unique solution of (3.4). By the maximum principle, $u_n \leq u_{n+1} \leq \overline{u}$ in ω, where $\overline{u}$ denotes a large solution of

$$\Delta u + \|q\|_\infty u = p_0 f(u) - \overline{r} \quad \text{in } \omega.$$

Thus $\lim_{n\to\infty} u_n(x) = u_\infty(x)$ exists and is a positive large solution of (3.5). Furthermore, every positive large solution of (3.5) dominates u_∞, that is, the solution u_∞ is the *minimal large solution*. This follows from the definition of u_∞ and the maximum principle. □

Lemma 3.4 *If $0 \not\equiv \Phi \in C^{0,\mu}(\partial\Omega)$ is a nonnegative function and $b > 0$ on $\partial\Omega$, then the boundary value problem*

$$\begin{cases} \Delta u + au = b(x)f(u) & \text{in } \Omega, \\ u > 0 & \text{in } \Omega, \\ u = \Phi & \text{on } \partial\Omega, \end{cases} \tag{3.6}$$

has a solution if and only if $a \in (-\infty, \lambda_{\infty,1})$. Moreover, in this case, the solution is unique.

Proof. The first part follows exactly in the same way as the proof of Theorem 3.1 (necessary condition).

For the sufficient condition, fix $a < \lambda_{\infty,1}$ and let $\lambda_{\infty,1} > \lambda_* > \max\{a, \lambda_1(\mu_0)\}$. Let u_* be the unique positive solution of (E_a) with $a = \lambda_*$.

Let Ω_i $(i = 1,2)$ be subdomains of Ω such that $\Omega_0 \subset\subset \Omega_1 \subset\subset \Omega_2 \subset\subset \Omega$ and $\Omega \setminus \overline{\Omega}_1$ is smooth.

We define $u_+ \in C^2(\Omega)$ as a positive function in Ω such that $u_+ \equiv u_\infty$ on $\Omega \setminus \Omega_2$ and $u_+ \equiv u_*$ on Ω_1. Here u_∞ denotes a positive large solution of (3.5) for $p(x) = b(x)$, $r(x) = 0$, $q(x) = a$ and $\omega = \Omega \setminus \overline{\Omega}_1$. So, since $b_0 := \inf_{\Omega_2 \setminus \Omega_1} b$ is positive, it is easy to check that if $C > 0$ is large enough then $\overline{v}_\Phi = Cu_+$ satisfies

$$\begin{cases} \Delta \overline{v}_\Phi + a\overline{v}_\Phi \leq b(x) f(\overline{v}_\Phi) & \text{in } \Omega\,, \\ \overline{v}_\Phi = \infty & \text{on } \partial\Omega\,, \\ \overline{v}_\Phi \geq \max\limits_{\partial\Omega} \Phi & \text{in } \Omega\,. \end{cases}$$

Let $\underline{v}_\Phi$ be the unique classical solution of the problem

$$\begin{cases} \Delta \underline{v}_\Phi = |a|\underline{v}_\Phi + \|b\|_\infty f(\underline{v}_\Phi) & \text{in } \Omega, \\ \underline{v}_\Phi > 0 & \text{in } \Omega, \\ \underline{v}_\Phi = \Phi & \text{on } \partial\Omega\,. \end{cases}$$

It is clear that $\underline{v}_\Phi$ is a positive subsolution of (3.6) and $\underline{v}_\Phi \leq \max_{\partial\Omega} \Phi \leq \overline{v}_\Phi$ in Ω. Therefore, by the sub-super solution method, problem (3.6) has at least a solution v_Φ between $\underline{v}_\Phi$ and $\overline{v}_\Phi$. Next, the uniqueness of the solution to (3.6) can be obtained by using essentially the same technique as in [30, Theorem 1] or [26, Appendix II]. □

Proof of Theorem 3.1 continued. Fix $a \in (-\infty, \lambda_{\infty,1})$. Two cases may occur:

Case 1: $b > 0$ on $\partial\Omega$. Denote by v_n the unique solution of (3.6) with $\Phi \equiv n$. For $\Phi \equiv 1$, set $v := \underline{v}_\Phi$ and $V := \overline{v}_\Phi$, where $\underline{v}_\Phi$ and $\overline{v}_\Phi$ are defined in the proof of Lemma 3.4. The sub and supersolutions method combined with the uniqueness of the solution of (3.6) shows that $v \leq v_n \leq v_{n+1} \leq V$ in Ω. Hence $v_\infty(x) := \lim_{n\to\infty} v_n(x)$ exists and is a positive large solution of (3.2).

Case 2: $b \geq 0$ on $\partial\Omega$. Let z_n $(n \geq 1)$ be the unique solution of (3.4) for $p \equiv b + 1/n$, $r \equiv 0$, $q \equiv a$, $\Phi \equiv n$ and $\omega = \Omega$. By the maximum principle, (z_n) is non-decreasing. Moreover, $\{z_n\}$ is uniformly bounded on every compact subdomain of Ω. Indeed, if $K \subset \Omega$ is an arbitrary compact set, then $d := \operatorname{dist}(K, \partial\Omega) > 0$. Choose

$\delta \in (0,d)$ small enough so that $\overline{\Omega}_0 \subset C_\delta$, where $C_\delta = \{x \in \Omega : \text{dist}(x, \partial\Omega) > \delta\}$. Since $b > 0$ on ∂C_δ, Case 1 allows us to define z_+ as a positive large solution of (3.2) for $\Omega = C_\delta$. A standard argument based on the maximum principle implies that $z_n \le z_+$ in C_δ, for all $n \ge 1$. So, $\{z_n\}$ is uniformly bounded on K. By the monotonicity of $\{z_n\}$, we conclude that $z_n \to \underline{z}$ in $L^\infty_{\text{loc}}(\Omega)$. Finally, standard elliptic regularity arguments lead to $z_n \to \underline{z}$ in $C^{2,\mu}(\Omega)$. This completes the proof of Theorem 3.1. □

Denote by $\mathscr{D}$ and $\mathscr{R}$ the boundary operators

$$\mathscr{D}u := u \qquad \text{and} \qquad \mathscr{R}u := \partial_\nu u + \beta(x)u,$$

where ν is the unit outward normal to $\partial\Omega$, and $\beta \in C^{1,\mu}(\partial\Omega)$ is nonnegative. Hence, $\mathscr{D}$ is the *Dirichlet* boundary operator and $\mathscr{R}$ is either the *Neumann* boundary operator, if $\beta \equiv 0$, or the *Robin* boundary operator, if $\beta \not\equiv 0$. Throughout this work, $\mathscr{B}$ can define any of these boundary operators.

Note that the Robin condition $\mathscr{R} = 0$ applies essentially to heat flow problems in a body with constant temperature in the surrounding medium. More generally, if α and β are smooth functions on $\partial\Omega$ such that $\alpha, \beta \ge 0$, $\alpha + \beta > 0$, then the boundary condition $Bu = \alpha\partial_\nu u + \beta u = 0$ represents the exchange of heat at the surface of the reactant by Newtonian cooling. Moreover, the boundary condition $Bu = 0$ is called an isothermal (Dirichlet) condition if $\alpha \equiv 0$, and it becomes an adiabatic (Neumann) condition if $\beta \equiv 0$. An intuitive meaning of the condition $\alpha + \beta > 0$ on $\partial\Omega$ is that, for the diffusion process described by problem (3.2), either the reflection phenomenon or the absorption phenomenon may occur at each point of the boundary.

We are now concerned with the following boundary blow-up problem

$$\begin{cases} \Delta u + au = b(x)f(u) & \text{in } \Omega \setminus \overline{\Omega}_0\,, \\ \mathscr{B}u = 0 & \text{on } \partial\Omega\,, \\ u = \infty & \text{on } \partial\Omega_0\,, \end{cases} \tag{3.7}$$

where $b > 0$ on $\partial\Omega$, while $\overline{\Omega}_0$ is nonempty, connected and with smooth boundary. Here, $u = \infty$ on $\partial\Omega_0$ means that $u(x) \to \infty$ as $x \in \Omega \setminus \overline{\Omega}_0$ and $d(x) := \text{dist}(x, \Omega_0) \to 0$.

The question of existence and uniqueness of positive solutions for problem (3.7) in the case of pure superlinear power in the nonlinearity is treated by Du and Huang [64]. Our next results extend their previous paper to the case of much more general nonlinearities of Keller–Osserman type.

In the following, by $(\tilde{A}_1)$ we mean that (A_1) is fulfilled and there exists

$$\lim_{u\to\infty} (F/f)'(u) := \gamma.$$

Then, $\gamma \geq 0$.

We prove

Theorem 3.5 *Let $(\tilde{A}_1)$ and (A_2) hold. Then, for any $a \in \mathbb{R}$, problem (3.7) has a minimal (resp. maximal) positive solution $\underline{U}_a$ (resp. $\overline{U}_a$).*

Proof. In proving Theorem 3.5 we rely on an appropriate comparison principle which allows us to prove that $\{u_n\}$ is nondecreasing, where u_n is the unique positive solution of problem (3.9) with $\Phi \equiv n$. The minimal positive solution of (3.7) will be obtained as the limit of the sequence $\{u_n\}$. Note that, since $b = 0$ on $\partial\Omega_0$, the main difficulty is related to the construction of an upper bound of this sequence which must fit to our general framework. Next, we deduce the maximal positive solution of (3.7) as the limit of the nonincreasing sequence $\{v_m\}_{m\geq m_1}$ provided m_1 is large so that $\Omega_{m_1} \subset\subset \Omega$. We denoted by v_m the minimal positive solution of (3.7) with Ω_0 replaced by

$$\Omega_m := \{x \in \Omega : d(x) < 1/m\}, \qquad m \geq m_1. \tag{3.8}$$

We start with the following auxiliary result (see Cîrstea and Rădulescu [44]).

Lemma 3.6 *Assume $b > 0$ on $\partial\Omega$. If (A_1) and (A_2) hold, then for any positive function $\Phi \in C^{2,\mu}(\partial\Omega_0)$ and $a \in \mathbb{R}$ the problem*

$$\begin{cases} \Delta u + au = b(x)f(u) & \text{in } \Omega\setminus\overline{\Omega}_0\,, \\ \mathscr{B}u = 0 & \text{on } \partial\Omega\,, \\ u = \Phi & \text{on } \partial\Omega_0\,, \end{cases} \tag{3.9}$$

has a unique positive solution.

We now come back to the proof of Theorem 3.5, which will be divided into two steps:

Step 1: Existence of the minimal positive solution for problem (3.7).

For any $n \geq 1$, let u_n be the unique positive solution of problem (3.9) with $\Phi \equiv n$. By the maximum principle, $u_n(x)$ increases with n for all $x \in \overline{\Omega}\setminus\overline{\Omega}_0$. Moreover, we prove

Lemma 3.7 *The sequence $(u_n(x))_n$ is bounded from above by some function $V(x)$ which is uniformly bounded on all compact subsets of $\overline{\Omega} \setminus \overline{\Omega}_0$.*

Proof. Let b^* be a C^2-function on $\overline{\Omega} \setminus \Omega_0$ such that

$$0 < b^*(x) \le b(x) \quad \forall x \in \overline{\Omega} \setminus \overline{\Omega}_0.$$

For x bounded away from $\partial\Omega_0$ it is not a problem to find such a function b^*. For x satisfying $0 < d(x) < \delta$ with $\delta > 0$ small such that $x \to d(x)$ is a C^2-function, we can take

$$b^*(x) = \int_0^{d(x)} \int_0^t [\min_{d(z)\ge s} b(z)]\, ds\, dt.$$

Let $g \in \mathscr{G}$ be a function such that (A_g) holds. Since $b^*(x) \to 0$ as $d(x) \searrow 0$, we deduce, by (A_1), the existence of some $\delta > 0$ such that for all $x \in \Omega$ with $0 < d(x) < \delta$ and $\xi > 1$

$$\frac{b^*(x) f(g(b^*(x))\xi)}{g''(b^*(x))\,\xi} > \sup_{\overline{\Omega}\setminus\Omega_0} |\nabla b^*|^2 + \frac{g'(b^*(x))}{g''(b^*(x))} \inf_{\overline{\Omega}\setminus\Omega_0} (\Delta b^*) + a \frac{g(b^*(x))}{g''(b^*(x))}.$$

Here, $\delta > 0$ is taken sufficiently small so that $g'(b^*(x)) < 0$ and $g''(b^*(x)) > 0$ for all x with $0 < d(x) < \delta$.

For $n_0 \ge 1$ fixed, define V^* as follows

(i) $V^*(x) = u_{n_0}(x) + 1$ for $x \in \overline{\Omega}$ and near $\partial\Omega$;
(ii) $V^*(x) = g(b^*(x))$ for x satisfying $0 < d(x) < \delta$;
(iii) $V^* \in C^2(\overline{\Omega} \setminus \overline{\Omega}_0)$ is positive on $\overline{\Omega} \setminus \overline{\Omega}_0$.

We show that for $\xi > 1$ large enough the upper bound of the sequence $(u_n(x))_n$ can be taken as $V(x) = \xi V^*(x)$. Since

$$\mathscr{B}V(x) = \xi\, \mathscr{B}V^*(x) \ge \xi \min\{1, \beta(x)\} \ge 0, \qquad \text{for all } x \in \partial\Omega$$

and

$$\lim_{d(x)\searrow 0} [u_n(x) - V(x)] = -\infty < 0,$$

to conclude that $u_n(x) \le V(x)$ for all $x \in \overline{\Omega} \setminus \overline{\Omega}_0$ it is sufficient to show that

$$-\Delta V(x) \ge aV(x) - b(x) f(V(x)), \qquad \forall x \in \Omega \setminus \overline{\Omega}_0. \tag{3.10}$$

For $x \in \Omega$ satisfying $0 < d(x) < \delta$ and $\xi > 1$ we have

$$
\begin{aligned}
&-\Delta V(x)-aV(x)+b(x)f(V(x))\\
=&-\xi\Delta g(b^*(x))-a\xi g(b^*(x))+b(x)f(g(b^*(x))\xi)\\
\geq&\xi g''(b^*)\left(-\frac{g'(b^*(x))}{g''(b^*(x))}\Delta b^*-|\nabla b^*|^2-a\frac{g(b^*(x))}{g''(b^*(x))}+b^*\frac{f(g(b^*(x))\xi)}{g''(b^*(x))\,\xi}\right)\\
>&0.
\end{aligned}
$$

For $x\in\Omega$ satisfying $d(x)\geq\delta$,

$$
-\Delta V(x)-aV(x)+b(x)f(V(x))=\xi\left(-\Delta V^*(x)-aV^*(x)+b(x)\frac{f(\xi V^*(x))}{\xi}\right),
$$

which is positive for ξ sufficiently large. It follows that (3.10) is fulfilled provided ξ is large enough. This finishes the proof of the lemma. □

By Lemma 3.7, $\underline{U}_a(x)\equiv\lim_{n\to\infty}u_n(x)$ exists, for any $x\in\overline{\Omega}\setminus\overline{\Omega}_0$. Moreover, $\underline{U}_a$ is a positive solution of (3.7). Using the maximum principle once more, we find that any positive solution u of (3.7) satisfies $u\geq u_n$ on $\overline{\Omega}\setminus\overline{\Omega}_0$, for all $n\geq 1$. Hence $\underline{U}_a$ is the minimal positive solution of (3.7).

Proof of Theorem 3.5 continued.

Step 2: Existence of the maximal positive solution for problem (3.7).

Lemma 3.8 *If Ω_0 is replaced by Ω_m defined in (3.8), then problem (3.7) has a minimal positive solution provided that (A_1) and (A_2) are fulfilled.*

Proof. The argument used here (more easier, since $b>0$ on $\overline{\Omega}\setminus\Omega_m$) is similar to that in Step 1. The only difference which appears in the proof (except the replacement of Ω_0 by Ω_m) is related to the construction of $V^*(x)$ for x near $\partial\Omega_m$. Here, we use our Theorem 3.1 which says that, for any $a\in\mathbb{R}$, there exists a positive large solution $u_{a,\infty}$ of problem (3.2) in the domain $\Omega\setminus\overline{\Omega}_m$. We define $V^*(x)=u_{a,\infty}(x)$ for $x\in\Omega\setminus\overline{\Omega}_m$ and near $\partial\Omega_m$. For $\xi>1$ and $x\in\Omega\setminus\overline{\Omega}_m$ near $\partial\Omega_m$ we have

$$
\begin{aligned}
-\Delta V(x)-aV(x)+b(x)f(V(x))&=-\xi\Delta V^*(x)-a\xi V^*(x)+b(x)f(\xi V^*(x))\\
&=b(x)[f(\xi V^*(x))-\xi f(V^*(x))]\geq 0.
\end{aligned}
$$

This completes the proof. □

Let v_m be the minimal positive solution for the problem considered in the statement of Lemma 3.8. By the maximum principle, $v_m\geq v_{m+1}\geq u$ on $\overline{\Omega}\setminus\overline{\Omega}_m$, where u is any positive solution of (3.7). Hence $\overline{U}_a(x):=\lim_{m\to\infty}v_m(x)\geq u(x)$. A regularity and compactness argument shows that $\overline{U}_a$ is a positive solution of (3.7). Consequently, $\overline{U}_a$ is the maximal positive solution. This concludes the proof of Theorem 3.5. □

The next question is whether one can conclude the uniqueness of positive solutions of problem (3.7). We recall first what is already known in this direction. When $f(u) = u^p$, $p > 1$, Du and Huang [64] proved the uniqueness of the solution to problem (3.7) and established its behavior near $\partial\Omega_0$, under the assumption

$$\lim_{d(x)\searrow 0} \frac{b(x)}{[d(x)]^\tau} = c \quad \text{for some positive constants } \tau, c > 0. \tag{3.11}$$

We shall give a general uniqueness result provided that b and f satisfy the following assumptions:

(B_1) $\lim_{d(x)\searrow 0} \frac{b(x)}{k(d(x))} = c$ for some constant $c > 0$, where $0 < k \in C^1(0, \delta_0)$ is increasing and satisfies

(B_2) $K(t) = \frac{\int_0^t \sqrt{k(s)}\,ds}{\sqrt{k(t)}} \in C^1[0, \delta_0)$, for some $\delta_0 > 0$.

Assume there exist $\zeta > 0$ and $t_0 \geq 1$ such that

(A_3) $f(\xi t) \leq \xi^{1+\zeta} f(t), \quad \forall \xi \in (0,1), \quad \forall t \geq t_0/\xi;$

(A_4) the mapping $(0,1] \ni \xi \longmapsto A(\xi) = \lim_{u\to\infty} \frac{f(\xi u)}{\xi f(u)}$ is a continuous positive function.

Our uniqueness result is

Theorem 3.9 *Assume the conditions $(\tilde{A}_1)$ with $\gamma \neq 0$, (A_3), (A_4), (B_1) and (B_2) hold. Then, for any $a \in \mathbb{R}$, problem (3.7) has a unique positive solution U_a. Moreover,*

$$\lim_{d(x)\searrow 0} \frac{U_a(x)}{h(d(x))} = \xi_0,$$

where h is defined by

$$\int_{h(t)}^{\infty} \frac{ds}{\sqrt{2F(s)}} = \int_0^t \sqrt{k(s)}\,ds, \quad \forall t \in (0, \delta_0) \tag{3.12}$$

and ξ_0 is the unique positive solution of $A(\xi) = \frac{K'(0)(1-2\gamma)+2\gamma}{c}$.

Remark 1 *(a)* $(A_1) + (A_3) \Rightarrow (A_2)$. Indeed, $\lim_{u\to\infty} \frac{f(u)}{u^{1+\zeta}} > 0$ since $\frac{f(t)}{t^{1+\zeta}}$ is nondecreasing for $t \geq t_0$.

(b) $K'(0)(1-2\gamma)+2\gamma \in (0,1]$ *when* $(\tilde{A}_1)$ *with* $\gamma \neq 0$, (A_2), (B_1) *and* (B_2) *hold.*
(c) The function $(0,\infty) \ni \xi \longmapsto A(\xi) \in (0,\infty)$ *is bijective when* (A_3) *and* (A_4) *hold (see Lemma 3.10).*

Among the nonlinearities f that satisfy the assumptions of Theorem 3.9 we note: (i) $f(u) = u^p$, $p > 1$; (ii) $f(u) = u^p \ln(u+1)$, $p > 1$; (iii) $f(u) = u^p \arctan u$, $p > 1$.

Proof of Theorem 3.9. By (A_4) we deduce that the mapping $(0,\infty) \ni \xi \longmapsto A(\xi) = \lim_{u\to\infty} \frac{f(\xi u)}{\xi f(u)}$ is a continuous positive function, since $A(1/\xi) = 1/A(\xi)$ for any $\xi \in (0,1)$. Moreover, we claim

Lemma 3.10 *The function* $A : (0,\infty) \to (0,\infty)$ *is bijective, provided that* (A_3) *and* (A_4) *are fulfilled.*

Proof. By the continuity of A, we see that the surjectivity of A follows if we prove that $\lim_{\xi \searrow 0} A(\xi) = 0$. To this aim, let $\xi \in (0,1)$ be fixed. Using (A_3) we find

$$\frac{f(\xi u)}{\xi f(u)} \leq \xi^{\zeta}, \quad \forall u \geq \frac{t_0}{\xi}$$

which yields $A(\xi) \leq \xi^{\zeta}$. Since $\xi \in (0,1)$ is arbitrary, it follows that $\lim_{\xi \searrow 0} A(\xi) = 0$.

We now prove that the function $\xi \longmapsto A(\xi)$ is increasing on $(0,\infty)$ which concludes our lemma. Let $0 < \xi_1 < \xi_2 < \infty$ be chosen arbitrarily. Using assumption (A_3) once more, we obtain

$$f(\xi_1 u) = f\left(\frac{\xi_1}{\xi_2}\xi_2 u\right) \leq \left(\frac{\xi_1}{\xi_2}\right)^{1+\zeta} f(\xi_2 u), \quad \forall u \geq t_0 \frac{\xi_2}{\xi_1}.$$

It follows that

$$\frac{f(\xi_1 u)}{\xi_1 f(u)} \leq \left(\frac{\xi_1}{\xi_2}\right)^{\zeta} \frac{f(\xi_2 u)}{\xi_2 f(u)}, \quad \forall u \geq t_0 \frac{\xi_2}{\xi_1}.$$

Passing to the limit as $u \to \infty$ we find

$$A(\xi_1) \leq \left(\frac{\xi_1}{\xi_2}\right)^{\zeta} A(\xi_2) < A(\xi_2),$$

which finishes the proof. □

Proof of Theorem 3.9 continued. Set

$$\Pi(\xi) = \lim_{d(x)\searrow 0} b(x) \frac{f(h(d(x))\xi)}{h''(d(x))\xi},$$

for any $\xi > 0$. Using (B_1) we find

$$\begin{aligned}\Pi(\xi) &= \lim_{d(x)\searrow 0} \frac{b(x)}{k(d(x))} \frac{k(d(x))f(h(d(x)))}{h''(d(x))} \frac{f(h(d(x))\xi)}{\xi f(h(d(x)))} \\ &= c \lim_{t\searrow 0} \frac{k(t)f(h(t))}{h''(t)} \lim_{u\to\infty} \frac{f(\xi u)}{\xi f(u)} \\ &= \frac{c}{K'(0)(1-2\gamma)+2\gamma} A(\xi).\end{aligned}$$

This and Lemma 3.10 imply that the function $\Pi : (0,\infty) \to (0,\infty)$ is bijective. Let ξ_0 be the unique positive solution of $\Pi(\xi) = 1$, that is

$$A(\xi_0) = \frac{K'(0)(1-2\gamma)+2\gamma}{c}.$$

For $\varepsilon \in (0,1/4)$ arbitrary, we denote $\xi_1 = \Pi^{-1}(1-4\varepsilon)$, respectively $\xi_2 = \Pi^{-1}(1+4\varepsilon)$.

We choose $\delta > 0$ small enough such that

(i) $\operatorname{dist}(x,\partial\Omega_0)$ is a C^2 function on the set $\{x \in \Omega : \operatorname{dist}(x,\partial\Omega_0) \le 2\delta\}$;

(ii) $\left| \frac{h'(s)}{h''(s)} \Delta d(x) + a \frac{h(s)}{h''(s)} \right| < \varepsilon$ and $h''(s) > 0$ for all $0 < s, d(x) < 2\delta$;

(iii) $(\Pi(\xi_2)-\varepsilon) \frac{h''(d(x))\,\xi_2}{f(h(d(x))\xi_2)} \le b(x) \le (\Pi(\xi_1)+\varepsilon) \frac{h''(d(x))\,\xi_1}{f(h(d(x))\xi_1)}$ if $0 < d(x) < 2\delta$;

(iv) $b(y) < (1+\varepsilon)b(x)$, for all x,y with $0 < d(y) < d(x) < 2\delta$.

Let $\sigma \in (0,\delta)$ be arbitrary. We define $\underline{v}_\sigma(x) = h(d(x)+\sigma)\xi_1$, for any x with $d(x)+\sigma < 2\delta$, respectively $\overline{v}_\sigma(x) = h(d(x)-\sigma)\xi_2$ for any x with $\sigma < d(x) < 2\delta$.

Using (ii), (iv) and the first inequality in (iii), when $\sigma < d(x) < 2\delta$, we obtain (since $|\nabla d(x)| \equiv 1$)

$$\begin{aligned}&-\Delta\overline{v}_\sigma(x) - a\overline{v}_\sigma(x) + b(x)f(\overline{v}_\sigma(x)) \\ &= \xi_2 \left[-h'(d(x)-\sigma)\Delta d - h''(d(x)-\sigma) - ah(d(x)-\sigma) + \frac{b(x)f(h(d(x)-\sigma)\xi_2)}{\xi_2} \right] \\ &= \xi_2 h''(d(x)-\sigma) \\ &\quad \times \left[-\frac{h'(d(x)-\sigma)}{h''(d(x)-\sigma)} \Delta d(x) - a \frac{h(d(x)-\sigma)}{h''(d(x)-\sigma)} - 1 + \frac{b(x)f(h(d(x)-\sigma)\xi_2)}{h''(d(x)-\sigma)\,\xi_2} \right] \\ &\ge \xi_2 h''(d(x)-\sigma) \left[-\frac{h'(d(x)-\sigma)}{h''(d(x)-\sigma)} \Delta d(x) - a \frac{h(d(x)-\sigma)}{h''(d(x)-\sigma)} - 1 + \frac{\Pi(\xi_2)-\varepsilon}{1+\varepsilon} \right] \\ &\ge 0,\end{aligned}$$

for all x satisfying $\sigma < d(x) < 2\delta$. Similarly, using (ii), (iv) and the second inequality in (iii), when $d(x)+\sigma < 2\delta$ we find

$$\begin{aligned}
&-\Delta \underline{v}_\sigma(x) - a\underline{v}_\sigma(x) + b(x)f(\underline{v}_\sigma(x)) \\
&\quad = \xi_1 h''(d(x)+\sigma) \\
&\quad\quad \times \left[-\frac{h'(d(x)+\sigma)}{h''(d(x)+\sigma)} \Delta d(x) - a\frac{h(d(x)+\sigma)}{h''(d(x)+\sigma)} - 1 + \frac{b(x)f(h(d(x)+\sigma)\xi_1)}{h''(d(x)+\sigma)\,\xi_1} \right] \\
&\quad \leq \xi_1 h''(d(x)+\sigma) \\
&\quad\quad \times \left[-\frac{h'(d(x)+\sigma)}{h''(d(x)+\sigma)} \Delta d(x) - a\frac{h(d(x)+\sigma)}{h''(d(x)+\sigma)} - 1 + (1+\varepsilon)(\Pi(\xi_1)+\varepsilon) \right] \\
&\quad \leq 0.
\end{aligned}$$

Define $\Omega_\delta \equiv \{x \in \Omega : \ d(x) < \delta\}$. Let $\omega \subset\subset \Omega_0$ be such that the first Dirichlet eigenvalue of $(-\Delta)$ in the smooth domain $\Omega_0 \setminus \overline{\omega}$ is strictly greater than a. Denote by w a positive large solution to the following problem

$$-\Delta w = aw - p(x)f(w) \qquad \text{in } \Omega_\delta\,,$$

where $p \in C^{0,\mu}(\overline{\Omega}_\delta)$ satisfies

$$0 < p(x) \leq b(x) \text{ for } x \in \overline{\Omega}_\delta \setminus \overline{\Omega}_0, \quad p(x) = 0 \text{ on } \overline{\Omega}_0 \setminus \omega \quad \text{and} \quad p(x) > 0 \text{ for } x \in \omega.$$

The existence of w is guaranteed by our Theorem 3.1.

Suppose that u is an arbitrary solution of (3.7) and let $v := u + w$. Then v satisfies

$$-\Delta v \geq av - b(x)f(v) \quad \text{in } \Omega_\delta \setminus \overline{\Omega}_0.$$

Since

$$v|_{\partial\Omega_0} = \infty > \underline{v}_\sigma|_{\partial\Omega_0} \quad \text{and} \quad v|_{\partial\Omega_\delta} = \infty > \underline{v}_\sigma|_{\partial\Omega_\delta},$$

we find

$$u + w \geq \underline{v}_\sigma \ \text{ on } \Omega_\delta \setminus \overline{\Omega}_0. \tag{3.13}$$

Similarly

$$\overline{v}_\sigma + w \geq u \ \text{ on } \Omega_\delta \setminus \overline{\Omega}_\sigma. \tag{3.14}$$

Letting $\sigma \to 0$ in (3.13) and (3.14), we deduce that

$$h(d(x))\,\xi_2 + 2w \geq u + w \geq h(d(x))\,\xi_1, \quad \forall x \in \Omega_\delta \setminus \overline{\Omega}_0.$$

Since w is uniformly bounded on $\partial\Omega_0$, it follows that

$$\xi_1 \le \liminf_{d(x)\searrow 0} \frac{u(x)}{h(d(x))} \le \limsup_{d(x)\searrow 0} \frac{u(x)}{h(d(x))} \le \xi_2. \tag{3.15}$$

Letting $\varepsilon \to 0$ in (3.15) and looking at the definition of ξ_1 respectively ξ_2 we find

$$\lim_{d(x)\searrow 0} \frac{u(x)}{h(d(x))} = \xi_0. \tag{3.16}$$

This behavior of the solution will be speculated in order to prove that problem (3.7) has a unique solution. Indeed, let u_1, u_2 be two positive solutions of (3.7). For any $\varepsilon > 0$, denote $\tilde{u}_i = (1+\varepsilon)\,u_i$, $i = 1,2$. By virtue of (3.16) we get

$$\lim_{d(x)\searrow 0} \frac{u_1(x) - \tilde{u}_2(x)}{h(d(x))} = \lim_{d(x)\searrow 0} \frac{u_2(x) - \tilde{u}_1(x)}{h(d(x))} = -\varepsilon\,\xi_0 < 0$$

which implies

$$\lim_{d(x)\searrow 0} [u_1(x) - \tilde{u}_2(x)] = \lim_{d(x)\searrow 0} [u_2(x) - \tilde{u}_1(x)] = -\infty.$$

On the other hand, since $\frac{f(u)}{u}$ is increasing for $u > 0$, we obtain

$$\begin{aligned} -\Delta \tilde{u}_i &= -(1+\varepsilon)\,\Delta u_i \\ &= (1+\varepsilon)\,(a u_i - b(x) f(u_i)) \\ &\ge a \tilde{u}_i - b(x) f(\tilde{u}_i) \quad \text{in } \Omega \setminus \overline{\Omega}_0 \end{aligned}$$

and

$$\mathscr{B}\tilde{u}_i = \mathscr{B}u_i = 0 \quad \text{on } \partial\Omega.$$

By the maximum principle,

$$u_1(x) \le \tilde{u}_2(x), \quad u_2(x) \le \tilde{u}_1(x), \qquad \text{for all } x \in \Omega \setminus \overline{\Omega}_0.$$

Letting $\varepsilon \to 0$, we obtain $u_1 \equiv u_2$. The proof of Theorem 3.9 is complete. □

3.1.1 A Karamata Regular Variation Theory Approach

The major purpose in this section is to advance innovative methods to study the uniqueness and asymptotic behavior of large solutions of (3.2). This approach is due to Cîrstea and Rădulescu [43, 45–48] and it relies essentially on the *regular variation theory* introduced by Karamata (see Bingham, Goldie, and Teugels [19], Karamata [137]), not only in the statement but in the proof as well. This enables us to obtain significant information about the qualitative behavior of the large solution to (3.2) in a general framework that removes previous restrictions in the literature.

Definition 3.11 *A positive measurable function R defined on $[D,\infty)$, for some $D>0$, is called regularly varying (at infinity) with index $q\in\mathbb{R}$ (written $R\in RV_q$) if for all $\xi>0$*

$$\lim_{u\to\infty} R(\xi u)/R(u)=\xi^q.$$

When the index of regular variation q is zero, we say that the function is slowly varying.

We remark that any function $R\in RV_q$ can be written in terms of a slowly varying function. Indeed, set $R(u)=u^qL(u)$. From the above definition we easily deduce that L varies slowly.

The canonical q-varying function is u^q. The functions $\ln(1+u)$, $\ln\ln(e+u)$, $\exp\{(\ln u)^\alpha\}$, $\alpha\in(0,1)$ vary slowly, as well as any measurable function on $[D,\infty)$ with positive limit at infinity.

In what follows L denotes an arbitrary slowly varying function and $D>0$ a positive number. For details on the properties below, we refer to Seneta [176].

Proposition 3.12 *(i) For any $m>0$, $u^mL(u)\to\infty$, $\quad u^{-m}L(u)\to 0$ as $u\to\infty$.*

(ii) Any positive C^1-function on $[D,\infty)$ satisfying $uL_1'(u)/L_1(u)\to 0$ as $u\to\infty$ is slowly varying. Moreover, if the above limit is $q\in\mathbb{R}$, then $L_1\in RV_q$.

(iii) Assume $R:[D,\infty)\to(0,\infty)$ is measurable and Lebesgue integrable on each finite subinterval of $[D,\infty)$. Then R varies regularly if and only if there exists $j\in\mathbb{R}$ such that

$$\lim_{u\to\infty}\frac{u^{j+1}R(u)}{\int_D^u x^jR(x)\,dx} \tag{3.17}$$

exists and is a positive number, say a_j+1. In this case, $R\in RV_q$ with $q=a_j-j$.

(iv) (Karamata Theorem, 1933). If $R\in RV_q$ is Lebesgue integrable on each finite subinterval of $[D,\infty)$, then the limit defined by (3.17) is $q+j+1$, for every $j>-q-1$.

Lemma 3.13 *Assume (A_1) holds. Then we have the equivalence*

$$a)\ f' \in RV_\rho \Longleftrightarrow b)\ \lim_{u\to\infty} uf'(u)/f(u) := \vartheta < \infty \Longleftrightarrow c)\ \lim_{u\to\infty} (F/f)'(u) := \gamma > 0.$$

Remark 2 *Let a) of Lemma 3.13 be fulfilled. Then the following assertions hold*

(i) ρ is nonnegative;

(ii) $\gamma = 1/(\rho+2) = 1/(\vartheta+1)$;

(iii) If $\rho \neq 0$, then (A_2) holds (use $\lim_{u\to\infty} f(u)/u^p = \infty$, $\forall p \in (1, 1+\rho)$). The converse implication is not necessarily true (take $f(u) = u\ln^4(u+1)$). However, there are cases when $\rho = 0$ and (A_2) fails so that (3.2) has **no** *large solutions. This is illustrated by $f(u) = u$ or $f(u) = u\ln(u+1)$.*

Inspired by the definition of γ, we denote by $\mathscr{K}$ the *Karamata class*, that is, the set of all positive, increasing C^1-functions k defined on $(0, \nu)$, for some $\nu > 0$, which satisfy

$$\lim_{t\to 0+} \left(\frac{\int_0^t k(s)\,ds}{k(t)} \right)^{(i)} := \ell_i,\ i = 0, 1.$$

It is easy to see that $\ell_0 = 0$ and $\ell_1 \in [0,1]$, for every $k \in \mathscr{K}$. Our next result gives examples of functions $k \in \mathscr{K}$ with $\lim_{t\to 0+} k(t) = 0$, for *every* $\ell_1 \in [0,1]$.

Lemma 3.14 *Let $S \in C^1[D,\infty)$ be such that $S' \in RV_q$ with $q > -1$. Hence the following hold:*

a) If $k(t) = \exp\{-S(1/t)\}$ $\forall t \le 1/D$, then $k \in \mathscr{K}$ with $\ell_1 = 0$.

b) If $k(t) = 1/S(1/t)$ $\forall t \le 1/D$, then $k \in \mathscr{K}$ with $\ell_1 = 1/(q+2) \in (0,1)$.

c) If $k(t) = 1/\ln S(1/t)$ $\forall t \le 1/D$, then $k \in \mathscr{K}$ with $\ell_1 = 1$.

Remark 3 *If $S \in C^1[D,\infty)$, then $S' \in RV_q$ with $q > -1$ if and only if for some $m > 0$, $C > 0$ and $B > D$ we have $S(u) = Cu^m \exp\left\{\int_B^u \frac{y(t)}{t}\,dt\right\}$, $\forall u \ge B$, where $y \in C[B,\infty)$ satisfies $\lim_{u\to\infty} y(u) = 0$. In this case, $S' \in RV_q$ with $q = m-1$. (This is a consequence of Proposition 3.12 (iii) and (iv).)*

Our main result is

Theorem 3.15 *Let (A_1) hold and $f' \in RV_\rho$ with $\rho > 0$. Assume $b \equiv 0$ on $\partial\Omega$ satisfies*

(B) $b(x) = ck^2(d(x)) + o(k^2(d(x)))$ as $d(x) \to 0$, for some constant $c > 0$ and $k \in \mathscr{K}$.

Then, for any $a \in (-\infty, \lambda_{\infty,1})$, (3.2) admits a unique large solution u_a. Moreover,

$$\lim_{d(x)\to 0} \frac{u_a(x)}{h(d(x))} = \xi_0, \tag{3.18}$$

where $\xi_0 = \left(\frac{2+\ell_1\rho}{c(2+\rho)}\right)^{1/\rho}$ *and h is defined by*

$$\int_{h(t)}^{\infty} \frac{ds}{\sqrt{2F(s)}} = \int_0^t k(s)\,ds, \quad \forall t \in (0,\nu). \tag{3.19}$$

By Remark 3, the assumption $f' \in RV_\rho$ with $\rho > 0$ holds if and only if there exist $p > 1$ and $B > 0$ such that $f(u) = Cu^p \exp\left\{\int_B^u \frac{y(t)}{t}\,dt\right\}$, for all $u \geq B$ (y as before and $p = \rho + 1$). If B is large enough ($y > -\rho$ on $[B,\infty)$), then $f(u)/u$ is increasing on $[B,\infty)$. Thus, to get the whole range of functions f for which our Theorem 3.15 applies we have only to "paste" a suitable smooth function on $[0,B]$ in accordance with (A_1). A simple way to do this is to define

$$f(u) = u^p \exp\{\int_0^u \frac{z(t)}{t}\,dt\} \quad \text{for all} \quad u \geq 0,$$

where $z \in C[0,\infty)$ is nonnegative such that $\lim_{t\to 0^+} z(t)/t \in [0,\infty)$ and $\lim_{u\to\infty} z(u) = 0$. Clearly, $f(u) = u^p$, $f(u) = u^p \ln(u+1)$, and $f(u) = u^p \arctan u$ ($p > 1$) fall into this category.

Lemma 3.14 provides a practical method to find functions k which can be considered in the statement of Theorem 3.15. Here are some examples:
(i) $k(t) = -1/\ln t$;
(ii) $k(t) = t^\alpha$;
(iii) $k(t) = \exp\{-1/t^\alpha\}$;
(iv) $k(t) = \exp\{-\ln(1+\frac{1}{t})/t^\alpha\}$;
(v) $k(t) = \exp\{-\left[\arctan\left(\frac{1}{t}\right)\right]/t^\alpha\}$;
(vi) $k(t) = t^\alpha/\ln(1+\frac{1}{t})$, for some $\alpha > 0$.

As we shall see, the uniqueness lies upon the crucial observation (3.18), which shows that all explosive solutions have the same boundary behavior. Note that the only case of Theorem 3.15 studied so far is $f(u) = u^p$ ($p > 1$) and $k(t) = t^\alpha$ ($\alpha > 0$) (see García-Melián, Letelier-Albornoz, and Sabina de Lis [82]). For related results on the uniqueness of explosive solutions (mainly in the cases $b \equiv 1$ and $a = 0$) we refer to Bandle and Marcus [12], Loewner and Nirenberg [138], Marcus and Véron [142].

Proof of Lemma 3.13. From Proposition 3.12 (iv) and Remark 2 (i) we deduce $a) \Longrightarrow b)$ and $\vartheta = \rho + 1$. Conversely, $b) \Longrightarrow a)$ follows by Proposition 3.12 (iii) since $\vartheta \geq 1$ cf. (A_1).

$b) \Longrightarrow c)$. Indeed, $\lim_{u\to\infty} uf(u)/F(u) = 1 + \vartheta$, which yields

$$\frac{\vartheta}{1+\vartheta} = \lim_{u\to\infty}\left[1 - \left(\frac{F}{f}\right)'(u)\right] = 1 - \gamma.$$

$c) \Longrightarrow b)$. Choose $s_1 > 0$ such that

$$\left(\frac{F}{f}\right)'(u) \geq \frac{\gamma}{2}, \quad \text{for all} \quad u \geq s_1.$$

So,

$$\left(\frac{F}{f}\right)(u) \geq \frac{(u-s_1)\gamma}{2} + \left(\frac{F}{f}\right)(s_1) \quad \text{for all } u \geq s_1.$$

Passing to the limit as $u \to \infty$, we find $\lim_{u\to\infty} F(u)/f(u) = \infty$. Thus,

$$\lim_{u\to\infty} \frac{uf(u)}{F(u)} = \frac{1}{\gamma}.$$

Since

$$1-\gamma := \lim_{u\to\infty} \frac{F(u)f'(u)}{f^2(u)},$$

we obtain

$$\lim_{u\to\infty} \frac{uf'(u)}{f(u)} = \frac{1-\gamma}{\gamma}.$$

This finishes the proof of the lemma. □

Proof of Lemma 3.14. Since $\lim_{u\to\infty} uS'(u) = \infty$ (cf. Proposition 3.12 (i)), from the Karamata theorem we deduce

$$\lim_{u\to\infty} \frac{uS'(u)}{S(u)} = q + 1 > 0.$$

Therefore, in any of the cases $a)$, $b)$, $c)$, $\lim_{t\to 0^+} k(t) = 0$ and k is an increasing C^1-function on $(0, \nu)$, for $\nu > 0$ sufficiently small.

$a)$ It is clear that

$$\lim_{t\to 0^+} \frac{tk'(t)}{k(t)\ln k(t)} = \lim_{t\to 0^+} \frac{-S'(1/t)}{tS(1/t)} = -(q+1).$$

By l'Hospital's rule we deduce

$$\ell_0 = \lim_{t\to 0^+} \frac{k(t)}{k'(t)} = 0 \quad \text{and} \quad \lim_{t\to 0^+} \frac{\left(\int_0^t k(s)\,ds\right) \ln k(t)}{tk(t)} = -\frac{1}{q+1}.$$

So,

$$1-\ell_1 := \lim_{t\to 0^+} \frac{\left(\int_0^t k(s)\,ds\right) k'(t)}{k^2(t)} = 1.$$

b) We see that

$$\lim_{t\to 0^+} \frac{tk'(t)}{k(t)} = \lim_{t\to 0^+} \frac{S'(1/t)}{tS(1/t)} = q+1.$$

By l'Hospital's rule,

$$\ell_0 = 0 \quad \text{and} \quad \lim_{t\to 0^+} \frac{\int_0^t k(s)\,ds}{tk(t)} = \frac{1}{q+2}.$$

So,

$$\ell_1 = 1 - \lim_{t\to 0^+} \frac{\int_0^t k(s)\,ds}{tk(t)} \frac{tk'(t)}{k(t)} = \frac{1}{q+2}.$$

c) We have

$$\lim_{t\to 0^+} \frac{tk'(t)}{k^2(t)} = \lim_{t\to 0^+} \frac{S'(1/t)}{tS(1/t)} = q+1.$$

By l'Hospital's rule,

$$\lim_{t\to 0^+} \frac{\int_0^t k(s)\,ds}{tk(t)} = 1.$$

Thus,

$$\ell_0 = 0 \quad \text{and} \quad \ell_1 = 1 - \lim_{t\to 0^+} \frac{\int_0^t k(s)\,ds}{t} \frac{tk'(t)}{k^2(t)} = 1.$$

This finishes the proof of our lemma. □

Proof of Theorem 3.15. Fix $a \in (-\infty, \lambda_{\infty,1})$. By Theorem 3.1, problem (3.2) has at least a large solution.

If we prove that (3.18) holds for an *arbitrary* large solution u_a of (3.2), then the uniqueness follows easily. Indeed, if u_1 and u_2 are two arbitrary large solutions of (3.2), then (3.18) yields $\lim_{d(x)\to 0^+} \frac{u_1(x)}{u_2(x)} = 1$. Hence, for any $\varepsilon \in (0,1)$, there exists $\delta = \delta(\varepsilon) > 0$ such that

$$(1-\varepsilon)u_2(x) \le u_1(x) \le (1+\varepsilon)u_2(x), \quad \forall x \in \Omega \text{ with } 0 < d(x) \le \delta. \tag{3.20}$$

Choosing eventually a smaller $\delta > 0$, we can assume that

$$\overline{\Omega}_0 \subset C_\delta \quad \text{where } C_\delta := \{x \in \Omega :\ d(x) > \delta\}.$$

It is clear that u_1 is a positive solution of the boundary value problem

$$\Delta\phi + a\phi = b(x)f(\phi) \quad \text{in } C_\delta, \qquad \phi = u_1 \quad \text{on } \partial C_\delta. \tag{3.21}$$

By (A_1) and (3.20), we see that $\phi^- = (1-\varepsilon)u_2$ (resp. $\phi^+ = (1+\varepsilon)u_2$) is a positive subsolution (resp. supersolution) of (3.21). By the sub and supersolutions method, (3.21) has a positive solution ϕ_1 satisfying

$$\phi^- \le \phi_1 \le \phi^+ \quad \text{in } C_\delta.$$

Since $b > 0$ on $\overline{C}_\delta \setminus \overline{\Omega}_0$, we deduce that (3.21) has a *unique* positive solution, that is, $u_1 \equiv \phi_1$ in C_δ. This yields

$$(1-\varepsilon)u_2(x) \le u_1(x) \le (1+\varepsilon)u_2(x) \quad \text{in } C_\delta,$$

so that (3.20) holds in Ω. Passing to the limit $\varepsilon \to 0^+$, we conclude that $u_1 \equiv u_2$.

In order to prove (3.18) we state some useful properties about h:

(h_1) $h \in C^2(0,\nu)$, $\lim_{t\to 0^+} h(t) = \infty$ (straightforward from (3.19)).

(h_2) $\lim_{t\to 0^+} \dfrac{h''(t)}{k^2(t)f(h(t)\xi)} = \dfrac{1}{\xi^{\rho+1}}\dfrac{2+\rho\ell_1}{2+\rho}$, $\forall \xi > 0$ (so, $h'' > 0$ on $(0,2\delta)$, for $\delta > 0$ small enough).

(h_3) $\lim_{t\to 0^+} h(t)/h''(t) = \lim_{t\to 0^+} h'(t)/h''(t) = 0$.

We check (h_2) for $\xi = 1$ only, since $f \in RV_{\rho+1}$. Clearly,

$$h'(t) = -k(t)\sqrt{2F(h(t))}$$

and

$$h''(t) = k^2(t)f(h(t))\left(1 - 2\frac{k'(t)\left(\int_0^t k(s)\,ds\right)}{k^2(t)}\frac{\sqrt{F(h(t))}}{f(h(t))\int_{h(t)}^\infty [F(s)]^{-1/2}ds}\right), \tag{3.22}$$

for all $0 < t < \nu$. We see that $\lim_{u\to\infty}\sqrt{F(u)}/f(u) = 0$. Thus, from l'Hospital's rule and Lemma 3.13 we infer that

$$\lim_{u\to\infty}\frac{\sqrt{F(u)}}{f(u)\int_u^\infty [F(s)]^{-1/2}ds} = \frac{1}{2} - \gamma = \frac{\rho}{2(\rho+2)}. \tag{3.23}$$

Using (3.22) and (3.23) we derive (h_2) and also

$$\lim_{t\to 0^+} \frac{h'(t)}{h''(t)} = \frac{-2(2+\rho)}{2+\ell_1\rho} \lim_{t\to 0^+} \frac{\int_0^t k(s)\,ds}{k(t)} \lim_{u\to\infty} \frac{\sqrt{F(u)}}{f(u)\int_u^\infty [F(s)]^{-1/2} ds} = \frac{-\rho\ell_0}{2+\ell_1\rho} = 0. \tag{3.24}$$

From (h_1) and (h_2), $\lim_{t\to 0^+} h'(t) = -\infty$. So, l'Hospital's rule and (3.24) yield $\lim_{t\to 0^+} h(t)/h'(t) = 0$. This and (3.24) lead to $\lim_{t\to 0^+} h(t)/h''(t) = 0$ which proves (h_3).

Proof of (3.18). Fix $\varepsilon \in (0, c/2)$. Since $b \equiv 0$ on $\partial\Omega$ and (B) holds, we take $\delta > 0$ so that

(i) $d(x)$ is a C^2-function on the set $\{x \in \mathbb{R}^N :\ d(x) < 2\delta\}$;
(ii) k^2 is increasing on $(0, 2\delta)$;
(iii) $(c-\varepsilon)k^2(d(x)) < b(x) < (c+\varepsilon)k^2(d(x))$, $\forall x \in \Omega$ with $0 < d(x) < 2\delta$;
(iv) $h''(t) > 0\ \forall t \in (0, 2\delta)$ (from (h_2)).

Let $\sigma \in (0, \delta)$ be arbitrary. We define

$$\xi^{\pm} = \left[\frac{2+\ell_1\rho}{(c\mp 2\varepsilon)(2+\rho)}\right]^{1/\rho} \quad \text{and } v_\sigma^-(x) = h(d(x)+\sigma)\xi^-,$$

for all x with $d(x)+\sigma < 2\delta$ resp. $v_\sigma^+(x) = h(d(x)-\sigma)\xi^+$, for all x with $\sigma < d(x) < 2\delta$.

Using (i)–(iv), when $\sigma < d(x) < 2\delta$ we obtain (since $|\nabla d(x)| \equiv 1$)

$$\begin{aligned}\Delta v_\sigma^+ &+ a v_\sigma^+ - b(x) f(v_\sigma^+) \\ &\le \xi^+ h''(d(x)-\sigma)\left(\frac{h'(d(x)-\sigma)}{h''(d(x)-\sigma)}\Delta d(x) + a\frac{h(d(x)-\sigma)}{h''(d(x)-\sigma)} + 1\right.\\ &\qquad\left. -(c-\varepsilon)\frac{k^2(d(x)-\sigma) f(h(d(x)-\sigma)\xi^+)}{h''(d(x)-\sigma)\xi^+}\right).\end{aligned}$$

Similarly, when $d(x)+\sigma < 2\delta$ we find

$$\begin{aligned}\Delta v_\sigma^- &+ a v_\sigma^- - b(x) f(v_\sigma^-) \\ &\ge \xi^- h''(d(x)+\sigma)\left(\frac{h'(d(x)+\sigma)}{h''(d(x)+\sigma)}\Delta d(x) + a\frac{h(d(x)+\sigma)}{h''(d(x)+\sigma)} + 1\right.\\ &\qquad\left. -(c+\varepsilon)\frac{k^2(d(x)+\sigma) f(h(d(x)+\sigma)\xi^-)}{h''(d(x)+\sigma)\xi^-}\right).\end{aligned}$$

Using (h_2) and (h_3) we see that, by diminishing δ, we can assume

$$\Delta v_\sigma^+(x) + a v_\sigma^+(x) - b(x) f(v_\sigma^+(x)) \le 0 \quad \forall x \text{ with } \sigma < d(x) < 2\delta;$$

$$\Delta v_{\sigma}^{-}(x) + a v_{\sigma}^{-}(x) - b(x) f(v_{\sigma}^{-}(x)) \geq 0 \quad \forall x \text{ with } d(x) + \sigma < 2\delta.$$

Let Ω_1 and Ω_2 be smooth bounded domains such that $\Omega \subset\subset \Omega_1 \subset\subset \Omega_2$ and the first Dirichlet eigenvalue of $(-\Delta)$ in the domain $\Omega_1 \setminus \overline{\Omega}$ is greater than a. Let $p \in C^{0,\mu}(\overline{\Omega}_2)$ satisfy $0 < p(x) \leq b(x)$ for $x \in \Omega \setminus C_{2\delta}$, $p = 0$ on $\overline{\Omega}_1 \setminus \Omega$ and $p > 0$ on $\Omega_2 \setminus \overline{\Omega}_1$. Denote by w a positive large solution of

$$\Delta w + a w = p(x) f(w) \quad \text{in } \Omega_2 \setminus \overline{C}_{2\delta}.$$

The existence of w is ensured by Theorem 3.1.

Suppose that u_a is an arbitrary large solution of (3.2) and let $v := u_a + w$. Then v satisfies

$$\Delta v + a v - b(x) f(v) \leq 0 \quad \text{in } \Omega \setminus \overline{C}_{2\delta}.$$

Since $v_{|\partial\Omega} = \infty > v_{\sigma|\partial\Omega}^{-}$ and $v_{|\partial C_{2\delta}} = \infty > v_{\sigma|\partial C_{2\delta}}^{-}$, the maximum principle implies

$$u_a + w \geq v_{\sigma}^{-} \quad \text{on } \Omega \setminus \overline{C}_{2\delta}. \tag{3.25}$$

Similarly,

$$v_{\sigma}^{+} + w \geq u_a \quad \text{on } C_{\sigma} \setminus \overline{C}_{2\delta}. \tag{3.26}$$

Letting $\sigma \to 0$ in (3.25) and (3.26), we deduce

$$h(d(x))\xi^{+} + 2w \geq u_a + w \geq h(d(x))\xi^{-} \quad \text{for all } x \in \Omega \setminus \overline{C}_{2\delta}.$$

Since w is uniformly bounded on $\partial\Omega$, we have

$$\xi^{-} \leq \liminf_{d(x)\to 0} \frac{u_a(x)}{h(d(x))} \leq \limsup_{d(x)\to 0} \frac{u_a(x)}{h(d(x))} \leq \xi^{+}.$$

Letting $\varepsilon \to 0^{+}$ we obtain (3.18). This concludes the proof of Theorem 3.15. □

Bandle and Marcus proved in [13] that the blow-up rate of the unique large solution of (3.2) depends on the curvature of the boundary of Ω. Our purpose in what follows is to refine the blow-up rate of u_a near $\partial\Omega$ by giving the second term in its expansion near the boundary. This is a more subtle question which represents the goal of more recent literature (see García-Melián, Letelier-Albornoz, and Sabina de Lis [82] and the references therein). The following is very general and, as a novelty, it relies on the Karamata regular variation theory.

Recall that $\mathscr{K}$ denotes the set of all positive increasing C^1-functions k defined on $(0,\nu)$, for some $\nu > 0$, which satisfy

$$\lim_{t\searrow 0}\left[\frac{\int_0^t k(s)ds}{k(t)}\right]^{(i)} := \ell_i, \quad i=1,2.$$

We also recall that RV_q ($q \in \mathbb{R}$) is the set of all positive measurable functions

$$Z:[A,\infty)\to\mathbb{R}, A>0, \quad \text{satisfying} \quad \lim_{u\to\infty}\frac{Z(\xi u)}{Z(u)} = \xi^q \ \text{ for all } \xi > 0.$$

Define by NRV_q the class of functions

$$f:[B,\infty)\to\mathbb{R}, B>0, \quad f(u) = Cu^q \exp\left\{\int_B^u \frac{\phi(t)}{t}\,dt\right\},$$

where $C>0$ is a constant and $\phi \in C[B,\infty)$ satisfies $\lim_{t\to\infty}\phi(t)=0$. The Karamata representation theorem shows that $NRV_q \subset RV_q$.

For any $\zeta > 0$, set $\mathscr{K}_{0,\zeta}$ the subset of $\mathscr{K}$ with

$$\ell_1 = 0 \quad \text{and} \quad \lim_{t\searrow 0}\left[\frac{\int_0^t k(s)ds}{k(t)}\right]' := L_\star \in \mathbb{R}.$$

It can be proven that $\mathscr{K}_{0,\zeta} \equiv \mathscr{R}_{0,\zeta}$, where $\mathscr{R}_{0,\zeta}$ is the class of all functions $k: [d_1,\infty)\to\mathbb{R}$, $d_1>0$ such that

$$k(u^{-1}) = d_0 u\,[\Lambda(u)]^{-1}\exp\left[-\int_{d_1}^u (s\Lambda(s))^{-1}ds\right], d_0>0,$$

where $\Lambda \in C^1[d_1,\infty)$ is a positive function such that

$$\lim_{u\to\infty}\Lambda(u) = \lim_{u\to\infty} u\Lambda'(u) = 0 \quad \text{and} \quad \lim_{u\to\infty} u^{\zeta+1}\Lambda'(u) = \ell_\star \in \mathbb{R}.$$

Define

$$\mathscr{F}_{\rho\eta} = \left\{f \in NRV_{\rho+1}\ (\rho>0):\ \phi \in RV_\eta \text{ or } -\phi \in RV_\eta\right\},\ \eta \in (-\rho-2,0];$$
$$\mathscr{F}_{\rho 0,\tau} = \{f \in \mathscr{F}_{\rho 0}:\ \lim_{u\to\infty}(\ln u)^\tau \phi(u) = \ell^\star \in \mathbb{R}\},\ \tau \in (0,\infty).$$

The following result establishes a precise asymptotic estimate in the neighborhood of the boundary.

Theorem 3.16 *Assume that*

$$b(x) = k^2(d)(1+\tilde{c}d^\theta + o(d^\theta)) \quad as \quad d(x)\to 0, \tag{3.27}$$

where $k \in \mathscr{R}_{0,\zeta}$, $\theta > 0$ and $\tilde{c} \in \mathbb{R}$. Suppose that f fulfills (A_1) and one of the following growth conditions at infinity:

(i) $f(u) = Cu^{\rho+1}$ *in a neighbourhood of infinity;*

(ii) $f \in \mathscr{F}_{\rho\eta}$ *with* $\eta \neq 0$*;*

(iii) $f \in \mathscr{F}_{\rho 0,\tau_1}$ *with* $\tau_1 = \varpi/\zeta$*, where* $\varpi = \min\{\theta, \zeta\}$.

Then, for any $a \in (-\infty, \lambda_{\infty,1})$, the unique positive solution u_a of (3.2) satisfies

$$u_a(x) = \xi_0 h(d)(1 + \chi d^{\varpi} + o(d^{\varpi})) \quad \text{as } d(x) \to 0, \tag{3.28}$$

where $\xi_0 = [2(2+\rho)^{-1}]^{1/\rho}$ and h is defined by

$$\int_{h(t)}^{\infty} \frac{ds}{\sqrt{2F(s)}} = \int_0^t k(s)\,ds,$$

for $t > 0$ small enough. The expression for χ is

$$\chi = \begin{cases} \chi_1 & \text{if (i) or (ii) holds} \\ \chi_1 - \dfrac{\ell^\star}{\rho}\left(-\dfrac{\rho \ell_\star}{2}\right)^{\tau_1}\left[\dfrac{1}{\rho+2} + \ln \xi_0\right] & \text{if } f \text{ obeys (iii),} \end{cases}$$

where

$$\chi_1 := -\frac{(1+\zeta)\ell_\star}{2\zeta}\,\text{Heaviside}\,(\theta - \zeta) - \frac{\tilde{c}}{\rho}\,\text{Heaviside}\,(\zeta - \theta).$$

Proof. We first state two auxiliary results. Their proofs are straightforward and we shall omit them.

Lemma 3.17 *Assume (3.27) and $f \in NRV_{\rho+1}$ satisfies (A_1). Then h has the following properties:*

(i) $h \in C^2(0,\nu)$, $\lim_{t\searrow 0} h(t) = \infty$ *and* $\lim_{t\searrow 0} h'(t) = -\infty$*;*

(ii) $\lim_{t\searrow 0} h''(t)/[k^2(t) f(h(t)\xi)] = (2+\rho\ell_1)/[\xi^{\rho+1}(2+\rho)]$, $\forall \xi > 0$*;*

(iii) $\lim_{t\searrow 0} h(t)/h''(t) = \lim_{t\searrow 0} h'(t)/h''(t) = \lim_{t\searrow 0} h(t)/h'(t) = 0$*;*

(iv) $\lim_{t\searrow 0} h'(t)/[t h''(t)] = -\rho\ell_1/(2+\rho\ell_1)$ *and*
$\lim_{t\searrow 0} h(t)/[t^2 h''(t)] = \rho^2\ell_1^2/[2(2+\rho\ell_1)]$*;*

(v) $\lim_{t\searrow 0} h(t)/[t h'(t)] = \lim_{t\searrow 0} [\ln t]/[\ln h(t)] = -\rho\ell_1/2$*;*

(vi) *If* $\ell_1 = 0$*, then* $\lim_{t\searrow 0} t^j h(t) = \infty$*, for all* $j > 0$*;*

(vii) $\lim_{t\searrow 0} 1/[t^{\zeta} \ln h(t)] = -\rho\ell_\star/2$ *and*
$\lim_{t\searrow 0} h'(t)/[t^{\zeta+1} h''(t)] = \rho\ell_\star/(2\zeta)$, $\forall k \in \mathscr{R}_{0,\zeta}$.

Let $\tau > 0$ be arbitrary. For any $u > 0$, define

$$T_{1,\tau}(u) = \{\rho/[2(\rho+2)] - \Xi(u)\}(\ln u)^{\tau}$$

and

$$T_{2,\tau}(u) = \{f(\xi_0 u)/[\xi_0 f(u)] - \xi_0^{\rho}\}(\ln u)^{\tau}.$$

Note that if $f(u) = Cu^{\rho+1}$, for u in a neighborhood V_∞ of infinity, then $T_{1,\tau}(u) = T_{2,\tau}(u) = 0$ for each $u \in V_\infty$.

Lemma 3.18 *Assume* (A_1) *and* $f \in \mathscr{F}_{\rho\eta}$. *The following hold:*

(i) *If* $f \in \mathscr{F}_{\rho 0,\tau}$, *then*

$$\lim_{u\to\infty} T_{1,\tau}(u) = -\frac{\ell^\star}{(\rho+2)^2} \quad \text{and} \quad \lim_{u\to\infty} T_{2,\tau}(u) = \xi_0^{\rho}\ell^\star \ln \xi_0.$$

(ii) *If* $f \in \mathscr{F}_{\rho\eta}$ *with* $\eta \neq 0$, *then*

$$\lim_{u\to\infty} T_{1,\tau}(u) = \lim_{u\to\infty} T_{2,\tau}(u) = 0.$$

Fix $\varepsilon \in (0,1/2)$. We can find $\delta > 0$ such that $d(x)$ is of class C^2 on $\{x \in \mathbb{R}^N : d(x) < \delta\}$, k is nondecreasing on $(0,\delta)$, and $h'(t) < 0 < h''(t)$ for all $t \in (0,\delta)$. A straightforward computation shows that

$$\lim_{t\searrow 0} \frac{t^{1-\theta}k'(t)}{k(t)} = \infty, \qquad \text{for every } \theta > 0.$$

Using now (3.27), it follows that we can diminish $\delta > 0$ such that

$$(0,\delta) \ni t \longmapsto k^2(t)\left[1+(\tilde{c}-\varepsilon)t^{\theta}\right] \quad \text{is increasing}$$

and

$$1+(\tilde{c}-\varepsilon)d^{\theta} < b(x)/k^2(d) < 1+(\tilde{c}+\varepsilon)d^{\theta}, \tag{3.29}$$

for all $x \in \Omega$ with $d = d(x) \in (0,\delta)$. We define $u^{\pm}(x) = \xi_0 h(d)(1+\chi_\varepsilon^{\pm} d^{\varpi})$, with $d \in (0,\delta)$, where $\chi_\varepsilon^{\pm} = \chi \pm \varepsilon\,[1+\text{Heaviside}\,(\zeta-\theta)]/\rho$. Take $\delta > 0$ small enough such that $u^{\pm}(x) > 0$, for each $x \in \Omega$ with $d \in (0,\delta)$. By the Lagrange mean value theorem, we obtain

$$f(u^{\pm}(x)) = f(\xi_0 h(d)) + \xi_0 \chi_\varepsilon^{\pm} d^{\varpi} h(d) f'(\Upsilon^{\pm}(d)),$$

where

$$\Upsilon^{\pm}(d) = \xi_0 h(d)(1 + \lambda^{\pm}(d)\chi_\varepsilon^{\pm} d^{\varpi}), \qquad \text{for some } \lambda^{\pm}(d) \in [0,1].$$

We claim that

$$\lim_{d \searrow 0} f(\Upsilon^{\pm}(d))/f(\xi_0 h(d)) = 1. \tag{3.30}$$

Fix $\sigma \in (0,1)$ and $M > 0$ such that $|\chi_\varepsilon^{\pm}| < M$. Choose $\mu^\star > 0$ so that $|(1 \pm Mt)^{\rho+1} - 1| < \sigma/2$, for all $t \in (0, 2\mu^\star)$. Let $\mu_\star \in (0, (\mu^\star)^{1/\varpi})$ be such that, for every $x \in \Omega$ with $d \in (0, \mu_\star)$

$$\left| f(\xi_0 h(d)(1 \pm M\mu^\star))/f(\xi_0 h(d)) - (1 \pm M\mu^\star)^{\rho+1} \right| < \sigma/2.$$

Hence,

$$\begin{aligned} 1 - \sigma &< (1 - M\mu^\star)^{\rho+1} - \frac{\sigma}{2} \\ &< \frac{f(\Upsilon^{\pm}(d))}{f(\xi_0 h(d))} \\ &< (1 + M\mu^\star)^{\rho+1} + \frac{\sigma}{2} \\ &< 1 + \sigma, \end{aligned}$$

for every $x \in \Omega$ with $d \in (0, \mu_\star)$. This proves (3.30).

Step 1: There exists $\delta_1 \in (0, \delta)$ so that

$$\Delta u^+ + a u^+ - k^2(d)[1 + (\tilde{c} - \varepsilon)d^{\theta}] f(u^+) \leq 0 \quad \text{for all } x \in \Omega \text{ with } d \in (0, \delta_1)$$

and

$$\Delta u^- + a u^- - k^2(d)[1 + (\tilde{c} + \varepsilon)d^{\theta}] f(u^-) \geq 0 \quad \text{for all } x \in \Omega \text{ with } d \in (0, \delta_1).$$

Indeed, for every $x \in \Omega$ with $d \in (0, \delta)$, we have

$$\begin{aligned} &\Delta u^{\pm} + a u^{\pm} - k^2(d)\left[1 + (\tilde{c} \mp \varepsilon)d^{\theta}\right] f(u^{\pm}) \\ &= \xi_0 d^{\varpi} h''(d) \Big[a\chi_\varepsilon^{\pm} \tfrac{h(d)}{h''(d)} + \chi_\varepsilon^{\pm} \Delta d \tfrac{h'(d)}{h''(d)} + 2\varpi \chi_\varepsilon^{\pm} \tfrac{h'(d)}{d h''(d)} + \varpi \chi_\varepsilon^{\pm} \Delta d \tfrac{h(d)}{d h''(d)} \\ &\quad + \varpi(\varpi - 1)\chi_\varepsilon^{\pm} \tfrac{h(d)}{d^2 h''(d)} + \Delta d \tfrac{h'(d)}{d^{\varpi} h''(d)} + \tfrac{a h(d)}{d^{\varpi} h''(d)} + \textstyle\sum_{j=1}^{4} \mathscr{S}_j^{\pm}(d) \Big] \end{aligned}$$

where, for any $t \in (0, \delta)$, we denote

$$\begin{aligned}
\mathscr{S}_1^{\pm}(t) &= (-\tilde{c}\pm\varepsilon)t^{\theta-\varpi}k^2(t)\frac{f(\xi_0 h(t))}{\xi_0 h''(t)},\\
\mathscr{S}_2^{\pm}(t) &= \chi_\varepsilon^{\pm}\left(1-k^2(t)h(t)\frac{f'(\Upsilon^{\pm}(t))}{h''(t)}\right),\\
\mathscr{S}_3^{\pm}(t) &= (-\tilde{c}\pm\varepsilon)\chi_\varepsilon^{\pm}t^{\theta}k^2(t)h(t)\frac{f'(\Upsilon^{\pm}(t)}{h''(t)},\\
\mathscr{S}_4^{\pm}(t) &= t^{-\varpi}\left(1-k^2(t)\frac{f(\xi_0 h(t))}{\xi_0 h''(t)}\right).
\end{aligned}$$

By Lemma 3.17 (ii), we find

$$\lim_{t\searrow 0} k^2(t)\frac{f(\xi_0 h(t))}{\xi_0 h''(t)} = 1,$$

which yields

$$\lim_{t\searrow 0}\mathscr{S}_1^{\pm}(t) = (-\tilde{c}\pm\varepsilon)\text{Heaviside}\,(\zeta-\theta).$$

Using (3.30), we next obtain

$$\lim_{t\searrow 0} k^2(t)h(t)\frac{f'(\Upsilon^{\pm}(t))}{h''(t)} = \rho+1.$$

Hence, $\lim_{t\searrow 0}\mathscr{S}_2^{\pm}(t) = -\rho\chi_\varepsilon^{\pm}$ and $\lim_{t\searrow 0}\mathscr{S}_3^{\pm}(t) = 0$. Using the expression of h'', we derive

$$\mathscr{S}_4^{\pm}(t) = \frac{k^2(t)f(h(t))}{h''(t)}\sum_{i=1}^{3}\mathscr{S}_{4,i}(t) \quad \text{for all } t\in(0,\delta),$$

where

$$\begin{aligned}
\mathscr{S}_{4,1}(t) &= 2\frac{\Xi(h(t))}{t^{\varpi}}\left[\frac{\int_0^t k(s)\,ds}{k(t)}\right]',\\
\mathscr{S}_{4,2}(t) &= 2\frac{T_{1,\tau_1}(h(t))}{[t^{\zeta}\ln h(t)]^{\tau_1}}\\
\mathscr{S}_{4,3}(t) &= -\frac{T_{2,\tau_1}(h(t))}{[t^{\zeta}\ln h(t)]^{\tau_1}}.
\end{aligned}$$

Since $\mathscr{R}_{0,\zeta}\equiv\mathscr{K}_{0,\zeta}$, we find

$$\lim_{t\searrow 0}\mathscr{S}_{4,1}(t) = -\frac{(1+\zeta)\rho\ell_\star}{\zeta(\rho+2)}\text{Heaviside}\,(\theta-\zeta).$$

Cases (i), (ii). By Lemma 3.17 (vii) and Lemma 3.18 (ii), we find

$$\lim_{t\searrow 0} S_{4,2}(t) = \lim_{t\searrow 0} S_{4,3}(t) = 0.$$

In view of Lemma 3.17 (ii), we derive that

$$\lim_{t\searrow 0} S_4^{\pm}(t) = -\frac{(1+\zeta)\rho\ell_\star}{2\zeta}\,\text{Heaviside}\,(\theta-\zeta).$$

Case (iii). By Lemma 3.17 (vii) and Lemma 3.18 (i) we find

$$\lim_{t\searrow 0} S_{4,2}(t) = -\frac{2\ell^\star}{(\rho+2)^2}\left(-\frac{\rho\ell_\star}{2}\right)^{\tau_1} \quad \text{and} \quad \lim_{t\searrow 0} S_{4,3}(t) = -\frac{2\ell^\star}{\rho+2}\left(-\frac{\rho\ell_\star}{2}\right)^{\tau_1}\ln\xi_0.$$

Using Lemma 3.17 (ii) once more, we arrive at

$$\lim_{t\searrow 0} S_4^{\pm}(t) = -\frac{(1+\zeta)\rho\ell_\star}{2\zeta}\,\text{Heaviside}\,(\theta-\zeta) - \ell^\star\left(-\frac{\rho\ell_\star}{2}\right)^{\tau_1}\left[\frac{1}{\rho+2}+\ln\xi_0\right].$$

Note that in each of the cases (i)–(iii), the definition of $\chi_\varepsilon^{\pm}$ yields

$$\lim_{t\searrow 0}\sum_{j=1}^{4}\mathscr{S}_j^{+}(t) = -\varepsilon < 0 \quad \text{and} \quad \lim_{t\searrow 0}\sum_{j=1}^{4}\mathscr{S}_j^{-}(t) = \varepsilon > 0.$$

By Lemma 3.17 (vii), we have

$$\lim_{t\searrow 0}\frac{h'(t)}{t^{\varpi}h''(t)} = 0.$$

But

$$\lim_{t\searrow 0}\frac{h(t)}{h'(t)} = 0, \quad \text{so} \quad \lim_{t\searrow 0}\frac{h(t)}{t^{\varpi}h''(t)} = 0.$$

Thus, using Lemma 3.17 [(iii), (iv)], relation (3.31) concludes our Step 1.

Step 2: There exists M^+, $\delta^+ > 0$ such that

$$u_a(x) \le u^+(x) + M^+ \quad \text{for all } x\in\Omega \text{ with } 0<d<\delta^+.$$

Define

$$(0,\infty)\ni u \longmapsto \Psi_x(u) = au - b(x)f(u), \quad \text{for all } x\in\Omega \text{ with } d\in(0,\delta_1).$$

Clearly, $\Psi_x(u)$ is decreasing when $a\le 0$. Suppose $a\in(0,\lambda_{\infty,1})$. Obviously,

$$(0,\infty)\ni t\longmapsto \frac{f(t)}{t}\in(f'(0),\infty) \quad \text{is bijective.}$$

Let $\delta_2\in(0,\delta_1)$ be such that $b(x)<1$, $\forall x$ with $d\in(0,\delta_2)$. Let u_x define the unique positive solution of $b(x)f(u)/u = a+f'(0)$, $\forall x$ with $d\in(0,\delta_2)$. Hence, for any x with $d\in(0,\delta_2)$, $u\to\Psi_x(u)$ is decreasing on (u_x,∞). But

$$\lim_{d(x)\searrow 0} b(x)\frac{f(u^+(x))}{u^+(x)} = +\infty$$

which follows by using $\lim_{d(x)\searrow 0} u^+(x)/h(d) = \xi_0$, (A_1) and Lemma 3.17 [(ii) and (iii)]. So, for δ_2 small enough, $u^+(x) > u_x$, $\forall x$ with $d \in (0, \delta_2)$.

Fix $\sigma \in (0, \delta_2/4)$ and set $\mathcal{N}_\sigma := \{x \in \Omega : \sigma < d(x) < \delta_2/2\}$. We define

$$u^*_\sigma(x) = u^+(d-\sigma, s) + M^+,$$

where (d,s) are the local coordinates of $x \in \mathcal{N}_\sigma$. We choose $M^+ > 0$ large enough to have

$$u^*_\sigma(\delta_2/2, s) \geq u_a(\delta_2/2, s) \quad \text{for all } (\sigma, s) \in (0, \delta_2/4) \times \partial\Omega.$$

Using (3.29) and Step 1, we find

$$\begin{aligned}
-\Delta u^*_\sigma(x) &\geq au^+(d-\sigma,s) - [1 + (\tilde{c} - \varepsilon)(d-\sigma)^\theta]k^2(d-\sigma)f(u^+(d-\sigma,s)) \\
&\geq au^+(d-\sigma,s) - [1 + (\tilde{c} - \varepsilon)d^\theta]k^2(d)f(u^+(d-\sigma,s)) \\
&\geq \Psi_x(u^+(d-\sigma,s)) \\
&\geq \Psi_x(u^*_\sigma) = au^*_\sigma(x) - b(x)f(u^*_\sigma(x)) \quad \text{in } \mathcal{N}_\sigma.
\end{aligned}$$

Thus, by the maximum principle, $u_a \leq u^*_\sigma$ in $\mathcal{N}_\sigma$, $\forall \sigma \in (0, \delta_2/4)$. Letting $\sigma \to 0$, we have proved Step 2.

Step 3: There exists M^-, $\delta^- > 0$ such that $u_a(x) \geq u^-(x) - M^-$, for all $x \in \Omega$ with $0 < d < \delta^-$.

For every $r \in (0, \delta)$, define $\Omega_r = \{x \in \Omega : 0 < d(x) < r\}$. We will prove that for $\lambda > 0$ sufficiently small, $\lambda u^-(x) \leq u_a(x)$, $\forall x \in \Omega_{\delta_2/4}$. Indeed, fix arbitrarily $\sigma \in (0, \delta_2/4)$. Define

$$v^*_\sigma(x) = \lambda u^-(d+\sigma, s) \quad \text{for } x = (d,s) \in \Omega_{\delta_2/2}.$$

We choose $\lambda \in (0,1)$ small enough such that

$$v^*_\sigma(\delta_2/4, s) \leq u_a(\delta_2/4, s) \quad \text{for all } (\sigma, s) \in (0, \delta_2/4) \times \partial\Omega.$$

Using (3.29), Step 1 and (A_1), we find

$$\begin{aligned}
\Delta v^*_\sigma(x) + av^*_\sigma(x) &\geq \lambda k^2(d+\sigma)[1 + (\tilde{c} + \varepsilon)(d+\sigma)^\theta]f(u^-(d+\sigma,s)) \\
&\geq k^2(d)[1 + (\tilde{c} + \varepsilon)d^\theta]f(\lambda u^-(d+\sigma,s)) \geq bf(v^*_\sigma),
\end{aligned}$$

for all $x=(d,s)\in\Omega_{\delta_2/4}$, that is v^*_σ is a subsolution of $\Delta u+au=b(x)f(u)$ in $\Omega_{\delta_2/4}$. By the maximum principle, we conclude that $v^*_\sigma\le u_a$ in $\Omega_{\delta_2/4}$. Letting $\sigma\to 0$, we find $\lambda u^-(x)\le u_a(x)$, $\forall x\in\Omega_{\delta_2/4}$.

Since $\lim_{d\searrow 0}u^-(x)/h(d)=\xi_0$, by using (A_1) and Lemma 3.17 [(ii), (iii)], we can easily obtain

$$\lim_{d\searrow 0}k^2(d)\frac{f(\lambda^2u^-(x))}{u^-(x)}=\infty.$$

So, there exists $\tilde\delta\in(0,\delta_2/4)$ such that

$$k^2(d)[1+(\tilde c+\varepsilon)d^\theta]f(\lambda^2u^-)/u^-\ge\lambda^2|a|,\quad\forall x\in\Omega\text{ with }0<d\le\tilde\delta.\tag{3.31}$$

By Lemma 3.17 [(i) and (v)], we deduce that $u^-(x)$ decreases with d when $d\in(0,\tilde\delta)$ (if necessary, $\tilde\delta>0$ is diminished). Choose $\delta_*\in(0,\tilde\delta)$, close enough to $\tilde\delta$, such that

$$h(\delta_*)(1+\chi_\varepsilon^-\delta_*^\varpi)/[h(\tilde\delta)(1+\chi_\varepsilon^-\tilde\delta^\varpi)]<1+\lambda.\tag{3.32}$$

For each $\sigma\in(0,\tilde\delta-\delta_*)$, we define $z_\sigma(x)=u^-(d+\sigma,s)-(1-\lambda)u^-(\delta_*,s)$. We prove that z_σ is a subsolution of $\Delta u+au=b(x)f(u)$ in Ω_{δ_*}. Using (3.32), we have

$$z_\sigma(x)\ge u^-(\tilde\delta,s)-(1-\lambda)u^-(\delta_*,s)>0\quad\text{ for all }x=(d,s)\in\Omega_{\delta_*}.$$

By (3.29) and Step 1, z_σ is a subsolution of $\Delta u+au=b(x)f(u)$ in Ω_{δ_*} if

$$k^2(d+\sigma)[1+(\tilde c+\varepsilon)(d+\sigma)^\theta]\left[f(u^-(d+\sigma,s))-f(z_\sigma(d,s))\right]\ge a(1-\lambda)u^-(\delta_*,s),\tag{3.33}$$

for all $(d,s)\in\Omega_{\delta_*}$. Applying the Lagrange mean value theorem and (A_1), we infer that (3.33) is a consequence of

$$k^2(d+\sigma)[1+(\tilde c+\varepsilon)(d+\sigma)^\theta]\frac{f(z_\sigma(d,s))}{z_\sigma(d,s)}\ge|a|,\quad\text{ for all }(d,s)\in\Omega_{\delta_*}.$$

This inequality holds by virtue of (3.31), (3.32) and the decreasing character of u^- with d.

On the other hand,

$$z_\sigma(\delta_*,s)\le\lambda u^-(\delta_*,s)\le u_a(x)\quad\text{ for all }x=(\delta_*,s)\in\Omega.$$

Clearly,

$$\limsup_{d\to 0}(z_\sigma-u_a)(x)=-\infty$$

and $b > 0$ in Ω_{δ_*}. Thus, by the maximum principle,

$$z_\sigma \le u_a \quad \text{in } \Omega_{\delta_*}, \text{ for all } \sigma \in (0, \tilde{\delta} - \delta_*).$$

Letting $\sigma \to 0$, we conclude the assertion of Step 3.

By Steps 2 and 3 we have

$$\chi_\varepsilon^+ \ge \left\{-1 + \frac{u_a(x)}{\xi_0 h(d)}\right\} d^{-\varpi} - \frac{M^+}{\xi_0 d^{\varpi} h(d)} \quad \text{if } x \in \Omega, d = d(x) \in (0, \delta^+),$$

and

$$\chi_\varepsilon^- \le \left\{-1 + \frac{u_a(x)}{\xi_0 h(d)}\right\} d^{-\varpi} + \frac{M^-}{\xi_0 d^{\varpi} h(d)} \quad \text{if } x \in \Omega, d = d(x) \in (0, \delta^-).$$

Passing to the limit as $d \to 0$ and using Lemma 3.17 (vi), we obtain

$$\chi_\varepsilon^- \le \liminf_{d \to 0} \left\{-1 + \frac{u_a(x)}{\xi_0 h(d)}\right\} d^{-\varpi}$$

and

$$\limsup_{d \to 0} \left\{-1 + \frac{u_a(x)}{\xi_0 h(d)}\right\} d^{-\varpi} \le \chi_\varepsilon^+.$$

Letting $\varepsilon \to 0$, we conclude our proof. □

3.2 Keller–Osserman Condition Revisited

In this section we are concerned with the existence, uniqueness, and numerical approximation of boundary blow-up solutions for elliptic PDEs as $\Delta u = f(u)$, where f satisfies a Keller–Osserman type condition. We characterize existence of such solutions for nonmonotone f. As an example, we construct an infinite family of boundary blow-up solutions for the equation $\Delta u = u^2(1 + \cos u)$ on a ball. We prove uniqueness (on balls) when f is increasing and convex in a neighborhood of infinity and we discuss and perform some numerical computations to approximate such boundary blow-up solutions. Our approach relies on the methods developed in [65].

3.2.1 *Setting of the Problem*

Let f be a nonnegative function defined on $[0,+\infty)$ such that $f(0)=0$. We assume, for the sake of simplicity that f is a C^1 function. Considering Ω a smooth bounded domain of $\mathbb{R}^D$, $D \geq 1$, we seek $u > 0$ a smooth function such that

$$\begin{cases} \Delta u = f(u) & \text{in } \Omega, \\ u = +\infty & \text{on } \partial\Omega, \end{cases} \tag{3.34}$$

where the boundary condition is to be understood as :

$$\lim_{x \to x_0} u(x) = +\infty \quad \forall x_0 \in \partial\Omega.$$

To prove existence of such a boundary blow-up solution, it is classically assumed that f is a nondecreasing function with suitable growth rate at infinity, as demonstrated independently by Keller [115] and Osserman [155].

In this chapter we study existence, asymptotic behavior, uniqueness and numerical approximation of solutions of (3.34), when f may exhibit *nonmonotone* behavior.

3.2.1.1 Existence Results

Existence of solutions of (3.34) is closely related to the following growth conditions : for $s \in [0,+\infty)$, let $F(s) = \int_0^s f(t)\,dt$ and define $\Phi : (0,+\infty) \to (0,+\infty]$ by

$$\Phi(\alpha) = \frac{1}{\sqrt{2}} \int_\alpha^\infty \frac{ds}{\sqrt{F(s)-F(\alpha)}},$$

where we let by convention $\Phi(\alpha) = +\infty$, whenever the integral is divergent or $F(s) = F(\alpha)$ on a set of positive measure.

Definition 3.19 We say that f satisfies the *Keller–Osserman condition* whenever

$$\exists \alpha > 0 \quad \Phi(\alpha) < \infty. \tag{3.35}$$

We say that f satisfies the *Sharpened Keller–Osserman condition* whenever

$$\liminf_{\alpha \to \infty} \Phi(\alpha) = 0. \tag{3.36}$$

Clearly, the Sharpened Keller–Osserman condition implies the classical one. It turns out that both conditions are equivalent:

Proposition 3.20 *Assume* (3.35) *holds for some* $\alpha > 0$. *Then* (3.36) *holds.*

We point out that in general, $\lim_{\alpha\to\infty}\Phi(\alpha)$ may not exist: in the special case $f(u) = u^2(1+\cos u)$, $\limsup_{\alpha\to\infty}\Phi(\alpha) = +\infty$. However, (3.36) still holds.

Proof of Proposition 3.20. Assume that f satisfies the Keller–Osserman condition. Up to translation, we may always assume that

$$\int_0^{+\infty} \frac{dt}{\sqrt{F(t)}} < +\infty. \tag{3.37}$$

Consider the change of variable $u = F(t)$. Then, letting $g(u) = (F^{-1})'(u)$, (3.37) reads also

$$\int_0^{+\infty} \frac{g(u)}{\sqrt{u}} du < +\infty, \tag{3.38}$$

whereas (3.36) can be rewritten as

$$\liminf_{\beta\to+\infty} \int_\beta^{+\infty} \frac{g(u)}{\sqrt{u-\beta}} du = 0. \tag{3.39}$$

First step: we claim that

$$\limsup_{\beta\to+\infty} \int_{2\beta}^{+\infty} \frac{g(u)}{\sqrt{u-\beta}} du = 0. \tag{3.40}$$

Observe that $u \le 2(u-\beta)$ and then

$$\int_{2\beta}^{+\infty} \frac{g(u)}{\sqrt{u-\beta}} du \le \sqrt{2} \int_{2\beta}^{+\infty} \frac{g(u)}{\sqrt{u}} du.$$

Second step: it remains to prove that

$$\liminf_{\beta\to+\infty} \int_\beta^{2\beta} \frac{g(u)}{\sqrt{u-\beta}} du = 0. \tag{3.41}$$

We argue by contradiction. Let us observe that

$$\int_\beta^{2\beta} \frac{g(u)}{\sqrt{u-\beta}} du = \frac{1}{2} \int_0^{\sqrt{\beta}} g(u^2+\beta) du.$$

Let us assume that there exists $C > 0$ such that for any β

$$0 < C \leq \int_0^\beta g(u^2+\beta^2)du. \tag{3.42}$$

Integrate this for β between 0 and R

$$0 < CR \leq \int_0^R \int_0^R g(u^2+\beta^2)\chi_{u\leq\beta}dud\beta. \tag{3.43}$$

The right-hand side of this inequality is bounded by an integral on a half disc of radius R. By symmetry and using polar coordinates

$$0 < CR \leq \pi \int_0^R g(r^2)rdr = \frac{\pi}{2}\int_0^{R^2} g(s)ds. \tag{3.44}$$

Remember that g is the derivative of F^{-1}. Thus,

$$0 < CR \leq \frac{\pi}{2}F^{-1}(R^2). \tag{3.45}$$

Setting $\xi = F^{-1}(R^2)$ this leads to $F(\xi) \leq C\xi^2$. This contradicts the Keller–Osserman condition (3.35). □

Next, we consider the special case $f(u) = u^2(1+\cos u)$.

Proposition 3.21 *Let* $f(u) = u^2(1+\cos u)$. *Then,*

$$\limsup_{\alpha\to\infty} \Phi(\alpha) = \infty,$$
$$\liminf_{\alpha\to\infty} \Phi(\alpha) = 0.$$

Proof. Set $\alpha = (2k+1)\pi$. For t close to α, $F(t) - F(\alpha) \sim \alpha^2(t-\alpha)^2$. Therefore $\Phi(\alpha) = +\infty$. In particular, $\limsup_{\alpha\to\infty} \Phi(\alpha) = \infty$. □

With these definitions in mind, our main results concerning existence read as follows.

Theorem 3.22 *The following statements are equivalent.*

- f *satisfies the Keller–Osserman condition* (3.35).
- f *satisfies the Sharpened Keller–Osserman condition* (3.36).
- *There exists a ball* $\Omega = B_R$ *such that* (3.34) *admits (at least) a positive boundary blow-up.*
- *Given any (smooth bounded) domain* Ω, (3.34) *admits (at least) a positive boundary blow-up solution.*

Theorem 3.22 is a straightforward consequence of Proposition 3.20 and the following two theorems.

Theorem 3.23 *f satisfies the Keller–Osserman condition if and only if* (3.34) *admits (at least) a positive boundary blow-up solution on some ball.*

Theorem 3.24 *f satisfies the Sharpened Keller–Osserman condition if and only if* (3.34) *admits (at least) a positive boundary blow-up solution on any (smooth bounded) domain Ω.*

In particular, Theorem 3.24 implies existence of boundary blow-up solutions for functions such as $f(u) = u^2(1+\cos u)$.

3.2.1.2 Asymptotic Behavior

The blow-up rate of solutions of (3.34) is determined implicitly by the following theorem.

Theorem 3.25 *Assume Ω satisfies uniform interior and exterior sphere conditions on its boundary. Assume* (3.36) *holds and let u denote any positive solution of* (3.34). *Then,*

$$\lim_{x \to x_0} \frac{\displaystyle\int_{u(x)}^{\infty} \frac{dt}{\sqrt{2F(t)}}}{\delta(x)} = 1,$$

where $\delta(x) = \mathrm{dist}(x, \partial\Omega)$.

3.2.1.3 Uniqueness Results

In view of the maximum principle, it seems natural to only assume that f is nondecreasing in order to obtain uniqueness. To the best of our knowledge, no proof (or counter-example) of such a statement has been given yet. Extra requirements such as the convexity of f or the monotony of $f(u)/u$ are needed in the proofs found in the literature (see e.g. [143]). When the domain is a ball, we relax such assumptions as follows:

Theorem 3.26 *Assume that f is nondecreasing and that f is convex in a neighborhood of $+\infty$. Assume also that f satisfies the Keller–Osserman condition* (3.35). *Then on any ball $B(0,R)$ there exists a unique boundary blow-up solution of* (3.34).

Remark 4 The same result also holds if f is nondecreasing, $f(u)/u$ is nondecreasing in a neighbourhood of $+\infty$ and (3.35) holds.

In dimension $D = 1$, a necessary and sufficient condition for uniqueness can be derived. Namely, we have the following

Proposition 3.27 *Assume that f satisfies the Sharpened Keller–Osserman condition. Then* (3.34) *admits a unique solution on $\Omega = (-R,R)$ if and only if the equation*

$$\Phi(\alpha) = R$$

admits exactly one solution.

As a straightforward consequence, we obtain

Corollary 3.28 *Assume that f satisfies the Sharpened Keller–Osserman condition.* (3.34) *admits a unique solution on any domain $\Omega = (-R,R)$ if and only if $\Phi : (0,\infty) \to (0,\infty)$ is one-to-one.*

Remark 5 In particular, if f is nondecreasing, one can easily show that Φ is one-to-one.

3.2.2 Minimality Principle

We restate the well-known sub and supersolution method (see [50] and [51]) and derive elementary but important corollaries.

Proposition 3.29 *Consider Ω a bounded domain of $\mathbb{R}^D$ such that all boundary points are regular, $f \in C(\mathbb{R})$ and $g \in C(\partial\Omega)$. Assume there exist two functions $\underline{u}$, $\overline{u} \in C(\bar{\Omega})$ such that $\underline{u} \le \overline{u}$ and*

$$\begin{cases} \Delta \underline{u} \ge f(\underline{u}) & \text{in } \mathscr{D}'(\Omega), \qquad (\text{resp. } \Delta \overline{u} \le f(\overline{u}) \text{ in } \mathscr{D}'(\Omega)) \\ \underline{u} \le g & \text{on } \partial\Omega \qquad (\text{resp. } \overline{u} \ge g \text{ on } \partial\Omega). \end{cases} \tag{3.46}$$

Then the problem

$$\begin{cases} \Delta u = f(u) & \text{in } \mathscr{D}'(\Omega), \\ u = g & \text{on } \partial\Omega, \end{cases} \tag{3.47}$$

possesses at least one solution $u \in C(\bar{\Omega}, \mathbb{R})$ such that $\underline{u} \le u \le \overline{u}$.

Corollary 3.30 *(Minimality Principle) Make the same assumptions as in Proposition 3.29. Then there exists a unique solution $u \in C(\bar{\Omega})$ of (3.47) such that $\underline{u} \leq u$ and $u|_\omega \leq \overline{v}$ for any open subset ω of Ω and any function $\overline{v} \in C(\bar{\omega})$ satisfying*

$$\begin{cases} \Delta\overline{v} \leq f(\overline{v}) & \text{in } \mathscr{D}'(\omega), \\ \overline{v} \geq \underline{u} & \text{in } \omega, \\ \overline{v} \geq u & \text{on } \partial\omega. \end{cases} \tag{3.48}$$

We call u the minimal solution of (3.47) *relative to $\underline{u}$.*

Proof. Let $\underline{u}, \overline{u}$ be the sub and supersolution given in the statement of Proposition 3.29. Let $(I, >)$ denote the set of all finite families containing $\overline{u}$ of supersolutions of (3.47) which stay above $\underline{u}$, ordered by inclusion : $i \in I$ if there exist $n \in \mathbb{N}$ and supersolutions $\overline{v}_k \in C(\bar{\Omega})$, $1 \leq k \leq n$ (that is, (3.46) holds when $\overline{u}$ is replaced by $\overline{v}_k$) with $\overline{v}_k \geq \underline{u}$, such that $i = \{\overline{u}, \overline{v}_1, \ldots, \overline{v}_n\}$.

I is nonempty since $\{\overline{u}\} \in I$. I is filtrating increasing, that is, if $i_1, i_2 \in I$ there exists $i_3 \in I$ such that $i_3 > i_1, i_2$ (take e.g. $i_3 = i_1 \cup i_2$). We prove that given $i = \{\overline{u}, \overline{v}_1, \ldots, \overline{v}_n\} \in I$ there exists a solution $u_i \in C(\bar{\Omega})$ of (3.47) such that $u_i \leq \overline{v}$ for all $\overline{v} \in i$. Let indeed u^0 denote the solution given by Proposition 3.29. Following [51], since $\underline{u} \leq u^0 \leq \overline{u}$, u^0 is also a solution of (3.47), when f is replaced by the truncation $f^0 \in C(\bar{\Omega} \times \mathbb{R})$ defined by

$$f^0(x,u) = \begin{cases} f(\underline{u}(x)) & \text{if } u < \underline{u}(x), \\ f(u) & \text{if } \underline{u}(x) \leq u \leq \overline{u}(x), \\ f(\overline{u}(x)) & \text{if } u > \overline{u}(x). \end{cases}$$

In [51], the authors prove that in fact *any* solution u of (3.47) with nonlinearity f^0 satisfies $\underline{u} \leq u \leq \overline{u}$ (and solves the problem with the original nonlinearity f). For convenience, we reproduce here the argument of Clément and Sweers : take any solution u of (3.47) with nonlinearity f^0. Assume by contradiction that $\Omega_+ := \{x \in \Omega \,:\, u(x) > \overline{u}(x)\}$ is nonempty. Working if necessary on a connected component of Ω_+, we may also assume that Ω_+ is connected. For $x \in \partial\Omega_+$, either $u(x) = \overline{u}(x)$ or $x \in \partial\Omega$, so that $u(x) = g(x) \leq \overline{u}(x)$. Hence,

$$\begin{cases} \Delta(\overline{u} - u) \leq f(\overline{u}) - f^0(x,u) = 0 & \text{in } \mathscr{D}'(\Omega_+), \\ \overline{u} - u \geq 0 & \text{on } \partial\Omega_+. \end{cases}$$

By the (weak) Maximum Principle, $\overline{u} \geq u$ in Ω_+, which is a contradiction. Hence, $u \leq \overline{u}$ and we can prove similarly that $u \geq \underline{u}$.

Define now the truncation $f^1 \in C(\bar{\Omega} \times \mathbb{R})$ of f^0 associated to $\overline{v}_1$ by :

$$f^1(x,u) = \begin{cases} f^0(x,\underline{u}(x)) & \text{if } u < \underline{u}(x), \\ f^0(x,u) & \text{if } \underline{u}(x) \leq u \leq \overline{v}_1(x), \\ f^0(x,\overline{v}_1(x)) & \text{if } u > \overline{v}_1(x). \end{cases}$$

Clearly, $\underline{u}$ and $\overline{v}_1$ are a sub and a supersolution of (3.47) with nonlinearity f^1. Applying Proposition 3.29 (which still holds for nonautonomous nonlinearities, see [51]), we can thus construct a solution u^1 of (3.47) with nonlinearity f^1, satisfying $\underline{u} \leq u^1 \leq \overline{v}_1$. Clearly, u^1 is a solution of the problem with nonlinearity f^0 and, as we mentioned earlier, we must have $u^1 \leq \overline{u}$. Repeating the process inductively, we obtain a solution $u_i := u^n$ such that $\underline{u} \leq u_i \leq \overline{u}, \overline{v}_1, \ldots, \overline{v}_n$.

Note that u_i may not be unique. Nevertheless, using the Axiom of Choice on the set of all such solutions, we can construct a well-defined generalized sequence $(u_i)_{i \in I}$, contained in the set K of all solutions u satisfying $\underline{u} \leq u \leq \overline{u}$.

By standard elliptic estimates, K is a compact subset of $C(\bar{\Omega})$ so there exists a generalized subsequence $\{u_{\phi(j)}\}_{j \in J}$ converging to a solution u of (3.47).

Choose now an arbitrary supersolution $\overline{v} \geq \underline{u}$ and let $i_1 := \{\overline{v}, \overline{u}\} \in I$. Given $\varepsilon > 0$, let $j_0 \in J$ such that $j > j_0 \Longrightarrow \|u_{\phi(j)} - u\|_\infty < \varepsilon$. Also choose $j_1 \in J$ such that $j > j_1 \Longrightarrow \phi(j) > i_1$. Finally pick $j_3 > j_1, j_2$. Then, for $j > j_3$,

$$u \leq \|u_{\phi(j)} - u\|_\infty + u_{\phi(j)} \leq \varepsilon + \overline{v}.$$

Letting $\varepsilon \to 0$, we conclude that $u \leq \overline{v}$ for any supersolution $\overline{v} \geq \underline{u}$. Clearly, u is the unique such solution.

It remains to prove that given any subdomain ω and any function $\overline{v} \in C(\bar{\omega})$ satisfying (3.48), $u \leq \overline{v}$. Fix such a function $\overline{v}$ and define $h^k \in C(\bar{\Omega} \times \mathbb{R})$, $k = 0, 1$, by

$$h^0(x,t) = \begin{cases} f(\underline{u}(x)) & \text{if } t < \underline{u}(x), \\ f(t) & \text{if } \underline{u}(x) \leq t \leq u(x), \\ f(u(x)) & \text{if } t > u(x). \end{cases}$$

and

$$h^1(x,t) = \begin{cases} h^1(x,\underline{u}(x)) & \text{if } t < \underline{u}(x), \\ h^1(x,t) & \text{if } x \in \Omega \setminus \omega \text{ or if } x \in \omega \text{ and } \underline{u}(x) \le t \le \overline{v}(x), \\ h^1(x,\overline{v}(x)) & \text{if } x \in \omega \text{ and } t > \overline{v}(x). \end{cases}$$

Working as before, we may solve (3.47) with nonlinearity h^1 and obtain a solution $\tilde{u}$ of (3.47) with nonlinearity f such that $\underline{u} \le \tilde{u} \le u$ and $\tilde{u}|_\omega \le \overline{v}$. Since $\tilde{u}$ is a supersolution of (3.47), we also have $u \le \tilde{u}$. Hence $u|_\omega = \tilde{u}|_\omega \le \overline{v}$. □

We present in what follows a short proof of Corollary 3.30 in the case where f is a locally Lipschitz function.

Proof. Uniqueness : Let u_1, u_2 be two such solutions. Choosing $\omega = \Omega$ and $\overline{v} = u_2$ in the statement of Corollary 3.30, we conclude that $u_1 \le u_2$. Reversing the roles of u_1 and u_2, we conclude that $u_1 = u_2$.
Existence : Let $\Lambda = \sup_{[\min \underline{u}, \max \overline{u}]} |f'|$, $u_0 = \underline{u}$ and for $k \ge 1$, define $u_k \in C(\bar{\Omega})$ inductively by

$$\begin{cases} \Delta u_k - \Lambda u_k = f(u_{k-1}) - \Lambda u_{k-1} & \text{in } \mathscr{D}'(\Omega), \\ u_k = g & \text{on } \partial\Omega. \end{cases}$$

Then it is known that the sequence $\{u_k\}$ is nondecreasing and converges to a solution $u \in C^2(\Omega) \cap C(\bar{\Omega})$ of (3.47), which satisfies in addition $\underline{u} \le u \le \overline{u}$.

Let $\overline{v} \in C(\bar{\omega})$ verify (3.48) and assume by contradiction that the set $\omega_1 := \{x \in \omega \,:\, \overline{v}(x) < u(x)\}$ is nonempty. Clearly ω_1 is open. Working if necessary with a connected component of ω_1, we assume that ω_1 is connected. We prove by induction that $\overline{v} \ge u_k$ in ω_1 for all $k \in \mathbb{N}$. Passing to the limit as $k \to \infty$, we then obtain a contradiction with the definition of ω_1.

By assumption, $\overline{v} \ge \underline{u} = u_0$ in ω_1. Given $k \ge 1$, assume that $\overline{v} \ge u_{k-1}$ in ω_1. In particular, we have that $\overline{v}(x) \in [\min \underline{u}, \max \overline{u}]$ for $x \in \omega_1$.

Observe that if $x \in \partial\omega_1$ then either $\overline{v}(x) = u(x)$, or $x \in \partial\omega$, whence $\overline{v}(x) \ge u(x)$. Since $u \ge u_k$, we conclude that $\overline{v} \ge u \ge u_k$ on $\partial\omega_1$. Hence,

$$\begin{cases} \Delta(\overline{v} - u_k) - \Lambda(\overline{v} - u_k) \le f(\overline{v}) - f(u_{k-1}) - \Lambda(\overline{v} - u_{k-1}) \le 0 & \text{in } \mathscr{D}'(\omega_1), \\ \overline{v} - u_k \ge 0 & \text{on } \partial\omega_1. \end{cases}$$

By the (weak) Maximum Principle, $\overline{v} \ge u_k$ in ω_1. □

Remark 6 Applying the Minimality Principle to (3.47) with nonlinearity $-f$, we also obtain the existence and uniqueness of a maximal solution relative to $\overline{u}$, defined in the straightforward way.

Remark 7 Assume $\Omega = B_R$ is a ball centered at the origin and g is a positive constant. If $\underline{u}$ is radial , one easily sees that the minimal solution u relative to $\underline{u}$ is radial : just apply the Minimality Principle 3.30 with $\overline{v}(x) = u(O(x))$, where $O \in \mathscr{O}_D$ is an arbitrary rotation of the Euclidean space. A well-known result of Gidas, Ni and Nirenberg [97] states that *any* solution $u > g$ is radially symmetric, provided f is e.g. locally Lipschitz.

Finally, letting $\phi(r) = u(x)$ for $r = |x|$, it follows from standard ODE theory that $\phi'(0) = 0$ and $\phi'(r) > 0$ in $(0,R)$.

Corollary 3.31 *(Minimality Principle for blow-up solutions) Let Ω be a bounded domain of $\mathbb{R}^D$ such that all boundary points are regular, $f \in C(\mathbb{R})$. Assume there exist a function $\underline{u} \in C(\bar{\Omega})$ such that $\Delta \underline{u} \geq f(\underline{u})$ in $\mathscr{D}'(\Omega)$ and a function $v \in C(\Omega)$ such that $\Delta v \leq f(v)$ in $\mathscr{D}'(\Omega)$, $\lim_{x \to x_0} v(x) = +\infty$ for all $x_0 \in \partial\Omega$ and $v \geq \underline{u}$.*

Then there exists a unique solution $u \in C(\Omega)$ of (3.34) *such that $\underline{u} \leq u$ and $u|_\omega \leq \overline{v}$ for any open subset $\omega \subset \Omega$ and any $\overline{v} \in C(\omega)$ satisfying*

$$\begin{cases} \Delta \overline{v} \leq f(\overline{v}) & \text{in } \mathscr{D}'(\omega), \\ \overline{v} \geq \underline{u} & \text{in } \omega, \\ \lim_{x \to x_0} \overline{v}(x) = +\infty & \text{for all } x_0 \in \partial\omega. \end{cases} \tag{3.49}$$

We call u the minimal solution of (3.34) *relative to $\underline{u}$.*

Proof. Clearly, there exists at most one such solution. Let now N denote any integer larger than $\|\underline{u}\|_{L^\infty(\Omega)}$ and let u_N denote the minimal solution relative to $\underline{u}$ of

$$\begin{cases} \Delta u_N = f(u_N) & \text{in } \mathscr{D}'(\Omega), \\ u_N = N & \text{on } \partial\Omega. \end{cases}$$

Using the Minimality Principle 3.30, one can easily show that the sequence (u_N) is nondecreasing and that $u_N \leq N$. Take any (smooth) open set $\omega \subset\subset \Omega$ such that $v \geq N$ on $\partial\omega$. By the Minimality Principle 3.30 again, we conclude that $v|_\omega \geq u_N|_\omega$. Since this holds for any such ω, we conclude that $v \geq u_N$ in Ω. In particular the sequence (u_N) is bounded on compact subsets of Ω and using elliptic regularity, we conclude that (u_N) converges to a blow-up solution u of (3.34) such that $\underline{u} \le u \le v$.

Now take $\omega \subset \Omega$ open and $\overline{v} \in C(\omega)$ satisfying (3.49). Take $\tilde{\omega} \subset\subset \omega$ such that $\overline{v} \ge N$ on $\partial\tilde{\omega}$. Applying the Minimality Principle 3.30, we have that $u_N|_{\tilde{\omega}} \le \overline{v}|_{\tilde{\omega}}$. Again, since $\tilde{\omega} \subset\subset \omega$ is arbitrary, we conclude that $u_N|_{\omega} \le \overline{v}$. Letting $N \to \infty$ yields the desired inequality. □

Remark 8 In contrast with Remark 6, there *does not* exist in general a maximal boundary blow-up solution of (3.34). See Sect. 3.2.10.3 for enlightening counter-examples.

3.2.3 Existence of Solutions on Some Ball

In this section, we prove that (3.35) implies the existence of a boundary blow-up solution on some ball. First, we state and prove a useful technical lemma.

Lemma 3.32 *Let $\phi \in C^2(0,R)$ be a nondecreasing function solving*

$$\phi'' + \frac{D-1}{r}\phi' = f(\phi) \quad \text{in } (0,R). \tag{3.50}$$

Then, given $0 < r_1 < r_2 < R$,

$$\frac{1}{\sqrt{2}}\int_{\phi(r_1)}^{\phi(r_2)} \frac{1}{\sqrt{F(s)-F(\phi(r_1))}}\,ds \ge \frac{1}{D-2}r_1\left(1-\left(\frac{r_1}{r_2}\right)^{D-2}\right),$$

if $D \ne 2$ and

$$\frac{1}{\sqrt{2}}\int_{\phi(r_1)}^{\phi(r_2)} \frac{1}{\sqrt{F(s)-F(\phi(r_1))}}\,ds \ge r_1 \ln\frac{r_2}{r_1},$$

if $D = 2$.

Proof. For $r \in (r_1, r_2)$, (3.50) is equivalent to

$$\frac{d}{dr}\left(r^{D-1}\phi'\right) = r^{D-1}f(\phi). \tag{3.51}$$

Multiplying the above equation by $r^{D-1}\phi'$ and integrating between r_1 and r, we obtain

$$\frac{1}{2}r^{2D-2}\left(\phi'(r)\right)^2 \geq \frac{1}{2}\left(r^{2D-2}\left(\phi'(r)\right)^2 - r_1^{2D-2}\left(\phi'(r_1)\right)^2\right)$$
$$= \int_{r_1}^{r} t^{2D-2} f(\phi(t))\phi'(t)\,dt$$
$$\geq r_1^{2D-2}\left(F(\phi(r)) - F(\phi(r_1))\right).$$

So,

$$\frac{1}{\sqrt{2}}\frac{\phi'(r)}{\sqrt{F(\phi(r)) - F(\phi(r_1))}} \geq \left(\frac{r_1}{r}\right)^{D-1}.$$

Integrating the above equation between r_1 and r_2, we obtain the desired result.□

Assume now (3.35) holds for some $\alpha > 0$. If $D \neq 2$, assume temporarily that $\Phi(\alpha) < \frac{1}{|D-2|}$. Applying Proposition 3.29 with $\underline{u} = 0$ and $\overline{u} = \alpha$, let u be the minimal solution relative to $\underline{u}$ of

$$\begin{cases} \Delta u = f(u) & \text{in } B_1, \\ u = \alpha & \text{on } \partial B_1. \end{cases}$$

Using Remark 7 and letting $\tilde{\alpha} = u(0)$, $\phi(r) := u(x)$ for $r = |x|$ solves (3.50), subject to the initial conditions $\phi(0) = \tilde{\alpha}$ and $\phi'(0) = 0$. ϕ can thus be extended on some maximal interval $(0,R)$. Assume temporarily that $R < \infty$. Then u is a boundary blow-up solution on B_R. Indeed, by definition of R, we must have either $\phi(R) = +\infty$ or $\phi'(R) = +\infty$. In the latter case, multiply (3.50) by ϕ' and integrate between 0 and r to obtain that $\frac{1}{2}(\phi')^2 \leq F(\phi)$. Hence $F(\phi(R)) = +\infty$, $\phi(R) = +\infty$ and u is a boundary blow-up solution. It remains to prove that $R < \infty$.

Assume by contradiction that $R = \infty$. Apply Lemma 3.32 between $r_1 = 1$ and $r_2 > 1$:

$$\Phi(\alpha) \geq \frac{1}{D-2}\left(1 - \left(\frac{1}{r_2}\right)^{D-2}\right),$$

if $D \geq 3$ and

$$\Phi(\alpha) \geq \ln r_2,$$

if $D = 2$. Letting r_2 converge to ∞, we obtain a contradiction if either $D = 2$ or $\Phi(\alpha) < \frac{1}{|D-2|}$.

If $D \neq 2$ and $\Phi(\alpha) \geq \frac{1}{|D-2|}$, choose $K > 0$ so large that $\frac{1}{K}\Phi(\alpha) < \frac{1}{|D-2|}$. The above proof provides a boundary blow-up solution u of (3.34) on some ball B_R, when f is replaced by $K^2 f$. $\tilde{u}(x) := u(x/K)$ is then a boundary blow-up solution of (3.34) with nonlinearity f on B_{RK}.□

Remark 9 Let B be a ball of radius R and assume $\underline{u} \in C(\bar{B})$ is such that $\Delta\underline{u} \geq f(\underline{u})$ in B. Assume (3.35) holds for some $\alpha \geq \sup_B \underline{u}$ and let $\overline{u} = \alpha$. Using Proposition 3.29, let u be the minimal solution relative to $\underline{u}$ of

$$\begin{cases} \Delta u = f(u) & \text{in } B, \\ u = \alpha & \text{on } \partial B. \end{cases}$$

Repeating the above proof, we conclude that u can be extended to a radially symmetric boundary blow-up solution on some ball $\tilde{B}$ of radius $\tilde{R} > R$, satisfying $u \geq \underline{u}$ in B.

3.2.4 Existence of Solutions on Small Balls

Assume (3.36) holds. By Theorem 3.23, (3.34) has a solution on some ball and we may define

$$R_0 := \inf\{R > 0 \; : \; (3.34) \text{ has a solution in } B_R\}.$$

We assume by contradiction that $R_0 > 0$. Let (β_n) be a sequence of real numbers increasing to infinity and satisfying

$$\lim_{\beta_n \to \infty} \Phi(\beta_n) = 0.$$

Applying Proposition 3.29 with $\underline{u} = 0$ and $\overline{u} = \beta_n$, let u_n be the minimal solution relative to $\underline{u}$ of

$$\begin{cases} \Delta u_n = f(u_n) & \text{in } B_{R_0/2}, \\ u_n = \beta_n & \text{on } \partial B_{R_0/2}. \end{cases}$$

By Remark 7, letting $\alpha_n = u_n(0)$, $\phi_n(r) := u_n(x)$ for $r = |x|$ solves (3.50) with initial conditions $\phi_n(0) = \alpha_n$ and $\phi_n'(0) = 0$. By definition of R_0, ϕ_n can be extended so that ϕ_n remains a solution of (3.50) in $(0, R_0)$. Now apply Lemma 3.32 with $r_1 = R_0/2$ and $r_2 = R_0$:

$$\Phi(\beta_n) \geq \frac{1}{D-2}\frac{R_0}{2}\left[1 - \left(\frac{1}{2}\right)^{D-2}\right],$$

if $D \geq 3$ and

$$\Phi(\beta_n) \geq \frac{R_0}{2}\ln(2),$$

if $D = 2$. Passing to the limit as $n \to \infty$, we obtain a contradiction in both cases. We have just proved that

$$\inf\{R > 0 \; : \; (3.34) \text{ has a solution in } B_R\} = 0. \tag{3.52}$$

Remark 10 Let B be a ball of radius R and assume $\underline{u} \in C(\bar{B})$ is such that $\Delta \underline{u} \geq f(\underline{u})$ in B. Using Remark 9 and working as above, one can show that $\inf\{\tilde{R} > R$: (3.34) has a solution u in $B_{\tilde{R}}$ such that $u \geq \underline{u}$ in $B\} = R$.

3.2.5 Existence of Solutions on Smooth Domains

We assume here that (3.36) is valid. Applying Proposition 3.29 with $\underline{u} = 0$ and $\bar{u} = N$, $N \in \mathbb{N}$, let u_N be the minimal solution relative to $\underline{u}$ of (3.47) with $g \equiv N$. For $x \in \Omega$, choose a ball $B(x,r) \subset \Omega$ such that there exists a boundary blow-up solution u_r on $B(x,r)$. This is always possible since (3.52) holds. Applying the Minimality Principle 3.30 with $\bar{v} = u_r$, we conclude that $0 \leq u_N \leq u_r$ in $B(x,r)$. In particular, the sequence (u_N) is uniformly bounded in $B(x,r/2)$.

Let K denote an arbitrary compact subset of Ω. Covering K by finitely many balls $B(x_i, r_i/2)$, we conclude that $\{u_N\}$ is uniformly bounded on K by a constant depending only on K and f. Applying the Minimality Principle 3.30 with $\bar{v} = u_{N+1}$, we can also infer that $\{u_N\}$ is a nondecreasing sequence. Using these two facts and elliptic regularity, we conclude that $\{u_N\}$ converges to a function u solving $\Delta u = f(u)$ in Ω.

Fix a point $x_0 \in \partial\Omega$ and an arbitrary sequence (x_k) in Ω converging to x_0. Then, since $u \geq u_N$,

$$\liminf_{k \to \infty} u(x_k) \geq \liminf_{k \to \infty} u_N(x_k) = N.$$

Letting N converge to infinity, we conclude that u is a boundary blow-up solution of (3.34) in Ω.

Proof of Theorem 3.23 continued. By Sect. 3.2.3, we know that if (3.35) holds, there exists a blow-up solution on some ball. Conversely, assume that $u > 0$ solves (3.34) on some ball B of radius R centered at the origin. By Corollary 3.31, we may always assume that u is the minimal solution relative to $\underline{u} = 0$ of (3.34). In particular u is radial and we define $\phi(r) = u(x)$ for $r = |x|$, so that ϕ solves (3.51) in $(0, R)$.

Multiplying (3.51) by $r^{D-1}\phi'$ and integrating between 0 and r, we obtain

$$\frac{1}{2}r^{2D-2}\phi'(r)^2 = \int_0^r t^{2D-2} f(\phi(t))\phi'(t)\,dt \le r^{2D-2}\left[F(\phi(r)) - F(\phi(0))\right].$$

Integrating once more between 0 and R,

$$0 \le \int_0^R \frac{\phi'(r)}{\sqrt{2\left[F(\phi(r)) - F(\phi(0))\right]}}\,dr \le R, \tag{3.53}$$

which implies (3.35) with $\alpha = \phi(0)$. □

Proof of Theorem 3.24 continued. By Sect. 3.2.5, we know that if (3.36) holds, there exists a blow-up solution on any domain. Conversely, given $n \in \mathbb{N}$, assume that $u_n > 0$ solves (3.34) on the ball B of radius $1/n$ centered at the origin. By Corollary 3.31, we may always assume that u_n is the minimal solution relative to $\underline{u} = 0$. In particular u_n is radial. Let now $\beta_n = u_n(0)$. We claim that (β_n) is unbounded. Taking a subsequence if necessary, we then have that $\lim_n \beta_n = \infty$ and (3.36) follows from (3.53) applied with $R = 1/n$.

It remains to prove that (β_n) is unbounded. If not, up to a subsequence, (β_n) converges to some $\beta \ge 0$. By (3.53) applied with $R = 1/n$, we have

$$0 \le \int_{\beta_n}^{\infty} \frac{dt}{\sqrt{2\left[F(t) - F(\beta_n)\right]}}\,dr \le 1/n.$$

By Fatou's lemma, we conclude that

$$\int_{\beta}^{\infty} \frac{dt}{\sqrt{2\left[F(t) - F(\beta)\right]}}\,dr = 0,$$

which is not possible. □

3.2.6 Blow-Up Rate of Radially Symmetric Solutions

Proposition 3.33 *Assume that f satisfies the Keller–Osserman condition* (3.35). *Assume ϕ is a radially symmetric and monotone boundary blow-up solution on the unit ball. Then, for $r \sim 1$,*

$$\int_{\phi(r)}^{+\infty} \frac{dt}{\sqrt{F(t)}} \sim \sqrt{2}(1-r). \tag{3.54}$$

Proof. Multiplying (3.51) by $r^{D-1}\phi'$ and integrating by parts, we easily obtain that given $r \in (0,1)$,

$$\frac{(\phi')^2(r)}{2} = F(\phi(r)) - G_\phi(r), \tag{3.55}$$

where

$$G_\phi(r) = \frac{2D-2}{r}\int_0^r \left(\frac{s}{r}\right)^{2D-1} F(\phi(s))ds.$$

We claim that

$$G_\phi(r) = o(F(\phi(r))), \qquad \text{as } r \to 1. \tag{3.56}$$

Let indeed $\varepsilon > 0$. Then, since F is nondecreasing,

$$\begin{aligned}\frac{G_\phi(r)}{F(\phi(r))} =& \frac{2D-2}{r}\int_0^{1-\varepsilon} \left(\frac{s}{r}\right)^{2D-1} \frac{F(\phi(s))}{F(\phi(r))}ds \\ &+ \frac{2D-2}{r}\int_{1-\varepsilon}^{r} \left(\frac{s}{r}\right)^{2D-1} \frac{F(\phi(s))}{F(\phi(r))}ds. \\ \leq & C\frac{F(\phi(1-\varepsilon))}{F(\phi(r))} + C\varepsilon.\end{aligned}$$

Letting $r \to 1$ and then $\varepsilon \to 0$, we obtain the desired result. Returning to (3.55), we obtain

$$1 - \frac{\phi'}{\sqrt{2F(\phi)}} = 1 - \left[1 - \frac{G_\phi}{F(\phi)}\right]^{1/2}.$$

Combining this with (3.56), it follows that for $r \sim 1$,

$$1 - \frac{\phi'}{\sqrt{2F(\phi)}} \sim \frac{G_\phi}{2F(\phi)}$$

and, integrating between r and 1,

$$(1-r) - \int_{\phi(r)}^{\infty} \frac{dt}{\sqrt{2F(t)}} \sim \int_r^1 \frac{G_\phi(s)}{2F(\phi(s))}ds = o(1),$$

which implies (3.54).

3.2.7 Blow-Up Rate of Solutions on Smooth Domains

Let u be a blow-up solution on a domain Ω, which satisfies an interior and an exterior sphere condition at any boundary point. Fix $x_0 \in \partial\Omega$ and let $B_R \subset \Omega$ denote a small ball which is tangent to $\partial\Omega$ at x_0. Fix $\eta \in (0,1)$. Let $\underline{u} := u|_{B_{\eta R}}$. By Remark 9, there exists a radial boundary blow-up solution v defined on some ball $\tilde{B} \supset B_{\eta R}$, such that $v \geq \underline{u}$ in $B_{\eta R}$. Let $K > 0$ such that $K\tilde{B} = B_R$ and let $v_K(x) := v((x - x_1)/K + x_1)$, where x_1 is the center of B_R. Then v_K solves

$$\begin{cases} \Delta v_K = \dfrac{1}{K^2} f(v_K) & \text{in } B_R, \\ v_K = +\infty & \text{on } \partial B_R. \end{cases}$$

Since $u(x_1) \leq v(x_1) = v_K(x_1)$, Proposition 3.33 implies that

$$K \int_{u(x_1)}^{+\infty} \frac{dt}{\sqrt{F(t)}} \geq K \int_{v_K(x_1)}^{+\infty} \frac{dt}{\sqrt{F(t)}} \sim \sqrt{2}R.$$

Letting $R \to 0$, we then have

$$K \liminf_{x \to x_0} \frac{\int_{u(x)}^{+\infty} \frac{dt}{\sqrt{F(t)}}}{\delta(x)} \geq \sqrt{2}.$$

By Remark 10, we may take K arbitrarily close to $1/\eta$. Also, $0 < \eta < 1$ was chosen arbitrarily, so letting $K, \eta \to 1$, we finally obtain

$$\liminf_{x \to x_0} \frac{\int_{u(x)}^{+\infty} \frac{dt}{\sqrt{F(t)}}}{\delta(x)} \geq \sqrt{2}.$$

Choose another ball $B_{R'} \subset \mathbb{R}^N \setminus \bar{\Omega}$ which is tangent to $\partial\Omega$ at x_0 and a concentric ball $B_{R''}$ with $R'' > R'$ so large that $\Omega \subset B_{R''}$. Finally, let $A = B_{R''} \setminus B_{R'}$. Let v denote the minimal boundary blow-up solution (relative to $\underline{u} = 0$) on A. By the Minimality Principle 3.31, we deduce that $u \geq v$ in Ω. Applying Proposition 3.33 (which still holds on an annulus) with v, we conclude that

$$\limsup_{x \to x_0} \frac{\int_{u(x)}^{+\infty} \frac{dt}{\sqrt{F(t)}}}{\delta(x)} \leq \limsup_{x \to x_0} \frac{\int_{v(x)}^{+\infty} \frac{dt}{\sqrt{F(t)}}}{\delta(x)} \leq \sqrt{2}.$$

This finishes the proof of Theorem 3.25. More can be said about the asymptotic behavior of solutions provided F satisfies some extra growth assumption.

Lemma 3.34 *Let u, v denote two radially symmetric boundary blow-up solutions defined on the unit ball B. Assume there exist $\beta > 0$ and $M > 0$ such that $\frac{F(v)}{v^2} \geq \beta^2 \frac{F(u)}{u^2}$ whenever $M \leq u \leq v$. Then $u(r) \sim v(r)$ on ∂B.*

Proof. We recall from the proof of Proposition 3.33 that for ϕ a radially symmetric boundary blow-up solution and for $r \sim 1$,

$$(1-r)-\int_{\phi(r)}^{\infty}\frac{dt}{\sqrt{2F(t)}}\leq C\int_r^1\frac{G_\phi(s)}{F(\phi(s))}ds\leq C\int_r^1\frac{\phi(s)}{\sqrt{F(\phi(s))}}ds.$$

Using that $\frac{\dot{\phi(s)}}{\sqrt{2F(\phi(s))}}\sim 1$, we then obtain that, introducing $K(r)$ such that $1-r=\int_{K(r)}^{\infty}\frac{dt}{\sqrt{2F(t)}}$,

$$\int_{K(r)}^{\phi(r)}\frac{dt}{\sqrt{2F(t)}}\leq C\int_{\phi(r)}^{\infty}\frac{t}{F(t)}dt. \tag{3.57}$$

Since F is increasing, we thus obtain

$$\left(1-\frac{K(r)}{\phi(r)}\right)\leq C\frac{\sqrt{F(\phi(r))}}{\phi(r)}\int_{\phi(r)}^{\infty}\frac{t}{F(t)}dt. \tag{3.58}$$

Since $\frac{F(v)}{v^2}\geq\beta^2\frac{F(u)}{u^2}$ for $u\leq v$ large enough, (3.58) implies that

$$\left(1-\frac{K(r)}{\phi(r)}\right)\leq\frac{C}{\beta}\int_{\phi(r)}^{\infty}\frac{dt}{\sqrt{F(t)}}.$$

The classical Keller–Osserman (3.35) condition gives the result. □

Corollary 3.35 *Assume either that f is convex on some interval $[a,+\infty)$ or that $f(t)/t$ is nondecreasing on $[a,+\infty)$. Then the result of the previous lemma holds.*

Proof. Assume f is convex in $[a,+\infty)$ and let $G(t)=F(t+a)-F(a)-f(a)t$. Then $G(0)=G'(0)=0$ and G' is convex in $\mathbb{R}^+$. So $G(t)/t^2$ is nondecreasing, that is, $t\to\frac{F(t+a)-L(t)}{t^2}$ is nondecreasing, where $L(t)=F(a)+f(a)t$ is affine. Observe that $\lim_{t\to\infty}F(t)/t^2=+\infty$ since the Keller–Osserman condition (3.35) implies

$$\frac{u}{2\sqrt{F(u)}}\leq\int_{u/2}^{u}\frac{dt}{\sqrt{F(t)}}=o(1)\qquad\text{as } u\to\infty.$$

It follows that there exists $\beta>0$ such that $\frac{F(v)}{v^2}\geq\beta^2\frac{F(u)}{u^2}$ for $u\leq v$ large enough and we may apply Lemma 3.34. The case where $f(t)/t$ is nondecreasing on $[a,+\infty)$ is similar, so we skip it. □

3.2.8 A Uniqueness Result

We start with the following auxiliary result.

Lemma 3.36 *Assume that f is nondecreasing on $[0,+\infty)$ and convex in a neighborhood of $+\infty$ (say $[a,+\infty)$). Consider two radially symmetric boundary blow-up solutions such that $u(r) \le v(r)$ on $B(0,R)$, then in fact $u = v$ everywhere.*

Proof. Set $R = 1$ for the sake of simplicity. Since f is nondecreasing, either $u(0) = v(0)$ (and then $u = v$ everywhere) or $u(r) < v(r)$ everywhere since $\partial(r^{D-1}(\dot{v} - \dot{u})) = r^{D-1}(f(v) - f(u))$ so the map $r \to v(r) - u(r)$ is nondecreasing. Assume then $u(0) < v(0)$.

Let $\varepsilon > 0$. Consider the set $\omega_\varepsilon = \{r \in [0,1); \forall s < r,\ (1+\varepsilon)u(s) < v(s)\}$. If ε is small enough, $0 \in \omega_\varepsilon$. Due to Lemma 3.34, $R = 1 \notin \omega_\varepsilon$ since $u \sim v$ close to the boundary. Then introduce $r_\varepsilon^0 = \sup \omega_\varepsilon$, which satisfies $0 < r_\varepsilon^0 < 1$. We now have

$$v(0) - u(0) \le v(r_\varepsilon^0) - u(r_\varepsilon^0) = \varepsilon u(r_\varepsilon^0). \tag{3.59}$$

Then either r_ε^0 converges towards $R = 1$ when $\varepsilon \to 0$ or, letting $\varepsilon \to 0$, $u(0) = v(0)$ and the proof is complete.

Introduce now a such that f is convex on $[a,+\infty)$. Introduce R_0 such that $u(r) \ge a$ for $r \ge R_0$. Then for ε small enough $r_\varepsilon^0 > R_0$. Set $w(r) = (1+\varepsilon)u(r)$. Then, on the annulus $R_0 < s < r_\varepsilon^0$, using the convexity

$$\begin{aligned}\Delta(v-w) &= f(v) - (1+\varepsilon)f(u)\\ &\ge f(v) - (1+\varepsilon)\left[f(w) - f(a)\frac{u-a}{w-a} - (1+\varepsilon)f(a)\right]\\ &\ge f(v) - f(w) + \frac{\varepsilon}{w-a}(af(w) - wf(a)).\end{aligned} \tag{3.60}$$

Observe now that the map $X \to \frac{Xf(a) - af(X)}{X-a}$ is majorized by some constant C for $X \ge a$. Then introducing χ that satisfies $-\Delta\chi = 1$ with homogeneous Dirichlet condition at $R = 1$ ($\chi(1) = 0$)

$$\Delta(v - w - C\varepsilon\chi) \ge 0 \tag{3.61}$$

and by the maximum principle, for any r in $R_0 < r < r_\varepsilon^0$

$$v(r) - w(r) - C\varepsilon\chi(r) \le \max(-C\varepsilon\chi(r_\varepsilon^0), v(R_0) - w(R_0) - C\varepsilon\chi(R_0)). \tag{3.62}$$

Then letting $\varepsilon \to 0$ we obtain that for any fixed r such that $R_0 < r < 1$,

$$v(r) - u(r) \le v(R_0) - u(R_0). \tag{3.63}$$

Since the map $r \to v(r) - u(r)$ is nondecreasing, this implies that $v(r) - u(r)$ is *constant* on $[R_0, 1)$. By standard ODE theory, this implies that v and u coincide everywhere on $[0,1]$. □

Remark 11 Since f is nondecreasing, there exist $\underline{U}, \overline{U}$ the minimal and the maximal boundary blow-up solutions of the problem (the latter can be obtained e.g. as the monotone limit of $u(R)$ as $R \to 1^-$, where $u(R)$ denotes the minimal boundary blow-up solution on B_R). Clearly both $\underline{U}$ and $\overline{U}$ are radial and they coincide by the previous lemma. Since any solution u of the problem must stay between $\underline{U}$ and $\overline{U}$, Theorem 3.26 follows. Alternatively, according to a result of Poretta and Veron [160], any boundary blow-up solution is radially symmetric if f is convex in a neighborhood of $+\infty$, whence again Theorem 3.26 follows from the previous lemma.

Remark 12 The previous lemma is still valid if we substitute the assumption $\frac{f(u)}{u}$ increasing in a neighborhood of infinity to the convexity assumption. Since the proofs are easier they are left as an exercise to the reader.

3.2.9 Discrete Equations

We are concerned with finite difference approximations of (3.34) when $D = 1$ or $D = 2$ on a cube or a ball. After introducing some notation, we observe that both the maximum principle and the minimality principle extend to the case of finite difference operators. We conclude this section with some theoretical error estimates, assuming that f is a nondecreasing function.

3.2.9.1 Finite Differences

To begin with, consider the interval $[-1, 1]$ or the unit square $[-1, 1]^2$. Consider a uniform grid Ω_h with mesh size $h = \frac{1}{L}$ for some integer L. The nodes on the grid are respectively jh if D=1, $-L \le j \le L$, or (ih, jh) if $D = 2$, with $-L \le i, j \le L$.

The discrete Laplace operator is then defined on each point/node of the grid respectively by

$$(\Delta_h U)_j = \frac{1}{h^2}(-2U_j + U_{j+1} + U_{j-1}), \tag{3.64}$$

if $D = 1$ and

$$(\Delta_h U)_{i,j} = \frac{1}{h^2}(-4U_{i,j} + U_{i,j+1} + U_{i,j-1} + U_{i+1,j} + U_{i-1,j}), \tag{3.65}$$

if $D = 2$. In the above, U is a vector in $\mathbb{R}^{2L+1}$ (or $\mathbb{R}^{(2L+1)2}$ in 2D) with components $U_j \simeq u(jh)$ (or $U_{i,j} \simeq u(ih, jh)$). If $D = 2$, we then solve $\Delta_h U = f(U)$, that is, $(\Delta_h U)_{i,j} = f(U_{i,j})$ for all interior nodes (i, j) and set $U_{\pm L,j} = U_{i,\pm L} = N$ at all boundary nodes, where N is a fixed large constant. We work accordingly when $D = 1$.

It is standard to prove that the matrix Δ_h has positive inverse, that is, the entries of the inverse matrix are positive. Therefore, the maximum principle is valid (see [40]). Actually, if U satisfies $\Delta_h U \le 0$ on the interior nodes of the grid and $U \ge 0$ on the boundary, then $U \ge 0$ everywhere. Here and throughout this section we write $U \ge 0$ if and only if $U_{i,j} \ge 0$ for all (i, j) nodes of the grid. We shall use the same notation for B a matrix: $B \ge 0$ if and only if the entries of B are all nonnegative.

When working on the unit ball, we use a slightly modified scheme. Focussing on radially symmetric functions, we approximate the equation

$$\frac{1}{r}\frac{\partial}{\partial r}\left(r\frac{\partial u}{\partial r}\right) = f \qquad \text{for } r \in (0,1). \tag{3.66}$$

Discretize $[0,1]$ by setting $Lh = 1$, $0 \le j \le L$. At $j = L$ set $U_L = N$ (boundary condition). For $0 < j < L$ solve

$$\frac{1}{r}\partial(r\partial u) \simeq \frac{1}{jh}D^+(jhD^-U) = \frac{-2U_j + U_{j+1} + U_{j-1}}{h^2} + \frac{U_{j+1} - U_j}{jh^2} = F_j, \tag{3.67}$$

where $(D^+U)_j = \frac{U_{j+1}-U_j}{h}$, $(D^-U)_j = \frac{U_j - U_{j-1}}{h}$. It remains to define the equation at $j = 0$. For that purpose, we use the symmetry property $u(h) = u(-h)$ and the approximation $\frac{\dot{u}(0)}{0} = \ddot{u}(0)$ to set

$$\frac{4}{h^2}(-U_0 + U_1) = F_0.$$

This approximation of the Laplace operator satisfies the maximum principle. Indeed, it can be easily checked that if $F_j \ge 0$ then $j \to U_j$ is increasing. The maximum principle follows promptly.

3.2.9.2 Computing an Approximation

We aim to solve the following problem

$$\begin{cases} \Delta_h u_h = f(u_h) & \text{in } \Omega, \\ u_h = N & \text{on } \partial\Omega, \end{cases} \tag{3.68}$$

for N large enough.

We expect that u_h is an approximation for u, the minimal boundary blow-up solution corresponding to $\underline{u} = 0$. As in Proposition 3.29, u_h is obtained by monotone iteration, starting from the discrete subsolution 0. We claim that Proposition 3.29 and Corollary 3.30 are valid for the finite difference approximation. The proof follows the guidelines of the continuous case and is left as an exercise for the reader.

An approximation of the solution to problem (3.68) is recursively obtained by the following discrete iterative scheme:

Consider $u^k \in \mathbb{R}^{(2L+1)D}$, where $L = \frac{1}{h}$, recursively defined by

$$u^0 = 0$$

and for $k \geq 0$, u^{k+1} solves

$$\begin{cases} (\Delta_h - \Lambda_N Id)u^{k+1} = f(u^k) - \Lambda_N u^k & \text{in } \Omega_h, \\ u^{k+1} = N & \text{on } \partial\Omega_h, \end{cases} \tag{3.69}$$

where $\Lambda_N = \sup_{[0,N]} f'$.

Therefore the error between u, the minimal boundary blow-up solution, and u^k the k-th iterate of (3.69) can be split as follows:

$$I_h(u) - u^k = I_h(u - u_N) + (I_h(u_N) - u_h) + (u_h - u^k) \tag{3.70}$$

where I_h is the interpolation operator defined by $I_h(u)_i = u(x_i)$ when $D = 1$ (respectively by $I_h(u)_a = u(a)$ when $D = 2$ for a node $a = (ih, jh)$ on the grid), and u_N is the solution of

$$\begin{cases} \Delta u = f(u) & \text{in } \Omega, \\ u = N & \text{on } \partial\Omega. \end{cases} \tag{3.71}$$

3.2.9.3 Error Estimate

Throughout this section, we assume that f is a convex increasing function and that $\Omega = [-1,1]^D$. We first bound from above the rate of convergence of the algorithm (3.69).

Lemma 3.37 *Let $u^k \in \mathbb{R}^{(2L+1)D}$ be given by* (3.69) *and $u_h = \lim_{k\to\infty} u^k$. Then, there exist constants $C = C(D)$, $\mu = \mu(D) > 0$ and $\lambda = \lambda(f) < \Lambda_N$ such that*

$$\|u^k - u_h\|_{\ell^\infty} \leq C\left(1 - \frac{\lambda}{\Lambda_N}\right)^k \min\left(1, \frac{1}{h^{D/2}}\left[1 - \frac{\mu}{\Lambda_N} + \frac{\mu^2}{\Lambda_N^2}\right]^k\right) \|u^0 - u_h\|_{\ell^\infty}. \quad (3.72)$$

Proof. In the sequel let us denote by a a node of the grid (that is $a = ih$ in 1D or $a = (ih, jh)$ in 2D). (u^k) is a nondecreasing sequence in $\mathbb{R}^{(2L+1)D}$ (that is, $u_a^k \leq u_a^{k+1}$ for each node a). By the mean value theorem, there exists $\theta_a \in (u_a^k, (u_h)_a)$ such that

$$f(u_a^k) - f((u_h)_a) = f'(\theta_a)(u_a^k - (u_h)_a). \quad (3.73)$$

Therefore

$$\left|\left((u_h)_a - \frac{f((u_h)_a)}{\Lambda_N}\right) - \left((u^k)_a - \frac{f((u^k)_a)}{\Lambda_N}\right)\right| \leq (1 - \frac{\lambda}{\Lambda_N})((u_h)_a - u_a^k), \quad (3.74)$$

where $\lambda = \inf f'$.

On the other hand

$$u_h - u^k = \left(Id - \frac{\Delta_h}{\Lambda_N}\right)^{-1}\left(\left(u_h - \frac{f(u_h)}{\Lambda_N}\right) - \left(u^k - \frac{f(u^k)}{\Lambda_N}\right)\right), \quad (3.75)$$

where $f(u)$ denotes the vector with components $f(u)_a = f(u_a)$. The key argument is to observe that the matrix $Id - \frac{\Delta_h}{\Lambda_N}$ satisfies the maximum principle. Therefore,

$$0 \leq u_h - u^{k+1} \leq \left(1 - \frac{\lambda}{\Lambda_N}\right)\left(Id - \frac{\Delta_h}{\Lambda_N}\right)^{-1}(u_h - u^k), \quad (3.76)$$

where inequalities hold component by component. We thus obtain that

$$\|u_h - u^k\|_{\ell^\infty} \leq \left(1 - \frac{\lambda}{\Lambda_N}\right)^k \left\|\left(Id - \frac{\Delta_h}{\Lambda_N}\right)^{-k}\right\|_{\mathscr{L}(\ell^\infty)} \|u_h - u^0\|_{\ell^\infty}. \quad (3.77)$$

On the one hand, the maximum principle implies that

$$\left\| \left(Id - \frac{\Delta_h}{\Lambda_N} \right)^{-k} \right\|_{\mathscr{L}(\ell^\infty)} \leq C, \tag{3.78}$$

for some constant C depending only on the dimension D.

On the other hand, since the spectrum of $-\Delta_h$ lies in a segment $[\mu, \frac{\mu}{h^2}]$ (see [40]) and $\frac{1}{\sqrt{(2/h+1)D}} \| \cdot \|_{\ell^2} \leq \| \cdot \|_{\ell^\infty} \leq \| \cdot \|_{\ell^2}$ in $\mathbb{R}^{(2L+1)D}$, we get

$$||(Id - \frac{\Delta_h}{\Lambda_N})^{-k}||_{\mathscr{L}(\ell^\infty)} \leq Ch^{-D/2} ||(Id - \frac{\Delta_h}{\Lambda_N})^{-k}||_{\mathscr{L}(\ell^2)} = Ch^{-D/2} \left(\frac{1}{1 + \frac{\mu}{\Lambda_N}} \right)^k. \tag{3.79}$$

Using that $\frac{1}{1+\frac{\mu}{\Lambda_N}} \leq 1 - \frac{\mu}{\Lambda_N} + \frac{\mu^2}{\Lambda_N^2}$ and collecting (3.77), (3.78) and (3.79), the proof is complete. □

We now provide an upper bound for $u_h - I_h(u_N)$.

Lemma 3.38 *Assume f is convex. Let u_h be the solution of (3.68) and u_N be the solution of (3.71). Then,*

$$\|u_h - I_h(u_N)\|_{\ell^\infty} \leq Ch^2 \alpha(N, f) \tag{3.80}$$

where $\alpha(N, f) = \|u_N^{(4)}\|_{L^\infty}$ is a constant depending only on N and f.

Proof. For the sake of simplicity we will denote $I_h(u_N)$ by u_N; this introduces no confusion.

We write down the proof for the 2D problem, leaving the 1D case as an exercise for the reader. Let $a = (ih, jh)$ be a node on the grid. By the mean value theorem, for each node a, there exist ξ, η in $\mathbb{R}^2$ such that

$$|\xi - a| < h \text{ and } |\eta - a| < h, \tag{3.81}$$

and

$$\Delta_h(u_N)_a - (\Delta u_N)_a = ch^2 (\frac{\partial^4 u}{\partial x^4}(\xi) + \frac{\partial^4 u}{\partial y^4}(\eta)). \tag{3.82}$$

Therefore

$$\begin{cases} \Delta_h(u_h - u_N)_a = f(u_h)_a - f(u_N)_a - ch^2 \left[\frac{\partial^4 u}{\partial x^4}(\xi) + \frac{\partial^4 u}{\partial y^4}(\eta) \right] & \text{in } \Omega_h, \\ (u_h - u_N)_a = 0 & \text{on } \partial\Omega_h. \end{cases} \tag{3.83}$$

Consider $\omega = u_h - u_N$, then working as in the previous lemma we obtain

$$w_a = \left[(\Delta_h + \Lambda_N)^{-1} \left(Id - D\left(\frac{f'(\theta)}{\Lambda_N}\right)\right) \Lambda_N \right] ch^2 \left(\frac{\partial^4 u}{\partial x^4}(\xi) + \frac{\partial^4 u}{\partial y^4}(\eta) \right), \quad (3.84)$$

where $\theta_a \in (u_N(a), (u_h)_a)$.

We therefore obtain

$$\|w_a\|_{\ell^\infty} \leq \frac{1}{\Lambda_N} \frac{4}{1+\frac{4}{\Lambda_N}} \|\Delta_h^{-1}\|_{\mathscr{L}(\ell^\infty)} Ch^2 \|u_N^{(4)}\|_{L^\infty}, \quad (3.85)$$

where $\|u_N^{(4)}\|_{L^\infty} = \max_{|\alpha| \leq 4}(\|\partial^\alpha u_N\|_{L^\infty})$.□

Remark 13 When $D = 1$ and $f(u) = u^p$ with $p \geq 2$, the constant $\alpha(N, f)$ is given by

$$\alpha(N, f) = \|u_N^{(4)}\|_{L^\infty} = \frac{p(3p-1)}{p+1} N^{2p-1}.$$

3.2.9.4 Error Estimate for $\|u - u_N\|$

Assume here that the Sharpened Keller–Osserman condition (3.36) is valid. Consider then a sequence (α_N) such that $\Phi(\alpha_N)$ converges towards 0. Consider the minimal solution u_N of

$$\begin{cases} \Delta u_N = f(u_N) & \text{in } \Omega = B(0,1), \\ u_N = \alpha_N & \text{on } \partial\Omega. \end{cases} \quad (3.86)$$

Then one may wonder how u_N approximates the minimal boundary blow-up solution u defined on the unit ball. We first state a qualitative result.

Proposition 3.39 *There exists R_N such that u_N is a boundary blow-up solution on $B(0, R_N)$. Moreover*

$$R_N - 1 \sim \frac{1}{\sqrt{2}} \int_{\alpha_N}^{+\infty} \frac{dt}{\sqrt{F(t)}}. \quad (3.87)$$

Remark 14 This proposition shows that when we plot the approximation u_N, we plot in fact a boundary blow-up solution on a ball that is close to the unit ball.

Let us proceed to the proof in the case where $D \geq 3$. The cases $D = 1, 2$ are very similar and so omitted. Assume that u_N extends to $\mathbb{R}^D$. Then by Lemma 3.32

$$\frac{1}{\sqrt{2}} \int_{u_N(1)}^{u_N(r)} \frac{dt}{\sqrt{F(t) - F(u_N(1))}} \geq \frac{1}{D-2}\left(1 - \left(\frac{1}{r}\right)^{D-2}\right). \quad (3.88)$$

Here we have used that u_N is radially symmetric. Therefore since $u_N(1) = \alpha_N$, $\Phi(\alpha_N) \geq \frac{1}{D-2}(1-(\frac{1}{r})^{D-2})$. $N \to +\infty$ leads to a contradiction. The estimate (3.87) comes from Proposition 3.33. □

Remark 15 Observe that if $f(u) = u^p$, then $u_N(r) = R_N^{-\frac{2}{p-1}} u(rR_N)$. In that case, $0 \leq u(r) - u_N(r) \leq C(R_N - 1)(\dot{u}(r) + u(r))$. The inequality is sharp for some numerical constant C. To prove the estimate in a more general context, we need extra hypotheses.

Definition 3.40 Consider $g : [0,+\infty) \to [0,+\infty)$ a function. We say that g is *strongly increasing* if the function

$$\rho(\lambda) = \inf_{u \geq 0} \frac{g(\lambda u)}{g(u)}$$

is a C^1 increasing function on $[1,+\infty)$ that satisfies $\dot{\rho}(1) \neq 0$.

A strongly increasing function is increasing in the usual sense. $g(u) = u^p$, $p > 0$ is strongly increasing. $g(u) = \ln(u+1)$ is not.

We now state and prove

Proposition 3.41 *Assume that $\frac{f(u)}{u}$ is strongly increasing. Then*

$$0 \leq u(r) - u_N(r) \leq C(R_N - 1)(\dot{u}(r) + u(r)).$$

Proof. $v(r) = \lambda u_N(rR_N)$ is a blow-up function on the unit ball. We have

$$\Delta v = \lambda R_N^2 f\left(\frac{v}{\lambda}\right) \leq \frac{R_N^2}{\rho(\lambda)} f(v). \tag{3.89}$$

For N large enough, we choose λ_N close to 1 such that $\rho(\lambda_N) = R_N^2$. Then v is a blow-up supersolution to (3.34). Since u is the minimal blow-up solution, then $u(r) \leq v(r)$. Therefore

$$\begin{aligned} 0 \leq& u(r) - u_N(r) \leq u(r) - \frac{1}{\lambda_N} v\left(\frac{r}{R_N}\right) \\ \leq& (R_N - 1)\dot{u}(r) + \left(1 - \frac{1}{\lambda_N}\right) u(rR_N^{-1}) \leq C(R_N - 1)(\dot{u}(r) + u(r)), \end{aligned} \tag{3.90}$$

since $(1 - \frac{1}{\lambda_N}) \leq 2\frac{R_N - 1}{\dot{\rho}(1)}$. □

3.2.10 Numerical Computations

In this section, we present some numerical results obtained with our method of approximation.

Remark 16 At this stage, we would like to point out that our method is self-contained, and does not use the knowledge of the boundary blow-up behavior of the solution. In fact, as in [119], one can introduce another approximate problem such as taking $\Omega_\varepsilon \subset \Omega$ where $\text{dist}(\Omega, \Omega_\varepsilon) \leq \varepsilon$, and solve the problem

$$\begin{cases} \Delta u_\varepsilon = f(u_\varepsilon) & \text{in } \Omega_\varepsilon, \\ u_\varepsilon(x) = K(x) & \text{on } \partial\Omega_\varepsilon, \end{cases} \tag{3.91}$$

where

$$\frac{1}{\sqrt{2}} \int_{K(x)}^{+\infty} \frac{dt}{\sqrt{F(t)}} = \text{dist}(x, \partial\Omega).$$

We discuss our numerical results successively on three examples:

- $f(u) = u^2$ (that is, $\frac{f(u)}{u}$ is increasing),
- $f(u) = u^2(2 + \cos u)$,
- $f(u) = u^2(1 + \cos u)$.

3.2.10.1 Case $f(u) = u^2$

Since $\frac{f(u)}{u}$ is increasing, on any domain we have a unique boundary blow-up solution (see e.g. [2] and references therein).

We see in Fig. 3.1 that Φ is a strictly decreasing function.

For the sake of completeness, we plot in 1D the values of $u_h(0)$ for several values of h and we compare them to theoretical results (see Fig. 3.2 and Proposition 3.41).

Remark 17 The value of $u(0)$ is obtained by solving

$$\frac{1}{\sqrt{2}} \int_{u(0)}^{+\infty} \frac{dt}{\sqrt{F(t) - F(u(0))}} = 1.$$

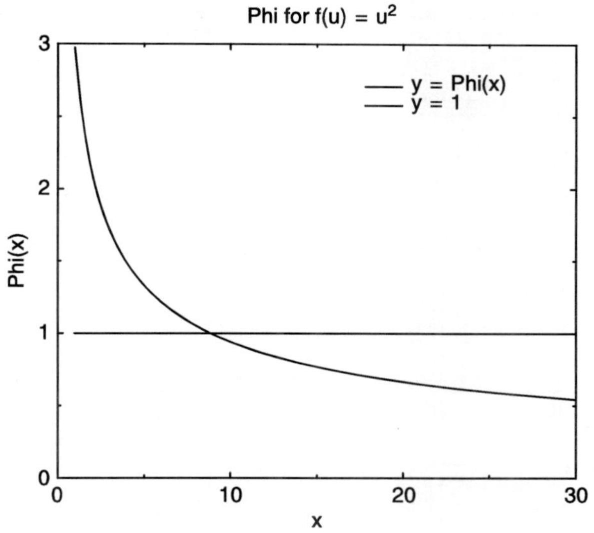

Fig. 3.1 Φ when $f(u) = u^2$

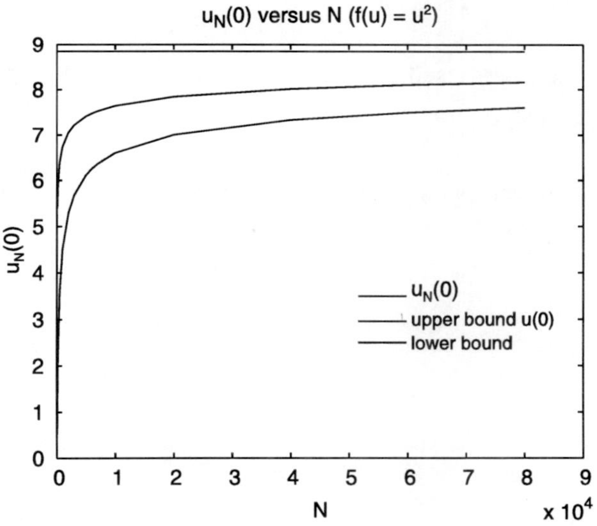

Fig. 3.2 Convergence of $u_h(0)$ to $u(0)$ versus h

3.2.10.2 Case $f(u) = u^2(2 + \cos u)$

This function f satisfies the Sharpened Keller–Osserman condition (3.36).

Figure 3.3 shows that $\Phi(\alpha)$ tends to 0 when α tends to $+\infty$ (then the Sharpened Keller–Osserman condition is valid). Note that Φ is not a decreasing function; for

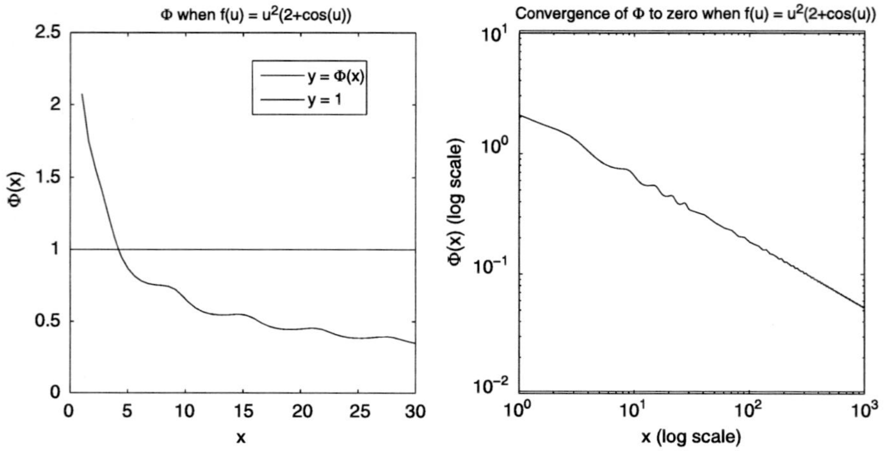

Fig. 3.3 Φ when $f(u) = u^2(2+\cos(u))$

instance for $c \approx 0.49$ there exist $\alpha \neq \beta$ such that $\Phi(\alpha) = \Phi(\beta) = c$. Therefore, at least in 1D, uniqueness does not hold.

Remark: The uniqueness result for $f(u) = u^2(2+\cos u)$ in $B(0,1) \subset \mathbb{R}^D$ is still an open question for $D \geq 2$.

Figure 3.4 shows an approximation of the minimal boundary blow-up radial solution on $B(0,1) \subset \mathbb{R}^2$.

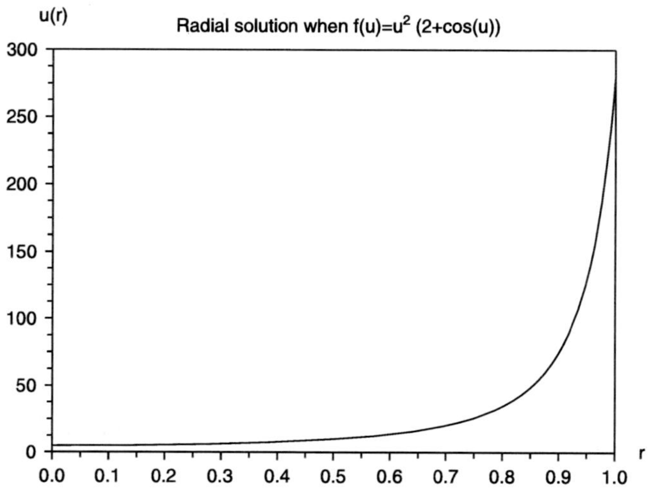

Fig. 3.4 Solution on the disk when $f(u) = u^2(2+\cos u)$

3.2.10.3 Case $f(u) = u^2(1+\cos u)$

We first state

Proposition 3.42 *The function $f(u) = u^2(1+\cos u)$ satisfies the Sharpened Keller–Osserman condition* (3.36). *Moreover,*

$$\lim_{\alpha \to (2k+1)\pi} \Phi(\alpha) = +\infty.$$

For a proof, see Proposition 3.21.

Figure 3.5 shows that for any domain, there exist an infinite sequence of boundary blow-up solutions.

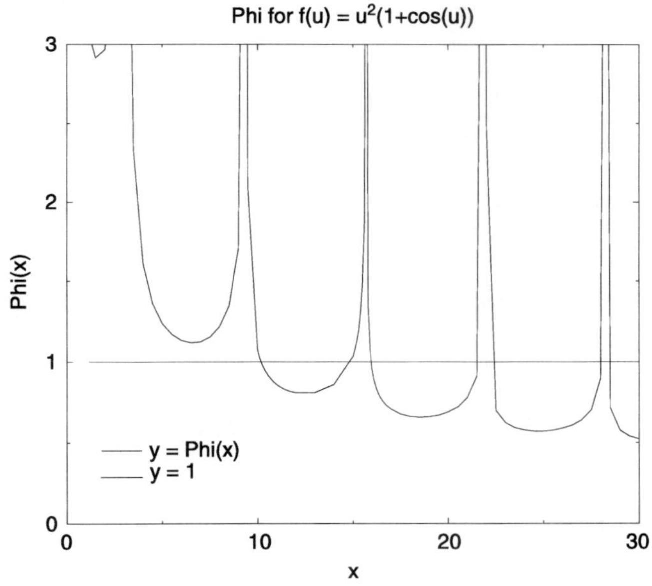

Fig. 3.5 Φ when $f(u) = u^2(1+\cos(u))$

When $D = 1$, this follows from the fact that $\Phi(\alpha) = R$ admits an infinite number of solutions α. When $D \geq 2$, fix an integer m and observe that $\alpha_m = (2m+1)\pi$ is a subsolution. Let u_m denote the minimal boundary blow-up solution relative to $\underline{u} = \alpha_m$. Infinitely many u_m must be distinct. Indeed, choose m_1 such that $\alpha_{m_1} > u_0(0)$. Then $u_{m_1}(0) \geq \alpha_{m_1} > u_0(0)$. Repeating this process inductively yields infinitely many distinct solutions u_{m_k}.

We plot in Fig. 3.6 approximations of different boundary blow-up solutions on the interval $[-1,1]$.

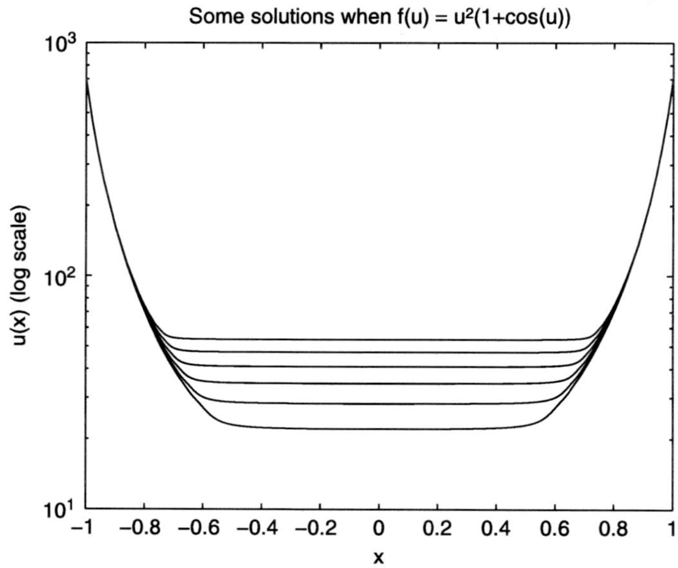

Fig. 3.6 Some solutions when $f(u) = u^2(1+\cos u)$

Figure 3.7 shows two radial approximations of different boundary blow-up solutions on the unit ball in $\mathbb{R}^2$.

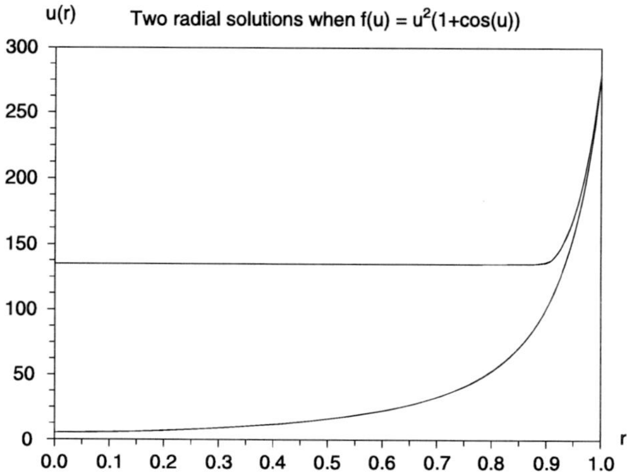

Fig. 3.7 Solutions on the disk when $f(u) = u^2(1+\cos u)$

3.3 Entire Large Solutions

Let $f \in C^1[0,\infty)$ be a nonnegative function such that $f(0) = 0$. f is assumed to be positive at infinity, in the sense that

$$\text{there exists } a \in \mathbb{R} \quad \text{s.t.} \quad f(a) > 0 \quad \text{and} \quad f(t) \geq 0 \quad \text{for } t > a. \tag{3.92}$$

and f is superlinear in the sense that

$$\int^{+\infty} \frac{ds}{\sqrt{F(s)}} < +\infty, \tag{KO}$$

where $F(s) = \int_a^s f(t)dt$.

In this section we are interested in the qualitative properties of solutions to

$$\begin{cases} \Delta u = \rho(|x|)f(u), u \geq 0 & \text{in } \mathbb{R}^D, \\ u(x) \to 0 & \text{as } |x| \to +\infty, \end{cases} \tag{3.93}$$

where $D \geq 3$ is a positive integer and $\rho \in C[0,\infty)$ is a positive function such that

$$\int_0^\infty r\rho(r)dr < +\infty. \tag{3.94}$$

Such solutions are called *entire large solutions* (in short *ELS*).

3.3.1 A Useful Result: Bounded Entire Solutions

In this section we are concerned with bounded entire solutions associated with (3.93).

Proposition 3.43 *Assume* (KO) *and (3.94). Then, for any* $0 \leq \beta \leq \infty$, *there exists a radially symmetric function* $w_\beta \in C^2(\mathbb{R}^D)$ *such that*

$$\begin{cases} \Delta w_\beta \leq \rho(|x|)f(w_\beta) & \text{in } \mathbb{R}^D, \\ \lim\limits_{|x|\to\infty} w_\beta(x) = \beta. \end{cases} \tag{3.95}$$

Moreover, the family $\{w_\beta\}_{\beta\in[0,+\infty]}$ *is increasing in* β *and* $\lim_{\beta\to\infty} w_\beta = w_\infty$.

Proof. Integrating by parts and using (3.94) we have

$$\int_0^r t^{1-D}\int_0^t s^{D-1}\rho(s)dsdt = \frac{1}{D-2}\left[\int_0^r s\rho(s)ds - r^{2-D}\int_0^r s^{D-1}\rho(s)ds\right]$$
$$< \frac{1}{D-2}\int_0^\infty s\rho(s)ds.$$

Hence, for all $x\in\mathbb{R}^D$ we can define

$$U(x)=\int_{|x|}^\infty t^{1-D}\int_0^t s^{D-1}\rho(s)dsdt.$$

It follows that

$$-\Delta U(x)=\rho(|x|)\quad \text{in } \mathbb{R}^D \qquad \text{and} \qquad \lim_{|x|\to+\infty}U(x)=0.$$

Let $\bar f\in C^1[0,\infty)$ be an increasing function such that

$$\bar f\geq f,\quad \bar f(0)=0 \quad \text{and} \quad \bar f>0 \text{ in } (0,\infty).$$

Before we proceed with the construction of w_β let us first show that

$$\int^\infty \frac{1}{\bar f(s)}ds<+\infty. \tag{3.96}$$

Indeed, let $\bar F(t)=\int_a^t \bar f(s)ds$, using the fact that $\bar f$ is increasing we have

$$0=\lim_{t\to\infty}\int_t^\infty\frac{1}{\sqrt{\bar F(s)}}ds\geq \lim_{t\to\infty}\int_t^{2t}\frac{1}{\sqrt{\bar F(s)}}ds\geq\lim_{t\to\infty}\frac{t}{\sqrt{\bar F(2t)}}.$$

Therefore, for $t>a$ large enough we have

$$\bar F(t)\geq t^2. \tag{3.97}$$

On the other hand,

$$\bar F(t)=\int_a^t \bar f(s)ds\leq t\bar f(t)\quad \text{for all } t>a. \tag{3.98}$$

Now, using (3.97) and (3.98) we deduce

$$\int_t^\infty\frac{1}{\bar f(s)}ds\leq\int_t^\infty\frac{s}{\bar F(s)}ds\leq\int_t^\infty\frac{1}{\sqrt{\bar F(s)}}ds\to 0\quad \text{as } t\to\infty,$$

so (3.96) follows. As a consequence of (3.96) we derive that for all $0<\beta\leq\infty$ the mapping

$$(0,\beta) \ni t \mapsto \int_t^\beta \frac{ds}{\bar{f}(s)} \in (0,\infty)$$

is bijective. Therefore, for any $0 < \beta \leq \infty$, there is a unique

$$w_\beta : \mathbb{R}^D \to (0,\beta)$$

such that

$$\int_{w_\beta(x)}^\beta \frac{ds}{\bar{f}(s)} = U(x) \quad \text{for all } x \in \mathbb{R}^D. \tag{3.99}$$

Since $U(x)$ is radial, w_β is radial too. Furthermore, w_β is increasing with respect to β and $|x|$. This proves the limit equality as $\beta \to \infty$ in (3.95). Now,

$$\nabla U(x) = -\frac{1}{\bar{f}(w_\beta(x))}\nabla w_\beta(x) \quad \text{in } \mathbb{R}^D$$

and

$$\rho(|x|) = -\Delta U = \frac{1}{\bar{f}(w_\beta)}\Delta w_\beta - \frac{\bar{f}'(w_\beta)}{\bar{f}^2(w_\beta)}|\nabla w_\beta|^2 \leq \frac{1}{\bar{f}(w_\beta)}\Delta w_\beta \quad \text{in } \mathbb{R}^D.$$

Hence w_β satisfies (3.95). □

3.3.2 Existence of an Entire Large Solution

We are now in a position to derive the existence of solutions to (3.93). Our main result in this sense is the following.

Theorem 3.44 *Assume that (3.92), (KO) and (3.94) hold.*

(a) There exists a minimal ELS u_0 of (3.93) in the sense that any ELS u of (3.93) satisfies $u \geq u_0$ in $\mathbb{R}^D$. Furthermore, u_0 is radially symmetric.

(b) If $r \mapsto \rho(r)$ is decreasing, then for any ELS u of (3.93) there exists a radial ELS $\bar{u}$ such that $u \leq \bar{u}$ in $\mathbb{R}^D$.

Proof. We shall perform the proof of Theorem 3.44 in three steps.

Step 1: Any ELS u of (3.93) satisfies $u \geq w_\infty$ in $\mathbb{R}^D$, where w_∞ is the function defined in Proposition 3.43 (given by (3.99)) for $\beta = \infty$.

Let u be an arbitrary ELS of (3.93). Define

$$\mathscr{E} = \{\beta \geq 0 : w_\beta \leq u \text{ in } \mathbb{R}^D\}$$

and remark first that $\mathscr{E}$ is nonempty. Indeed, if $m := \inf_{\mathbb{R}^D} u \geq 0$ then $w_m \leq m \leq u$ in $\mathbb{R}^D$ where w_m is the function defined by (3.99) for $\beta = m$. Also, if $\beta \in \mathscr{E}$ then $[0,\beta] \subset \mathscr{E}$.

Lemma 3.45 $\mathscr{E} = [0,\infty)$.

Proof. Assume by contradiction that $\beta_0 = \sup \mathscr{E} < +\infty$. Then, $w_{\beta_0} \leq u$ in $\mathbb{R}^D$. We claim that there exists $x_0 \in \mathbb{R}^D$ such that $w_{\beta_0}(x_0) = u(x_0)$. Indeed, if this is not the case, we have $w_{\beta_0} < u$ in $\mathbb{R}^D$. Since $\lim_{|x|\to\infty}(u(x) - w_{\beta_0}(x)) = \infty$, it follows that there exists $c > 0$ such that

$$w_{\beta_0} + c < u \quad \text{in } \mathbb{R}^D. \tag{3.100}$$

In addition, from (3.99) we have

$$\int_{w_{\beta_0}}^{\beta_0} \frac{ds}{\bar{f}(s)} = \int_{w_\beta}^{\beta} \frac{ds}{\bar{f}(s)}, \qquad \text{for all } \beta > \beta_0.$$

This yields

$$\int_{\beta_0}^{\beta} \frac{ds}{\bar{f}(s)} = \int_{w_{\beta_0}}^{w_\beta} \frac{ds}{\bar{f}(s)}, \qquad \text{for all } \beta > \beta_0.$$

Since $\bar{f}$ is increasing, we find

$$|w_\beta - w_{\beta_0}| \leq \frac{\bar{f}(w_\beta)}{\bar{f}(w_{\beta_0})}|\beta - \beta_0| \leq \frac{\bar{f}(\beta)}{\bar{f}(w_{\beta_0})}|\beta - \beta_0| \quad \text{for all } \beta > \beta_0.$$

In particular,

$$|w_{\beta_0+\varepsilon} - w_{\beta_0}| \leq \varepsilon \frac{\bar{f}(\beta_0+\varepsilon)}{\bar{f}(w_{\beta_0})}.$$

Thus, for small values of $\varepsilon > 0$ we find $w_{\beta_0+\varepsilon} \leq w_{\beta_0} + c/2$ in $\mathbb{R}^D$. Combining this last estimate with (3.100) we obtain $w_{\beta_0+\varepsilon} + c/2 < u$ in $\mathbb{R}^D$, which contradicts the definition of β_0.

Hence, there exists $x_0 \in \mathbb{R}^D$ such that $w_{\beta_0}(x_0) = u(x_0)$. We fix now $R > |x_0|$ and let

$$\Lambda := \sup\{f'(s) : \min_{B_R} w_{\beta_0} \leq s \leq \max_{B_R} u\}.$$

Then $g(t) = f(t) - \Lambda t$ is a nonincreasing function on $[\min_{B_R} w_{\beta_0}, \max_{B_R} u]$ and

$$\Delta(u - w_{\beta_0}) - \Lambda(u - w_{\beta_0}) \leq g(u) - g(w_{\beta_0}) \leq 0 \quad \text{in } B_R,$$

$$0 \le u - w_{\beta_0} \quad \text{on } \partial B_R, \quad u(x_0) = w_{\beta_0}(x_0).$$

By the strong maximum principle it follows that $u \equiv w_{\beta_0}$ in B_R. Since $R > |x_0|$ was arbitrarily chosen, this yields $u \equiv w_{\beta_0}$. But this is clearly a contradiction since u is an ELS to (3.93), while w_{β_0} is bounded in $\mathbb{R}^D$.

Therefore $\mathscr{E} = [0, \infty)$. □

By Lemma 3.45 it follows that $w_\beta \le u$ in $\mathbb{R}^D$ for all $\beta > 0$. It remains now to pass to the limit with $\beta \to \infty$ to reach the conclusion in Step 1.

Step 2: There exists a radial ELS u_0 to (3.93) such that for any ELS u to (3.93), we have $u \ge u_0$.

For any $R > 0$ denote by u_R the unique minimal solution relative to w_∞, to the problem

$$\begin{cases} \Delta u_R = \rho(|x|) f(u_R) & \text{in } B_R, \\ u_R = w_\infty & \text{on } \partial B_R. \end{cases}$$

Also, by Proposition 3.31 there exists U_R a unique minimal boundary blow-up solution (in short BBUS) relative to w_∞, solving

$$\begin{cases} \Delta U_R = \rho(|x|) f(U_R) & \text{in } B_R, \\ U_R = +\infty & \text{on } \partial B_R. \end{cases} \tag{3.101}$$

Since w_∞ is radially symmetric, so is u_R. Moreover, for all $R > 0$ we have

$$w_\infty \le u_R \le u_{R+1} \le U_R \quad \text{in } B_R. \tag{3.102}$$

Indeed, the first two inequalities follow from the minimality principle for u_R as stated in Proposition 3.29 whereas the last inequality is obtained as follows. We first consider $0 < R' < R$ such that

$$u_{R+1} < U_R \quad \text{in } B_R \setminus B_{R'}.$$

Again by Proposition 3.29 and the minimality of u_{R+1} relative to w_∞ we find $u_{R+1} \le U_R$ in $B_{R'}$. Hence, $u_{R+1} \le U_R$ in B_R and (3.102) follows.

Next, by (3.102) and the Arzela–Ascoli theorem, there exists a subsequence of $\{u_R\}$ (still denoted $\{u_R\}$) that converges uniformly to some u_0 on each compact subset of $\mathbb{R}^D$. By standard elliptic regularity it follows that $u_0 = \lim_{R \to +\infty} u_R$ is a radial solution of (3.93) and by (3.102) we also have $w_\infty \le u_0$ in $\mathbb{R}^D$. Thus, u_0 is a radial ELS of (3.93). Now, if u is any ELS of (3.93), from Step 1 we have $w_\infty \le u$ in $\mathbb{R}^D$. Thus, in virtue of the minimality of u_R relative to w_∞ we have $u_R \le u$ in B_R. This yields $u_0 = \lim_{R\to+\infty} u_R \le u$ in $\mathbb{R}^D$.

Step 3: Assume that $r \mapsto \rho(r)$ is nonincreasing and u is an ELS of (3.93). Then, there exists $\overline{u}$ a radial ELS of (3.93) such that $u \le \overline{u}$.

Let u be an arbitrary ELS of (3.93). We fix $R > 0$ and let $N = N(u,R) \ge 1$ be such that $\max_{B_R} u < N$. By Proposition 3.29, for all $n \ge N$ there exists a minimal solution u_R^n relative to u of the problem

$$\begin{cases} \Delta u_R^n = \rho(|x|) f(u_R^n) & \text{in } B_R, \\ u_R^n = n & \text{on } \partial B_R. \end{cases}$$

By minimality arguments we have $u \le u_R^n < n$ in B_R. Therefore, the function

$$v := n - u_R^n$$

satisfies

$$\begin{cases} -\Delta v = \rho(|x|) f(n-v),\ v > 0 & \text{in } B_R, \\ v = 0 & \text{on } \partial B_R. \end{cases}$$

Since $r \mapsto \rho(r)$ is nonincreasing, Theorem 1.6 implies that v and so u_R^n are radially symmetric functions.

Let U_R be the BBUS of (3.101) which is minimal relative to u. As in Step 2 one can easily see that

$$u \le u_R^n \le U_R \quad \text{in } B_R. \tag{3.103}$$

Applying further Proposition 3.29 we find

$$u_R^n \le u_R^{n+1}, \quad u_{R+1}^n \le u_R^n \quad \text{in } B_R, \text{ for all } n \ge N. \tag{3.104}$$

Using the first inequality in (3.104) together with (3.103) we obtain

$$u \le \tilde{u} := \lim_{n\to\infty} u_R^n \le U_R \quad \text{in } B_R.$$

Remark that $\tilde{u}$ is radially symmetric. Also, by elliptic regularity we derive that $\tilde{u}$ is a BBUS of (3.101). Thus, by the minimality of U_R it follows that $\tilde{u} \equiv U_R$, so U_R is radially symmetric.

From the second inequality in (3.104) and (3.103) we obtain

$$U_R = \lim_{n\to\infty} u_R^n \geq \lim_{n\to\infty} u_{R+1}^n = U_{R+1} \geq u \quad \text{in } B_R,$$

so $\{U_R\}$ is nonincreasing. Therefore, $\overline{u} := \lim_{R\to\infty} U_R$ satisfies $\overline{u}$ is radial, $\overline{u} \geq u$ in $\mathbb{R}^D$ and by standard elliptic arguments $\overline{u}$ is a solution of (3.93). This completes the proof of Theorem 3.44. □

The uniqueness does not hold in general due to the lack of monotonicity of f. If f is strongly oscillating, our problem (3.93) exhibits infinitely many solutions. More precisely we have

Corollary 3.46 *Assume that (3.94) holds and f satisfies* (3.92), (KO) *and that there exists a sequence $\{t_k\} \subset (0,\infty)$ such that $f(t_k) = 0$ for all $k \geq 1$ and* $\lim_{k\to\infty} = \infty$.

Then (3.93) has infinitely many solutions.

Proof. Let $f_k(t) = f(t+t_k)$, $t \geq 0$, $k \geq 1$. Then $f_k \in C^1[0,\infty)$, $f_k(0) = 0$, $f_k \geq 0$ and f_k satisfies (KO). Therefore, by Theorem 3.44 there exists an ELS v_k of $\Delta v_k = \rho(|x|) f_k(v_k)$ in $\mathbb{R}^D$. Now, if $u_k = v_k + t_k$ we have that u_k is an ELS of (3.93) and $u_k \geq t_k$. Since $\{t_k\}$ is unbounded, it follows that we have infinitely many solutions of (3.93). □

It follows from Theorem 3.44 that any ELS u to (3.93) is sandwiched between two radial ELS U, V to the same equation:

$$U(|x|) \leq u(x) \leq V(|x|), \qquad \text{for all } x \in \mathbb{R}^D.$$

The lower bound U can be chosen to be universal for all ELS to (3.93). However, there need not exist a maximal ELS to (3.93) as we see in the next result.

Corollary 3.47 *Consider the equation*

$$\Delta u = \rho(|x|) u^2 (1 + \cos u), u > 0 \quad \text{in } \mathbb{R}^D, \tag{3.105}$$

where ρ satisfies (3.94). Then:

(a) Equation (3.105) has infinitely many ELS.
(b) Equation (3.105) has a minimal ELS.
(c) Equation (3.105) has no maximal ELS.

Proof. This follows from Corollary 3.46 by taking $f(t) = t^2(1+\cos t)$ and $t_k = (2k+1)\pi$, $k \geq 1$. Finally, if U would be a maximal ELS of (3.105) then $U \geq u_k \geq t_k = (2k+1)\pi$ in $\mathbb{R}^D$, a contradiction. □

3.3.3 Uniqueness of Solution

In this section, we establish the uniqueness of radial solution under the hypothesis that f is nondecreasing. In view of Theorem 3.44 this implies the uniqueness in general of a solution to entire large solutions (ELS) of (3.93). More precisely we have:

Theorem 3.48 *Assume that f is nondecreasing and satisfies (KO) and (3.92). Assume also that $\rho(r) = Cr^{2-D}$ for some $r \geq r_0$.*

Then (3.93) has a unique ELS.

Proof. By the results in Theorem (3.44) it is enough to prove the uniqueness of a radial ELS to (3.93). We start with the following result.

Lemma 3.1. *Let u_1, u_2 be two ELS of* (3.93) *such that*

$$\lim_{|x|\to\infty} (u_1 - u_2)(x) = 0.$$

Then, $u_1 = u_2$.

Proof. Set $w = u_1 - u_2$ which verifies

$$\begin{cases} \Delta w = \rho(|x|)\dfrac{f(u_1) - f(u_2)}{u_1 - u_2} w & \text{in } \mathbb{R}^D, \\ w(x) \to 0 & \text{as } x \to \infty. \end{cases}$$

Fix $\varepsilon > 0$ and let $R > 0$ be large enough such that $w(x) < \varepsilon$ for all $|x| > R$. Then

$$\begin{cases} \Delta w = a(x)w & \text{in } B_R, \\ w \leq \varepsilon & \text{on } \partial B_R, \end{cases}$$

where

$$a(x) := \begin{cases} \rho(|x|)\dfrac{f(u_1(x)) - f(u_2(x))}{u_1(x) - u_2(x)} & \text{if } u_1(x) \neq u_2(x), \\ \rho(|x|)f'(u_1(x)) & \text{if } u_1(x) = u_2(x). \end{cases}$$

Since f is nondecreasing, a is nonnegative. By the Strong Maximum Principle, we have $w \le \varepsilon$ in B_R. Hence, $u_1 - u_2 \le \varepsilon$ on $\mathbb{R}^D$. Since $\varepsilon > 0$ was arbitrarily chosen, we have $u_1 \le u_2$ in $\mathbb{R}^D$. In the same way we obtain $u_2 \le u_1$ in $\mathbb{R}^D$ so $u_1 \equiv u_2$. □

Consider u_1, u_2 two radial ELS of (3.93) and let $t = r^{2-D}$. Then $v_i(t) := u_i(r)$ satisfies

$$\begin{cases} \dfrac{d^2 v_i}{dt^2}(t) = C_0 f(v_i(t)) & \text{for all } 0 < t \le t_0 = r_0^{1/(2-D)}, \\ v_i(t) \to +\infty, & \text{as } t \to 0^+. \end{cases} \tag{3.106}$$

We multiply (3.106) by v_i' and we integrate over $[t, t_0]$. We obtain, for $i = 1, 2$

$$v_i'^2(t) = 2c_0(F(v_i(t)) + C_i) \quad \text{for all } 0 < t \le t_0,$$

where $C_i = v_i'^2(t_0) - F(v_i(t_0))$, $i = 1, 2$. Thus,

$$\frac{-v_1'}{\sqrt{F(v_1) + C_1}} = \frac{-v_2'}{\sqrt{F(v_2) + C_2}}.$$

Integrating between 0 and t we find

$$\int_{v_1(t)}^{\infty} \frac{ds}{\sqrt{F(s) + C_1}} = \int_{v_2(t)}^{\infty} \frac{ds}{\sqrt{F(s) + C_2}}.$$

Without loss of generality, we assume that $v_2 \ge v_1$. Then,

$$\int_{v_1(t)}^{v_2(t)} \frac{ds}{\sqrt{F(s) + C_1}} = \int_{v_2(t)}^{\infty} \frac{\sqrt{F(s) + C_1} - \sqrt{F(s) + C_2}}{\sqrt{(F(s) + C_1)(F(s) + C_2)}} ds.$$

Since F is increasing, we have

$$\frac{v_2(t) - v_1(t)}{\sqrt{F(v_2(t)) + C_1}} \le \int_{v_2(t)}^{\infty} \frac{C}{F(s)^{3/2}} ds \le \frac{C}{\sqrt{F(v_2(t))}} \int_{v_2(t)}^{\infty} \frac{1}{F(s)} ds.$$

This implies

$$0 \le v_2(t) - v_1(t) \le C \int_{v_2(t)}^{\infty} \frac{1}{F(s)} ds \to 0 \text{ as } t \to 0^+.$$

This implies $u_2(r) - u_1(r) \to 0$ as $r \to \infty$. By Lemma 3.1 we now obtain $u_1 \equiv u_2$ so (3.93) has a unique solution. □

3.4 Elliptic Equations with Absorption

Consider the problem

$$\begin{cases} \Delta u + q(x)|\nabla u|^a = p(x)f(u) & \text{in } \Omega\,, \\ u \geq 0,\ u \not\equiv 0 & \text{in } \Omega\,, \end{cases} \tag{3.107}$$

where $\Omega \subset \mathbb{R}^N$ ($N \geq 3$) is a smooth domain (bounded or possibly unbounded) with compact (possibly empty) boundary. We assume that $a \leq 2$ is a positive real number, p,q are nonnegative functions such that $p \not\equiv 0$, $p,q \in C^{0,\alpha}(\overline{\Omega})$ if Ω is bounded, and $p,q \in C^{0,\alpha}_{\mathrm{loc}}(\Omega)$ otherwise. Throughout this section (see [91]) we assume that the nonlinearity f fulfills the following conditions:

$(f1)$ $\quad f \in C^1[0,\infty)$, $f' \geq 0$, $f(0) = 0$ and $f > 0$ on $(0,\infty)$.

$(f2)$ $\quad \displaystyle\int_1^\infty [F(t)]^{-1/2}\,dt < \infty, \quad$ where $\displaystyle F(t) = \int_0^t f(s)\,ds.$

$(f3)$ $\quad \dfrac{F(t)}{f^{2/a}(t)} \to 0 \quad$ as $\quad t \to \infty$.

Cf. Véron [199], f is called an absorption term. The above conditions hold provided that $f(t) = t^k$, $k > 1$ and $0 < a < \frac{2r}{r+1} (< 2)$, or $f(t) = e^t - 1$, or $f(t) = e^t - t$ and $a < 2$. We observe that by $(f1)$ and $(f3)$ it follows that $f/F^{a/2} \geq \beta > 0$ for t large enough, that is, $(F^{1-a/2})' \geq \beta > 0$ for t large enough which yields $0 < a \leq 2$. We also deduce that conditions $(f2)$ and $(f3)$ imply $\displaystyle\int_1^\infty f^{-1/a}(t)dt < \infty$.

We are mainly interested in finding properties of *large (explosive) solutions* of (3.107), that is solutions u satisfying $u(x) \to \infty$ as $\mathrm{dist}\,(x,\partial\Omega) \to 0$ (if $\Omega \not\equiv \mathbb{R}^N$), or $u(x) \to \infty$ as $|x| \to \infty$ (if $\Omega = \mathbb{R}^N$). In the latter case the solution is called an *entire large (explosive)* solution.

Problems of this type appear in stochastic control theory and were first studied by Lasry and Lions [129]. The corresponding parabolic equation was considered in Quittner [167] and in Galaktionov and Vázquez [80]. In terms of the dynamic programming approach, an explosive solution of (3.107) corresponds to a value function (or Bellman function) associated to an infinite exit cost (see Lasry and Lions [129]).

Bandle and Giarrusso [11] studied the existence of a large solution of problem (3.107) in the case $p \equiv 1$, $q \equiv 1$ and Ω bounded. Lair and Wood [126] studied the sublinear case corresponding to $p \equiv 1$, while Cîrstea and Rădulescu [42] proved the existence of large solutions to (3.107) in the case $q \equiv 0$.

As observed by Bandle and Giarrusso [11], the simplest case is $a = 2$, which can be reduced to a problem without gradient term. Indeed, if u is a solution of (3.107) for $q \equiv 1$, then the function $v = e^u$ (Gelfand transformation) satisfies

$$\begin{cases} \Delta v = p(x) v f(\ln v) & \text{in } \Omega, \\ v(x) \to +\infty & \text{if } \operatorname{dist}(x, \partial\Omega) \to 0. \end{cases}$$

We shall therefore mainly consider the case where $0 < a < 2$.

The main results in this section are due to Ghergu, Niculescu, and Rădulescu [87]. These results generalize those obtained by Cîrstea and Rădulescu [42] in the case of the presence of a convection (gradient) term.

Our first result concerns the existence of a large solution to problem (3.107) when Ω is bounded.

Theorem 3.49 *Suppose that Ω is bounded and assume that p satisfies*
$(p1)$ *for every $x_0 \in \Omega$ with $p(x_0) = 0$, there exists a domain $\Omega_0 \ni x_0$ such that $\overline{\Omega_0} \subset \Omega$ and $p > 0$ on $\partial\Omega_0$.*

Then problem (3.107) has a positive large solution.

A crucial role in the proof of the above result is played by the following auxiliary result (see Ghergu, Niculescu, and Rădulescu [87]).

Lemma 3.50 *Let Ω be a bounded domain. Assume that $p, q \in C^{0,\alpha}(\overline{\Omega})$ are non-negative functions, $0 < a < 2$ is a real number, f satisfies $(f1)$ and $g : \partial\Omega \to (0, \infty)$ is continuous. Then the boundary value problem*

$$\begin{cases} \Delta u + q(x)|\nabla u|^a = p(x) f(u), & \text{in } \Omega, \\ u = g, & \text{on } \partial\Omega, \\ u \geq 0,\ u \not\equiv 0, & \text{in } \Omega \end{cases} \tag{3.108}$$

has a classical solution. If p is positive, then the solution is unique.

Sketch of the proof of Theorem 3.49. By Lemma 3.50, the boundary value problem

$$\begin{cases} \Delta v_n + q(x)|\nabla v_n|^a = \left(p(x) + \dfrac{1}{n}\right) f(v_n), & \text{in } \Omega, \\ v_n = n, & \text{on } \partial\Omega, \\ v_n \geq 0,\ v_n \not\equiv 0, & \text{in } \Omega \end{cases}$$

has a unique positive solution, for any $n \geq 1$. Next, by the maximum principle, the sequence $\{v_n\}$ is nondecreasing and is bounded from below in Ω by a positive function.

To conclude the proof, it is sufficient to show that

(a) for all $x_0 \in \Omega$ there exists an open set $\mathscr{O} \subset\subset \Omega$ which contains x_0 and $M_0 = M_0(x_0) > 0$ such that $v_n \leq M_0$ in $\mathscr{O}$ for all $n \geq 1$;

(b) $\lim_{x\to\partial\Omega} v(x) = \infty$, where $v(x) = \lim_{n\to\infty} v_n(x)$.

We observe that the statement (a) shows that the sequence (v_n) is uniformly bounded on every compact subset of Ω. Standard elliptic regularity arguments (see Gilbarg and Trudinger [99]) show that v is a solution of problem (3.107). Then, by (b), it follows that v is a large solution of problem (3.107).

To prove (a) we distinguish two cases:

Case $p(x_0) > 0$. By the continuity of p, there exists a ball $B = B(x_0, r) \subset\subset \Omega$ such that

$$m_0 := \min\{p(x);\ x \in \overline{B}\} > 0.$$

Let w be a positive solution of the problem

$$\begin{cases} \Delta w + q(x)|\nabla w|^a = m_0 f(w), & \text{in } B\\ w(x) \to \infty, & \text{as } x \to \partial B. \end{cases}$$

The existence of w follows by considering the problem

$$\begin{cases} \Delta w_n + q(x)|\nabla w_n|^a = m_0 f(w_n), & \text{in } B\\ w_n = n, & \text{on } \partial B. \end{cases}$$

The maximum principle implies $w_n \leq w_{n+1} \leq \theta$, where

$$\begin{cases} \Delta \theta + \|q\|_{L^\infty}|\nabla \theta|^a = m_0 f(\theta), & \text{in } B\\ \theta(x) \to \infty, & \text{as } x \to \partial B. \end{cases}$$

Standard arguments show that $v_n \leq w$ in B. Furthermore, w is bounded in $\overline{B(x_0, r/2)}$. Setting $M_0 = \sup_{\mathscr{O}} w$, where $\mathscr{O} = B(x_0, r/2)$, we obtain (a).

Case $p(x_0) = 0$. Our hypothesis $(p1)$ and the boundedness of Ω imply the existence of a domain $\mathscr{O} \subset\subset \Omega$ which contains x_0 such that $p > 0$ on $\partial\mathscr{O}$. The above case shows that for any $x \in \partial\mathscr{O}$ there exist a ball $B(x, r_x)$ strictly contained in Ω and a constant $M_x > 0$ such that $v_n \leq M_x$ on $B(x, r_x/2)$, for any $n \geq 1$. Since $\partial\mathscr{O}$ is compact, it follows that it may be covered by a finite number of such balls, say $B(x_i, r_{x_i}/2)$,

$i = 1, \cdots, k_0$. Setting $M_0 = \max\{M_{x_1}, \cdots, M_{x_{k_0}}\}$ we have $v_n \le M_0$ on $\partial\mathscr{O}$, for any $n \ge 1$. Applying the maximum principle we obtain $v_n \le M_0$ in $\mathscr{O}$ and (a) follows.

Let z be the unique function satisfying $-\Delta z = p(x)$ in Ω and $z = 0$, on $\partial\Omega$. Moreover, by the maximum principle, we have $z > 0$ in Ω. We first observe that for proving (b) it is sufficient to show that

$$\int_{v(x)}^{\infty} \frac{dt}{f(t)} \le z(x) \quad \text{for any } x \in \Omega. \tag{3.109}$$

By [42, Lemma 1], the left-hand side of (3.109) is well defined in Ω. We choose $R > 0$ so that $\overline{\Omega} \subset B(0,R)$ and fix $\varepsilon > 0$. Since $v_n = n$ on $\partial\Omega$, let $n_1 = n_1(\varepsilon)$ be such that

$$n_1 > \frac{1}{\varepsilon(N-3)(1+R^2)^{-1/2} + 3\varepsilon(1+R^2)^{-5/2}}, \tag{3.110}$$

and

$$\int_{v_n(x)}^{\infty} \frac{dt}{f(t)} \le z(x) + \varepsilon(1+|x|^2)^{-1/2} \qquad \forall\, x \in \partial\Omega, \forall\, n \ge n_1. \tag{3.111}$$

In order to prove (3.109), it is enough to show that

$$\int_{v_n(x)}^{\infty} \frac{dt}{f(t)} \le z(x) + \varepsilon(1+|x|^2)^{-1/2} \qquad \forall\, x \in \Omega,\ \forall n \ge n_1. \tag{3.112}$$

Indeed, taking $n \to \infty$ in (3.112) we deduce (3.109), since $\varepsilon > 0$ is arbitrarily chosen. Assume now, by contradiction, that (3.112) fails. Then

$$\max_{x \in \overline{\Omega}} \left\{ \int_{v_n(x)}^{\infty} \frac{dt}{f(t)} - z(x) - \varepsilon(1+|x|^2)^{-1/2} \right\} > 0.$$

Using (3.111) we see that the point where the maximum is achieved must lie in Ω. A straightforward computation shows that at this point, say x_0, we have

$$0 \ge \Delta \left(\int_{v_n(x)}^{\infty} \frac{dt}{f(t)} - z(x) - \varepsilon(1+|x|^2)^{-1/2} \right)_{|x=x_0} > 0.$$

This contradiction shows that inequality (3.111) holds and the proof of Theorem 3.49 is complete. □

Similar arguments based on the maximum principle and the approximation of large balls $B(0,n)$ imply the following existence result.

Theorem 3.51 *Assume that $\Omega = \mathbb{R}^N$ and that problem (3.107) has at least one solution. Suppose that p satisfies the condition*

$(p1)'$ *There exists a sequence of smooth bounded domains $(\Omega_n)_{n\geq 1}$ such that $\overline{\Omega_n} \subset \Omega_{n+1}$, $\mathbb{R}^N = \cup_{n=1}^{\infty}\Omega_n$, and $(p1)$ holds in Ω_n, for any $n \geq 1$.*

Then there exists a classical solution U of (3.107) which is a maximal solution if p is positive.

Assume that p verifies the additional condition

$$(p2) \qquad \int_0^\infty r\Phi(r)\,dr < \infty, \quad \text{where } \Phi(r) = \max\{p(x): \ |x| = r\}.$$

Then U is an entire large solution of (3.107).

We now consider the case in which $\Omega \neq \mathbb{R}^N$ and Ω is unbounded. We say that a large solution u of (3.107) is *regular* if u tends to zero at infinity. In [141, Theorem 3.1] Marcus proved for this case (and if $q = 0$) the existence of regular large solutions to problem (3.107) by assuming that there exist $\gamma > 1$ and $\beta > 0$ such that

$$\liminf_{t\to 0} f(t)t^{-\gamma} > 0 \qquad \text{and} \qquad \liminf_{|x|\to\infty} p(x)|x|^{\beta} > 0.$$

The large solution constructed in Marcus [141] is the *smallest* large solution of problem (3.107). In the next result we show that problem (3.107) admits a *maximal* classical solution U and that U blows-up at infinity if $\Omega = \mathbb{R}^N \setminus \overline{B(0,R)}$.

Theorem 3.52 *Suppose that $\Omega \neq \mathbb{R}^N$ is unbounded and that problem (3.107) has at least a solution. Assume that p satisfies condition $(p1)'$ in Ω. Then there exists a classical solution U of problem (3.107) which is a maximal solution if p is positive.*

If $\Omega = \mathbb{R}^N \setminus \overline{B(0,R)}$ and p satisfies the additional condition $(p2)$, with $\Phi(r) = 0$ for $r \in [0,R]$, then the solution U of (3.107) is a large solution that blows-up at infinity.

We refer to Ghergu, Niculescu and Rădulescu [87] for complete proofs of Theorems 3.51 and 3.52.

A useful observation is given in the following property.

Remark 18 *Assume that $p \in C(\mathbb{R}^N)$ is a nonnegative and nontrivial function which satisfies $(p2)$. Let f be a function satisfying assumption $(f1)$. Then condition*

$$\int_1^\infty \frac{dt}{f(t)} < \infty \tag{3.113}$$

is necessary for the existence of entire large solutions to (3.107).

Indeed, let u be an entire large solution of problem (3.107). Define

$$\bar{u}(r) = \frac{1}{\omega_N r^{N-1}} \int_{|x|=r} \left(\int_{a_0}^{u(x)} \frac{dt}{f(t)} \right) dS = \frac{1}{\omega_N} \int_{|\xi|=1} \left(\int_{a_0}^{u(r\xi)} \frac{dt}{f(t)} \right) dS,$$

where ω_N denotes the surface area of the unit sphere in $\mathbb{R}^N$ and a_0 is chosen such that $a_0 \in (0, u_0)$, where $u_0 = \inf_{\mathbb{R}^N} u > 0$. By the divergence theorem we have

$$\bar{u}'(r) = \frac{1}{\omega_N r^{N-1}} \int_{B(0,r)} \Delta \left(\int_{a_0}^{u(x)} \frac{dt}{f(t)} \right) dx.$$

Since u is a positive classical solution it follows that

$$|\bar{u}'(r)| \leq Cr \to 0 \qquad \text{as } r \to 0\,.$$

On the other hand

$$\omega_N \left(R^{N-1}\bar{u}'(R) - r^{N-1}\bar{u}'(r) \right) = \int_r^R \left(\int_{|x|=z} \Delta \left(\int_{a_0}^{u(x)} \frac{dt}{f(t)} \right) dS \right) dz.$$

Dividing by $R - r$ and taking $R \to r$ we find

$$\begin{aligned}
\omega_N (r^{N-1}\bar{u}'(r))' &= \int_{|x|=r} \Delta \left(\int_{a_0}^{u(x)} \frac{dt}{f(t)} \right) dS = \int_{|x|=r} \operatorname{div} \left(\frac{1}{f(u(x))} \nabla u(x) \right) dS \\
&= \int_{|x|=r} \left[\left(\frac{1}{f} \right)' (u(x)) \cdot |\nabla u(x)|^2 + \frac{1}{f(u(x))} \Delta u(x) \right] dS \\
&\leq \int_{|x|=r} \frac{p(x) f(u(x))}{f(u(x))} dS \leq \omega_N r^{N-1} \Phi(r).
\end{aligned}$$

The above inequality yields by integration

$$\bar{u}(r) \leq \bar{u}(0) + \int_0^r \sigma^{1-N} \left(\int_0^\sigma \tau^{N-1} \Phi(\tau)\, d\tau \right) d\sigma \qquad \text{for all } r \geq 0. \tag{3.114}$$

On the other hand, according to $(p2)$, for all $r > 0$ we have

$$\begin{aligned} &\int_0^r \sigma^{1-N} \left(\int_0^\sigma \tau^{N-1} \Phi(\tau)\, d\tau \right) d\sigma \\ &\qquad = \frac{1}{2-N} r^{2-N} \int_0^r \tau^{N-1} \Phi(\tau)\, d\tau - \frac{1}{2-N} \int_0^r \sigma \Phi(\sigma)\, d\sigma \\ &\qquad \leq \frac{1}{N-2} \int_0^\infty r \Phi(r)\, dr < \infty. \end{aligned}$$

So, by (3.114), $\bar{u}(r) \leq \bar{u}(0) + K$, for all $r \geq 0$. The last inequality implies that $\bar{u}$ is bounded and assuming that (3.113) is not fulfilled it follows that u cannot be a large solution. □

We point out that the hypothesis $(p2)$ on p is essential in the statement of Remark 18. Indeed, let us consider $f(t) = t$, $p \equiv 1$, $\alpha \in (0,1)$, $q(x) = 2^{\alpha-2} \cdot |x|^\alpha$, $a = 2 - \alpha \in (1,2)$. Then the corresponding problem has the entire large solution $u(x) = |x|^2 + 2N$, but (3.113) is not fulfilled.

3.5 Lack of the Keller–Osserman Condition

We have already seen that if f is smooth and increasing on $[0,\infty)$ such that $f(0) = 0$ and $f > 0$ in $(0,\infty)$, then the problem

$$\begin{cases} \Delta u = f(u) & \text{in } \Omega, \\ u > 0 & \text{in } \Omega, \\ u = +\infty & \text{on } \partial\Omega \end{cases}$$

has a solution if and only if the Keller–Osserman condition $\int_1^\infty [F(t)]^{-1/2} dt < \infty$ is fulfilled, where $F(t) = \int_0^t f(s)\, ds$. In particular, this implies that f must have a superlinear growth. In this section we are concerned with the problem

$$\begin{cases} \Delta u + |\nabla u| = p(x) f(u) & \text{in } \Omega, \\ u \geq 0 & \text{in } \Omega, \end{cases} \tag{3.115}$$

where $\Omega \subset \mathbb{R}^N$ $(N \geq 3)$ is either a smooth bounded domain or the whole space. Our main assumption on f is that it has a *sublinear* growth, so we cannot expect that problem (3.115) admits a blow-up boundary solution. Our main purpose in this section is to establish a necessary and sufficient condition on the variable potential $p(x)$ for the existence of an entire large solution.

Throughout this section we assume that p is a nonnegative function such that $p \in C^{0,\alpha}(\overline{\Omega})$ $(0 < \alpha < 1)$ if Ω is bounded, and $p \in C^{0,\alpha}_{\text{loc}}(\mathbb{R}^N)$ otherwise. The nondecreasing nonlinearity $f \in C^{0,\alpha}_{\text{loc}}[0,\infty)$ fulfills $f(0) = 0$ and $f > 0$ on $(0,\infty)$. We also assume that f is sublinear at infinity, in the sense that $\Lambda := \sup_{s\geq 1} \frac{f(s)}{s} < \infty$.

If Ω is bounded we prove the following nonexistence result.

Theorem 3.53 *Suppose that $\Omega \subset \mathbb{R}^N$ is a smooth bounded domain. Then problem (3.115) has no positive large solution in Ω.*

Proof. Suppose by contradiction that problem (3.115) has a positive large solution u and define $v(x) = \ln(1 + u(x))$, $x \in \Omega$. It follows that v is positive and $v(x) \to \infty$ as $\operatorname{dist}(x, \partial\Omega) \to 0$. We have

$$\Delta v = \frac{1}{1+u}\Delta u - \frac{1}{(1+u)^2}|\nabla u|^2 \qquad \text{in } \Omega$$

and so

$$\Delta v \leq p(x)\frac{f(u)}{1+u} \leq \|p\|_\infty \frac{f(u)}{1+u} \leq A \qquad \text{in } \Omega,$$

for some constant $A > 0$. Therefore

$$\Delta(v(x) - A|x|^2) < 0, \qquad \text{for all } x \in \Omega.$$

Let $w(x) = v(x) - A|x|^2$, $x \in \Omega$. Then $\Delta w < 0$ in Ω. Moreover, since Ω is bounded, it follows that $w(x) \to \infty$ as $\operatorname{dist}(x, \partial\Omega) \to 0$.

Let $M > 0$ be arbitrary. We claim that $w \geq M$ in Ω. For all $\delta > 0$, we set

$$\Omega_\delta = \{x \in \Omega\,;\ \operatorname{dist}(x, \partial\Omega) > \delta\}.$$

Since $w(x) \to \infty$ as $\operatorname{dist}(x, \partial\Omega) \to 0$, we can choose $\delta > 0$ such that

$$w(x) \geq M \qquad \text{for all } x \in \Omega \setminus \Omega_\delta. \tag{3.116}$$

On the other hand,

$$\begin{aligned} -\Delta(w(x)-M) &> 0 \quad \text{in } \Omega_\delta, \\ w(x)-M &\geq 0 \quad \text{on } \partial\Omega_\delta. \end{aligned}$$

By the maximum principle we get $w(x)-M \geq 0$ in Ω_δ. So, by (3.116), $w \geq M$ in Ω. Since $M>0$ is arbitrary, it follows that $w \geq n$ in Ω, for all $n \geq 1$. Obviously, this is a contradiction and the proof is now complete. □

Next, we consider the problem (3.115) when $\Omega = \mathbb{R}^N$. For all $r \geq 0$ we set

$$\phi(r) = \max_{|x|=r} p(x), \qquad \psi(r) = \min_{|x|=r} p(x), \qquad \text{and} \qquad h(r) = \phi(r) - \psi(r).$$

We suppose that

$$\int_0^\infty rh(r)\Psi(r)dr < \infty, \tag{3.117}$$

where

$$\Psi(r) = \exp\left(\Lambda_N \int_0^r s\psi(s)ds\right), \qquad \Lambda_N = \frac{\Lambda}{N-2}.$$

Obviously, if p is radial then $h \equiv 0$ and (3.117) occurs. Assumption (3.117) shows that the variable potential $p(x)$ has a slow variation. An example of nonradial potential for which (3.117) holds is $p(x) = \dfrac{1+|x_1|^2}{(1+|x_1|^2)(1+|x|^2)+1}$. In this case $\phi(r) = \dfrac{r^2+1}{(r^2+1)^2+1}$ and $\psi(r) = \dfrac{1}{r^2+2}$. If $\Lambda_N = 1$, by direct computation we get $rh(r)\Psi(r) = O\left(r^{-2}\right)$ as $r \to \infty$ and so (3.117) holds.

Theorem 3.54 *Assume $\Omega = \mathbb{R}^N$ and p satisfies (3.117). Then problem (3.115) has a positive entire large solution if and only if*

$$\int_1^\infty e^{-t}t^{1-N} \int_0^t e^s s^{N-1}\psi(s)dsdt = \infty. \tag{3.118}$$

Proof. Several times in the proof of Theorem 3.54 we shall apply the following elementary inequality:

$$\int_0^r e^{-t}t^{1-N} \int_0^t e^s s^{N-1} g(s)dsdt \leq \frac{1}{N-2}\int_0^r tg(t)dt, \qquad \forall\, r>0, \tag{3.119}$$

for any continuous function $g: [0,\infty) \to [0,\infty)$. The proof follows easily by integration by parts.

Necessary condition. Suppose that (3.117) fails and (3.115) has a positive entire large solution u. We claim that

$$\int_1^\infty e^{-t}t^{1-N}\int_0^t e^s s^{N-1}\phi(s)dsdt < \infty. \tag{3.120}$$

We first recall that $\phi = h + \psi$. Thus

$$\int_1^\infty e^{-t}t^{1-N}\int_0^t e^s s^{N-1}\phi(s)dsdt = \int_1^\infty e^{-t}t^{1-N}\int_0^t e^s s^{N-1}\psi(s)dsdt$$
$$+\int_1^\infty e^{-t}t^{1-N}\int_0^t e^s s^{N-1}h(s)dsdt.$$

By virtue of (3.119) we find

$$\int_1^\infty e^{-t}t^{1-N}\int_0^t e^s s^{N-1}\phi(s)dsdt \le \int_1^\infty e^{-t}t^{1-N}\int_0^t e^s s^{N-1}\psi(s)dsdt + \frac{\int_0^\infty th(t)dt}{N-2}$$
$$\le \int_1^\infty e^{-t}t^{1-N}\int_0^t e^s s^{N-1}\psi(s)dsdt + \frac{\int_0^\infty th(t)\Psi(t)dt}{N-2}.$$

Since $\int_1^\infty e^{-t}t^{1-N}\int_0^t e^s s^{N-1}\psi(s)dsdt < \infty$, by (3.117) we deduce (3.120).

Now, let $\bar{u}$ be the spherical average of u, that is,

$$\bar{u}(r) = \frac{1}{\omega_N r^{N-1}}\int_{|x|=r} u(x)d\sigma_x, \qquad r \ge 0,$$

where ω_N is the surface area of the unit sphere in $\mathbb{R}^N$. Since u is a positive entire large solution of (3.107) it follows that $\bar{u}$ is positive and $\bar{u}(r) \to \infty$ as $r \to \infty$. With the change of variable $x \to ry$, we have

$$\bar{u}(r) = \frac{1}{\omega_N}\int_{|y|=1} u(ry)d\sigma_y, \qquad r \ge 0$$

and

$$\bar{u}'(r) = \frac{1}{\omega_N}\int_{|y|=1} \nabla u(ry)\cdot y d\sigma_y, \qquad r \ge 0. \tag{3.121}$$

Hence

$$\bar{u}'(r) = \frac{1}{\omega_N} \int\limits_{|y|=1} \frac{\partial u}{\partial r}(ry)\,d\sigma_y = \frac{1}{\omega_N r^{N-1}} \int\limits_{|x|=r} \frac{\partial u}{\partial r}(x)\,d\sigma_x,$$

that is

$$\bar{u}'(r) = \frac{1}{\omega_N r^{N-1}} \int\limits_{B(0,R)} \Delta u(x)\,dx, \qquad \text{for all } r \geq 0. \tag{3.122}$$

Due to the gradient term $|\nabla u|$ in (3.107), we cannot infer that $\Delta u \geq 0$ in $\mathbb{R}^N$ and so we cannot expect that $\bar{u}' \geq 0$ in $[0,\infty)$. We define the auxiliary function

$$U(r) = \max_{0 \leq t \leq r} \bar{u}(t), \qquad r \geq 0. \tag{3.123}$$

Then U is positive and nondecreasing. Moreover, $U \geq \bar{u}$ and $U(r) \to \infty$ as $r \to \infty$.

The assumptions $(f1)$ and $(f2)$ yield $f(t) \leq \Lambda(1+t)$, for all $t \geq 0$. So, by (3.121) and (3.122) we have

$$\begin{aligned}
\bar{u}'' + \frac{N-1}{r}\bar{u}' + \bar{u}' &\leq \frac{1}{\omega_N r^{N-1}} \int\limits_{|x|=r} [\Delta u(x) + |\nabla u|(x)]\,d\sigma_x \\
&= \frac{1}{\omega_N r^{N-1}} \int\limits_{|x|=r} p(r) f(u(x))\,d\sigma_x \\
&\leq \Lambda \phi(r) \frac{1}{\omega_N r^{N-1}} \int\limits_{|x|=r} (1+u(x))\,d\sigma_x \\
&= \Lambda \phi(r)(1+\bar{u}(r)) \leq \Lambda \phi(r)(1+U(r)),
\end{aligned}$$

for all $r \geq 0$. It follows that

$$\left(r^{N-1} e^r \bar{u}'\right)' \leq \Lambda e^r r^{N-1} \phi(r)(1+U(r)), \qquad \text{for all } r \geq 0.$$

So, for all $r \geq r_0 > 0$,

$$\bar{u}(r) \leq \bar{u}(r_0) + \Lambda \int_{r_0}^{r} e^{-t} t^{1-N} \int_0^t e^s s^{N-1} \phi(s)(1+U(s))\,ds\,dt.$$

The monotonicity of U implies

$$\bar{u}(r) \leq \bar{u}(r_0) + \Lambda(1+U(r)) \int_{r_0}^{r} e^{-t} t^{1-N} \int_0^t e^s s^{N-1} \phi(s)\,ds\,dt, \tag{3.124}$$

for all $r \geq r_0 \geq 0$. By (3.120) we can choose $r_0 \geq 1$ such that

$$\int_{r_0}^{\infty} e^{-t} t^{1-N} \int_0^t e^s s^{N-1} \phi(s) ds dt < \frac{1}{2\Lambda}. \tag{3.125}$$

Thus (3.124) and (3.125) yield

$$\bar{u}(r) \leq \bar{u}(r_0) + \frac{1}{2}(1 + U(r)), \qquad \text{for all } r \geq r_0. \tag{3.126}$$

By the definition of U and $\lim_{r \to \infty} \bar{u}(r) = \infty$, we find $r_1 \geq r_0$ such that

$$U(r) = \max_{r_0 \leq t \leq r} \bar{u}(r), \qquad \text{for all } r \geq r_1. \tag{3.127}$$

Considering now (3.126) and (3.127) we obtain

$$U(r) \leq \bar{u}(r_0) + \frac{1}{2}(1 + U(r)), \qquad \text{for all } r \geq r_1.$$

Hence

$$U(r) \leq 2\bar{u}(r_0) + 1, \qquad \text{for all } r \geq r_1.$$

This means that U is bounded, so u is also bounded, a contradiction. It follows that (3.107) has no positive entire large solutions.

Sufficient condition. We need the following auxiliary comparison result.

Lemma 3.55 *Assume that (3.117) and (3.118) hold. Then the equations*

$$\Delta v + |\nabla v| = \phi(|x|) f(v) \qquad \Delta w + |\nabla w| = \psi(|x|) f(w) \tag{3.128}$$

have positive entire large solution such that

$$v \leq w \qquad \text{in } \mathbb{R}^N. \tag{3.129}$$

Proof. Radial solutions of (3.128) satisfy

$$v'' + \frac{N-1}{r} v' + |v'| = \phi(r) f(v)$$

and

$$w'' + \frac{N-1}{r} w' + |w'| = \psi(r) f(w).$$

Assuming that v' and w' are nonnegative, we deduce

$$\left(e^r r^{N-1} v'\right)' = e^r r^{N-1} \phi(r) f(v)$$

and

$$\left(e^r r^{N-1} w'\right)' = e^r r^{N-1} \psi(r) f(w).$$

Thus any positive solutions v and w of the integral equations

$$v(r) = 1 + \int_0^r e^{-t} t^{1-N} \int_0^t e^s s^{N-1} \phi(s) f(v(s)) ds dt, \qquad r \geq 0, \tag{3.130}$$

$$w(r) = b + \int_0^r e^{-t} t^{1-N} \int_0^t e^s s^{N-1} \psi(s) f(w(s)) ds dt, \qquad r \geq 0, \tag{3.131}$$

provide a solution of (3.128), for any $b > 0$. Since $w \geq b$, it follows that $f(w) \geq f(b) > 0$ which yields

$$w(r) \geq b + f(b) \int_0^r e^{-t} t^{1-N} \int_0^t e^s s^{N-1} \psi(s) ds dt, \qquad r \geq 0.$$

By (3.118), the right-hand side of this inequality goes to $+\infty$ as $r \to \infty$. Thus $w(r) \to \infty$ as $r \to \infty$. With a similar argument we find $v(r) \to \infty$ as $r \to \infty$.

Let $b > 1$ be fixed. We first show that (3.131) has a positive solution. Similarly, (3.130) has a positive solution.

Let $\{w_k\}$ be the sequence defined by $w_1 = b$ and

$$w_{k+1}(r) = b + \int_0^r e^{-t} t^{1-N} \int_0^t e^s s^{N-1} \psi(s) f(w_k(s)) ds dt, \qquad k \geq 1. \tag{3.132}$$

We remark that $\{w_k\}$ is a nondecreasing sequence. To get the convergence of $\{w_k\}$ we will show that $\{w_k\}$ is bounded from above on bounded subsets. To this aim, we fix $R > 0$ and we prove that

$$w_k(r) \leq b e^{Mr}, \qquad \text{for any } 0 \leq r \leq R, \text{ and for all } k \geq 1, \tag{3.133}$$

where $M \equiv \Lambda_N \max_{t \in [0,R]} t\psi(t)$.

We achieve (3.133) by induction. We first notice that (3.133) is true for $k = 1$. Furthermore, the assumption $(f2)$ and the fact that $w_k \geq 1$ lead us to $f(w_k) \leq \Lambda w_k$, for all $k \geq 1$. So, by (3.132),

$$w_{k+1}(r) \leq b + \Lambda \int_0^r e^{-t} t^{1-N} \int_0^t e^s s^{N-1} \psi(s) w_k(s) ds dt, \qquad r \geq 0.$$

Using now (3.119) (for $g(t) = \psi(t)w_k(t)$) we deduce

$$w_{k+1}(r) \le b + \Lambda_N \int_0^r t\psi(t)w_k(t)dt, \qquad \forall\, r \in [0,R].$$

The induction hypothesis yields

$$w_{k+1}(r) \le b + bM \int_0^r e^{Mt}dt = be^{Mr}, \qquad \forall\, r \in [0,R].$$

Hence, by induction, the sequence $\{w_k\}$ is bounded in $[0,R]$, for any $R > 0$. It follows that $w(r) = \lim_{k\to\infty} w_k(r)$ is a positive solution of (3.131). In a similar way we conclude that (3.130) has a positive solution on $[0,\infty)$.

The next step is to show that the constant b may be chosen sufficiently large so that (3.129) holds. More exactly, if

$$b > 1 + K\Lambda_N \int_0^\infty sh(s)\Psi(s)ds, \tag{3.134}$$

where $K = \exp\left(\Lambda_N \int_0^\infty th(t)dt\right)$, then (3.129) occurs.

We first prove that the solution v of (3.130) satisfies

$$v(r) \le K\Psi(r), \qquad \forall\, r \ge 0. \tag{3.135}$$

Since $v \ge 1$, from $(f2)$ we have $f(v) \le \Lambda v$. We use this fact in (3.130) and then we apply the estimate (3.119) for $g = \phi$. It follows that

$$v(r) \le 1 + \Lambda_N \int_0^r s\phi(s)v(s)ds, \qquad \forall\, r \ge 0. \tag{3.136}$$

By Gronwall's inequality we obtain

$$v(r) \le \exp\left(\Lambda_N \int_0^r s\phi(s)ds\right), \qquad \forall\, r \ge 0,$$

and, by (3.136),

$$v(r) \le 1 + \Lambda_N \int_0^r s\phi(s)\exp\left(\Lambda_N \int_0^s t\phi(t)dt\right)ds, \qquad \forall\, r \ge 0.$$

Hence

$$v(r) \leq 1 + \int_0^r \left(\exp\left(\Lambda_N \int_0^s t\phi(t)dt \right) \right)' ds, \qquad \forall\, r \geq 0,$$

that is

$$v(r) \leq \exp\left(\Lambda_N \int_0^r t\phi(t)dt \right), \qquad \forall\, r \geq 0. \tag{3.137}$$

Inserting $\phi = h + \psi$ in (3.137) we have

$$v(r) \leq e^{\Lambda_N \int_0^r th(t)dt} \Psi(r) \leq K\Psi(r), \qquad \forall\, r \geq 0,$$

so (3.135) follows.

Since $b > 1$ it follows that $v(0) < w(0)$. Then there exists $R > 0$ such that $v(r) < w(r)$, for any $0 \leq r \leq R$. Set

$$R_\infty = \sup\{\, R > 0 \,|\, v(r) < w(r),\ \forall\, r \in [0,R] \,\}.$$

In order to conclude our proof, it remains to show that $R_\infty = \infty$. Suppose the contrary. Since $v \leq w$ on $[0,R_\infty]$ and $\phi = h + \psi$, from (3.130) we deduce

$$\begin{aligned} v(R_\infty) =& 1 + \int_0^{R_\infty} e^{-t} t^{1-N} \int_0^t e^s s^{N-1} h(s) f(v(s)) ds dt \\ &+ \int_0^{R_\infty} e^{-t} t^{1-N} \int_0^t e^s s^{N-1} \psi(s) f(v(s)) ds dt. \end{aligned}$$

So, by (3.119),

$$v(R_\infty) \leq 1 + \frac{1}{N-2} \int_0^{R_\infty} th(t) f(v(t)) dt + \int_0^{R_\infty} e^{-t} t^{1-N} \int_0^t e^s s^{N-1} \psi(s) f(w(s)) ds dt.$$

Taking into account that $v \geq 1$ and the assumption $(f2)$, it follows that

$$v(R_\infty) \leq 1 + K\Lambda_N \int_0^{R_\infty} th(t)\Psi(t)dt + \int_0^{R_\infty} e^{-t} t^{1-N} \int_0^t e^s s^{N-1} \psi(s) f(w(s)) ds dt.$$

Now, using (3.134) we obtain

$$v(R_\infty) < b + \int_0^{R_\infty} e^{-t} t^{1-N} \int_0^t e^s s^{N-1} \psi(s) f(w(s))\,ds\,dt = w(R_\infty).$$

Hence $v(R_\infty) < w(R_\infty)$. Therefore, there exists $R > R_\infty$ such that $v < w$ on $[0,R]$, which contradicts the maximality of R_∞. This contradiction shows that inequality (3.129) holds and the proof of Lemma 3.55 is now complete. □

Proof of Theorem 3.54 continued. Suppose that (3.118) holds. For all $k \geq 1$ we consider the problem

$$\begin{cases} \Delta u_k + |\nabla u_k| = p(x) f(u_k) & \text{in } B(0,k), \\ u_k(x) = w(k) & \text{on } \partial B(0,k). \end{cases} \tag{3.138}$$

Then v and w defined by (3.130) and (3.131) are positive sub and supersolutions of (3.138). So this problem has at least a positive solution u_k and

$$v(|x|) \leq u_k(x) \leq w(|x|) \qquad \text{in } B(0,k), \text{ for all } k \geq 1.$$

By Theorem 14.3 in Gilbarg and Trudinger [99], the sequence $\{\nabla u_k\}$ is bounded on every compact set in $\mathbb{R}^N$. Hence the sequence $\{u_k\}$ is bounded and equicontinuous on compact subsets of $\mathbb{R}^N$. So, by the Arzela–Ascoli theorem, the sequence $\{u_k\}$ has a uniform convergent subsequence, $\{u_k^1\}$ on the ball $B(0,1)$. Let $u^1 = \lim_{k\to\infty} u_k^1$. Then $\{f(u_k^1)\}$ converges uniformly to $f(u^1)$ on $B(0,1)$ and, by (3.138), the sequence $\{\Delta u_k^1 + |\nabla u_k^1|\}$ converges uniformly to $pf(u^1)$. Since the sum of the Laplace and Gradient operators is a closed operator, we deduce that u^1 satisfies (3.107) on $B(0,1)$.

Now, the sequence $\{u_k^1\}$ is bounded and equicontinuous on the ball $B(0,2)$, so it has a convergent subsequence $\{u_k^2\}$. Let $u^2 = \lim\limits_{k\to\infty} u_k^2$ on $B(0,2)$ and suppose u^2 satisfies (3.107) on $B(0,2)$. Proceeding in the same way, we construct a sequence $\{u^n\}$ so that u^n satisfies (3.107) on $B(0,n)$ and $u^{n+1} = u^n$ on $B(0,n)$ for all n. Moreover, the sequence $\{u^n\}$ converges in $L^\infty_{\rm loc}(\mathbb{R}^N)$ to the function u defined by

$$u(x) = u^m(x), \qquad \text{for } x \in B(0,m).$$

Since $v \leq u^n \leq w$ on $B(0,n)$ it follows that $v \leq u \leq w$ on $\mathbb{R}^N$, and u satisfies (3.107). From $v \leq u$ we deduce that u is a positive entire large solution of (3.107). This completes the proof. □

Chapter 4
Singular Lane–Emden–Fowler Equations and Systems

Do not go where the path may lead, go instead where there is no path and leave a trail.

Ralph Waldo Emerson (1803–1882)

4.1 Bifurcation Problems for Singular Elliptic Equations

In this section we study the bifurcation problem

$$\begin{cases} -\Delta u = \lambda f(u) + a(x)g(u) & \text{in } \Omega, \\ u > 0 & \text{in } \Omega, \\ u = 0 & \text{on } \partial\Omega, \end{cases} \tag{P_λ}$$

where $\lambda \in \mathbb{R}$ is a parameter and $\Omega \subset \mathbb{R}^N$ $(N \geq 2)$ is a bounded domain with smooth boundary $\partial\Omega$. The main feature of this boundary value problem is the presence of the "smooth" nonlinearity f combined with the "singular" nonlinearity g. More exactly, we assume that $0 < f \in C^{0,\beta}[0,\infty)$ and $0 \leq g \in C^{0,\beta}(0,\infty)$ $(0 < \beta < 1)$ fulfill the hypotheses

($f1$) f is nondecreasing on $(0,\infty)$ while $f(s)/s$ is nonincreasing for $s > 0$.

($g1$) g is nonincreasing on $(0,\infty)$ with $\lim_{s \searrow 0} g(s) = +\infty$.

($g2$) there exist $C_0, \eta_0 > 0$ and $\alpha \in (0,1)$ so that $g(s) \leq C_0 s^{-\alpha}$, $\forall s \in (0,\eta_0)$.

M. Ghergu and V. Rădulescu, *Nonlinear PDEs*, Springer Monographs in Mathematics,
DOI 10.1007/978-3-642-22664-9_4, © Springer-Verlag Berlin Heidelberg 2012

The assumption $(g2)$ implies the following Keller–Osserman-type growth condition around the origin

$$\int_0^1 \left(\int_0^t g(s)ds \right)^{-1/2} dt < +\infty. \tag{4.1}$$

As proved by Bénilan, Brezis and Crandall in [14], condition (4.1) is equivalent to the *property of compact support*, that is, for any $h \in L^1(\mathbb{R}^N)$ with compact support, there exists a unique $u \in W^{1,1}(\mathbb{R}^N)$ with compact support such that $\Delta u \in L^1(\mathbb{R}^N)$ and

$$-\Delta u + g(u) = h \qquad \text{a.e. in } \mathbb{R}^N.$$

In many papers (see, e.g., Dalmasso [56], Kusano and Swanson [125]) the potential $a(x)$ is assumed to depend "almost" radially on x, in the sense that $C_1\, p(|x|) \leq a(x) \leq C_2\, p(|x|)$, where C_1, C_2 are positive constants and $p(|x|)$ is a positive function satisfying some integrability condition. We do not impose any growth assumption on a, but we suppose that the variable potential $a(x)$ satisfies $a \in C^{0,\beta}(\overline{\Omega})$ and $a > 0$ in Ω.

If $\lambda = 0$ this equation is called the Lane–Emden–Fowler equation and arises in the boundary-layer theory of viscous fluids (see Wong [213]). Problems of this type, as well as the associated evolution equations, describe naturally certain physical phenomena. For example, super-diffusivity equations of this type have been proposed by de Gennes [62] as a model for long range Van der Waals interactions in thin films spreading on solid surfaces.

Our purpose is to study the effect of the asymptotically linear perturbation $f(u)$ in (P_λ), as well as to describe the set of values of the positive parameter λ such that problem (P_λ) admits a solution. In this case, we also prove a uniqueness result. Due to the singular character of (P_λ), we can not expect to find solutions in $C^2(\overline{\Omega})$. However, under the above assumptions we will show that (P_λ) has solutions in the class

$$\mathscr{E} := \{u \in C^2(\Omega) \cap C^{1,1-\alpha}(\overline{\Omega});\ \Delta u \in L^1(\Omega)\}.$$

We first observe that, in view of the assumption $(f1)$, there exists

$$m := \lim_{s \to \infty} \frac{f(s)}{s} \in [0, \infty).$$

This number plays a crucial role in our analysis. More precisely, the existence of the solutions to (P_λ) will be separately discussed for $m > 0$ and $m = 0$. Let $a_* = \min_{x \in \overline{\Omega}} a(x)$.

Theorem 4.1 *Assume* $(f1)$, $(g1)$, $(g2)$ *and* $m = 0$. *If* $a_* > 0$ *(resp.* $a_* = 0$*), then* (P_λ) *has a unique solution* $u_\lambda \in \mathscr{E}$ *for all* $\lambda \in \mathbb{R}$ *(resp.* $\lambda \geq 0$*) with the properties:*

(i) u_λ *is strictly increasing with respect to* λ.
(ii) *there exist two positive constant* $c_1, c_2 > 0$ *depending on* λ *such that* $c_1 d(x) \leq u_\lambda \leq c_2 d(x)$ *in* Ω.

The bifurcation diagram in the "sublinear" case $m = 0$ is depicted in Fig. 4.1.

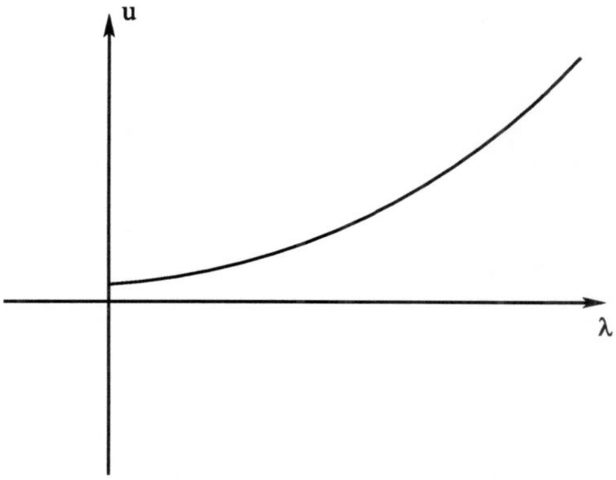

Fig. 4.1 The "sublinear" case $m = 0$

Proof. We first recall the following existence result that we need in the proof.

Lemma 4.2 (Shi and Yao [180]). *Let* $F : \overline{\Omega} \times (0, \infty) \to \mathbb{R}$ *be a Hölder continuous function with exponent* $\beta \in (0,1)$ *on each compact subset of* $\overline{\Omega} \times (0, \infty)$ *which satisfies*

(F1) $\limsup_{s \to +\infty} \left(s^{-1} \max_{x \in \overline{\Omega}} F(x,s) \right) < \lambda_1$.
(F2) *for each* $t > 0$*, there exists a constant* $D(t) > 0$ *such that*

$$F(x,r) - F(x,s) \geq -D(t)(r-s), \quad \textit{for } x \in \overline{\Omega} \textit{ and } r \geq s \geq t.$$

(F3) *there exists* $\eta_0 > 0$ *and an open subset* $\Omega_0 \subset \Omega$ *such that*

$$\min_{x \in \overline{\Omega}} F(x,s) \geq 0 \quad \textit{for } s \in (0, \eta_0),$$

and

$$\lim_{s \searrow 0} \frac{F(x,s)}{s} = +\infty \quad \textit{uniformly for } x \in \Omega_0.$$

Then for any nonnegative function $\phi_0 \in C^{2,\beta}(\partial\Omega)$*, the problem*

$$\begin{cases} -\Delta u = F(x,u) & \text{in } \Omega, \\ u > 0 & \text{in } \Omega, \\ u = \phi_0 & \text{on } \partial\Omega, \end{cases}$$

has at least one positive solution $u \in C^{2,\beta}(G) \cap C(\overline{\Omega})$*, for any compact set* $G \subset \Omega \cup \{x \in \partial\Omega;\ \phi_0(x) > 0\}$.

Lemma 4.3 (Shi and Yao [180]). *Let* $F : \overline{\Omega} \times (0,\infty) \to \mathbb{R}$ *be a continuous function such that the mapping* $(0,\infty) \ni s \longmapsto \dfrac{F(x,s)}{s}$ *is strictly decreasing at each* $x \in \Omega$*. Assume that there exists* $v,\ w \in C^2(\Omega) \cap C(\overline{\Omega})$ *such that*

(a) $\Delta w + F(x,w) \le 0 \le \Delta v + F(x,v)$ *in* Ω*.*
(b) $v, w > 0$ *in* Ω *and* $v \le w$ *on* $\partial\Omega$*.*
(c) $\Delta v \in L^1(\Omega)$*.*

Then $v \le w$ *in* Ω*.*

Now, we are ready to give the proof of Theorem 4.1. This will be divided into four steps.

Step 1: Existence of solutions to problem (P_λ).

For any $\lambda \in \mathbb{R}$, define the function

$$\Phi_\lambda(x,s) = \lambda f(s) + a(x)g(s), \quad (x,s) \in \overline{\Omega} \times (0,\infty). \tag{4.2}$$

Taking into account the assumptions of Theorem 4.1, it follows that Φ_λ verifies the hypotheses of Lemma 4.2 for $\lambda \in \mathbb{R}$ if $a_* > 0$ and $\lambda \ge 0$ if $a_* = 0$. Hence, for λ in the above range, (P_λ) has at least one solution $u_\lambda \in C^{2,\beta}(\Omega) \cap C(\overline{\Omega})$.

Step 2: Uniqueness of solution.

Fix $\lambda \in \mathbb{R}$ (resp. $\lambda \ge 0$) if $a_* > 0$ (resp. $a_* = 0$). Let u_λ be a solution of (P_λ). Denote $\lambda^- = \min\{0,\lambda\}$ and $\lambda^+ = \max\{0,\lambda\}$. We claim that $\Delta u_\lambda \in L^1(\Omega)$. Since $a \in C^{0,\beta}(\overline{\Omega})$, by [99, Theorem 6.14], there exists a unique nonnegative solution $\zeta \in C^{2,\beta}(\overline{\Omega})$ of

$$\begin{cases} -\Delta \zeta = a(x) & \text{in } \Omega, \\ \zeta = 0 & \text{on } \partial\Omega. \end{cases}$$

By the weak maximum principle (see e.g., [99, Theorem 2.2]), $\zeta > 0$ in Ω. Moreover, we are going to prove that

(a) $z(x) := c\zeta(x)$ is a subsolution of (P_λ), for $c > 0$ small enough.
(b) $z(x) \geq c_1 d(x)$ in $\overline{\Omega}$, for some positive constant $c_1 > 0$.
(c) $u_\lambda \geq z$ in $\overline{\Omega}$.

Therefore, by (b) and (c), $u_\lambda \geq c_1 d(x)$ in $\overline{\Omega}$. Using $(g2)$, we obtain $g(u_\lambda) \leq Cd^{-\alpha}(x)$ in Ω, where $C > 0$ is a constant. So, $g(u_\lambda) \in L^1(\Omega)$. This implies

$$\Delta u_\lambda \in L^1(\Omega).$$

Proof of (a). Using $(f1)$ and $(g1)$, we have

$$\begin{aligned}\Delta z(x) + \Phi_\lambda(x,z) &= -ca(x) + \lambda f(c\zeta) + a(x)g(c\zeta)\\ &\geq -ca(x) + \lambda^- f(c\|\zeta\|_\infty) + a(x)g(c\|\zeta\|_\infty)\\ &\geq ca(x)\left[\frac{g(c\|\zeta\|_\infty)}{2c} - 1\right] + f(c\|\zeta\|_\infty)\left[a_*\frac{g(c\|\zeta\|_\infty)}{2f(c\|\zeta\|_\infty)} + \lambda^-\right]\end{aligned}$$

for each $x \in \Omega$. Since $\lambda < 0$ corresponds to $a_* > 0$, using $\lim_{t\searrow 0} g(t) = +\infty$ and $\lim_{t\to 0} f(t) \in (0,\infty)$, we can find $c > 0$ small enough that

$$\Delta z + \Phi_\lambda(x,z) \geq 0, \quad \forall x \in \Omega.$$

This concludes (a).

Proof of (b). Since $\zeta \in C^{2,\beta}(\overline{\Omega})$, $\zeta > 0$ in Ω and $\zeta = 0$ on $\partial\Omega$, by Lemma 3.4 in Gilbarg and Trudinger [99], we have

$$\frac{\partial\zeta}{\partial\nu}(y) < 0, \quad \forall y \in \partial\Omega.$$

Therefore, there exists a positive constant c_0 such that

$$\frac{\partial\zeta}{\partial\nu}(y) := \lim_{x\in\Omega, x\to y} \frac{\zeta(y) - \zeta(x)}{|x-y|} \leq -c_0, \quad \forall y \in \partial\Omega.$$

So, for each $y \in \Omega$, there exists $r_y > 0$ such that

$$\frac{\zeta(x)}{|x-y|} \geq \frac{c_0}{2}, \quad \forall x \in B_{r_y}(y) \cap \Omega. \tag{4.3}$$

Using the compactness of $\partial\Omega$, we can find a finite number k of balls $B_{r_{y_i}}(y_i)$ such that $\partial\Omega \subset \cup_{i=1}^k B_{r_{y_i}}(y_i)$. Moreover, we can assume that for small $d_0 > 0$,

$$\{x \in \Omega : \ d(x) < d_0\} \subset \cup_{i=1}^k B_{r_{y_i}}(y_i).$$

Therefore, by (4.3) we obtain

$$\zeta(x) \geq \frac{c_0}{2} d(x), \quad \forall x \in \Omega \text{ with } d(x) < d_0.$$

This fact, combined with $\zeta > 0$ in Ω, shows that for some constant $\tilde{c} > 0$

$$\zeta(x) \geq \tilde{c} d(x), \quad \forall x \in \Omega.$$

Thus, (b) follows by the definition of z.

Proof of (c). We distinguish two cases:

Case 1. $\lambda \geq 0$. We see that Φ_λ verifies the hypotheses in Theorem 1.2. Since

$$\begin{aligned}
&\Delta u_\lambda + \Phi_\lambda(x, u_\lambda) \leq 0 \leq \Delta z + \Phi_\lambda(x, z) \quad \text{in } \Omega,\\
&u_\lambda, z > 0 \quad \text{in } \Omega,\\
&u_\lambda = z \quad \text{on } \partial\Omega,\\
&\Delta z \in L^1(\Omega),
\end{aligned}$$

by Theorem 1.2 it follows that $u_\lambda \geq z$ in $\overline{\Omega}$.

Now, if u_1 and u_2 are two solutions of (P_λ), we can use Theorem 1.2 in order to deduce that $u_1 = u_2$.

Case 2. $\lambda < 0$ (corresponding to $a_* > 0$). Let $\varepsilon > 0$ be fixed. We prove that

$$z \leq u_\lambda + \varepsilon(1 + |x|^2)^\tau \quad \text{in } \Omega, \tag{4.4}$$

where $\tau < 0$ is chosen such that $\tau|x|^2 + 1 > 0$, $\forall x \in \Omega$. This is always possible since $\Omega \subset \mathbb{R}^N$ $(N \geq 2)$ is bounded.

We argue by contradiction. Suppose that there exists $x_0 \in \Omega$ such that $u_\lambda(x_0) + \varepsilon(1 + |x_0|)^\tau < z(x_0)$. Then $\min_{x \in \overline{\Omega}}\{u_\lambda(x) + \varepsilon(1 + |x|^2)^\tau - z(x)\} < 0$ is achieved at some point $x_1 \in \Omega$. Since $\Phi_\lambda(x, z)$ is nonincreasing in z, we have

$$\begin{aligned}
0 &\geq -\Delta[u_\lambda(x) - z(x) + \varepsilon(1 + |x|^2)^\tau]|_{x=x_1}\\
&= \Phi_\lambda(x_1, u_\lambda(x_1)) - \Phi_\lambda(x_1, z(x_1)) - \varepsilon\Delta[(1 + |x|^2)^\tau]|_{x=x_1}
\end{aligned}$$

$$\geq -\varepsilon\Delta[(1+|x|^2)^\tau]|_{x=x_1} = -2\varepsilon\tau(1+|x_1|^2)^{\tau-2}[(N+2\tau-2)|x_1|^2+N]$$
$$\geq -4\varepsilon\tau(1+|x_1|^2)^{\tau-2}(\tau|x_1|^2+1) > 0.$$

This contradiction proves (4.4). Passing to the limit $\varepsilon \to 0$, we obtain (c).

In a similar way we can prove that (P_λ) has a unique solution.

Step 3: Dependence on λ.

We fix $\lambda_1 < \lambda_2$, where $\lambda_1, \lambda_2 \in \mathbb{R}$ if $a_* > 0$ resp. $\lambda_1, \lambda_2 \in [0,\infty)$ if $a_* = 0$. Let $u_{\lambda_1}, u_{\lambda_2}$ be the corresponding solutions of (P_{λ_1}) and (P_{λ_2}) respectively.

If $\lambda_1 \geq 0$, then Φ_{λ_1} verifies the hypotheses in Theorem 1.2. Furthermore, we have

$$\begin{aligned}
&\Delta u_{\lambda_2} + \Phi_{\lambda_1}(x,u_{\lambda_2}) \leq 0 \leq \Delta u_{\lambda_1} + \Phi_{\lambda_1}(x,u_{\lambda_1}) \quad \text{in } \Omega,\\
&u_{\lambda_1}, u_{\lambda_2} > 0 \quad \text{in } \Omega,\\
&u_{\lambda_1} = u_{\lambda_2} \quad \text{on } \partial\Omega,\\
&\Delta u_{\lambda_1} \in L^1(\Omega).
\end{aligned}$$

Again by Theorem 1.2, we conclude that $u_{\lambda_1} \leq u_{\lambda_2}$ in $\overline{\Omega}$. Moreover, by the maximum principle, $u_{\lambda_1} < u_{\lambda_2}$ in Ω.

Let $\lambda_2 \leq 0$; we show that $u_{\lambda_1} \leq u_{\lambda_2}$ in $\overline{\Omega}$. Indeed, supposing the contrary, there exists $x_0 \in \Omega$ such that $u_{\lambda_1}(x_0) > u_{\lambda_2}(x_0)$. We conclude now that $\max_{x\in\overline{\Omega}}\{u_{\lambda_1}(x) - u_{\lambda_2}(x)\} > 0$ is achieved at some point in Ω. At that point, say $\bar{x}$, we have

$$0 \leq -\Delta(u_{\lambda_1} - u_{\lambda_2})(\bar{x}) = \Phi_{\lambda_1}(\bar{x}, u_{\lambda_1}(\bar{x})) - \Phi_{\lambda_2}(\bar{x}, u_{\lambda_2}(\bar{x})) < 0,$$

which is a contradiction. It follows that $u_{\lambda_1} \leq u_{\lambda_2}$ in $\overline{\Omega}$, and by the maximum principle we have $u_{\lambda_1} < u_{\lambda_2}$ in Ω.

If $\lambda_1 < 0 < \lambda_2$, then $u_{\lambda_1} < u_0 < u_{\lambda_2}$ in Ω. This finishes the proof of Step 3.

Step 4: Regularity of the solution.ï¿½

Fix $\lambda \in \mathbb{R}$ and let $u_\lambda \in C^2(\Omega) \cap C(\overline{\Omega})$ be the unique solution of (P_λ). An important result in our approach is the following estimate

$$c_1 d(x) \leq u_\lambda(x) \leq c_2 d(x), \qquad \text{for all } x \in \Omega, \tag{4.5}$$

where c_1, c_2 are positive constants. The first inequality in (4.5) was established in Step 2. For the second one, we apply an idea found in Gui and Lin [107].

Using the smoothness of $\partial\Omega$, we can find $\delta \in (0,1)$ such that for all $x_0 \in \Omega_\delta := \{x \in \Omega\,;\, d(x) \le \delta\}$, there exists $y \in \mathbb{R}^N \setminus \overline{\Omega}$ with $d(y, \partial\Omega) = \delta$ and $d(x_0) = |x_0 - y| - \delta$.

Let $K > 1$ be such that $\operatorname{diam}(\Omega) < (K-1)\delta$ and let w be the unique solution of the Dirichlet problem

$$\begin{cases} -\Delta w = \lambda^+ f(w) + g(w) & \text{in } B_K(0) \setminus \overline{B_1}(0), \\ w > 0 & \text{in } B_K(0) \setminus \overline{B_1}(0), \\ w = 0 & \text{on } \partial(B_K(0) \setminus \overline{B_1}(0)), \end{cases} \tag{4.6}$$

where $B_r(0)$ is the open ball in $\mathbb{R}^N$ of radius r and centered at the origin. By uniqueness, w is radially symmetric. Hence $w(x) = \tilde{w}(|x|)$ and

$$\begin{cases} \tilde{w}'' + \dfrac{N-1}{r}\tilde{w}' + \lambda^+ f(\tilde{w}) + g(\tilde{w}) = 0 & \text{for } r \in (1,K), \\ \tilde{w} > 0 & \text{in } (1,K), \\ \tilde{w}(1) = \tilde{w}(K) = 0. \end{cases} \tag{4.7}$$

Integrating in (4.7) we have

$$\begin{aligned} \tilde{w}'(t) &= \tilde{w}'(a)a^{N-1}t^{1-N} - t^{1-N}\int_a^t r^{N-1}\left[\lambda^+ f(\tilde{w}(r)) + g(\tilde{w}(r))\right] dr, \\ &= \tilde{w}'(b)b^{N-1}t^{1-N} + t^{1-N}\int_t^b r^{N-1}\left[\lambda^+ f(\tilde{w}(r)) + g(\tilde{w}(r))\right] dr, \end{aligned}$$

where $1 < a < t < b < K$. Since $g(\tilde{w}) \in L^1(1,K)$, we deduce that both $\tilde{w}'(1)$ and $\tilde{w}'(K)$ are finite, so $\tilde{w} \in C^2(1,K) \cap C^1[1,K]$. Furthermore,

$$w(x) \le C\min\{K - |x|, |x| - 1\}, \quad \text{for any } x \in B_K(0) \setminus B_1(0). \tag{4.8}$$

Let us fix $x_0 \in \Omega_\delta$. Then we can find $y_0 \in \mathbb{R}^N \setminus \overline{\Omega}$ with $d(y_0, \partial\Omega) = \delta$ and $d(x_0) = |x_0 - y| - \delta$. Thus, $\Omega \subset B_{K\delta}(y_0) \setminus B_\delta(y_0)$. Define $v(x) = cw\left(\dfrac{x - y_0}{\delta}\right)$, $x \in \overline{\Omega}$. We show that v is a supersolution of (P_λ), provided that c is large enough. Indeed, if $c > \max\{1, \delta^2\|a\|_\infty\}$, then for all $x \in \Omega$ we have

$$\begin{aligned} \Delta v + \lambda f(v) + a(x)g(v) \le \frac{c}{\delta^2}&\left(\tilde{w}''(r) + \frac{N-1}{r}\tilde{w}'(r)\right) \\ &+ \lambda^+ f(c\tilde{w}(r)) + a(x)g(c\tilde{w}(r)), \end{aligned}$$

where $r = \frac{|x-y_0|}{\delta} \in (1,K)$. Using the assumption $(f1)$ we get $f(c\tilde{w}) \leq cf(\tilde{w})$ in $(1,K)$. The above relations lead us to

$$\begin{aligned}\Delta v + \lambda f(v) + a(x)g(v) &\leq \frac{c}{\delta^2}\left(\tilde{w}'' + \frac{N-1}{r}\tilde{w}'\right) + \lambda^+ cf(\tilde{w}) + \|a\|_\infty g(\tilde{w}) \\ &\leq \frac{c}{\delta^2}\left(\tilde{w}'' + \frac{N-1}{r}\tilde{w}' + \lambda^+ f(\tilde{w}) + g(\tilde{w})\right) \\ &= 0.\end{aligned}$$

Since $\Delta u_\lambda \in L^1(\Omega)$, with a similar proof as in Step 2 we get $u_\lambda \leq v$ in Ω. This combined with (4.8) yields

$$u_\lambda(x_0) \leq v(x_0) \leq C\min\{K - \frac{|x_0-y_0|}{\delta}, \frac{|x_0-y_0|}{\delta} - 1\} \leq \frac{C}{\delta}d(x_0).$$

Hence $u_\lambda \leq \frac{C}{\delta}d(x)$ in Ω_δ and the last inequality in (4.5) follows.

Let G be the Green's function associated with the Laplace operator in Ω. Then, for all $x \in \Omega$ we have

$$u_\lambda(x) = -\int_\Omega G(x,y)\,[\lambda f(u_\lambda(y)) + a(y)g(u_\lambda(y))]\,dy,$$

and

$$\nabla u_\lambda(x) = -\int_\Omega G_x(x,y)\,[\lambda f(u_\lambda(y)) + a(y)g(u_\lambda(y))]\,dy.$$

If $x_1, x_2 \in \Omega$, using $(g2)$ we obtain

$$\begin{aligned}|\nabla u_\lambda(x_1) - \nabla u_\lambda(x_2)| \leq |\lambda| &\int_\Omega |G_x(x_1,y) - G_x(x_2,y)| \cdot f(u_\lambda(y))dy \\ &+ \tilde{c}\int_\Omega |G_x(x_1,y) - G_x(x_2,y)| \cdot u_\lambda^{-\alpha}(y)dy.\end{aligned}$$

Now, taking into account that $u_\lambda \in C(\overline{\Omega})$, by the standard regularity theory (see Gilbarg and Trudinger [99]) we get

$$\int_\Omega |G_x(x_1,y) - G_x(x_2,y)| \cdot f(u_\lambda(y)) \leq \tilde{c}_1|x_1 - x_2|.$$

On the other hand, with the same proof as in [107, Theorem 1], we deduce

$$\int_\Omega |G_x(x_1,y) - G_x(x_2,y)| \cdot u_\lambda^{-\alpha}(y) \leq \tilde{c}_2|x_1 - x_2|^{1-\alpha}.$$

The above inequalities imply $u_\lambda \in C^2(\Omega) \cap C^{1,1-\alpha}(\overline{\Omega})$. The proof of Theorem 4.1 is now complete. □

Next, consider the case $m > 0$. The results in this case are different from those presented in Theorem 4.1. A careful examination of (P_λ) reveals the fact that the singular term $g(u)$ is not significant. Actually, the conclusions are close to those established in Mironescu and Rădulescu [144, Theorem A], where an elliptic problem associated to an asymptotically linear function is studied.

Let λ_1 be the first Dirichlet eigenvalue of $(-\Delta)$ in Ω and $\lambda^* = \dfrac{\lambda_1}{m}$. Our result in this case is the following.

Theorem 4.4 *Assume $(f1)$, $(g1)$, $(g2)$ and $m > 0$. Then the following hold.*

(i) *If $\lambda \geq \lambda^*$, then (P_λ) has no solutions in $\mathscr{E}$.*
(ii) *If $a_* > 0$ (resp. $a_* = 0$) then (P_λ) has a unique solution $u_\lambda \in \mathscr{E}$ for all $-\infty < \lambda < \lambda^*$ (resp. $0 < \lambda < \lambda^*$) with the properties:*
  (ii1) *u_λ is strictly increasing with respect to λ.*
  (ii2) *there exist two positive constants $c_1, c_2 > 0$ depending on λ such that $c_1 d(x) \leq u_\lambda \leq c_2 d(x)$ in Ω.*
  (ii3) *$\lim_{\lambda \nearrow \lambda^*} u_\lambda = +\infty$, uniformly on compact subsets of Ω.*

The bifurcation diagram in the "linear" case $m > 0$ is depicted in Fig. 4.2.

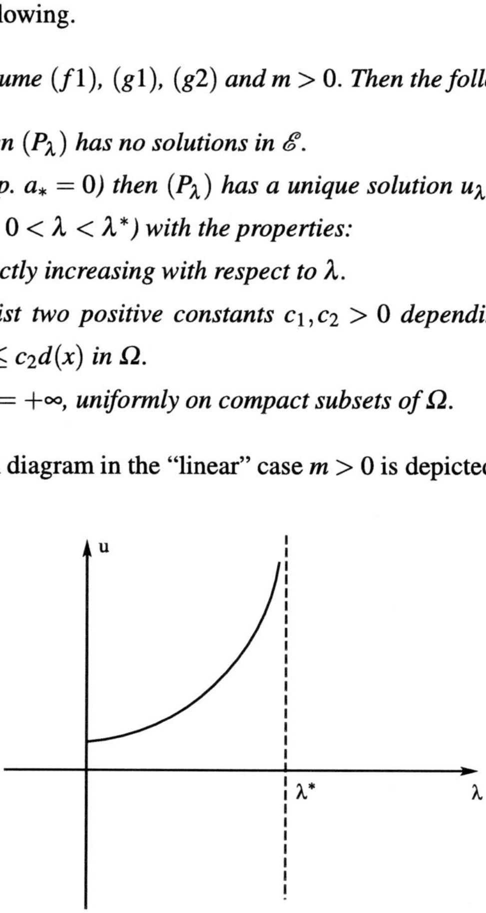

Fig. 4.2 The "linear" case $m > 0$

Proof. (i) Let ϕ_1 be the first eigenfunction of the Laplace operator in Ω with Dirichlet boundary condition. Arguing by contradiction, let us suppose that there exists $\lambda \geq \lambda^*$ such that (P_λ) has a solution $u_\lambda \in \mathscr{E}$.

Multiplying by ϕ_1 in (P_λ) and then integrating over Ω we get

$$-\int_\Omega \phi_1 \Delta u_\lambda = \lambda \int_\Omega f(u_\lambda)\phi_1 + \int_\Omega a(x)g(u_\lambda)\phi_1. \tag{4.9}$$

Since $\lambda \geq \dfrac{\lambda_1}{m}$, in view of the assumption $(f1)$ we get $\lambda f(u_\lambda) \geq \lambda_1 u_\lambda$ in Ω. Using this fact in (4.9) we obtain

$$-\int_\Omega \phi_1 \Delta u_\lambda > \lambda_1 \int_\Omega u_\lambda \phi_1.$$

The regularity of u_λ yields $-\int_\Omega u_\lambda \Delta\phi_1 > \lambda_1 \int_\Omega u_\lambda \phi_1$. This is clearly a contradiction since $-\Delta\phi_1 = \lambda_1\phi_1$ in Ω. Hence (P_λ) has no solutions in $\mathscr{E}$ for any $\lambda \geq \lambda^*$.

(ii) From now on, the proof of the existence, uniqueness and regularity of solution is the same as in Theorem 4.1.

(ii3) In what follows we shall apply some ideas developed in Mironescu and Rădulescu [144]. Due to the special character of our problem, we will be able to prove that, in certain cases, L^2 boundedness implies H_0^1 boundedness!

Let $u_\lambda \in \mathscr{E}$ be the unique solution of (P_λ) for $0 < \lambda < \lambda^*$. We prove that $\lim_{\lambda \nearrow \lambda^*} u_\lambda = +\infty$, uniformly on compact subsets of Ω. Suppose the contrary. Since $\{u_\lambda\}_{0<\lambda<\lambda^*}$ is a sequence of nonnegative super-harmonic functions in Ω, by Theorem 4.1.9 in Hörmander [111], there exists a subsequence of $\{u_\lambda\}_{\lambda<\lambda^*}$ (still denoted by $\{u_\lambda\}_{\lambda<\lambda^*}$) which is convergent in $L^1_{\text{loc}}(\Omega)$.

We first prove that $\{u_\lambda\}_{\lambda<\lambda^*}$ is bounded in $L^2(\Omega)$. We argue by contradiction. Suppose that $\{u_\lambda\}_{\lambda<\lambda^*}$ is not bounded in $L^2(\Omega)$. Thus, up to a subsequence we have $u_\lambda = M(\lambda)w_\lambda$, where

$$M(\lambda) = ||u_\lambda||_{L^2(\Omega)} \to \infty \text{ as } \lambda \nearrow \lambda^* \text{ and } w_\lambda \in L^2(\Omega),\ \ ||w_\lambda||_{L^2(\Omega)} = 1. \tag{4.10}$$

Using $(f1)$, $(g2)$ and the monotonicity assumption on g, we deduce the existence of $A, B, C, D > 0$ $(A > m)$ such that

$$f(t) \leq At + B, \quad g(t) \leq Ct^{-\alpha} + D, \qquad \text{for all } t > 0. \tag{4.11}$$

This implies

$$\frac{1}{M(\lambda)}(\lambda f(u_\lambda) + a(x)g(u_\lambda)) \to 0 \quad \text{in } L^1_{\text{loc}}(\Omega) \text{ as } \lambda \nearrow \lambda^*$$

that is,

$$-\Delta w_\lambda \to 0 \quad \text{in } L^1_{\text{loc}}(\Omega) \text{ as } \lambda \nearrow \lambda^*. \tag{4.12}$$

By Green's first identity, we have

$$\int_\Omega \nabla w_\lambda \cdot \nabla\phi\, dx = -\int_\Omega \phi\, \Delta w_\lambda\, dx = -\int_{\text{Supp}\,\phi} \phi\, \Delta w_\lambda\, dx \quad \forall \phi \in C_0^\infty(\Omega). \tag{4.13}$$

Using (4.12) we derive that

$$\begin{aligned}\left|\int_{\text{Supp}\,\phi} \phi\, \Delta w_\lambda\, dx\right| &\le \int_{\text{Supp}\,\phi} |\phi||\Delta w_\lambda|\, dx \\ &\le \|\phi\|_{L^\infty} \int_{\text{Supp}\,\phi} |\Delta w_\lambda|\, dx \to 0 \quad \text{as } \lambda \nearrow \lambda^*.\end{aligned} \tag{4.14}$$

Combining (4.13) and (4.14), we arrive at

$$\int_\Omega \nabla w_\lambda \cdot \nabla \phi\, dx \to 0 \text{ as } \lambda \nearrow \lambda^*, \quad \forall \phi \in C_0^\infty(\Omega). \tag{4.15}$$

By definition, the sequence $\{w_\lambda\}_{0<\lambda<\lambda^*}$ is bounded in $L^2(\Omega)$.

We claim that $\{w_\lambda\}_{\lambda<\lambda^*}$ is bounded in $H_0^1(\Omega)$. Indeed, using (4.11) and Hölder's inequality, we have

$$\begin{aligned}\int_\Omega |\nabla w_\lambda|^2 &= -\int_\Omega w_\lambda \Delta w_\lambda = \frac{-1}{M(\lambda)}\int_\Omega w_\lambda \Delta u_\lambda \\ &= \frac{1}{M(\lambda)}\int_\Omega [\lambda w_\lambda f(u_\lambda) + a(x) g(u_\lambda) w_\lambda] \\ &\le \frac{\lambda}{M(\lambda)}\int_\Omega w_\lambda (A u_\lambda + B) + \frac{\|a\|_\infty}{M(\lambda)}\int_\Omega w_\lambda (C u_\lambda^{-\alpha} + D) \\ &= \lambda A \int_\Omega w_\lambda^2 + \frac{\|a\|_\infty C}{M(\lambda)^{1+\alpha}}\int_\Omega w_\lambda^{1-\alpha} + \frac{\lambda B + \|a\|_\infty D}{M(\lambda)}\int_\Omega w_\lambda \\ &\le \lambda^* A + \frac{\|a\|_\infty C}{M(\lambda)^{1+\alpha}}|\Omega|^{(1+\alpha)/2} + \frac{\lambda B + \|a\|_\infty D}{M(\lambda)}|\Omega|^{1/2}.\end{aligned}$$

From the above estimates, it is easy to see that $\{w_\lambda\}_{\lambda<\lambda^*}$ is bounded in $H_0^1(\Omega)$, so the claim is proved. Then, there exists $w \in H_0^1(\Omega)$ such that (up to a subsequence)

$$w_\lambda \rightharpoonup w \quad \text{weakly in } H_0^1(\Omega) \text{ as } \lambda \nearrow \lambda^* \tag{4.16}$$

and, because $H_0^1(\Omega)$ is compactly embedded in $L^2(\Omega)$,

$$w_\lambda \to w \quad \text{strongly in } L^2(\Omega) \text{ as } \lambda \nearrow \lambda^*. \tag{4.17}$$

On the one hand, by (4.10) and (4.17), we derive that $\|w\|_{L^2(\Omega)} = 1$. Furthermore, using (4.15) and (4.16), we infer that

$$\int_\Omega \nabla w \cdot \nabla \phi \, dx = 0, \quad \forall \phi \in C_0^\infty(\Omega).$$

Since $w \in H_0^1(\Omega)$, using the above relation and the definition of $H_0^1(\Omega)$, we get $w = 0$. This contradiction shows that $\{u_\lambda\}_{\lambda<\lambda^*}$ is bounded in $L^2(\Omega)$. As above for w_λ, we can derive that u_λ is bounded in $H_0^1(\Omega)$. So, there exists $u^* \in H_0^1(\Omega)$ such that, up to a subsequence,

$$\begin{cases} u_\lambda \rightharpoonup u^* \text{ weakly in } H_0^1(\Omega) \text{ as } \lambda \nearrow \lambda^*, \\ u_\lambda \to u^* \text{ strongly in } L^2(\Omega) \text{ as } \lambda \nearrow \lambda^*, \\ u_\lambda \to u^* \text{ a.e. in } \Omega \text{ as } \lambda \nearrow \lambda^*. \end{cases} \tag{4.18}$$

Now we can proceed to get a contradiction. Multiplying by ϕ_1 in (P_λ) and integrating over Ω we have

$$-\int_\Omega \varphi_1 \Delta u_\lambda = \lambda \int_\Omega f(u_\lambda)\varphi_1 + \int_\Omega a(x)g(u_\lambda)\varphi_1, \quad \text{for all } 0 < \lambda < \lambda^*. \tag{4.19}$$

On the other hand, by $(f1)$ it follows that $f(u_\lambda) \geq m u_\lambda$ in Ω, for all $0 < \lambda < \lambda^*$. Combining this with (4.19) we obtain

$$\lambda_1 \int_\Omega u_\lambda \varphi_1 \geq \lambda m \int_\Omega u_\lambda \varphi_1 + \int_\Omega a(x)g(u_\lambda)\varphi_1, \quad \text{for all } 0 < \lambda < \lambda^*. \tag{4.20}$$

Notice that by $(g1)$, (4.18) and the monotonicity of u_λ with respect to λ we can apply the Lebesgue convergence theorem to find

$$\int_\Omega a(x)g(u_\lambda)\varphi_1 \, dx \to \int_\Omega a(x)g(u^*)\varphi_1 \, dx \text{ as } \lambda \nearrow \lambda_1.$$

Passing to the limit in (4.20) as $\lambda \nearrow \lambda^*$, and using (4.18), we get

$$\lambda_1 \int_\Omega u^* \varphi_1 \geq \lambda_1 \int_\Omega u^* \varphi_1 + \int_\Omega a(x)g(u^*)\varphi_1. \tag{4.21}$$

Hence $\int_\Omega a(x)g(u^*)\varphi_1 = 0$, which is a contradiction. This fact shows that $\lim_{\lambda \nearrow \lambda^*} u_\lambda = +\infty$, uniformly on compact subsets of Ω. This ends the proof. □

4.2 Lane–Emden–Fowler Systems with Negative Exponents

In this section we study the elliptic system

$$\begin{cases} -\Delta u = u^{-p}v^{-q}\,,\ u>0 & \text{in } \Omega, \\ -\Delta v = u^{-r}v^{-s}\,,\ v>0 & \text{in } \Omega, \\ \quad u=v=0 & \text{on } \partial\Omega, \end{cases} \tag{4.22}$$

where $\Omega \subset \mathbb{R}^N$ $(N \geq 1)$ is a bounded domain with C^2 boundary, $p,s \geq 0$ and $q,r>0$. By solution of (4.22) we understand a pair (u,v) with $u,v \in C^2(\Omega)\cap C(\overline{\Omega})$ such that $u,v>0$ in Ω and satisfies (4.22) pointwise.

The first motivation for the study of system (4.22) comes from the so-called *Lane–Emden* equation (see [68, 77, 128])

$$-\Delta u = u^p \quad \text{in } B_R(0),\ R>0, \tag{4.23}$$

subject to Dirichlet boundary condition. In astrophysics, the exponent p is called the *polytropic index* and positive radially symmetric solutions of (4.23) are used to describe the structure of the polytropic stars (we refer the interested reader to the book by Chandrasekhar [37] for an account on the above equation as well as for various mathematical techniques to describe the behavior of the solution to the Lane–Emden equation).

Systems of type (4.22) with $p,s \leq 0$ and $q,s<0$ have received considerable attention in the last decade (see, e.g., [33, 49, 72, 75, 146, 166, 171, 177, 178, 183, 216] and the references therein). It has been shown that for such range of exponents system (4.22) has a rich mathematical structure. Various techniques such as the moving plane method, Pohozaev-type identities, and rescaling arguments have been developed and suitably adapted to deal with (4.22) in this case.

Recently, there has been some interest in systems of type (4.22) where not all the exponents are negative. In [85, 93, 94] the system (4.22) is considered under the hypothesis $p,r<0<q,s$. This corresponds to the singular Gierer–Meinhardt system arising in molecular biology. In [111] the authors provide a nice sub and supersolution device that applies to general systems both in cooperative and noncooperative settings. This method was then used to discuss singular counterparts of some well-known models such as Gierer–Meinhardt, Lotka–Volterra or predator–prey systems.

We shall be concerned with system (4.22) in case $p,s \geq 0$ and $q,r > 0$. This corresponds to the prototype equation (4.23) in which the polytropic index p is negative. For such range of exponents, the above-mentioned methods do not apply; another difficulties in dealing with system (4.22) come from the noncooperative character of our system and from the lack of a variational structure. In turn, our approach relies on the boundary behavior of solutions to (4.23) (with $p < 0$) or more generally, to singular elliptic problems of the type

$$\begin{cases} -\Delta u = k(\delta(x))u^{-p}\,,\ u > 0 & \text{in } \Omega, \\ u = 0 & \text{on } \partial\Omega, \end{cases} \tag{4.24}$$

where

$$\delta(x) = \operatorname{dist}(x, \partial\Omega), \quad x \in \overline{\Omega},$$

and $k : (0,\infty) \to (0,\infty)$ is a decreasing function such that $\lim_{t\searrow 0} k(t) = \infty$.

The approach we adopt here is inspired from [86] and can be used to study more general systems in the form

$$\begin{cases} -\mathscr{L}u = f(x,u,v)\,,\ u > 0 & \text{in } \Omega, \\ -\mathscr{L}v = g(x,u,v)\,,\ v > 0 & \text{in } \Omega, \\ u = v = 0 & \text{on } \partial\Omega, \end{cases}$$

where $\mathscr{L}$ is a second order differential operator not necessarily in divergence form and

$$f(x,u,v) = k_1(x)u^{-p}v^{-q}, \quad g(x,u,v) = k_2(x)u^{-r}v^{-s},$$

or

$$f(x,u,v) = k_{11}(x)u^{-p} + k_{12}(x)v^{-q}, \quad g(x,u,v) = k_{21}(x)u^{-r} + k_{22}(x)v^{-s},$$

with $k_i, k_{ij} : \Omega \to (0,\infty)$ $(i,j = 1,2)$ continuous functions that behave like

$$\delta(x)^{-a}\log^b\left(\frac{A}{\delta(x)}\right) \quad \text{near } \partial\Omega, \tag{4.25}$$

for some $A, a > 0$ and $b \in \mathbb{R}$.

4.2.1 Preliminary Results

In this section we collect some old and new results concerning problems of type (4.24). Note that the method of sub and supersolutions is also valid in the singular framework as explained in [95, Theorem 1.2.3]. Our first result is a straightforward comparison principle between subsolutions and supersolutions for singular elliptic equations.

Proposition 4.5 *Let $p \geq 0$ and $\phi : \Omega \to (0,\infty)$ be a continuous function. If $\underline{u}$ is a subsolution and $\overline{u}$ is a supersolution of*

$$\begin{cases} -\Delta u = \phi(x)u^{-p}\,,\ u > 0 & \text{in } \Omega, \\ \quad u = 0 & \text{on } \partial\Omega, \end{cases}$$

then $\underline{u} \leq \overline{u}$ in Ω.

Proof. If $p = 0$ the result follows directly from the maximum principle. Let now $p > 0$. Assume by contradiction that the set $\omega := \{x \in \Omega : \overline{u}(x) < \underline{u}(x)\}$ is not empty and let $w := \underline{u} - \overline{u}$. Then, w achieves its maximum on Ω at a point that belongs to ω. At that point, say x_0, we have

$$0 \leq -\Delta w(x_0) \leq \phi(x_0)[\underline{u}(x_0)^{-p} - \overline{u}(x_0)^{-p}] < 0,$$

which is a contradiction. Therefore, $\omega = \emptyset$, that is, $\underline{u} \leq \overline{u}$ in Ω. □

Proposition 4.6 *Let $u \in C^2(\Omega) \cap C(\overline{\Omega})$ be such that $u = 0$ on $\partial\Omega$ and*

$$0 \leq -\Delta u \leq c\delta(x)^{-a} \quad \text{in } \Omega,$$

where $0 < a < 2$ and $c > 0$. Then, $u \in C^{0,\gamma}(\overline{\Omega})$ for some $0 < \gamma < 1$. Furthermore, if $0 < a < 1$, then $u \in C^{1,1-a}(\overline{\Omega})$.

Proof. Let $\mathscr{G}$ denote the Green's function for the negative Laplace operator. Thus, for all $x \in \Omega$ we have

$$u(x) = -\int_\Omega \mathscr{G}(x,y)\Delta u(y)dy.$$

Let $x_1, x_2 \in \Omega$. Then

$$\begin{aligned}|u(x_1)-u(x_2)| &\le -\int_\Omega |\mathcal{G}(x_1,y)-\mathcal{G}(x_2,y)|\Delta u(y)dy\\ &\le c\int_\Omega |\mathcal{G}(x_1,y)-\mathcal{G}_x(x_2,y)|\delta(y)^{-a}dy.\end{aligned}$$

Next, using the method in [107, Theorem 1.1] we have

$$|u(x_1)-u(x_2)| \le C|x_1-x_2|^\gamma \quad \text{for some } 0<\gamma<1.$$

Hence $u\in C^{0,\gamma}(\overline{\Omega})$. Assume now $0<a<1$. Then,

$$\nabla u(x) = -\int_\Omega \mathcal{G}_x(x,y)\Delta u(y)dy \quad \text{for all } x\in\Omega,$$

and

$$\begin{aligned}|\nabla u(x_1)-\nabla u(x_2)| &\le -\int_\Omega |\mathcal{G}_x(x_1,y)-\mathcal{G}_x(x_2,y)|\Delta u(y)dy\\ &\le c\int_\Omega |\mathcal{G}_x(x_1,y)-\mathcal{G}_x(x_2,y)|\delta(y)^{-a}dy.\end{aligned}$$

The same technique as in [107, Theorem 1.1] yields

$$|\nabla u(x_1)-\nabla u(x_2)| \le C|x_1-x_2|^{1-a} \quad \text{for all } x_1,x_2\in\Omega.$$

Therefore $u\in C^{1,1-a}(\overline{\Omega})$. □

Proposition 4.7 *Let (u,v) be a solution of system (4.22). Then, there exists a constant $c>0$ such that*

$$u(x)\ge c\delta(x) \quad \textit{and} \quad v(x)\ge c\delta(x) \quad \textit{in } \Omega. \tag{4.26}$$

Proof. Let w be the solution of

$$\begin{cases} -\Delta w = 1\ ,\ w>0 & \text{in } \Omega,\\ w=0 & \text{on } \partial\Omega.\end{cases} \tag{4.27}$$

Using the smoothness of $\partial\Omega$, we have $w\in C^2(\overline{\Omega})$ and by Hopf's boundary point lemma (see [162]), there exists $c_0>0$ such that $w(x)\ge c_0\delta(x)$ in Ω. Since $-\Delta u\ge C=-\Delta(Cw)$ in Ω, for some constant $C>0$, by standard maximum principle we deduce $u(x)\ge Cw(x)\ge c\delta(x)$ in Ω and similarly $v(x)\ge c\delta(x)$ in Ω, where $c>0$ is a positive constant. □

Let (λ_1,φ_1) be the first eigenvalue/eigenfunction of $-\Delta$ in Ω. It is well known that $\lambda_1>0$ and $\varphi_1\in C^2(\overline{\Omega})$ has constant sign in Ω. Further, using the smoothness of Ω and normalizing φ_1 with a suitable constant, we can assume

$$c_0\delta(x) \le \varphi_1(x) \le \delta(x) \quad \text{in } \Omega, \tag{4.28}$$

for some $0 < c_0 < 1$. By Hopf's boundary point lemma we have $\frac{\partial \varphi_1}{\partial n} < 0$ on $\partial\Omega$, where n is the outer unit normal vector at $\partial\Omega$. Hence, there exists $\omega \subset\subset \Omega$ and $c > 0$ such that

$$|\nabla \varphi_1| > c \quad \text{in } \Omega \setminus \omega. \tag{4.29}$$

Theorem 4.8 *Let $p \ge 0$, $A > \operatorname{diam}(\Omega)$ and $k : (0,A) \to (0,\infty)$ be a decreasing function such that*

$$\int_0^A tk(t)dt = \infty.$$

Then, the inequality

$$\begin{cases} -\Delta u \ge k(\delta(x))u^{-p} \ , \ u > 0 & \text{in } \Omega, \\ u = 0 & \text{on } \partial\Omega, \end{cases} \tag{4.30}$$

has no solutions $u \in C^2(\Omega) \cap C(\overline{\Omega})$.

Proof. Suppose by contradiction that there exists a solution u_0 of (4.30). For any

$$0 < \varepsilon < A - \operatorname{diam}(\Omega)$$

we consider the perturbed problem

$$\begin{cases} -\Delta u = k(\delta(x) + \varepsilon)(u + \varepsilon)^{-p} \ , \ u > 0 & \text{in } \Omega, \\ u = 0 & \text{on } \partial\Omega. \end{cases} \tag{4.31}$$

Then, $\overline{u} = u_0$ is a supersolution of (4.31). Also, if w is the solution of problem (4.27) it is easy to see that $\underline{u} = cw$ is a subsolution of (4.31) provided $c > 0$ is small enough. Further, by Proposition 4.5 it follows that $\underline{u} \le \overline{u}$ in Ω. Thus, by the sub and supersolution method we deduce that problem (4.31) has a solution $u_\varepsilon \in C^2(\overline{\Omega})$ such that

$$cw \le u_\varepsilon \le u_0 \quad \text{in } \Omega. \tag{4.32}$$

Multiplying with φ_1 in (4.31) and then integrating over Ω we find

$$\lambda_1 \int_\Omega u_\varepsilon \varphi_1 dx = \int_\Omega k(\delta(x) + \varepsilon)(u_\varepsilon + \varepsilon)^{-p} \varphi_1 dx.$$

Using (4.32) we obtain

$$M := \lambda_1 \int_\Omega u_0 \varphi_1 dx \geq \lambda_1 \int_\Omega u_\varepsilon \varphi_1 dx \geq \int_\omega k(\delta(x)+\varepsilon)(u_0+\varepsilon)^{-p} \varphi_1 dx,$$

for all $\omega \subset\subset \Omega$. Passing to the limit with $\varepsilon \to 0$ in the above inequality and using (4.28) we find

$$M \geq \int_\omega k(\delta(x)) u_0^{-p} \varphi_1 dx \geq c_0 \|u_0\|_\infty^{-p} \int_\omega k(\delta(x))\delta(x) dx.$$

Since $\omega \subset\subset \Omega$ was arbitrary, we deduce

$$\int_\Omega k(\delta(x))\delta(x) dx < \infty.$$

Using the smoothness of $\partial\Omega$, the above condition yields $\int_0^A tk(t)dt < \infty$, which contradicts our assumption on k. Hence, (4.30) has no solutions. □

A direct consequence of Theorem 4.8 is the following result.

Corollary 4.9 *Let $p \geq 0$ and $q \geq 2$. Then, there are no functions $u \in C^2(\Omega) \cap C(\overline{\Omega})$ such that*

$$\begin{cases} -\Delta u \geq \delta(x)^{-q} u^{-p}, \ u > 0 & \text{in } \Omega, \\ u = 0 & \text{on } \partial\Omega. \end{cases}$$

Proposition 4.10 *Let $p \geq 0$ and $0 < q < 2$. There exists $c > 0$ and $A >$ diam(Ω) such that any supersolution $\overline{u}$ of*

$$\begin{cases} -\Delta u = \delta(x)^{-q} u^{-p}, \ u > 0 & \text{in } \Omega, \\ u = 0 & \text{on } \partial\Omega, \end{cases} \tag{4.33}$$

satisfies:

(i) $\overline{u}(x) \geq c\delta(x)$ in Ω, if $p+q<1$.
(ii) $\overline{u}(x) \geq c\delta(x) \log^{\frac{1}{1+p}}\left(\frac{A}{\delta(x)}\right)$ in Ω if $p+q=1$.
(iii) $\overline{u}(x) \geq c\delta(x)^{\frac{2-q}{1+p}}$ in Ω, if $p+q>1$.

A similar result holds for subsolutions of (4.33).

Proof. If $p > 0$ then the result follows from Theorem 3.5 in [66] (see also [95, Section 9]). If $p = 0$ we proceed as in [66, Theorem 3.5], namely, for $m > 0$ we show that the function

$$\underline{u}(x) = \begin{cases} m\varphi_1(x) & \text{if } q < 1, \\ m\varphi_1(x)\log\left(\dfrac{A}{\varphi_1(x)}\right) & \text{if } q = 1, A > \operatorname{diam}(\Omega), \\ m\varphi_1(x)^{2-q} & \text{if } q > 1, \end{cases}$$

satisfies $-\Delta \underline{u} \le \delta(x)^{-q}$ in Ω. Thus, the estimates in Proposition 4.10 follows from (4.28) and the maximum principle. □

Theorem 4.11 *Let $0 < a < 1$, $A > \operatorname{diam}(\Omega)$, $p \ge 0$ and $q > 0$ be such that $p+q=1$. Then, the problem*

$$\begin{cases} -\Delta u = \delta(x)^{-q}\log^{-a}\left(\dfrac{A}{\delta(x)}\right)u^{-p}\,,\ u > 0 & \text{in } \Omega, \\ u = 0 & \text{on } \partial\Omega, \end{cases} \tag{4.34}$$

has a unique solution u which satisfies

$$c_1\delta(x)\log^{\frac{1-a}{1+p}}\left(\frac{A}{\delta(x)}\right) \le u(x) \le c_2\delta(x)\log^{\frac{1-a}{1+p}}\left(\frac{A}{\delta(x)}\right) \quad \text{in } \Omega, \tag{4.35}$$

for some $c_1, c_2 > 0$.

Proof. Let

$$w(x) = \varphi_1(x)\log^b\left(\frac{A}{\varphi_1(x)}\right), \quad x \in \Omega,$$

where $b = \frac{1-a}{1+p} \in (0,1)$. A straightforward computation yields

$$\begin{aligned} -\Delta w =& \lambda_1\varphi_1\log^b\left(\frac{A}{\varphi_1(x)}\right) + b(|\nabla\varphi_1|^2 - \lambda_1\varphi_1^2)\varphi_1^{-1}\log^{b-1}\left(\frac{A}{\varphi_1(x)}\right) \\ &+ b(1-b)|\nabla\varphi_1|^2\varphi_1^{-1}\log^{b-2}\left(\frac{A}{\varphi_1(x)}\right) \quad \text{in } \Omega. \end{aligned}$$

Using (4.29) we can find $C_1, C_2 > 0$ such that

$$C_1\varphi_1^{-1}\log^{b-1}\left(\frac{A}{\varphi_1(x)}\right) \le -\Delta w \le C_2\varphi_1^{-1}\log^{b-1}\left(\frac{A}{\varphi_1(x)}\right) \quad \text{in } \Omega,$$

that is,

$$C_1\varphi_1^{-q}\log^{-a}\left(\frac{A}{\varphi_1(x)}\right)w^{-p} \le -\Delta w \le C_2\varphi_1^{-q}\log^{-a}\left(\frac{A}{\varphi_1(x)}\right)w^{-p} \quad \text{in } \Omega.$$

We now deduce that $\underline{u} = mw$ and $\overline{u} = Mw$ are respectively subsolution and supersolution of (4.34) for suitable $0 < m < 1 < M$. Hence, the problem (4.34) has a solution $u \in C^2(\Omega) \cap C(\overline{\Omega})$ such that

$$m\varphi_1 \log^{\frac{1-a}{1+p}}\left(\frac{A}{\varphi_1(x)}\right) \leq u \leq M \log^{\frac{1-a}{1+p}}\left(\frac{A}{\varphi_1(x)}\right) \quad \text{in } \Omega. \tag{4.36}$$

The uniqueness follows from Proposition 4.5 while the boundary behavior of u follows from (4.36) and (4.28). This finishes the proof. □

Corollary 4.12 *Let $C > 0$ and a, A, p, q be as in Theorem 4.11. Then, there exists $c > 0$ such that any solution u of*

$$\begin{cases} -\Delta u \geq C\delta(x)^{-q} \log^{-a}\left(\dfrac{A}{\delta(x)}\right) u^{-p}, \; u > 0 & \text{in } \Omega, \\ u = 0 & \text{on } \partial\Omega, \end{cases}$$

satisfies

$$u(x) \geq c\delta(x) \log^{\frac{1-a}{1+p}}\left(\frac{A}{\delta(x)}\right) \quad \text{in } \Omega.$$

Proposition 4.13 *Let $A > 3\text{diam}(\Omega)$ and $C > 0$. There exists $c > 0$ such that any solution $u \in C^2(\Omega) \cap C(\overline{\Omega})$ of*

$$\begin{cases} -\Delta u \geq C\delta^{-1}(x) \log^{-1}\left(\dfrac{A}{\delta(x)}\right), \; u > 0 & \text{in } \Omega, \\ u = 0 & \text{on } \partial\Omega, \end{cases}$$

satisfies

$$u(x) \geq c\delta(x) \log\left[\log\left(\frac{A}{\delta(x)}\right)\right] \quad \text{in } \Omega. \tag{4.37}$$

Proof. Let

$$w(x) = \varphi_1(x) \log\left[\log\left(\frac{A}{\varphi_1(x)}\right)\right], \quad x \in \Omega.$$

An easy computation yields

$$\begin{aligned} -\Delta w =& \lambda_1 \varphi_1 \log\left[\log\left(\frac{A}{\varphi_1(x)}\right)\right] + \frac{|\nabla\varphi_1|^2 - \lambda_1 \varphi_1^2}{\varphi_1 \log\left(\frac{A}{\varphi_1(x)}\right)} + \frac{|\nabla\varphi_1|^2}{\varphi_1 \log^2\left(\frac{A}{\varphi_1(x)}\right)} \\ \leq & \frac{c_0}{\varphi_1 \log\left(\frac{A}{\varphi_1(x)}\right)} \quad \text{in } \Omega, \end{aligned}$$

for some $c_0 > 0$. Using (4.28) we can find $m > 0$ small enough such that

$$-\Delta(mw) \leq \frac{C}{\delta(x) \log\left(\frac{A}{\delta(x)}\right)} \quad \text{in } \Omega.$$

Now by the maximum principle we deduce $u \geq mw$ in Ω and by (4.28) we obtain that u satisfies the estimate (4.37). □

Theorem 4.14 *Let $p \geq 0$, $A > \operatorname{diam}(\Omega)$ and $a \in \mathbb{R}$. Then, problem*

$$\begin{cases} -\Delta u = \delta(x)^{-2}\log^{-a}\left(\dfrac{A}{\delta(x)}\right)u^{-p}\,,\ u>0 & \text{in } \Omega,\\ u=0 & \text{on } \partial\Omega, \end{cases} \tag{4.38}$$

has solutions if and only if $a > 1$. Furthermore, if $a > 1$ then (4.41) has a unique solution u and there exist $c_1, c_2 > 0$ such that

$$c_1 \log^{\frac{1-a}{1+p}}\left(\frac{A}{\delta(x)}\right) \leq u(x) \leq c_2 \log^{\frac{1-a}{1+p}}\left(\frac{A}{\delta(x)}\right) \quad \text{in } \Omega. \tag{4.39}$$

Proof. Fix $B > A$ such that the function $k : (0,B) \to \mathbb{R}$, $k(t) = t^{-2}\log^{-a}\left(\frac{B}{t}\right)$ is decreasing on $(0,A)$. Then, any solution u of (4.38) satisfies

$$\begin{cases} -\Delta u \geq ck(\delta(x))u^{-p}\,,\ u>0 & \text{in } \Omega,\\ u=0 & \text{on } \partial\Omega, \end{cases}$$

where $c > 0$. By virtue of Theorem 4.8 we deduce $\int_0^A tk(t)dt < \infty$, that is, $a > 1$.

For $a > 1$, let

$$w(x) = \log^b\left(\frac{B}{\varphi_1(x)}\right), \quad x \in \Omega,$$

where $b = \frac{1-a}{1+p} < 0$. It is easy to see that

$$\begin{aligned} -\Delta w = &-b(|\nabla\varphi_1|^2 + \lambda_1\varphi_1^2)\varphi_1^{-2}\log^{b-1}\left(\frac{B}{\varphi_1(x)}\right)\\ &-b(b-1)|\nabla\varphi_1|^2\varphi_1^{-2}\log^{b-2}\left(\frac{B}{\varphi_1(x)}\right) \quad \text{in } \Omega. \end{aligned}$$

Choosing $B > 0$ large enough, we may assume

$$\log\left(\frac{B}{\varphi_1(x)}\right) \geq 2(1-b) \quad \text{in } \Omega. \tag{4.40}$$

Therefore, from (4.29) and (4.40) there exist $C_1, C_2 > 0$ such that

$$C_1\varphi_1^{-2}\log^{b-1}\left(\frac{B}{\varphi_1(x)}\right) \leq -\Delta w \leq C_2\varphi_1^{-2}\log^{b-1}\left(\frac{B}{\varphi_1(x)}\right) \quad \text{in } \Omega,$$

that is,

$$C_1\varphi_1^{-2}\log^{-a}\left(\frac{B}{\varphi_1(x)}\right)w^{-p}\le -\Delta w\le C_2\varphi_1^{-2}\log^{-a}\left(\frac{B}{\varphi_1(x)}\right)w^{-p}\quad \text{in } \Omega.$$

As before, from (4.28) it follows that $\underline{u}=mw$ and $\overline{u}=Mw$ are respectively subsolution and supersolution of (4.38) provided $m>0$ is small and $M>1$ is large enough. The rest of the proof is the same as for Theorem 4.11. □

Corollary 4.15 *Let $C>0$, $p\ge 0$, $A>\operatorname{diam}(\Omega)$ and $a>1$. Then, there exists $c>0$ such that any solution $u\in C^2(\Omega)\cap C(\overline{\Omega})$ of*

$$\begin{cases} -\Delta u\ge C\delta(x)^{-2}\log^{-a}\left(\dfrac{A}{\varphi_1(x)}\right)u^{-p}\,,\ u>0 & \text{in } \Omega,\\ u=0 & \text{on } \partial\Omega, \end{cases}\tag{4.41}$$

satisfies

$$u(x)\ge c\log^{\frac{1-a}{1+p}}\left(\frac{A}{\delta(x)}\right)\quad \text{in } \Omega.$$

4.2.2 Nonexistence of a Solution

Our first result concerning the study of (4.22) is the following.

Theorem 4.16 (Nonexistence) *Let $p,s\ge 0$, $q,r>0$ be such that one of the following conditions holds:*

(i) $r\min\left\{1,\frac{2-q}{1+p}\right\}\ge 2.$
(ii) $q\min\left\{1,\frac{2-r}{1+s}\right\}\ge 2.$
(iii) $p>\max\{1,r-1\}$, $2r>(1-s)(1+p)$ *and* $q(1+p-r)>(1+p)(1+s)$.
(iv) $s>\max\{1,q-1\}$, $2q>(1-p)(1+s)$ *and* $r(1+s-q)>(1+p)(1+s)$.

Then the system (4.22) has no solutions.

Remark that condition (i) in Theorem 4.16 restricts the range of the exponent q to the interval $(0,2)$ while in (iii) the exponent q can take any value greater than 2, provided we adjust the other three exponents p,r,s accordingly. The same remark applies for the exponent r from the above conditions (ii) and (iv).

Proof. Since the system (4.22) is invariant under the transform

$$(u,v,p,q,r,s)\to(v,u,s,r,q,p),$$

we only need to prove (i) and (iii).

(i) Assume that there exists (u,v) a solution of system (4.22). Note that from (i) we have $0<q<2$. Also, using Proposition 4.7, we can find $c>0$ such that (4.26) holds.

Case 1: $p+q<1$. From our hypothesis (i) we deduce $r\geq 2$. Using the estimates (4.26) in the first equation of the system (4.22) we find

$$\begin{cases} -\Delta u \leq c_1\delta(x)^{-q}u^{-p}\,,\ u>0 & \text{in } \Omega,\\ u=0 & \text{on } \partial\Omega, \end{cases} \tag{4.42}$$

for some $c_1>0$. From Proposition 4.10(i) we now deduce $u(x)\leq c_2\delta(x)$ in Ω, for some $c_2>0$. Using this last estimate in the second equation of (4.22) we find

$$\begin{cases} -\Delta v \geq c_3\delta(x)^{-r}v^{-s}\,,\ v>0 & \text{in } \Omega,\\ u=0 & \text{on } \partial\Omega, \end{cases} \tag{4.43}$$

where $c_3>0$. According to Corollary 4.9, this is impossible, since $r\geq 2$.

Case 2: $p+q>1$. From hypothesis (i) we also have $\frac{r(2-q)}{1+p}\geq 2$. In the same manner as above, u satisfies (4.42). Thus, by Proposition 4.10(iii), there exists $c_4>0$ such that

$$u(x)\leq c_4\delta(x)^{\frac{2-q}{1+p}} \quad \text{in } \Omega.$$

Using this estimate in the second equation of system (4.22) we obtain

$$\begin{cases} -\Delta v \geq c_5\delta(x)^{-\frac{r(2-q)}{1+p}}v^{-s}\,,\ v>0 & \text{in } \Omega,\\ u=0 & \text{on } \partial\Omega, \end{cases}$$

for some $c_5>0$, which is impossible in view of Corollary 4.9, since $\frac{r(2-q)}{1+p}\geq 2$.

Case 3: $p+q=1$. From (i) it follows that $r\geq 2$. As in the previous two cases, we easily find that u is a solution of (4.42), for some $c_1>0$. Using Proposition 4.10(ii), there exists $c_6>0$ such that

$$u(x)\leq c_6\delta(x)\log^{\frac{1}{1+p}}\left(\frac{A}{\delta(x)}\right) \quad \text{in } \Omega,$$

for some $A>3\operatorname{diam}(\Omega)$. Using this estimate in the second equation of (4.22) we obtain

$$\begin{cases} -\Delta v \geq c_7\delta(x)^{-r}\log^{-\frac{r}{1+p}}\left(\frac{A}{\delta(x)}\right)v^{-s}\,,\ v>0 & \text{in } \Omega,\\ u=0 & \text{on } \partial\Omega, \end{cases} \tag{4.44}$$

where c_7 is a positive constant. From Theorem 4.8 it follows that

$$\int_0^1 t^{1-r}\log^{-\frac{r}{1+p}}\left(\frac{A}{t}\right)dt<\infty.$$

Since $r\geq 2$, the above integral condition implies $r=2$. Now, using (4.44) (with $r=2$) and Corollary 4.15, there exists $c_8>0$ such that

$$v(x)\geq c_8\log^{\frac{p-1}{(1+p)(1+s)}}\left(\frac{A}{\delta(x)}\right)\quad \text{in } \Omega. \tag{4.45}$$

Using the estimate (4.45) in the first equation of system (4.22) we deduce

$$\begin{cases} -\Delta u\leq c_9\log^{\frac{q(1-p)}{(1+p)(1+s)}}\left(\frac{A}{\delta(x)}\right)u^{-p}\,,\; u>0 & \text{in } \Omega,\\ u=0 & \text{on } \partial\Omega, \end{cases} \tag{4.46}$$

for some $c_9>0$. Fix $0<a<1-p$. Then, from (4.46) we can find a constant $c_{10}>0$ such that u satisfies

$$\begin{cases} -\Delta u\leq c_{10}\delta(x)^{-a}u^{-p}\,,\; u>0 & \text{in } \Omega,\\ u=0 & \text{on } \partial\Omega. \end{cases}$$

By Proposition 4.10(i) (since $a+p<1$) we derive $u(x)\leq c_{11}\delta(x)$ in Ω, where $c_{11}>0$. Using this last estimate in the second equation of (4.22) we finally obtain (note that $r=2$):

$$\begin{cases} -\Delta v\geq c_{12}\delta(x)^{-2}v^{-s}\,,\; v>0 & \text{in } \Omega,\\ v=0 & \text{on } \partial\Omega, \end{cases}$$

which is impossible according to Corollary 4.9. Therefore, the system (4.22) has no solutions.

(iii) Suppose that the system (4.22) has a solution (u,v) and let $M=\max_{x\in\overline{\Omega}} v$. From the first equation of (4.22) we have

$$\begin{cases} -\Delta u\geq c_1u^{-p}\,,\; u>0 & \text{in } \Omega,\\ u=0 & \text{on } \partial\Omega, \end{cases}$$

where $c_1 = M^{-q} > 0$. Using Proposition 4.10(iii) there exists $c_2 > 0$ such that $u(x) \geq c_2\delta(x)^{\frac{2}{1+p}}$ in Ω. Combining this estimate with the second equation of (4.22) we find

$$\begin{cases} -\Delta v \leq c_3\delta(x)^{-\frac{2r}{1+p}} v^{-s}, \; v > 0 & \text{in } \Omega, \\ v = 0 & \text{on } \partial\Omega. \end{cases}$$

Since $\frac{2r}{1+p} + s > 1$, again by Proposition 4.10(iii) we obtain that the function v satisfies

$$v(x) \leq c_4\delta(x)^{\frac{2(1+p-r)}{(1+p)(1+s)}} \quad \text{in } \Omega,$$

for some $c_4 > 0$. Using the above estimate in the first equation of (4.22) we find $c_5 > 0$ such that

$$\begin{cases} -\Delta u \geq c_5\delta(x)^{-\frac{2q(1+p-r)}{(1+p)(1+s)}} u^{-p}, \; u > 0 & \text{in } \Omega, \\ u = 0 & \text{on } \partial\Omega, \end{cases}$$

which contradicts Corollary 4.9 since $q(1+p-r) > (1+p)(1+s)$. Thus, the system (4.22) has no solutions. This ends the proof of Theorem 4.16. □

4.2.3 Existence of a Solution

The existence of solutions to (4.22) is obtained under the following assumption on the exponents p, q, r, s:

$$(1+p)(1+s) - qr > 0. \tag{4.47}$$

We also introduce the quantities

$$\alpha = p + q\min\left\{1, \frac{2-r}{1+s}\right\}, \quad \beta = r + s\min\left\{1, \frac{2-q}{1+p}\right\}.$$

The above values of α and β are related to the boundary behavior of the solution to the singular elliptic problem (4.24) as explained in Proposition 4.10. Our existence result is as follows.

Theorem 4.17 *Let $p, s \geq 0$, $q, r > 0$ satisfy (4.47) and one of the following conditions:*

(i) $\alpha \leq 1$ and $r < 2$.
(ii) $\beta \leq 1$ and $q < 2$.
(iii) $p, s \geq 1$ and $q, r < 2$.

Then, the system (4.22) has at least one solution.

The proof of the existence is based on the Schauder's fixed point theorem in a suitably chosen closed convex subset of $C(\overline{\Omega}) \times C(\overline{\Omega})$ that contains all the functions having a certain rate of decay expressed in terms of the distance function $\delta(x)$ up to the boundary of Ω.

Proof. (i) We divide the proof into six cases according to the boundary behavior of singular elliptic problems of type (4.24), as described in Proposition 4.10.

Case 1 : $r+s>1$ and $\alpha = p + \frac{q(2-r)}{1+s} < 1$. By Proposition 4.10(i) and (iii) there exist $0<c_1<1<c_2$ such that:

- Any subsolution $\underline{u}$ and any supersolution $\overline{u}$ of the problem

$$\begin{cases} -\Delta u = \delta(x)^{-\frac{q(2-r)}{1+s}} u^{-p}\,,\ u>0 & \text{in } \Omega, \\ u=0 & \text{on } \partial\Omega, \end{cases} \tag{4.48}$$

  satisfies

$$\overline{u}(x) \ge c_1\delta(x) \quad \text{and} \quad \underline{u}(x) \le c_2\delta(x) \quad \text{in } \Omega. \tag{4.49}$$

- Any subsolution $\underline{v}$ and any supersolution $\overline{v}$ of the problem

$$\begin{cases} -\Delta v = \delta(x)^{-r} v^{-s}\,,\ v>0 & \text{in } \Omega, \\ v=0 & \text{on } \partial\Omega, \end{cases} \tag{4.50}$$

  satisfies

$$\overline{v}(x) \ge c_1\delta(x)^{\frac{2-r}{1+s}} \quad \text{and} \quad \underline{v}(x) \le c_2\delta(x)^{\frac{2-r}{1+s}} \quad \text{in } \Omega. \tag{4.51}$$

  We fix $0<m_1<1<M_1$ and $0<m_2<1<M_2$ such that

$$M_1^{\frac{r}{1+s}} m_2 \le c_1 < c_2 \le M_1 m_2^{\frac{q}{1+p}}, \tag{4.52}$$

and

$$M_2^{\frac{q}{1+p}} m_1 \le c_1 < c_2 \le M_2 m_1^{\frac{r}{1+s}}. \tag{4.53}$$

Note that the above choice of m_i, M_i $(i=1,2)$ is possible in view of (4.47). Set

$$\mathscr{A} = \left\{ (u,v) \in C(\overline{\Omega}) \times C(\overline{\Omega}) : \begin{array}{ll} m_1\delta(x) \le u(x) \le M_1\delta(x) & \text{in } \Omega \\ m_2\delta(x)^{\frac{2-r}{1+s}} \le v(x) \le M_2\delta(x)^{\frac{2-r}{1+s}} & \text{in } \Omega \end{array} \right\}.$$

For any $(u,v) \in \mathscr{A}$, we consider (Tu,Tv) the unique solution of the decoupled system

$$\begin{cases} -\Delta(Tu) = v^{-q}(Tu)^{-p}\,,\ Tu>0 & \text{in } \Omega,\\ -\Delta(Tv) = u^{-r}(Tv)^{-s}\,,\ Tv>0 & \text{in } \Omega,\\ \quad Tu = Tv = 0 & \text{on } \partial\Omega, \end{cases} \tag{4.54}$$

and define

$$\mathscr{F}: \mathscr{A} \to C(\overline{\Omega}) \times C(\overline{\Omega}) \quad \text{by} \quad \mathscr{F}(u,v) = (Tu,Tv) \quad \text{for any } (u,v) \in \mathscr{A}. \tag{4.55}$$

Thus, the existence of a solution to system (4.22) follows once we prove that $\mathscr{F}$ has a fixed point in $\mathscr{A}$. To this aim, we shall prove that $\mathscr{F}$ satisfies the conditions:

$$\mathscr{F}(\mathscr{A}) \subseteq \mathscr{A},\ \mathscr{F} \text{ is compact and continuous.}$$

Then, by Schauder's fixed point theorem we deduce that $\mathscr{F}$ has a fixed point in $\mathscr{A}$, which, by standard elliptic estimates, is a classical solution to (4.22).

Step 1: $\mathscr{F}(\mathscr{A}) \subseteq \mathscr{A}$. Let $(u,v) \in \mathscr{A}$. From

$$v(x) \le M_2\delta(x)^{\frac{2-r}{1+s}} \quad \text{in } \Omega,$$

it follows that Tu satisfies

$$\begin{cases} -\Delta(Tu) \ge M_2^{-q}\delta(x)^{-\frac{q(2-r)}{1+s}}(Tu)^{-p}\,,\ Tu>0 & \text{in } \Omega,\\ \quad Tu = 0 & \text{on } \partial\Omega. \end{cases}$$

Thus, $\overline{u} := M_2^{\frac{q}{1+p}}Tu$ is a supersolution of (4.48). By (4.49) and (4.53) we obtain

$$Tu = M_2^{-\frac{q}{1+p}}\overline{u} \ge c_1 M_2^{-\frac{q}{1+p}}\delta(x) \ge m_1\delta(x) \quad \text{in } \Omega.$$

From $v(x) \ge m_2\delta(x)^{\frac{2-r}{1+s}}$ in Ω and the definition of Tu we deduce that

$$\begin{cases} -\Delta(Tu) \le m_2^{-q}\delta(x)^{-\frac{q(2-r)}{1+s}}(Tu)^{-p}\,,\ Tu>0 & \text{in } \Omega,\\ \quad Tu = 0 & \text{on } \partial\Omega. \end{cases}$$

Thus, $\underline{u} := m_2^{\frac{q}{1+p}}Tu$ is a subsolution of problem (4.48). Hence, from (4.49) and (4.52) we obtain

$$Tu = m_2^{-\frac{q}{1+p}}\underline{u} \le c_2 m_2^{-\frac{q}{1+p}}\delta(x) \le M_1\delta(x) \quad \text{in } \Omega.$$

We have proved that Tu satisfies

$$m_1\delta(x) \le Tu \le M_1\delta(x) \text{ in } \Omega.$$

In a similar manner, using the definition of $\mathscr{A}$ and the properties of the sub and supersolutions of problem (4.50) we show that Tv satisfies

$$m_2\delta(x)^{\frac{2-r}{1+s}} \le Tv \le M_2\delta(x)^{\frac{2-r}{1+s}} \text{ in } \Omega.$$

Thus, $(Tu,Tv) \in \mathscr{A}$ for all $(u,v) \in \mathscr{A}$, that is, $\mathscr{F}(\mathscr{A}) \subseteq \mathscr{A}$.

Step 2: $\mathscr{F}$ is compact and continuous. Let $(u,v) \in \mathscr{A}$. Since $\mathscr{F}(u,v) \in \mathscr{A}$, we can find $0 < a < 2$ such that

$$0 \le -\Delta(Tu), -\Delta(Tv) \le c\delta(x)^{-a} \quad \text{in } \Omega,$$

for some positive constant $c > 0$. Using Proposition 4.6 we now deduce $Tu, Tv \in C^{0,\gamma}(\overline{\Omega})$ $(0 < \gamma < 1)$. Since the embedding $C^{0,\gamma}(\overline{\Omega}) \hookrightarrow C(\overline{\Omega})$ is compact, it follows that $\mathscr{F}$ is also compact.

It remains to prove that $\mathscr{F}$ is continuous. To this aim, let $\{(u_n,v_n)\} \subset \mathscr{A}$ be such that $u_n \to u$ and $v_n \to v$ in $C(\overline{\Omega})$ as $n \to \infty$. Using the fact that $\mathscr{F}$ is compact, there exists $(U,V) \in \mathscr{A}$ such that up to a subsequence we have

$$Tu_n \to U, \quad Tv_n \to V \quad \text{in } C(\overline{\Omega}) \quad \text{as } n \to \infty.$$

On the other hand, by standard elliptic estimates, the sequences $\{Tu_n\}$ and $\{Tv_n\}$ are bounded in $C^{2,\beta}(\overline{\omega})$ $(0 < \beta < 1)$ for any smooth open set $\omega \subset\subset \Omega$. Therefore, up to a diagonal subsequence, we have

$$Tu_n \to U, \quad Tv_n \to V \quad \text{in } C^2(\overline{\omega}) \quad \text{as } n \to \infty,$$

for any smooth open set $\omega \subset\subset \Omega$. Passing to the limit in the definition of Tu_n and Tv_n we find that (U,V) satisfies

$$\begin{cases} -\Delta U = v^{-q}U^{-p}\,,\ U > 0 & \text{in } \Omega,\\ -\Delta V = u^{-r}V^{-s}\,,\ V > 0 & \text{in } \Omega,\\ \quad U = V = 0 & \text{on } \partial\Omega. \end{cases}$$

By uniqueness of (4.54), it follows that $Tu = U$ and $Tv = V$. Hence

$$Tu_n \to Tu, \quad Tv_n \to Tv \quad \text{in } C(\overline{\Omega}) \quad \text{as } n \to \infty.$$

This proves that $\mathscr{F}$ is continuous.

We are now in a position to apply the Schauder's fixed point theorem. Thus, there exists $(u,v)\in\mathscr{A}$ such that $\mathscr{F}(u,v)=(u,v)$, that is, $Tu=u$ and $Tv=v$. By standard elliptic estimates, it follows that (u,v) is a solution of system (4.22).

The remaining five cases will be considered in a similar way. Due to the different boundary behavior of solutions described in Proposition 4.10, the set $\mathscr{A}$ and the constants c_1,c_2 have to be modified accordingly. We shall point out the way we choose these constants in order to apply Schauder's fixed point theorem.

Case 2: $r+s=1$ and $\alpha=p+q<1$. According to Proposition 4.10(i)–(ii) there exist $0<a<1$ and $0<c_1<1<c_2$ such that:

- Any subsolution $\underline{u}$ of the problem

$$\begin{cases} -\Delta u=\delta(x)^{-q}u^{-p}\,,\ u>0 & \text{in } \Omega,\\ u=0 & \text{on } \partial\Omega, \end{cases}$$

  satisfies

$$\underline{u}(x)\le c_2\delta(x)\quad \text{in } \Omega.$$

- Any supersolution $\overline{u}$ of the problem

$$\begin{cases} -\Delta u=\delta(x)^{-q(1-a)}u^{-p}\,,\ u>0 & \text{in } \Omega,\\ u=0 & \text{on } \partial\Omega, \end{cases}$$

  satisfies

$$\overline{u}(x)\ge c_1\delta(x)\quad \text{in } \Omega.$$

- Any subsolution $\underline{v}$ and any supersolution $\overline{v}$ of problem (4.50) satisfies the estimates

$$\underline{v}(x)\le c_2\delta(x)^{1-a}\quad\text{and}\quad \overline{v}(x)\ge c_1\delta(x)\quad \text{in } \Omega.$$

We now define

$$\mathscr{A}=\left\{(u,v)\in C(\overline{\Omega})\times C(\overline{\Omega}):\begin{array}{ll} m_1\delta(x)\le u(x)\le M_1\delta(x) & \text{in } \Omega\\ m_2\delta(x)\le v(x)\le M_2\delta(x)^{1-a} & \text{in } \Omega \end{array}\right\},$$

where $0<m_i<1<M_i$ $(i=1,2)$ satisfy (4.52), (4.53) and

$$m_2[\operatorname{diam}(\Omega)]^a<M_2. \tag{4.56}$$

We next define the operator $\mathscr{F}$ in the same way as in Case 1 by (4.54) and (4.55). The fact that $\mathscr{F}(\mathscr{A}) \subseteq \mathscr{A}$ and $\mathscr{F}$ is continuous and compact follows in the same manner.

Case 3: $r+s<1$ and $\alpha=p+q<1$. In the same manner we define

$$\mathscr{A}=\left\{(u,v)\in C(\overline{\Omega})\times C(\overline{\Omega}):\begin{array}{ll} m_1\delta(x)\le u(x)\le M_1\delta(x) & \text{in } \Omega\\ m_2\delta(x)\le v(x)\le M_2\delta(x) & \text{in } \Omega\end{array}\right\},$$

where $0<m_i<1<M_i$ $(i=1,2)$ satisfy (4.52)–(4.53) for suitable constants c_1 and c_2.

Case 4: $r+s<1$ and $\alpha=p+q=1$. The approach is the same as in Case 2 above if we interchange u with v in the initial system (4.22).

Case 5: $r+s>1$ and $\alpha=p+q=1$. Let $0<a<1$ be fixed such that $ar+s>1$. From Proposition 4.10(i), (iii), there exist $0<c_1<1<c_2$ such that:

- Any subsolution $\underline{u}$ of the problem

$$\begin{cases} -\Delta u=\delta(x)^{-\frac{q(2-ar)}{1+s}}u^{-p}\,,\ u>0 & \text{in } \Omega,\\ u=0 & \text{on } \partial\Omega,\end{cases}$$

  satisfies

$$\underline{u}(x)\le c_2\delta(x)^a \quad \text{in } \Omega.$$

- Any supersolution $\overline{u}$ of the problem

$$\begin{cases} -\Delta u=\delta(x)^{-\frac{q(2-r)}{1+s}}u^{-p}\,,\ u>0 & \text{in } \Omega,\\ u=0 & \text{on } \partial\Omega,\end{cases}$$

  satisfies

$$\overline{u}(x)\ge c_1\delta(x) \quad \text{in } \Omega.$$

- Any subsolution $\underline{v}$ of problem (4.50) satisfies

$$\underline{v}(x)\le c_2\delta(x)^{\frac{2-r}{1+s}} \quad \text{in } \Omega.$$

- Any supersolution $\overline{v}$ of the problem

$$\begin{cases} -\Delta v=\delta(x)^{-ar}v^{-s}\,,\ v>0 & \text{in } \Omega,\\ v=0 & \text{on } \partial\Omega,\end{cases}$$

satisfies

$$\overline{v}(x) \geq c_1 \delta(x)^{\frac{2-ar}{1+s}} \quad \text{in } \Omega.$$

We now define

$$\mathscr{A} = \left\{ (u,v) \in C(\overline{\Omega}) \times C(\overline{\Omega}) : \begin{array}{ll} m_1 \delta(x) \leq u(x) \leq M_1 \delta(x)^a & \text{in } \Omega \\ m_2 \delta(x)^{\frac{2-ar}{1+s}} \leq v(x) \leq M_2 \delta(x)^{\frac{2-r}{1+s}} & \text{in } \Omega \end{array} \right\},$$

where $0 < m_i < 1 < M_i$ $(i = 1,2)$ satisfy (4.52)–(4.53) in which the constants c_1, c_2 are those given above and

$$m_1 [\text{diam}(\Omega)]^{1-a} < M_1, \quad m_2 [\text{diam}(\Omega)]^{\frac{r(1-a)}{1+s}} < M_2.$$

Case 6: $r+s = 1$ and $\alpha = p+q = 1$. We proceed in the same manner as above by considering

$$\mathscr{A} = \left\{ (u,v) \in C(\overline{\Omega}) \times C(\overline{\Omega}) : \begin{array}{ll} m_1 \delta(x) \leq u(x) \leq M_1 \delta(x)^{1-a} & \text{in } \Omega \\ m_2 \delta(x) \leq v(x) \leq M_2 \delta(x)^{1-a} & \text{in } \Omega \end{array} \right\},$$

where $0 < a < 1$ is a fixed constant and m_i, M_i $(i = 1,2)$ satisfy (4.52)–(4.53) for suitable $c_1, c_2 > 0$ and

$$m_i [\text{diam}(\Omega)]^a < M_i, \quad i = 1,2.$$

(iii) Let

$$a = \frac{2(1+s-q)}{(1+p)(1+s)-qr}, \quad b = \frac{2(1+p-r)}{(1+p)(1+s)-qr}.$$

Then

$$(1+p)a + bq = 2, \quad ar + (1+s)b = 2. \tag{4.57}$$

Since $p + bq > 1$ and $s + ar > 1$, from Proposition 4.10(iii) and (4.57) above we can find $0 < c_1 < 1 < c_2$ such that

- Any subsolution $\underline{u}$ and any supersolution $\overline{u}$ of the problem

$$\begin{cases} -\Delta u = \delta(x)^{-bq} u^{-p}, \; u > 0 & \text{in } \Omega, \\ u = 0 & \text{on } \partial\Omega, \end{cases}$$

  satisfies

$$\overline{u}(x) \geq c_1 \delta(x)^a \quad \text{and} \quad \underline{u}(x) \leq c_2 \delta(x)^a \quad \text{in } \Omega.$$

- Any subsolution $\underline{v}$ and any supersolution $\overline{v}$ of the problem

$$\begin{cases} -\Delta v = \delta(x)^{-ar} v^{-s}\,,\ v > 0 & \text{in } \Omega, \\ \quad v = 0 & \text{on } \partial\Omega, \end{cases}$$

satisfies

$$\overline{v}(x) \geq c_1 \delta(x)^b \quad \text{and} \quad \underline{v}(x) \leq c_2 \delta(x)^b \quad \text{in } \Omega.$$

As before, we now define

$$\mathscr{A} = \left\{ (u,v) \in C(\overline{\Omega}) \times C(\overline{\Omega}) : \begin{array}{ll} m_1 \delta(x)^a \leq u(x) \leq M_1 \delta(x)^a & \text{in } \Omega \\ m_2 \delta(x)^b \leq v(x) \leq M_2 \delta(x)^b & \text{in } \Omega \end{array} \right\},$$

where $0 < m_1 < 1 < M_1$ and $0 < m_2 < 1 < M_2$ satisfy (4.52)–(4.53). This concludes the proof of Theorem 4.17. □

From Theorem 4.16(i)–(ii) and Theorem 4.17(i)–(ii) we have the following necessary and sufficient conditions for the existence of solutions to (4.22).

Corollary 4.18 *Let $p,s \geq 0$, $q,r > 0$ satisfy (4.47).*

(i) Assume $p+q \leq 1$. Then system (4.22) has solutions if and only if $r < 2$.
(ii) Assume $r+s \leq 1$. Then system (4.22) has solutions if and only if $q < 2$.

4.2.4 Regularity of Solution

A particular feature of system (4.22) is that it does not possess $C^2(\overline{\Omega})$ solutions. Indeed, due to the fact that $q,r < 0$ and to the homogeneous Dirichlet boundary condition imposed on u and v we have that $u^{-p}v^{-q}$ and $u^{-r}v^{-s}$ are unbounded around $\partial\Omega$, so there are no $C^2(\overline{\Omega})$ solutions of (4.22). In turn, $C^2(\Omega) \cap C^1(\overline{\Omega})$ may exist and our next result provides necessary and sufficient conditions in terms of p,q,r and s for the existence of such solutions.

Theorem 4.19 *Let $p,s \geq 0$, $q,r > 0$ satisfy (4.47). Then*

(i) System (4.22) has a solution (u,v) with $u \in C^1(\overline{\Omega})$ if and only if $\alpha < 1$ and $r < 2$.
(ii) System (4.22) has a solution (u,v) with $v \in C^1(\overline{\Omega})$ if and only if $\beta < 1$ and $q < 2$.

(iii) System (4.22) has a solution (u,v) with $u,v \in C^1(\overline{\Omega})$ if and only if $p+q<1$ and $r+s<1$.

Proof. (i) Assume first that the system (4.22) has a solution (u,v) with $u \in C^1(\overline{\Omega})$. Then, there exists $c>0$ such that $u(x) \le c\delta(x)$ in Ω. Using this fact in the second equation of (4.22), we derive that v satisfies the elliptic inequality (4.43) for some $c_3>0$. By Corollary 4.9 this entails $r<2$.

In order to prove that $\alpha<1$ we argue by contradiction. Suppose that $\alpha \ge 1$ and we divide our argument into three cases.

Case 1: $r+s>1$. Then, $\alpha = p+\frac{q(2-r)}{1+s} \ge 1$. From Proposition 4.7 we have $u(x) \ge c\delta(x)$ in Ω, for some $c>0$. Then v satisfies

$$\begin{cases} -\Delta v \le c_1\delta(x)^{-r}v^{-s}\,,\ v>0 & \text{in } \Omega,\\ u=0 & \text{on } \partial\Omega, \end{cases} \tag{4.58}$$

where $c_1>0$. Since $r<2$, from Proposition 4.10(iii) we find $v(x) \le c_2\delta(x)^{\frac{2-r}{1+s}}$ in Ω, for some $c_2>0$. Using this estimate in the first equation of system (4.22) we deduce

$$\begin{cases} -\Delta u \ge c_3\delta(x)^{-\frac{q(2-r)}{1+s}}u^{-p}\,,\ u>0 & \text{in } \Omega,\\ u=0 & \text{on } \partial\Omega, \end{cases} \tag{4.59}$$

where $c_3>0$. Now, if $\frac{q(2-r)}{1+s} \ge 2$, from Corollary 4.9 the above inequality is impossible. Assume next that $\frac{q(2-r)}{1+s}<2$.

If $\alpha>1$, from (4.28), (4.59) and Proposition 4.10(iii) we find

$$u(x) \ge c_4\delta(x)^{\tau} \ge c_4\varphi_1(x)^{\tau} \quad \text{in } \Omega, \tag{4.60}$$

where

$$\tau = \frac{2-\frac{q(2-r)}{1+s}}{1+p} \in (0,1) \quad \text{and} \quad c_4>0.$$

Fix $x_0 \in \partial\Omega$ and let n be the outer unit normal vector on $\partial\Omega$ at x_0. Using (4.60) and the fact that $0<\tau<1$ we have

$$\begin{aligned} \frac{\partial u}{\partial n}(x_0) &= \lim_{t\nearrow 0}\frac{u(x_0+tn)-u(x_0)}{t}\\ &\le c_4\lim_{t\nearrow 0}\frac{\varphi_1(x_0+tn)-\varphi_1(x_0)}{t}\varphi_1^{\tau-1}(x_0+tn)\\ &= c_4\frac{\partial\varphi_1}{\partial n}(x_0)\lim_{t\nearrow 0}\varphi_1^{\tau-1}(x_0+tn)\\ &= -\infty. \end{aligned}$$

Hence, $u \notin C^1(\overline{\Omega})$.

If $\alpha = 1$ we proceed in the same manner. From (4.59) and Proposition 4.10(ii) we deduce

$$u(x) \geq c_5 \delta(x) \log^{\frac{1}{1+p}} \left(\frac{A}{\delta(x)}\right) \geq c_6 \varphi_1(x) \log^{\frac{1}{1+p}} \left(\frac{A}{\varphi_1(x)}\right) \quad \text{in } \Omega,$$

where $c_5, c_6 > 0$. As before, we obtain $\frac{\partial u}{\partial n}(x_0) = -\infty$, $x_0 \in \partial\Omega$, which contradicts $u \in C^1(\overline{\Omega})$.

Case 2: $r + s < 1$. Then, $\alpha = p + q \geq 1$. As in Case 1, v fulfills (4.58) and by Proposition 4.10(i) we find $v(x) \leq c_7 \delta(x)$ in Ω, for some $c_7 > 0$. Thus, u satisfies

$$\begin{cases} -\Delta u \geq c_8 \delta(x)^{-q} u^{-p}, \; u > 0 & \text{in } \Omega, \\ u = 0 & \text{on } \partial\Omega, \end{cases}$$

where $c_8 > 0$. From Corollary 4.9 it follows that $q < 2$. Since $\alpha = p + q \geq 1$, it follows that u satisfies either the estimate (ii) (if $p + q = 1$) or the estimate (iii) (if $p + q > 1$) in Proposition 4.10. Proceeding in the same way as before we derive that the outer unit normal derivative of u on $\partial\Omega$ is $-\infty$, which is impossible.

Case 3: $r + s = 1$. This also yields $\alpha = p + q \geq 1$. As before v satisfies (4.58) and by Proposition 4.10(ii) we deduce

$$v(x) \leq c_9 \delta(x) \log^{\frac{1}{1+s}} \left(\frac{A}{\delta(x)}\right) \quad \text{in } \Omega,$$

where $c_9 > 0$. It follows that u satisfies

$$\begin{cases} -\Delta u \geq c_{10} \delta(x)^{-q} \log^{-\frac{q}{1+s}} \left(\frac{A}{\delta(x)}\right) u^{-p}, \; u > 0 & \text{in } \Omega, \\ u = 0 & \text{on } \partial\Omega, \end{cases} \tag{4.61}$$

where $c_{10} > 0$. If $q - b \geq 2$ the above inequality is impossible in the light of Corollary 4.9. Assume next that $q - b < 2$. If $\alpha = p + q > 1$, we fix $0 < b < \min\{q, p + q - 1\}$ and from (4.61) we have that u satisfies

$$\begin{cases} -\Delta u \geq c_{11} \delta(x)^{-(q-b)} u^{-p}, \; u > 0 & \text{in } \Omega, \\ u = 0 & \text{on } \partial\Omega, \end{cases}$$

for some $c_{11} > 0$. Now, since $p + q - b > 1$, from Proposition 4.10(iii) we find

$$u(x) \geq c_{12}\delta(x)^{\frac{2-(q-b)}{1+p}} \quad \text{in } \Omega,$$

where $c_{12} > 0$. Since $0 < \frac{2-(q-b)}{1+p} < 1$, we obtain as before that the normal derivative of u on $\partial\Omega$ is infinite, which is impossible.

It remains to consider the case $\alpha = p+q = 1$, that is, $p+q = r+s = 1$. First, if $q < 1+s$, that is, $q \neq 1$ and $s \neq 0$, by (4.61) and Corollary 4.12 we deduce

$$u(x) \geq c_{13}\delta(x)\log^{\frac{1+s-q}{(1+p)(1+s)}}\left(\frac{A}{\delta(x)}\right) \quad \text{in } \Omega,$$

for some $c_{13} > 0$. Proceeding as before we obtain $\frac{\partial u}{\partial n} = -\infty$ on $\partial\Omega$, which is impossible.

If $q = 1$ and $s = 0$ then we apply Proposition 4.13 to obtain

$$u(x) \geq c_{14}\delta(x)\log\left[\log\left(\frac{A}{\delta(x)}\right)\right] \quad \text{in } \Omega,$$

where $c_{14} > 0$. This also leads us to the same contradiction $\frac{\partial u}{\partial n} = -\infty$ on $\partial\Omega$. Thus, we have proved that if the system (4.22) has a solution (u,v) with $u \in C^1(\overline{\Omega})$ then $\alpha < 1$ and $r < 2$.

Conversely, assume now that $\alpha < 1$ and $r < 2$. By Theorem 4.17(i) (Cases 1,2 and 3) there exists a solution (u,v) of (4.22) such that

$$u(x) \geq c\delta(x) \quad \text{in } \Omega,$$

and

$$v(x) \geq c\delta(x) \quad \text{in } \Omega, \quad \text{if } r+s \leq 1,$$

or

$$v(x) \geq c\delta(x)^{\frac{2-r}{1+s}} \quad \text{in } \Omega, \quad \text{if } r+s > 1,$$

for some $c > 0$. Using the above estimates we find

$$-\Delta u = u^{-p}v^{-q} \leq C\delta(x)^{-\alpha} \quad \text{in } \Omega,$$

for some $C > 0$. By Proposition 4.6, we now deduce $u \in C^{1,1-\alpha}(\overline{\Omega})$. The proof of (ii) is similar.

(iii) Assume first that the system (4.22) has a solution (u,v) with $u,v \in C^1(\overline{\Omega})$. Then, there exists $c > 0$ such that $v(x) \leq c\delta(x)$ in Ω. Using this estimate in the first equation of (4.22) we find that

$$\begin{cases} -\Delta u \geq C\delta(x)^{-q}u^{-p}\ , \ u>0 & \text{in } \Omega, \\ u=0 & \text{on } \partial\Omega, \end{cases}$$

where C is a positive constant. If $p+q \geq 1$, then we combine the result in Proposition 4.10(ii)–(iii) with the techniques used above to deduce $\frac{\partial u}{\partial n} = -\infty$ on $\partial\Omega$, so $u \notin C^1(\overline{\Omega})$. Thus, $p+q<1$ and in a similar way we obtain $r+s<1$.

Assume now that $p+q<1$ and $r+s<1$. By Theorem 4.17(i) (Case 3) we have that (4.22) has a solution (u,v) such that $u(x), v(x) \geq c\delta(x)$ in Ω, for some $c>0$. This yields

$$-\Delta u \leq C\delta(x)^{-(p+q)} \quad \text{in } \Omega,$$

$$-\Delta v \leq C\delta(x)^{-(r+s)} \quad \text{in } \Omega,$$

where $C>0$. Now Proposition 4.6 implies $u, v \in C^1(\overline{\Omega})$. This concludes the proof. □

4.2.5 Uniqueness

Another feature of system (4.22) is that under some conditions on p,q,r,s it has a unique solution (see Theorem 4.20 below). This is a striking difference between our setting and the case $p,s \leq 0$ and $q,r<0$ largely investigated in the literature so far, where the uniqueness does not seem to occur. In our framework, the uniqueness is achieved from the boundary behavior of the solution to (4.22) deduced from the study of the prototype model (4.24).

Theorem 4.20 *Let $p,s \geq 0$, $q,r>0$ satisfy (4.47) and one of the following conditions:*

(i) $p+q<1$ and $r<2$.
(ii) $r+s<1$ and $q<2$.

Then, the system (4.22) has a unique solution.

Proof. We shall prove only (i); the case (ii) follows in the same manner.

Let (u_1,v_1) and (u_2,v_2) be two solutions of system (4.22). Using Proposition 4.7 there exists $c_1>0$ such that

$$u_i(x), v_i(x) \geq c_1\delta(x) \quad \text{in } \Omega,\ i=1,2. \tag{4.62}$$

Hence, u_i satisfies

$$\begin{cases} -\Delta u_i \le c_2 \delta(x)^{-q} u_i^{-p}, \; u_i > 0 & \text{in } \Omega, \\ u_i = 0 & \text{on } \partial\Omega, \end{cases}$$

for some $c_2 > 0$. By Proposition 4.10(i) and (4.62) there exists $0 < c < 1$ such that

$$c\delta(x) \le u_i(x) \le \frac{1}{c}\delta(x) \quad \text{in } \Omega,\; i = 1,2. \tag{4.63}$$

This means that we can find a constant $C > 1$ such that $Cu_1 \ge u_2$ and $Cu_2 \ge u_1$ in Ω.

We claim that $u_1 \ge u_2$ in Ω. Supposing the contrary, let

$$M = \inf\{A > 1 : Au_1 \ge u_2 \text{ in } \Omega\}.$$

By our assumption, we have $M > 1$. From $Mu_1 \ge u_2$ in Ω, it follows that

$$-\Delta v_2 = u_2^{-r} v_2^{-s} \ge M^{-r} u_1^{-r} v_2^{-s} \quad \text{in } \Omega.$$

Therefore v_1 is a solution and $M^{\frac{r}{1+s}} v_2$ is a supersolution of

$$\begin{cases} -\Delta w = u_1^{-r} w^{-s}, \; w > 0 & \text{in } \Omega, \\ w = 0 & \text{on } \partial\Omega. \end{cases}$$

By Proposition 4.5 we obtain

$$v_1 \le M^{\frac{r}{1+s}} v_2 \quad \text{in } \Omega.$$

The above estimate yields

$$-\Delta u_1 = u_1^{-p} v_1^{-q} \ge M^{-\frac{qr}{1+s}} u_1^{-p} v_2^{-q} \quad \text{in } \Omega.$$

It follows that u_2 is a solution and $M^{\frac{qr}{(1+p)(1+s)}} u_1$ is a supersolution of

$$\begin{cases} -\Delta w = v_2^{-q} w^{-p}, \; w > 0 & \text{in } \Omega, \\ w = 0 & \text{on } \partial\Omega. \end{cases}$$

By Proposition 4.5 we now deduce

$$M^{\frac{qr}{(1+p)(1+s)}} u_1 \ge u_2 \quad \text{in } \Omega.$$

Since $M > 1$ and $\frac{qr}{(1+p)(1+s)} < 1$, the above inequality contradicts the minimality of M. Hence, $u_1 \geq u_2$ in Ω. Similarly we deduce $u_1 \leq u_2$ in Ω, so $u_1 \equiv u_2$ which also yields $v_1 \equiv v_2$. Therefore, the system has a unique solution. This completes the proof of Theorem 4.20. □

4.3 Sublinear Lane–Emden Systems with Singular Data

Let $\Omega \subset \mathbb{R}^N$ be a bounded domain with $C^{2,\alpha}(0 < \alpha < 1)$ boundary. In this section we shall be concerned with the following elliptic system

$$\begin{cases} -\Delta u = \delta(x)^{-a} v^p, u > 0 & \text{in } \Omega, \\ -\Delta v = \delta(x)^{-b} u^q, v > 0 & \text{in } \Omega, \\ \quad u = v = 0 & \text{on } \partial\Omega, \end{cases} \tag{4.64}$$

where $\delta(x) = \text{dist}(x, \partial\Omega)$, $a, b \geq 0$ and $p, q \in \mathbb{R}$ satisfy $-1 < pq < 1$.

Our discussion splits up into four cases according to the sign of p and q.

4.3.1 *Case* $p > 0$ *and* $q > 0$

Theorem 4.21 *Assume that $p, q > 0$ satisfy $pq < 1$ and one of the following holds:*

(i) $a - p < 2$ and $b - q\min\{1, 2 - a + p\} < 1$.
(ii) $b - q < 2$ and $a - p\min\{1, 2 - b + q\} < 1$.

Then, the system (4.64) has at least one solution.

Proof. Before we proceed with the proof of Theorem 4.21 we state the following auxiliary result whose proof is similar to those in Sect. 4.2.1.

Lemma 4.22 *Let $0 < a < 2$ and $u \in C^2(\Omega) \cap C(\overline{\Omega})$ be such that*

$$0 \leq -\Delta u \leq \delta(x)^{-a}, \; u > 0 \quad \textit{in } \Omega, \quad \textit{and} \quad u = 0 \quad \textit{on } \partial\Omega.$$

Then:

(i) There exists $0 < \gamma < 1$ such that $u \in C^{0,\gamma}(\overline{\Omega})$. Moreover, if $0 < a < 1$ then $u \in C^{1,1-a}(\overline{\Omega})$.

(ii) If $0 < a < 1$ *then there exists* $c > 0$ *such that* $u \le c\delta(x)$ *in* Ω.

(iii) If $a = 1$ *and* $\tau > 1$ *then, there exist* $c > 0$ *and* $A > \operatorname{diam}(\Omega)$ *such that*

$$u \le c\delta(x)\log^{\tau}\left(\frac{A}{\delta(x)}\right) \quad \text{in } \Omega.$$

In particular, for any $0 < \varepsilon < 1$ *there exist* $c > 0$ *such that* $u \le c\delta(x)^{1-\varepsilon}$ *in* Ω.

(iv) If $1 < a < 2$ *then, there exists* $c > 0$ *such that* $u \le c\delta(x)^{2-a}$ *in* Ω.

We shall prove only (i), the proof of (ii) is similar. First, we fix $\tau > 0$ such that $(p+\tau)q < 1$. We divide our argument into three cases according to the boundary behavior of the solutions to some singular elliptic inequalities as described in Lemma 4.22.

Case 1: $1 < a - p < 2$. By Lemma 4.22, there exists $C > 0$ such that:

- Any function $w \in C^2(\Omega) \cap C(\overline{\Omega})$ such that

$$\begin{cases} -\Delta w \le \delta(x)^{-a+p}, \; w > 0 & \text{in } \Omega, \\ w = 0 & \text{on } \partial\Omega, \end{cases} \tag{4.65}$$

  satisfies

$$w(x) \le C\delta(x)^{2-a+p} \quad \text{in } \Omega. \tag{4.66}$$

- Any function $w \in C^2(\Omega) \cap C(\overline{\Omega})$ such that

$$\begin{cases} -\Delta w \le \delta(x)^{-b+q(2-a+p)}, \; w > 0 & \text{in } \Omega, \\ w = 0 & \text{on } \partial\Omega, \end{cases} \tag{4.67}$$

  satisfies

$$w(x) \le C\delta(x) \quad \text{in } \Omega. \tag{4.68}$$

We fix $M > 1$ with the property

$$C < \min\{M^{\tau}, M^{1-(p+\tau)q}\} \tag{4.69}$$

and define

$$\mathscr{A} := \left\{ (u,v) \in C(\overline{\Omega}) \times C(\overline{\Omega}) : \begin{array}{ll} 0 \le u \le M^{p+\tau}\delta(x)^{2-a+p} & \text{in } \Omega \\ 0 \le v \le M\delta(x) & \text{in } \Omega \end{array} \right\}.$$

For any $(u,v) \in \mathscr{A}$, we consider (Tu,Tv) the unique solution of

$$\begin{cases} -\Delta(Tu) = \delta(x)^{-a}v^p \,,\; Tu > 0 & \text{in } \Omega, \\ -\Delta(Tv) = \delta(x)^{-b}u^q \,,\; Tv > 0 & \text{in } \Omega, \\ \quad Tu = Tv = 0 & \text{on } \partial\Omega, \end{cases} \tag{4.70}$$

and define $\mathscr{F} : \mathscr{A} \to C(\overline{\Omega}) \times C(\overline{\Omega})$ by

$$\mathscr{F}(u,v) = (Tu,Tv) \quad \text{for any } (u,v) \in \mathscr{A}. \tag{4.71}$$

Remark that (4.70) has a unique solution, in other words, $\mathscr{F}$ is well defined. Indeed for the existence of Tu we remark that $0 \le \delta(x)^{-a}v^p \le c\delta(x)^{-a+p}$ in Ω. Therefore, by Lemma 4.22(iv) we find that $\underline{w} \equiv 0$ and $\overline{w} = A\varphi_1^{2-a}$, $A > 1$ large, are sub and supersolutions respectively. Therefore, there exists $Tu \in C^2(\Omega) \cap C(\overline{\Omega})$ such that $-\Delta(Tu) = \delta(x)^{-a}v^p$ in Ω and $Tu = 0$ on $\partial\Omega$. The uniqueness of Tu and the fact that $Tu > 0$ in Ω follows from the standard maximum principle. The existence and uniqueness of Tv is similar.

As in the previous section we next prove that $\mathscr{F}$ is compact and continuous and that $\mathscr{F}(\mathscr{A}) \subseteq \mathscr{A}$. By the Schauder fixed point theorem we then obtain that $\mathscr{F}$ has a fixed point which is a solution of (4.64).

Case 2: $a - p = 1$. We fix $\varepsilon > 0$ small enough such that $b - q(1-\varepsilon) < 1$. By Lemma 4.22(ii)–(iii), there exists $C > 0$ such that

- Any function $w \in C^2(\Omega) \cap C(\overline{\Omega})$ such that

$$\begin{cases} -\Delta w \le \delta(x)^{-1} \,,\; w > 0 & \text{in } \Omega, \\ \quad w = 0 & \text{on } \partial\Omega, \end{cases}$$

  satisfies

$$w(x) \le C\delta(x)^{1-\varepsilon} \quad \text{in } \Omega.$$

- Any function $w \in C^2(\Omega) \cap C(\overline{\Omega})$ such that

$$\begin{cases} -\Delta w \le \delta(x)^{-b+q(1-\varepsilon)} \,,\; w > 0 & \text{in } \Omega, \\ \quad w = 0 & \text{on } \partial\Omega, \end{cases}$$

  satisfies

$$w(x) \le C\delta(x) \quad \text{in } \Omega.$$

We next proceed as before by considering

$$\mathscr{A}=\left\{(u,v)\in C(\overline{\Omega})\times C(\overline{\Omega}):\begin{array}{ll}0\le u(x)\le M^{p+\tau}\delta(x)^{1-\varepsilon} & \text{in } \Omega\\ 0\le v(x)\le M\delta(x) & \text{in } \Omega\end{array}\right\},$$

where $M>1$ is a constant that fulfills (4.69).

Case 3: $a-p<1$. In this case $-b+q<1$ and by Lemma 4.22 there exists $C>0$ such that

- Any function $w\in C^2(\Omega)\cap C(\overline{\Omega})$ such that

$$\begin{cases}-\Delta w\le \delta(x)^{-a+p}\,,\ w>0 & \text{in } \Omega,\\ \quad w=0 & \text{on } \partial\Omega,\end{cases}$$

  satisfies

$$w(x)\le C\delta(x)\quad \text{in } \Omega.$$

- Any function $w\in C^2(\Omega)\cap C(\overline{\Omega})$ such that

$$\begin{cases}-\Delta w\le \delta(x)^{-b+q(1-\varepsilon)}\,,\ w>0 & \text{in } \Omega,\\ \quad w=0 & \text{on } \partial\Omega,\end{cases}$$

  satisfies

$$w(x)\le C\delta(x)\quad \text{in } \Omega.$$

We next proceed as in Case 1 by considering

$$\mathscr{A}=\left\{(u,v)\in C(\overline{\Omega})\times C(\overline{\Omega}):\begin{array}{ll}0\le u(x)\le M^{p+\tau}\delta(x) & \text{in } \Omega\\ 0\le v(x)\le M\delta(x) & \text{in } \Omega\end{array}\right\}.$$

This finishes the proof of Theorem 4.21. □

4.3.2 Case $p>0$ and $q<0$

In this section we shall be concerned with the case $p>0>q$. First, we obtain the following nonexistence result.

Theorem 4.23 *Assume $p>0>q$ and one of the following holds:*

(i) $a-p\ge 2$ *or* $b\ge 2$.
(ii) $0<a<1$ *and* $b-q\ge 2$.

(iii) $a = 1$ *and* $b > 1$ *and* $b \geq 2 + q$.

(iv) $1 < a < 2$ *and* $b - q(2 - a) \geq 2$.

Then, the system (4.64) has no solutions.

Proof. (i) Assume $a - p \geq 2$. By Lemma 4.7 we have $v \geq c\delta(x)$ in Ω, for some $c > 0$. Using this estimate in the first equation of (4.64) we find $-\Delta u \geq c_1 \delta(x)^{a-p}$ in Ω, for some $c_1 > 0$, which is impossible by Corollary 4.9.

If $b \geq 2$, we use the second equation of (4.64) to derive

$$-\Delta v \geq \|u\|_\infty^q \delta(x)^{-b} \quad \text{in } \Omega,$$

which is impossible according to Corollary 4.9.

(ii) From the first equation of (4.64) and $p > 0$ we deduce

$$-\Delta u \leq \|v\|_\infty^p \delta(x)^{-a} \quad \text{in } \Omega.$$

Since $0 < a < 1$, Lemma 4.22(ii) yields $u \leq c_0 \delta(x)$ in Ω, for some $c_0 > 0$. Using this estimate in the second equation of (4.64) we deduce $-\Delta v \geq c_1 \delta(x)^{-b+q}$ in Ω. Since $b - q \geq 2$, we arrive at a contradiction according to Corollary 4.9.

(iii) Let $\tau > 1$. Since $-\Delta u \leq \|v\|_\infty^p \delta(x)^{-1}$ in Ω, there exists $c > 0$ such that

$$u \leq c\delta(x) \log^\tau \left(\frac{A}{\delta(x)} \right) \quad \text{in } \Omega.$$

Thus,

$$-\Delta v = \delta(x)^{-b} u^q \geq C\delta(x)^{-b+q} \log^{\tau q} \left(\frac{A}{\delta(x)} \right) \quad \text{in } \Omega,$$

where C is a positive constant. By Theorem 4.8 we now deduce

$$\int_0^1 t^{-b+q+1} \log^{\tau q} \left(\frac{A}{t} \right) dt < \infty$$

which, in view of the fact that $b \geq q + 2$ and $b > 1$, yields $b - q = 2$ and $\tau q < -1$. Since $\tau > 1$ was arbitrary, it follows that $q \leq -1$ so $b = q + 2 \leq 1$, which is impossible.

(iv) As before, from the first equation of (4.64) we have

$$-\Delta u \leq \|v\|_\infty^p \delta(x)^{-a} \quad \text{in } \Omega.$$

From Lemma 4.22(iv) we find $u \leq c_2 \delta(x)^{2-a}$ in Ω, where c_2 is a positive constant. Using this last inequality in the second equation of (4.64) we derive

$-\Delta v \geq c_3 \delta(x)^{-b+q(2-a)}$ in Ω, which is impossible by Corollary 4.9. This ends the proof of Theorem 4.23. □

Theorem 4.24 *Let $p > 0 > q$ satisfy $pq > -1$.*

(i) If $b - q < 1$ then the system (4.64) has solutions if and only if $a - p < 2$.
(ii) If $0 < a < 1$ then the system (4.64) has solutions if and only if $b - q < 2$.

Furthermore, in both the above cases the system (4.64) has a unique solution.

Proof. (i) If $a - p \geq 2$, then by Theorem 4.23 there are no solutions of system (4.64). Suppose next that $a - p < 2$. The proof is similar to that for Theorem 4.21. Fix $\tau > 0$ such that $(p+\tau)q > -1$ and assume first that $1 < a - p < 2$. According to Proposition 4.10(i) and Lemma 4.22(iv) there exist $c_1 > c_2 > 0$ such that

- Any function $w \in C^2(\Omega) \cap C(\overline{\Omega})$ that satisfies (4.65) also fulfills

$$w(x) \leq c_1 \delta(x)^{2-a+p} \quad \text{in } \Omega.$$

- Any function $w \in C^2(\Omega) \cap C(\overline{\Omega})$ that satisfies (4.67) also fulfills

$$w(x) \leq c_1 \delta(x) \quad \text{in } \Omega.$$

- Any function $w \in C^2(\Omega) \cap C(\overline{\Omega})$ that satisfies

$$\begin{cases} -\Delta w \geq \delta(x)^{-a+p}\,, \; w > 0 & \text{in } \Omega, \\ w = 0 & \text{on } \partial\Omega, \end{cases}$$

  also has the property that

$$w(x) \geq c_2 \delta(x)^{2-a+p} \quad \text{in } \Omega.$$

- Any function $w \in C^2(\Omega) \cap C(\overline{\Omega})$ such that

$$\begin{cases} -\Delta w \geq \delta(x)^{-b+q(2-a+p)}\,, \; w > 0 & \text{in } \Omega, \\ w = 0 & \text{on } \partial\Omega, \end{cases}$$

  satisfies

$$w(x) \geq c_2 \delta(x) \quad \text{in } \Omega.$$

Now we fix $M > 1$ such that

$$\min\{M^{\tau}, M^{1+(p+\tau)q}\} > \max\{c_1, c_2^{-1}\}.$$

We apply Schauder's fixed point theorem for the mapping $\mathscr{F}$ defined by (4.70) and (4.71) where this time the set $\mathscr{A} \subset C(\overline{\Omega}) \times C(\overline{\Omega})$ is given by

$$\mathscr{A} = \left\{ (u,v) \in C(\overline{\Omega}) \times C(\overline{\Omega}) : \begin{array}{ll} \frac{1}{M^{p+\tau}}\delta(x)^{2-a+p} \leq u(x) \leq M^{p+\tau}\delta(x)^{2-a+p} & \text{in } \Omega \\ \frac{1}{M}\delta(x) \leq v(x) \leq M\delta(x) & \text{in } \Omega \end{array} \right\}.$$

If $a-p=1$ we fix $0<\varepsilon<1$ and proceed in the same fashion with the set

$$\mathscr{A} = \left\{ (u,v) \in C(\overline{\Omega}) \times C(\overline{\Omega}) : \begin{array}{ll} \frac{1}{M^{p+\tau}}\delta(x) \leq u(x) \leq M^{p+\tau}\delta(x)^{1-\varepsilon} & \text{in } \Omega \\ \frac{1}{M}\delta(x) \leq v(x) \leq M\delta(x) & \text{in } \Omega \end{array} \right\},$$

where $M>1$ is a suitably chosen constant. Finally, if $a-p<1$ we define the set $\mathscr{A}$ as

$$\mathscr{A} = \left\{ (u,v) \in C(\overline{\Omega}) \times C(\overline{\Omega}) : \begin{array}{ll} \frac{1}{M^{p+\tau}}\delta(x) \leq u(x) \leq M^{p+\tau}\delta(x) & \text{in } \Omega \\ \frac{1}{M}\delta(x) \leq v(x) \leq M\delta(x) & \text{in } \Omega \end{array} \right\}.$$

For the uniqueness, we first remark that any solution (u,v) of (4.64) satisfies $v \in C^2(\Omega)\cap C^1(\overline{\Omega})$. Indeed, by Proposition 4.7 there exists $c>0$ such that $u,v \geq c\delta(x)$ in Ω. Thus, $-\Delta v \leq c\delta(x)^{-b+q}$ in Ω so, by Lemma 4.22(i)–(ii) we have $v \in C^1(\overline{\Omega})$ and $v(x) \leq c_0\delta(x)$ in Ω for some $c_0>0$.

Let (u_1,v_1) and (u_2,v_2) be two solutions of (4.64). Using the above remark, there exists $0<m<1$ such that

$$m\delta(x) \leq v_i(x) \geq \frac{1}{m}\delta(x) \quad \text{in } \Omega,\ i=1,2. \tag{4.72}$$

Therefore, we can find a constant $C>1$ such that $Cv_1 \geq v_2$ and $Cv_2 \geq v_1$ in Ω. We claim that $v_1 \geq v_2$ in Ω. Supposing the contrary, let

$$M = \inf\{A>1 : Av_1 \geq v_2 \text{ in } \Omega\}.$$

By our assumption, we have $M>1$. From $Mv_1 \geq v_2$ in Ω, it follows that $-\Delta u_2 = \delta(x)^{-a}v_2^p \leq M^p\delta(x)^{-a}v_1^p$ in Ω. Hence $-\Delta(M^{-p}u_2) \leq \delta(x)^{-a}v_1^p = -\Delta u_1$ in Ω, which yields $M^{-p}u_2 \leq u_1$ in Ω. Using the last inequality we have

$$-\Delta v_1 = \delta(x)^{-b}u_1^q \leq M^{-pq}\delta(x)^{-b}u_2^q \quad \text{in } \Omega.$$

It follows that $-\Delta(M^{pq}v_1) \le \delta(x)^{-b}u_2^q = -\Delta v_2$ in Ω, which implies $M^{pq}v_1 \le v_2$ in Ω. Further we obtain $-\Delta u_2 = \delta(x)^{-a}v_2^p \ge M^{p^2q}\delta(x)^{-a}v_1^p$ in Ω. As before we now derive $M^{-p^2q}u_2 \ge u_1$ in Ω. Finally we have

$$-\Delta v_1 = \delta(x)^{-b}u_1^q \ge M^{-p^2q^2}\delta(x)^{-b}u_2^q \quad \text{in } \Omega.$$

It follows that

$$-\Delta(M^{p^2q^2}v_1) \ge \delta(x)^{-b}u_2^q = -\Delta v_2 \quad \text{in } \Omega.$$

Thus, $M^{p^2q^2}v_1 \ge v_2$ in Ω. Since $0 < p^2q^2 < 1$, the above inequality contradicts the definition of M. Thus, $v_1 \ge v_2$ in Ω and similarly we obtain $v_2 \ge v_1$ in Ω. Hence $v_1 \equiv v_2$ which yields $u_1 \equiv u_2$. The proof of (ii) is similar. □

4.3.3 *Case* $p < 0$ *and* $q < 0$

Theorem 4.25 *Let* $p, q < 0$ *satisfy* $pq < 1$ *and assume one of the following conditions holds:*

(i) $b - q < 1$ *and* $a - p < 2$.
(ii) $a - p < 1$ *and* $b - q < 2$.

Then, the system (4.64) has a unique solution.

Proof. The proof is similar to that for Theorem 4.24. We only point out the differences. We fix $\tau > 0$ such that $(p-\tau)q < 1$.

(i) If $1 < a - p < 2$ then we proceed as in the proof of Theorem 4.24 for the set $\mathscr{A} \subset C(\overline{\Omega}) \times C(\overline{\Omega})$ given by

$$\mathscr{A} = \left\{ (u,v) \in C(\overline{\Omega}) \times C(\overline{\Omega}) : \begin{array}{ll} M^{p-\tau}\delta(x)^{2-a+p} \le u(x) \le M^{-p+\tau}\delta(x)^{2-a+p} & \text{in } \Omega \\ \frac{1}{M}\delta(x) \le v(x) \le M\delta(x) & \text{in } \Omega \end{array} \right\}.$$

If $a - p = 1$ we fix $0 < \varepsilon < 1$ and proceed in the same way with the set

$$\mathscr{A} = \left\{ (u,v) \in C(\overline{\Omega}) \times C(\overline{\Omega}) : \begin{array}{ll} M^{p-\tau}\delta(x) \le u(x) \le M^{-p+\tau}\delta(x)^{1-\varepsilon} & \text{in } \Omega \\ \frac{1}{M}\delta(x) \le v(x) \le M\delta(x) & \text{in } \Omega \end{array} \right\},$$

where $M > 1$ is a suitably chosen constant. Finally, if $a - p < 1$ we define $\mathscr{A}$ as

$$\mathscr{A} = \left\{ (u,v) \in C(\overline{\Omega}) \times C(\overline{\Omega}) : \begin{array}{ll} M^{p-\tau}\delta(x) \le u(x) \le M^{-p+\tau}\delta(x) & \text{in } \Omega \\ \frac{1}{M}\delta(x) \le v(x) \le M\delta(x) & \text{in } \Omega \end{array} \right\}.$$

For the uniqueness, we first remark that any solution (u,v) of (4.64) satisfies $c_1\delta(x) \le v \le c_2\delta(x)$ in Ω. Indeed, by Proposition 4.7 there exists $c > 0$ such that $u,v \ge c\delta(x)$ in Ω. Then $-\Delta v \le c^q\delta(x)^{-b+q}$ in Ω so, according to Lemma 4.22(ii) we have $v \le c_0\delta(x)$ in Ω for some $c_0 > 0$.

Let (u_1,v_1) and (u_2,v_2) be two solutions of (4.64). By the above remark v_i $(i = 1,2)$, are both comparable to the distance function $\delta(x)$ up to the boundary. Therefore, we can find a constant $C > 1$ such that $Cv_1 \ge v_2$ and $Cv_2 \ge v_1$ in Ω. We next proceed in the same manner as in the proof of Theorem 4.24. The proof of (ii) is similar. □

4.3.4 Further Extensions: Superlinear Case

We want to point out here some features of the superlinear case $p,q > 0$ and $pq > 1$. In this setting, system (4.64) has a variational structure. More precisely, one can see (4.64) as a Hamiltonian system:

$$\begin{cases} -\Delta u = H_v(x,u,v) & \text{in } \Omega, \\ -\Delta v = H_u(x,u,v) & \text{in } \Omega, \\ u = v = 0 & \text{on } \partial\Omega, \end{cases}$$

where

$$H(x,u,v) = \frac{u^{p+1}}{(p+1)\delta(x)^a} + \frac{v^{q+1}}{(q+1)\delta(x)^b}.$$

The approach in this case is variational, it consists of using the fractional powers of the negative Laplace operator subject to homogeneous Dirichlet boundary conditions.

In fact, we deduce the existence of solutions for a more general system, namely

$$\begin{cases} -\Delta u = \delta(x)^{-a}|v|^{p-1}v & \text{in } \Omega, \\ -\Delta v = \delta(x)^{-b}|u|^{q-1}u & \text{in } \Omega, \\ u = v = 0 & \text{on } \partial\Omega. \end{cases} \tag{4.73}$$

Our main result concerning system (4.73) is the following.

Theorem 4.26 *Let $p,q>0$ satisfy $pq>1$ and*

$$\frac{1-a}{1+p}+\frac{1-b}{1+q}>\frac{N-2}{N}, \tag{4.74}$$

$$p<\frac{2N(1-a)}{N-4} \quad \text{and} \quad q<\frac{2N(1-b)}{N-4} \quad \text{if } N\geq 5. \tag{4.75}$$

Then, the system (4.73) has infinitely many solutions of which at least one is positive.

Proof. The proof is similar to that in [63, Theorem 1]; we only point out here the main differences. Consider the Laplace operator

$$-\Delta : H^2(\Omega)\cap H_0^1(\Omega)\subset L^2(\Omega)\to L^2(\Omega),$$

and denote by $\{\lambda_n, e_n\}$ the corresponding eigenvalues and eigenfunctions with

$$0<\lambda_1\leq\lambda_2\leq\cdots\leq\lambda_n\to\infty \quad \text{and} \quad \|e_n\|_2=1.$$

Thus, any $u\in H^2(\Omega)\cap H_0^1(\Omega)$ has the unique representation

$$u=\sum_{n\geq 1}a_n e_n \quad \text{where } a_n=\int_\Omega u e_n dx.$$

For any $0<s<1$ we define

$$E^s=\{u=\sum_{n\geq 1}a_n e_n\in L^2(\Omega): \sum_{n\geq 1}\lambda_n^{2s}a_n^2<\infty\}.$$

The s-power A^s of $-\Delta$ is defined as

$$A^s : E^s\subset L^2(\Omega)\to L^2(\Omega), \quad A^s=\sum_{n\geq 1}\lambda_n^s a_n e_n.$$

It turns out that E^s is a Hilbert space with the inner product

$$\langle u,v\rangle_{E^s}=\int_\Omega A^s u A^s v dx.$$

Moreover, E^s is a fractional Sobolev space (see [135]) and $E^s\subseteq H^{2s}(\Omega)$ for all $0<s<1$. Further, the embedding $E^s\hookrightarrow L^r(\Omega)$ is compact provided $\frac{1}{r}\geq\frac{1}{2}-\frac{2s}{N}$. Note that by Hölder inequality we have

$$\int_\Omega \frac{u^{q+1}}{\delta(x)^b}dx\leq \|u\|_{L^r(\Omega)}^{q+1}\Big(\int_\Omega \delta(x)^{\frac{-br}{r-(q+1)}}dx\Big)^{\frac{r-(q+1)}{r}}\leq C\|u\|_{L^r(\Omega)}^{q+1},$$

for all $u \in E^s$ and $r > (q+1)/(1-b)$. Thus the embedding

$$E^s \hookrightarrow L^{q+1}(\Omega, \delta(x)^{-b})$$

is compact. Using (4.74)–(4.75), one can find $0 < s,t < 1$ such that $s+t=1$ and

$$N-4t < \frac{2N(1-a)}{1+p} \quad \text{and} \quad N-4s < \frac{2N(1-b)}{1+q}.$$

Then the embeddings

$$E^s \hookrightarrow L^{q+1}(\Omega, \delta(x)^{-b}) \quad \text{and} \quad E^t \hookrightarrow L^{p+1}(\Omega, \delta(x)^{-a})$$

are compact.

Let $E = E^s \times E^t$. We first look for (s,t)-weak solutions to (4.73) in the following sense.

Definition 4.1. We say that (u,v) is an (s,t)-weak solution of system (4.73) if

$$\int_\Omega A^s u A^t \phi dx + \int_\Omega A^t v A^s \psi dx - \int_\Omega \frac{v^p}{\delta(x)^a} \phi dx - \int_\Omega \frac{u^q}{\delta(x)^b} \psi dx = 0,$$

for all $(\phi, \psi) \in E$.

It is easy to see that any (s,t)-weak solution of (4.73) is in fact a critical point of the functional

$$I : E \to \mathbb{R}, \quad I(u,v) = \int_\Omega A^s u A^t v dx - \int_\Omega H(x,u,v) dx.$$

Remark that if (u,v) is an (s,t)-weak solution of (4.73) then

$$u \in W^{2,p_1}(\Omega) \quad \text{and} \quad v \in W^{2,q_1}(\Omega),$$

for all $p_1, q_1 > 1$ that satisfy

$$p_1\left(p + \frac{2Na}{N-4t}\right) < \frac{2N}{N-4s} \quad \text{and} \quad q_1\left(q + \frac{2Nb}{N-4s}\right) < \frac{2N}{N-4s}.$$

From now on we employ step by step the same arguments as in [63] in order to deduce that system (4.73) has infinitely many solutions of which at least one is positive. □

Chapter 5
Singular Elliptic Inequalities in Exterior Domains

> The true sign of intelligence is not knowledge but imagination.
>
> Albert Einstein (1879–1955)

5.1 Introduction

In this chapter we study the existence and nonexistence of C^2 positive solutions $u(x)$ of the following semilinear elliptic inequality

$$-\Delta u \geq \varphi(\delta_K(x))f(u) \quad \text{in } \mathbb{R}^N \setminus K, \tag{5.1}$$

where K is a compact set in $\mathbb{R}^N$ ($N \geq 2$) and $\delta_K(x) = \text{dist}(x, \partial K)$. We assume that $\varphi \in C^{0,\gamma}(0,\infty)$ ($0 < \gamma < 1$) and $f \in C^1(0,\infty)$ are positive functions such that φ is nonincreasing and f is decreasing.

Elliptic equations or inequalities in unbounded domains have been subject to extensive study recently (see, e.g., [59,60,95,120,121,136,164,165] and the references therein). In [59, 60] the authors are concerned with elliptic problems with superlinear nonlinearities $f(t)$ in exterior domains. Large classes of elliptic inequalities in exterior or cone-like domains involving various types of differential operators are considered in [120, 121, 136, 164, 165]. In [190–194] elliptic inequalities are studied in a punctured neighborhood of the origin and asymptotic radial symmetry of solutions is investigated. The main novelty here is the presence of the distance func-

M. Ghergu and V. Rădulescu, *Nonlinear PDEs*, Springer Monographs in Mathematics,
DOI 10.1007/978-3-642-22664-9_5,

tion $\delta_K(x)$ to the boundary of the compact set K which, as we shall see, will play a significant role in the qualitative study of (5.1).

In our approach to (5.1) we shall distinguish between the case where K is *nondegenerate*, that is, K has at least one component which is the closure of a C^2 domain, and the case where K is *degenerate* which means that K reduces to a finite set of points.

We provide general nonexistence results of solutions to (5.1) for various types of compact sets $K \subset \mathbb{R}^N$, $N \geq 2$. Our study points out the role played by the geometry of K in the existence of C^2 positive solutions to (5.1). For instance, if K consists of finitely many components each of which is the closure of a C^2 domain, then (5.1) has solutions if and only if

$$\int_0^\infty r\varphi(r)dr < \infty. \tag{5.2}$$

In turn, if K consists of a finite number of points, then the existence of a C^2 positive solution to (5.1) depends on both φ and f. If $f(t) = t^{-p}$, $p > 0$, and K reduces to a single point (by translation one may consider $K = \{0\}$) we describe the solution set of

$$-\Delta u = \varphi(|x|)u^{-p} \quad \text{in } \mathbb{R}^N \setminus \{0\}. \tag{5.3}$$

For a large class of functions φ, we show that any C^2 positive solution of (5.3) (if it exists) is radially symmetric. Furthermore, the solution set of (5.3) consists of a two-parameter family of radially symmetric functions.

If (5.1) has solutions, we prove that it has a minimal C^2 positive solution $\tilde{u}$ in the sense of the usual order relation. Moreover, $\tilde{u}$ is a *ground-state* of (5.1), that is, $\tilde{u}(x) \to 0$ as $|x| \to \infty$. In some cases depending on the geometry of ∂K we prove that $\tilde{u}$ is continuous up to the boundary of K.

5.2 Some Elliptic Problems in Bounded Domains

In this section we obtain some results for related elliptic problems in bounded domains that will be further used in the sequel. We start with the following comparison result.

Lemma 5.1 *Let $\Omega \subset \mathbb{R}^N$ $(N \geq 2)$ be a nonempty open set and $g : \Omega \times (0,\infty) \to (0,\infty)$ be a continuous function such that $g(x,\cdot)$ is decreasing for all $x \in \Omega$. Assume* *that u, v are C^2 positive functions that satisfy*

$$\Delta u + g(x,u) \le 0 \le \Delta v + g(x,v) \quad \text{in } \Omega,$$

$$\lim_{x\in\Omega, x\to y} (v(x) - u(x)) \le 0 \quad \text{for all } y \in \partial^\infty \Omega.$$

Then $u \ge v$ in Ω. (Here $\partial^\infty \Omega$ stands for the Euclidean boundary $\partial\Omega$ if Ω is bounded and for $\partial\Omega \cup \{\infty\}$ if Ω is unbounded.)

Proof. Assume by contradiction that the set $\omega := \{x \in \Omega : u(x) < v(x)\}$ is not empty and let $w := v - u$. Since $\lim_{x\in\Omega, x\to y} w(x) \le 0$ for all $y \in \partial^\infty\Omega$, it follows that w is bounded from above and it achieves its maximum on Ω at a point that belongs to ω. At that point, say x_0, we have

$$0 \le -\Delta w(x_0) \le g(x_0, v(x_0)) - g(x_0, u(x_0)) < 0,$$

which is a contradiction. Therefore, $\omega = \emptyset$, that is, $u \ge v$ in Ω. □

Lemma 5.2 *Let $\Omega \subset \mathbb{R}^N$ ($N \ge 2$) be a bounded domain with C^2 boundary and let $g : \overline{\Omega} \times (0,\infty) \to (0,\infty)$ be a Hölder continuous function such that for all $x \in \overline{\Omega}$ we have $g(x,\cdot) \in C^1(0,\infty)$ and $g(x,\cdot)$ is decreasing. Then, for any $\phi \in C(\partial\Omega)$, $\phi \ge 0$, the problem*

$$\begin{cases} -\Delta u = g(x,u),\ u > 0 & \text{in } \Omega, \\ u = \phi(x) & \text{on } \partial\Omega, \end{cases} \tag{5.4}$$

has a unique solution $u \in C^2(\Omega) \cap C(\overline{\Omega})$.

Proof. For all $n \ge 1$ consider the following perturbed problem

$$\begin{cases} -\Delta u = g\Big(x, u + \dfrac{1}{n}\Big),\ u > 0 & \text{in } \Omega, \\ u = \phi(x) & \text{on } \partial\Omega. \end{cases} \tag{5.5}$$

It is easy to see that $\underline{u} \equiv 0$ is a subsolution. To construct a supersolution, let w be the solution of

$$\begin{cases} -\Delta w = 1,\ w > 0 & \text{in } \Omega, \\ w = 0 & \text{on } \partial\Omega. \end{cases}$$

Then $\overline{u} = Mw + ||\phi||_\infty + 1$ is a supersolution of (5.5) provided $M > 1$ is large enough. Thus, by the sub and supersolution method and Lemma 5.1, there exists a unique solution $u_n \in C^2(\Omega) \cap C(\overline{\Omega})$ of (5.5). Furthermore, since $g(x,\cdot)$ is decreasing, by

Lemma 5.1 we deduce

$$u_1 \le u_2 \le \cdots \le u_n \le \cdots \le \overline{u} \quad \text{in } \Omega, \tag{5.6}$$

$$u_n + \frac{1}{n} \ge u_{n+1} + \frac{1}{n+1} \quad \text{in } \Omega. \tag{5.7}$$

Hence $\{u_n(x)\}$ is increasing and bounded for all $x \in \Omega$. Letting $u(x) := \lim_{n\to\infty} u_n(x)$, a standard bootstrap argument (see [54], [99]) implies $u_n \to u$ in $C^2_{\text{loc}}(\Omega)$ so that passing to the limit in (5.5) we deduce $-\Delta u = g(x,u)$ in Ω. From (5.6) and (5.7) we obtain $u_n + 1/n \ge u \ge u_n$ in Ω, for all $n \ge 1$. This yields $u \in C(\overline{\Omega})$ and $u = \phi(x)$ on $\partial\Omega$. Therefore $u \in C^2(\Omega)\cap C(\overline{\Omega})$ is a solution of (5.4). The uniqueness follows from Lemma 5.1. □

Lemma 5.3 and Lemma 5.4 below extend the existence results obtained in [88, 89, 92].

Lemma 5.3 *Let $\Omega \subset \mathbb{R}^N$ $(N \ge 2)$ be a bounded domain with C^2 boundary. Also let $\varphi \in C^{0,\gamma}(0,\infty)$ $(0<\gamma<1)$ and $f \in C^1(0,\infty)$ be positive functions such that:*

(i) f is decreasing;
(ii) φ is nonincreasing and $\int_0^1 r\phi(r)dr < \infty$.

Then, the problem

$$\begin{cases} -\Delta u = \varphi(\delta_\Omega(x))f(u),\ u>0 & \text{in } \Omega,\\ u=0 & \text{on } \partial\Omega, \end{cases} \tag{5.8}$$

has a unique solution $u \in C^2(\Omega)\cap C(\overline{\Omega})$. Furthermore, there exist $c_1, c_2 > 0$ and $0 < r_0 < 1$ such that the unique solution u of (5.8) satisfies

$$c_1 \le \frac{u(x)}{H(\delta_\Omega(x))} \le c_2 \quad \text{in } \{x\in\Omega : 0<\delta_\Omega(x)<r_0\}, \tag{5.9}$$

where $H : [0,1] \to (0,\infty)$ is the unique solution of

$$\begin{cases} -H''(t) = \varphi(t)f(H(t)),\ H(t)>0 & 0<t<1,\\ H(0)=H(1)=0. \end{cases} \tag{5.10}$$

The existence of a solution to (5.10) follows from [3, Theorem 2.1].

Proof. Let (λ_1, e_1) be the first eigenvalue and the first eigenfunction of $-\Delta$ in Ω subject to Dirichlet boundary condition. It is well known that e_1 has constant sign in Ω so that normalizing, we may assume that $e_1 > 0$ in Ω. Also, since Ω has a C^2 boundary, we have $\partial e_1/\partial \nu < 0$ on $\partial\Omega$ and

$$C_1\delta_\Omega(x) \le e_1(x) \le C_2\delta_\Omega(x) \quad \text{in } \Omega, \tag{5.11}$$

where ν is the outward unit normal at $\partial\Omega$ and C_1, C_2 are two positive constants. We claim that there exist $M > 1$ and $c > 0$ such that $\overline{u} = MH(ce_1)$ is a supersolution of (5.8). First, since the solution H of (5.10) is positive and concave, we can find $0 < a < 1$ such that $H' > 0$ on $(0,a]$. Let $c > 0$ be such that

$$ce_1(x) \le \min\{a, \delta_\Omega(x)\} \quad \text{in } \Omega.$$

Then

$$\begin{aligned} -\Delta\overline{u} &= -Mc^2H''(ce_1)|\nabla e_1|^2 + Mc\lambda_1 e_1 H'(ce_1) \\ &= Mc^2\varphi(ce_1)f(H(ce_1))|\nabla e_1|^2 + Mc\lambda_1 e_1 H'(ce_1) \\ &\ge Mc^2\varphi(\delta_\Omega(x))f(\overline{u})|\nabla e_1|^2 + Mc\lambda_1 e_1 H'(ce_1) \quad \text{in } \Omega. \end{aligned} \tag{5.12}$$

Since $e_1 > 0$ in Ω and $\partial e_1/\partial\nu < 0$ on $\partial\Omega$, we can find $d > 0$ and a subdomain $\omega \subset\subset \Omega$ such that

$$|\nabla e_1| > d \quad \text{in } \Omega \setminus \omega.$$

Therefore, from (5.12) we obtain

$$-\Delta\overline{u} \ge Mc^2d^2\varphi(\delta_\Omega(x))f(\overline{u}) \quad \text{in } \Omega\setminus\omega, \qquad -\Delta\overline{u} \ge Mc\lambda_1 e_1 H'(ce_1) \quad \text{in } \omega. \tag{5.13}$$

Now, we choose $M > 0$ large enough such that

$$Mc^2d^2 > 1 \quad \text{and} \quad Mc\lambda_1 e_1 H'(ce_1) \ge \varphi(\delta_\Omega(x))f(\overline{u}) \quad \text{in } \omega. \tag{5.14}$$

Note that the last relation in (5.14) is possible since in ω the right side of the inequality is bounded and the left side is bounded away from zero. Thus, from (5.13) and (5.14), $\overline{u}$ is a supersolution for (5.8). Similarly, we can choose $m > 0$ small enough such that $\underline{u} = mH(ce_1)$ is a subsolution of (5.8). Therefore, by the sub and supersolution method we find a solution $u \in C^2(\Omega)\cap C(\overline{\Omega})$ such that $\underline{u} \le u \le \overline{u}$ in Ω. The uniqueness follows from Lemma 5.1. In order to prove the boundary estimate (5.9), note first that $ce_1 \le \delta_\Omega(x)$ in Ω so

$$u(x) \le \overline{u}(x) \le MH(\delta_\Omega(x)) \quad \text{in } \{x \in \Omega : 0 < \delta_\Omega(x) < a\}.$$

On the other hand, since H is concave and $H(0) = 0$, we easily derive that $t \to H(t)/t$ is decreasing on $(0,1)$. Also we can assume $cC_1 < 1$. Thus,

$$u(x) \ge mH(ce_1) \ge mH(cC_1\delta_\Omega(x)) \ge mcC_1H(\delta_\Omega(x)),$$

for all $x \in \Omega$ with $0 < \delta_\Omega(x) < 1$. The proof of Lemma 5.3 is now complete. □

Lemma 5.4 *Let $K \subset \mathbb{R}^N$ $(N \ge 2)$ be a compact set, $\Omega \subset \mathbb{R}^N$ be a bounded domain such that $K \subset \Omega$ and $\Omega \setminus K$ is connected and has C^2 boundary. Let φ and f be as in Lemma 5.3. Then, there exists a unique solution $u \in C^2(\Omega \setminus K) \cap C(\overline{\Omega} \setminus \mathrm{int}(K))$ of the problem*

$$\begin{cases} -\Delta u = \varphi(\delta_K(x))f(u),\ u > 0 & \text{in } \Omega \setminus K, \\ u = 0 & \text{on } \partial(\Omega \setminus K). \end{cases} \tag{5.15}$$

Furthermore, there exist $c_1, c_2 > 0$ and $0 < r_0 < 1$ such that the unique solution u of (5.15) satisfies

$$c_1 \le \frac{u(x)}{H(\delta_K(x))} \le c_2 \quad \text{in } \{x \in \Omega \setminus K : 0 < \delta_K(x) < r_0\}, \tag{5.16}$$

where H is the unique solution of (5.10).

Proof. According to Lemma 5.3 there exists $v \in C^2(\Omega \setminus K) \cap C(\overline{\Omega \setminus K})$ such that

$$\begin{cases} -\Delta v = \varphi(\delta_{\Omega\setminus K}(x))f(v),\ v > 0 & \text{in } \Omega \setminus K, \\ v = 0 & \text{on } \partial(\Omega \setminus K), \end{cases}$$

which further satisfies

$$c_1 \le \frac{v(x)}{H(\delta_{\Omega\setminus K}(x))} \le c_2 \quad \text{in } \{x \in \Omega \setminus K : 0 < \delta_{\Omega\setminus K}(x) < \rho_0\}, \tag{5.17}$$

for some $0 < \rho_0 < 1$ and $c_1, c_2 > 0$. Since $\delta_K(x) \ge \delta_{\Omega\setminus K}(x)$ for all $x \in \Omega \setminus K$ and φ is nonincreasing, it is easy to see that $\overline{u} = v$ is a supersolution of (5.15). Also it is not difficult to see that $\underline{u} = mw$ is a subsolution to (5.15) for $m > 0$ sufficiently small, where w satisfies

$$\begin{cases} -\Delta w = 1,\ w > 0 & \text{in } \Omega \setminus K, \\ w = 0 & \text{on } \partial(\Omega \setminus K). \end{cases}$$

Using Lemma 5.1 we have $\underline{u} \le \overline{u}$ in $\Omega \setminus K$. Therefore, there exists a solution $u \in C^2(\Omega \setminus K) \cap C(\overline{\Omega} \setminus \mathrm{int}(K))$ of (5.15). As before, the uniqueness follows from Lemma 5.1. In order to prove (5.16), let $0 < r_0 < \rho_0$ be small such that

$$\omega := \{x \in \Omega \setminus K : 0 < \delta_K(x) < r_0\} \subset\subset \Omega \quad \text{and} \quad \delta_{\Omega \setminus K}(x) = \delta_K(x) \quad \text{for all } x \in \omega.$$

Then, from (5.17) we have

$$u \le \overline{u} \le c_2 H(\delta_K(x)) \quad \text{in } \omega.$$

For the remaining part of (5.16), let $M > 1$ be such that $Mu \ge v$ on $\partial\omega \setminus \partial K$. Also

$$-\Delta(Mu) = M\varphi(\delta_K(x))f(u) \ge \varphi(\delta_K(x))f(Mu) \quad \text{in } \omega.$$

By Lemma 5.1 we have $Mu \ge v$ in ω and from (5.17) we obtain the first inequality in (5.16). This concludes the proof. □

The following result is a direct consequence of Lemma 5.4.

Lemma 5.5 *Let $K_1, K_2, L \subset \mathbb{R}^N$ $(N \ge 2)$ be three compact sets (see Fig. 5.1) such that*

$$K_1 \cap L = \emptyset, \quad K_2 \subset \mathrm{int}(L), \quad K_1, L \text{ are the closure of } C^2 \text{ domains.}$$

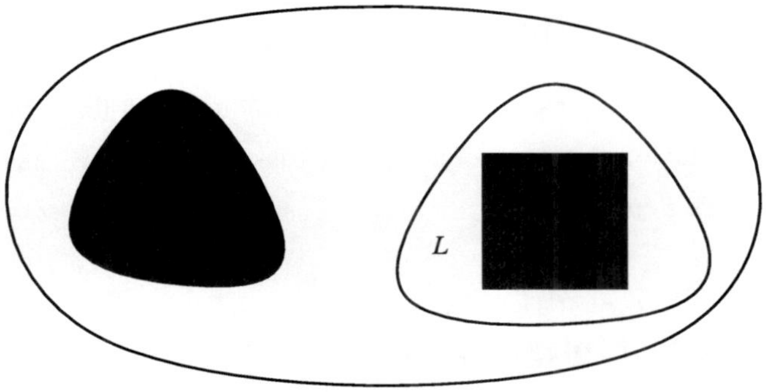

Fig. 5.1 The compact sets K_1, K_2 and L

Also let $\Omega \subset \mathbb{R}^N$ be a bounded domain with C^2 boundary such that $K_1 \cup L \subset \Omega$ and $\Omega \setminus (K_1 \cup L)$ is connected. Let φ, f be as in Lemma 5.3. Then, there exists a unique solution

$$u \in C^2(\Omega \setminus (K_1 \cup L)) \cap C(\overline{\Omega} \setminus \mathrm{int}(K_1 \cup L))$$

of the problem

$$\begin{cases} -\Delta u = \varphi(\delta_{K_1 \cup K_2}(x)) f(u),\ u > 0 & \text{in } \Omega \setminus (K_1 \cup L), \\ u = 0 & \text{on } \partial(\Omega \setminus (K_1 \cup L)). \end{cases} \tag{5.18}$$

Furthermore, there exist $c_1, c_2 > 0$ *and* $0 < r_0 < 1$ *such that the unique solution u of problem (5.18) satisfies*

$$c_1 \leq \frac{u(x)}{H(\delta_{K_1}(x))} \leq c_2 \quad \text{in } \{x \in \Omega \setminus (K_1 \cup L) : 0 < \delta_{K_1}(x) < r_0\}, \tag{5.19}$$

where H is the unique solution of (5.10).

Proof. By Lemma 5.4 there exists a unique $v \in C^2(\Omega \setminus (K_1 \cup L)) \cap C(\overline{\Omega} \setminus \mathrm{int}(K_1 \cup L))$ such that

$$\begin{cases} -\Delta v = \varphi(\delta_{K_1 \cup L}(x)) f(v),\ v > 0 & \text{in } \Omega \setminus (K_1 \cup L), \\ v = 0 & \text{on } \partial(\Omega \setminus (K_1 \cup L)). \end{cases}$$

Since $\delta_{K_1 \cup L}(x) \leq \delta_{K_1 \cup K_2}(x)$ in $\Omega \setminus (K_1 \cup L)$ and φ is nonincreasing, we derive that $\overline{u} = v$ is a supersolution of (5.18). As a subsolution we use $\underline{u} = mw$ where m is sufficiently small and w satisfies

$$\begin{cases} -\Delta w = 1,\ w > 0 & \text{in } \Omega \setminus (K_1 \cup L), \\ w = 0 & \text{on } \partial(\Omega \setminus (K_1 \cup L)). \end{cases}$$

Therefore, problem (5.18) has a solution u. The uniqueness follows from Lemma 5.1 while the asymptotic behavior of u around K_1 is obtained in the same manner as in Lemma 5.4. This ends the proof. □

5.3 An Equivalent Integral Condition

Several times in this section we shall use the following elementary result that provides an equivalent integral condition to (5.2).

Lemma 5.6 *Let* $N \geq 3$ *and* $\varphi : (0, \infty) \to [0, \infty)$ *be a continuous function.*

(i) $\int_0^1 r\varphi(r)dr < \infty$ *if and only if* $\int_0^1 t^{1-N} \int_0^t s^{N-1} \varphi(s) ds dt < \infty$.

(ii) $\int_1^\infty r\varphi(r)dr < \infty$ *if and only if* $\int_1^\infty t^{1-N}\int_1^t s^{N-1}\varphi(s)dsdt < \infty$.

(iii) $\int_0^\infty r\varphi(r)dr < \infty$ *if and only if* $\int_0^\infty t^{1-N}\int_0^t s^{N-1}\varphi(s)dsdt < \infty$.

Proof. We prove only (i). The proof of (ii) is similar, while (iii) follows from (i)–(ii).

Assume first that $\int_0^1 r\varphi(r)dr < \infty$. Integrating by parts we have

$$\begin{aligned}\int_0^1 t^{1-N}\int_0^t s^{N-1}\varphi(s)dsdt &= -\frac{1}{N-2}\int_0^1 \left(t^{2-N}\right)'\int_0^t s^{N-1}\varphi(s)dsdt\\ &= \frac{1}{N-2}\left(\int_0^1 t\varphi(t)dt - \int_0^1 t^{N-1}\varphi(t)dt\right)\\ &\le \frac{1}{N-2}\int_0^1 t\varphi(t)dt < \infty.\end{aligned}$$

Conversely, for $0 < \varepsilon < 1/2$ we have

$$\begin{aligned}\int_\varepsilon^1 t^{1-N}\int_0^t s^{N-1}\varphi(s)dsdt &= \frac{1}{N-2}\left(\int_\varepsilon^1 t\varphi(t)dt - \int_0^1 t^{N-1}\varphi(t)dt + \varepsilon^{2-N}\int_0^\varepsilon t^{N-1}\varphi(t)dt\right)\\ &\ge \frac{1}{N-2}\left(\int_\varepsilon^1 t\varphi(t)dt - \int_\varepsilon^1 t^{N-1}\varphi(t)dt\right)\\ &= \frac{1}{N-2}\int_\varepsilon^1 (1-t^{N-2})t\varphi(t)dt\\ &\ge \frac{1}{N-2}\left(1-\left(\frac{1}{2}\right)^{N-2}\right)\int_\varepsilon^{1/2} t\varphi(t)dt.\end{aligned}$$

Passing to the limit with $\varepsilon \searrow 0$ we deduce $\int_0^1 t\varphi(t)dt < \infty$. This concludes the proof of Lemma 5.6. □

5.4 The Nondegenerate Case

5.4.1 Nonexistence Results

We present some nonexistence results that hold in a more general setting for f and φ.

Theorem 5.7 *Let* $\varphi : (0,\infty) \to [0,\infty)$ *and* $f : (0,\infty) \to (0,\infty)$ *be continuous functions such that:*

(i) $\liminf_{t\searrow 0} f(t) > 0$.
(ii) $\varphi(r)$ *is monotone for* r *large.*
(iii) $\int_1^\infty r\varphi(r)dr = \infty$.

Then, for any compact set $K \subset \mathbb{R}^N$ $(N \geq 3)$ there does not exist a C^2 positive solution $u(x)$ of (5.1).

Proof. It is easy to construct a C^1 function $g : [0,\infty) \to (0,\infty)$ such that $g < f$ in $(0,\infty)$ and g' is negative and nondecreasing. Therefore, we may assume $f : [0,\infty) \to (0,\infty)$ is of class C^1 and f' is negative and nondecreasing.

Suppose for contradiction that $u(x)$ is a C^2 positive solution of (5.1). By translation, we may assume that $0 \in K$. Choose $r_0 > 0$ such that

$$K \subset B_{r_0/2}(0), \quad \varphi(r_0/2) > 0, \quad \text{and} \quad \varphi \text{ is monotone on } [r_0/2,\infty).$$

Define $\psi : [r_0/2,\infty) \to (0,\infty)$ by

$$\psi(r) = \min_{r_0/2 \leq \rho \leq r} \varphi(\rho) = \begin{cases} \varphi(r) & \text{if } \varphi \text{ is nonincreasing for } r \geq r_0/2, \\ \varphi(r_0/2) & \text{if } \varphi \text{ is nondecreasing for } r \geq r_0/2. \end{cases}$$

Then $\int_{r_0}^{\infty} r\psi(r)dr = \infty$. Also, since $r_0/2 \leq \delta_K(x) \leq |x|$ for all $x \in \mathbb{R}^N \setminus B_{r_0}(0)$, we have

$$\varphi(\delta_K(x)) \geq \psi(|x|) \quad \text{for all } x \in \mathbb{R}^N \setminus B_{r_0}(0).$$

Thus, the solution u of (5.1) satisfies

$$-\Delta u \geq \psi(|x|)f(u) \quad \text{in } \mathbb{R}^N \setminus B_{r_0}(0). \tag{5.20}$$

Averaging (5.20) and using Jensen's inequality, we get

$$-\left(\bar{u}''(r) + \frac{n-1}{r}\bar{u}'(r)\right) \geq \psi(r)f(\bar{u}(r)) \quad \text{for all } r \geq r_0. \tag{5.21}$$

Here $\bar{u}$ is the spherical average of u, that is

$$\bar{u}(r) = \frac{1}{\sigma_N r^{N-1}} \int_{\partial B_r(0)} u(x)\, d\sigma(x), \tag{5.22}$$

where σ denotes the surface area measure in $\mathbb{R}^N$ and $\sigma_N = \sigma(\partial B_1(0))$.

Making in (5.21) the change of variables $\bar{u}(r) = v(\rho)$, $\rho = r^{2-N}$ we get

$$-v''(\rho) \geq \frac{1}{(N-2)^2}\rho^{2(N-1)/(2-N)}\psi(\rho^{1/(2-N)})f(v(\rho)) \quad \text{for all } 0 < \rho \leq \rho_0,$$

where $\rho_0 = r_0^{2-N}$. Since v is concave down and positive, v is bounded for $0 < \rho \le \rho_0$. Hence $f(v(\rho)) \ge (N-2)^2 C$ for some positive constant C. Consequently

$$-v''(\rho) \ge C\rho^{2(N-1)/(2-N)}\psi(\rho^{1/(2-N)}) \quad \text{for all } 0 < \rho \le \rho_0.$$

Integrating this inequality twice we get

$$\begin{aligned}
\infty &> \int_0^{\rho_0} v'(\rho)\,d\rho - \rho_0 v'(\rho_0)\\
&\ge C\int_0^{\rho_0}\int_\rho^{\rho_0} \bar\rho^{2(N-1)/(2-N)}\psi(\bar\rho^{1/(2-N)})\,d\bar\rho\,d\rho\\
&= C\int_0^{\rho_0} \bar\rho^{1+2(N-1)/(2-N)}\psi(\bar\rho^{1/(2-N)})\,d\bar\rho\\
&= (N-2)C\int_{r_0}^{\infty} r\psi(r)\,dr = \infty.
\end{aligned}$$

This contradiction completes the proof. □

Theorem 5.8 *Let $\varphi : (0,\infty) \to [0,\infty)$ and $f : (0,\infty) \to (0,\infty)$ be continuous functions such that*

(i) $\liminf_{t\searrow 0} f(t) > 0$.
(ii) $\int_0^1 r\varphi(r)\,dr = \infty$.

Then there does not exist a C^2 positive solution $u(x)$ of

$$-\Delta u \ge \varphi(\delta_\Omega(x))f(u) \quad \text{in } \{x \in \mathbb{R}^N \setminus \overline{\Omega} : 0 < \delta_\Omega(x) < 2r_0\}, \quad N \ge 2, \tag{5.23}$$

where Ω is a C^2 bounded domain in $\mathbb{R}^N$ and $r_0 > 0$.

For the proof of Theorem 5.8 we shall use the following lemma concerning the geometry of a C^2 bounded domain. One can prove it using standard methods of differential geometry.

Lemma 5.9 *Let Ω be a C^2 bounded domain in $\mathbb{R}^N$, $N \ge 2$, such that $\mathbb{R}^N \setminus \Omega$ is connected. Then, there exists $r_0 > 0$ such that*

(i) $\Omega_r := \{x \in \mathbb{R}^N : \mathrm{dist}(x,\Omega) < r\}$ *is a C^1 domain for each $0 < r \le r_0$.*
(ii) For $0 \le r \le r_0$ the function $T(\cdot,r) : \partial\Omega \to \mathbb{R}^N$ defined by $T(\xi,r) = \xi + r\eta_\xi$, where η_ξ is the outward unit normal to $\partial\Omega$ at ξ, is a C^1 diffeomorphism from $\partial\Omega$ onto $\partial\Omega_r$ (onto $\partial\Omega$ if $r = 0$) whose volume magnification factor (that is,

the absolute value of its Jacobian determinant) $J(\cdot,r):\partial\Omega\to(0,\infty)$ *is continuous on* $\partial\Omega$ *and* C^∞ *with respect to* r.

(iii) *If* $\eta_{T(\xi,r)}$ *is the unit outward normal to* $\partial\Omega_r$ *at* $T(\xi,r)$ *then* $\eta_{T(\xi,r)}$ *and* η_ξ *are equal (but have different base points) for* $\xi\in\partial\Omega$ *and* $0\le r\le r_0$.

Proof. [of Theorem 5.8] Without loss of generality we can assume $\mathbb{R}^N\setminus\Omega$ is connected. Suppose for contradiction that $u(x)$ is a C^2 positive solution of (5.23). By decreasing r_0 if necessary, the conclusion of Lemma 5.9 holds.

Lemma 5.10 *The function*

$$g(r)=\int_{\partial\Omega_r}u(x)\,d\sigma(x),\quad 0<r\le r_0,$$

is continuously differentiable and there exists a positive constant C such that

$$\left|g'(r)-\int_{\partial\Omega_r}\frac{\partial u}{\partial\eta}\,d\sigma(x)\right|\le Cg(r)\quad \textit{for all } 0<r\le r_0,$$

where η *is the outward unit normal to* $\partial\Omega_r$.

Proof. By Lemma 5.9 we have

$$g(r)=\int_{\partial\Omega}u(\xi+r\eta_\xi)J(\xi,r)\,d\sigma(\xi)\quad \text{for all } 0<r\le r_0,$$

and thus

$$\begin{aligned}g'(r)&=\int_{\partial\Omega}\left[\frac{\partial}{\partial r}\Big(u(\xi+r\eta_\xi)\Big)\right]J(\xi,r)\,d\sigma(\xi)+\int_{\partial\Omega}u(\xi+r\eta_\xi)J_r(\xi,r)\,d\sigma(\xi)\\&=\int_{\partial\Omega_r}\frac{\partial u}{\partial\eta}(x)\,d\sigma(x)+\int_{\partial\Omega_r}u(x)\frac{J_r(\xi,r)}{J(\xi,r)}\,d\sigma(x),\end{aligned}\tag{5.24}$$

for all $0<r\le r_0$, where in the last integral $\xi=x-r\eta_\xi\in\partial\Omega$. Since, by Lemma 5.9, $J(\xi,r)$ is positive and continuous for $\xi\in\partial\Omega$ and $0\le r\le r_0$ and $J_r(\xi,r)$ is continuous there, we see that Lemma 5.10 follows from (5.24). □

We now come back to the proof of Theorem 5.8. For $0<r\le r_0$ we have

$$\begin{aligned}0&\le\int_{\Omega_{r_0}\setminus\Omega_r}-\Delta u(x)\,dx=\int_{\partial\Omega_r}\frac{\partial u}{\partial\eta}\,d\sigma(x)-\int_{\partial\Omega_{r_0}}\frac{\partial u}{\partial\eta}\,d\sigma(x)\\&\le g'(r)+Cg(r)+C\end{aligned}\tag{5.25}$$

for some positive constant C by Lemma 5.10. Hence

$$\left(e^{Cr}(g(r)+1)\right)' \geq 0 \quad \text{for all } 0 < r \leq r_0,$$

and integrating this inequality over $[r, r_0]$ we obtain

$$g(r) \leq e^{C(r_0-r)}(g(r_0)+1) - 1 \leq C_1 \quad \text{for all } 0 < r \leq r_0 \tag{5.26}$$

and for some $C_1 > 0$. Thus

$$U(r) := \frac{1}{|\partial\Omega_r|}\int_{\partial\Omega_r} u(x)\,d\sigma(x) = \frac{g(r)}{|\partial\Omega_r|}$$

is bounded for $0 < r \leq r_0$. Consequently, by the assumption (i) of f, it follows that

$$|\partial\Omega_\rho| f(U(\rho)) \geq \varepsilon > 0 \quad \text{for all } 0 < \rho \leq r_0. \tag{5.27}$$

As in the proof of Theorem 5.7, we may assume that $f : [0,\infty) \to (0,\infty)$ is of class C^1 and f' is negative and nondecreasing. From (5.23), (5.25)–(5.27) and Jensen's inequality we now obtain

$$\begin{aligned} g'(r) + C_2 &\geq \int_{\Omega_{r_0}\setminus\Omega_r} -\Delta u\,dx \\ &\geq \int_r^{r_0} \varphi(\rho)\int_{\partial\Omega_\rho} f(u(x))\,d\sigma(x)\,d\rho \\ &\geq \int_r^{r_0} \varphi(\rho)|\partial\Omega_\rho| f(U(\rho))\,d\rho \\ &\geq \varepsilon\int_r^{r_0} \varphi(\rho)\,d\rho \quad \text{for all } 0 < r \leq r_0. \end{aligned}$$

Integrating over $[r, r_0]$ in the above estimate we find

$$\begin{aligned} g(r_0) - g(r) + C_2 r_0 &\geq \varepsilon\int_r^{r_0}\int_s^{r_0} \varphi(\rho)\,d\rho\,ds \\ &= \varepsilon\int_r^{r_0} (\rho - r)\varphi(\rho)\,d\rho \to \varepsilon\int_0^{r_0} \rho\varphi(\rho)\,d\rho = \infty \quad \text{as } r \searrow 0, \end{aligned}$$

which contradicts $g > 0$ and completes the proof. □

5.4.2 Existence Results

In this section we obtain existence results for (5.1) in the nondegenerate case on K. We prove that for a large class of compact sets, condition (5.2) is sufficient for (5.1) to have C^2 positive solutions.

In the sequel, unless otherwise stated, we assume that φ and f satisfy the same hypotheses as in Lemma 5.3, that is, $\varphi \in C^{0,\gamma}(0,\infty)$ $(0<\gamma<1)$ and $f \in C^1(0,\infty)$ are positive functions such that φ is nondecreasing and f is decreasing. Our first result in this sense concerns the case where K consists of a finite number of components each of which is of class C^2. We have

Theorem 5.11 *Let K be a compact set in $\mathbb{R}^N$ $(N \geq 3)$ having a finite number of components each of which is the closure of a C^2 domain. Then (5.1) has C^2 positive solutions if and only if (5.2) holds. Furthermore, if (5.2) is fulfilled, then there exists a minimal solution $\tilde{u}$ of (5.1) such that*

$$\tilde{u} \in C^2(\mathbb{R}^N \setminus K) \cap C(\mathbb{R}^N \setminus \mathrm{int}(K))$$

and

$$\begin{cases} -\Delta \tilde{u} = \varphi(\delta_K(x)) f(\tilde{u}),\ \tilde{u} > 0 & \text{in } \mathbb{R}^N \setminus K, \\ \tilde{u} = 0 & \text{on } \partial K, \\ \tilde{u}(x) \to 0 & \text{as } |x| \to \infty. \end{cases} \tag{5.28}$$

In addition, there exist $c_1, c_2 > 0$ and $0 < r_0 < 1$ such that $\tilde{u}$ satisfies

$$c_1 \leq \frac{\tilde{u}(x)}{H(\delta_K(x))} \leq c_2 \quad \text{in } \{x \in \mathbb{R}^N \setminus K : 0 < \delta_K(x) < r_0\}, \tag{5.29}$$

where H is the unique solution of (5.10).

Proof. The necessity of (5.2) follows from Theorems 5.7 and 5.8. To prove the sufficiency of (5.2), assume first that $\mathbb{R}^N \setminus K$ is connected.

We fix $0 < \rho < R$ such that $K \subset B_\rho(0)$. By Lemma 5.4 there exists

$$u \in C^2(B_\rho(0) \setminus K) \cap C(\overline{B}_\rho(0) \setminus \mathrm{int}(K))$$

such that

$$\begin{cases} -\Delta u = \varphi(\delta_K(x)) f(u),\ u > 0 & \text{in } B_\rho(0) \setminus K, \\ u = 0 & \text{on } \partial(B_\rho(0) \setminus K). \end{cases} \tag{5.30}$$

We next construct a solution v of (5.1) in a neighborhood of infinity. To this aim, let

$$A(r) := \int_r^\infty t^{1-N} \int_R^t s^{N-1} \varphi(s-\rho)\,ds\,dt \quad \text{for all } r \geq R.$$

Since $\int_R^\infty r\varphi(r-\rho)dr < \infty$, by Lemma 5.6 we have that A is well defined for all $r \geq R$. Also, it is easy to check that

$$-\Delta A(|x|) = \varphi(|x|-\rho) \quad \text{in } \mathbb{R}^N \setminus B_R(0).$$

Since the mapping

$$[0,\infty) \ni t \longmapsto \int_0^t \frac{1}{f(s)}ds \in [0,\infty)$$

is bijective, we can define $v : \mathbb{R}^N \setminus B_R(0) \to (0,\infty)$ implicitly as the unique solution of

$$\int_0^{v(x)} \frac{1}{f(t)}dt = A(|x|) \quad \text{for all } x \in \mathbb{R}^N \setminus B_R(0). \tag{5.31}$$

Then, using the properties of A we deduce that $v \in C^2(\mathbb{R}^N \setminus B_R(0))$, $v > 0$ and $v(x) \to 0$ as $|x| \to \infty$. Further from (5.31) we obtain

$$\nabla A(|x|) = \frac{1}{f(v)}\nabla v \quad \text{in } \mathbb{R}^N \setminus B_R(0),$$

and

$$\varphi(|x|-\rho) = -\Delta A(|x|) = \frac{f'(v)}{f^2(v)}|\nabla v|^2 - \frac{1}{f(v)}\Delta v \quad \text{in } \mathbb{R}^N \setminus B_R(0).$$

Since f is decreasing, we have $f' \leq 0$ which implies

$$-\Delta v \geq \varphi(|x|-\rho)f(v) \quad \text{in } \mathbb{R}^N \setminus B_R(0).$$

Therefore, $v \in C^2(\mathbb{R}^N \setminus B_R(0))$ satisfies

$$\begin{cases} -\Delta v \geq \varphi(\delta_K(x))f(v),\ v > 0 & \text{in } \mathbb{R}^N \setminus B_R(0), \\ v(x) \to 0 & \text{as } |x| \to \infty. \end{cases} \tag{5.32}$$

Let now $0 < \rho_0 < \rho$ be such that $K \subset B_{\rho_0}(0)$ and let u, v be the solutions of (5.30) and (5.32) respectively. Consider

$$w : (B_{\rho_0}(0) \setminus \text{int}(K)) \cup (\mathbb{R}^N \setminus B_R(0)) \to [0,\infty),$$

defined by

$$w(x) = u(x) \text{ if } x \in B_{\rho_0}(0) \setminus \text{int}(K), \quad w(x) = v(x) \text{ if } x \in \mathbb{R}^N \setminus B_R(0).$$

Let W be a positive C^2 extension of w to $\mathbb{R}^N \setminus K$. We claim that there exists $M > 0$ large enough such that

$$U(x) = W(x) + M(1+|x|^2)^{(2-N)/2}, \quad x \in \mathbb{R}^N \setminus \text{int}(K) \tag{5.33}$$

satisfies (5.1). Indeed, since $(1+|x|^2)^{(2-N)/2}$ is superharmonic, this condition is already satisfied in $B_{\rho_0}(0) \setminus K$ and $\mathbb{R}^N \setminus B_R(0)$. In $B_R(0) \setminus B_{\rho_0}(0)$ we use the fact that $-\Delta(1+|x|^2)^{(2-N)/2}$ is positive and bounded away from zero. Therefore we have constructed a solution $U \in C^2(\mathbb{R}^N \setminus K) \cap C(\mathbb{R}^N \setminus \text{int}(K))$ of (5.1) that tends to zero at infinity.

Let us prove the existence of a minimal solution $\tilde{u}$ of (5.1). According to Lemma 5.4, for any $n \geq 1$ there exists a unique

$$u_n \in C^2(B_{R+n}(0) \setminus K) \cap C(\overline{B}_{R+n}(0) \setminus \text{int}(K))$$

such that

$$\begin{cases} -\Delta u_n = \varphi(\delta_K(x)) f(u_n),\ u_n > 0 & \text{in } B_{R+n}(0) \setminus K, \\ u_n = 0 & \text{on } \partial(B_{R+n}(0) \setminus K). \end{cases} \tag{5.34}$$

We extend $u_n = 0$ on $\mathbb{R}^N \setminus B_{R+n}(0)$ and by Lemma 5.1 we have that $\{u_n\}$ is a nondecreasing sequence of functions and $u_n \leq U$ in $\mathbb{R}^N \setminus K$. Let

$$\tilde{u}(x) := \lim_{n \to \infty} u_n(x) \quad \text{for all } x \in \mathbb{R}^N \setminus \text{int}(K).$$

By standard elliptic arguments, we have $\tilde{u} \in C^2(\mathbb{R}^N \setminus K)$ and

$$-\Delta \tilde{u} = \varphi(\delta_K(x)) f(\tilde{u}) \quad \text{in } \mathbb{R}^N \setminus K.$$

We next prove that $\tilde{u}$ vanishes continuously on ∂K.

Let u_1 be the unique solution of (5.34) with $n = 1$ and $\omega := \{x \in \mathbb{R}^N \setminus K : 0 < \delta_K(x) < 1\}$. Since both u_1 and $\tilde{u}$ are continuous and positive on $\partial\omega \setminus K$, we can find $M > 1$ such that $Mu_1 \geq \tilde{u}$ on $\partial\omega \setminus K$. Now, using the fact that the sequence $\{u_n\}$ is nondecreasing, this also yields $Mu_1 \geq u_n$ on $\partial\omega \setminus K$, for all $n \geq 1$. The above inequality also holds true on ∂K (since u_1 and u_n are zero there). Therefore $Mu_1 \geq u_n$ on $\partial\omega$ for all $n \geq 1$, which by the comparison result in Lemma 2.1 (note that the function Mu_1 satisfies (5.1) in ω) gives

$$Mu_1 \geq u_n \quad \text{in } \omega,$$

for all $n \geq 1$. Passing to the limit with $n \to \infty$ in the above estimate, we obtain $Mu_1 \geq \tilde{u}$ in ω and since u_1 vanishes continuously on ∂K, so does $\tilde{u}$.

The boundary behavior of $\tilde{u}$ near K follows from the fact that $u_1 \leq \tilde{u} \leq Mu_1$ in ω and u_1 satisfies (5.16). From Lemma 5.1 we obtain that any solution u of (5.1) satisfies $u \geq u_n$ in $\mathbb{R}^N \setminus K$ which implies $u \geq \tilde{u}$. Hence, $\tilde{u}$ is the minimal solution of (5.1).

Assume now that $\mathbb{R}^N \setminus K$ is not connected. We shall construct a solution to (5.1) by considering each component of $\mathbb{R}^N \setminus K$. Note that since K is compact, $\mathbb{R}^N \setminus K$ has only one unbounded component on which we proceed as above. Since φ satisfies (5.2), by Lemma 5.3, on each bounded component of $\mathbb{R}^N \setminus K$ we construct a solution of $-\Delta u = \varphi(\delta_K(x))f(u)$ that vanishes continuously on its boundary and has the behavior described by (5.9). This completes the proof of Theorem 5.11. □

Remark 19 *The approach in Theorem 5.11 can be used to study the inequality (5.1) in some cases where the compact set K consists of infinitely many components, all of them with C^2 boundary. For instance, it is easy to see that the same arguments apply for compact sets K of the form*

$$K = B_1(0) \cup \bigcup_{n \geq 1} \left\{ x \in \mathbb{R}^N : 1 + \frac{1}{2n+1} < |x| < 1 + \frac{1}{2n} \right\}$$

or

$$K = \partial B_1(0) \cup \bigcup_{n \geq 1} \partial B_{1+1/n}(0).$$

Remark 20 *The existence of a positive ground state solution in the exterior of a compact set is a particular feature of the case $N \geq 3$. Such solutions do not exist in dimension $N = 2$. Indeed, suppose that u is a C^2 positive solution of*

$$-\Delta u \geq 0 \quad \text{in } \mathbb{R}^2 \setminus K, \quad u(x) \to 0 \quad \text{as } |x| \to \infty,$$

where $K \subset \mathbb{R}^2$ is a compact set, not necessarily with smooth boundary. Choose $r_0 > 0$ such that $K \subset B_{r_0}(0)$ and let $m = \min_{|x|=r_0} u(x) > 0$. For each $r_1 > r_0$ define

$$v_{r_1} : \mathbb{R}^2 \setminus \{0\} \to \mathbb{R}, \quad v_{r_1}(x) = \frac{m(\log r_1 - \log|x|)}{\log r_1 - \log r_0}.$$

Then

$$v_{r_1} \text{ is harmonic in } \mathbb{R}^2 \setminus \{0\}, \quad v_{r_1} = m \text{ on } \partial B_{r_0}(0), \quad v_{r_1} = 0 \text{ on } \partial B_{r_1}(0).$$

Let $w_{r_1}(x) = u(x) - v_{r_1}(x)$, $x \in \mathbb{R}^N \setminus B_{r_0}(0)$. *Thus,*

$$-\Delta w_{r_1} = -\Delta u \geq 0 \quad \text{in } \overline{B}_{r_1}(0) \setminus B_{r_0}(0), \quad w_{r_1} \geq 0 \quad \text{on } \partial B_{r_1}(0) \cup \partial B_{r_0}(0).$$

By the maximum principle it follows that $w_{r_1} \geq 0$ *in* $\overline{B}_{r_1}(0) \setminus B_{r_0}(0)$, *that is* $u(x) \geq v_{r_1}(x)$ *in* $\overline{B}_{r_1}(0) \setminus B_{r_0}(0)$.

Let now $x \in \mathbb{R}^2 \setminus \overline{B}_{r_0}(0)$ *be fixed. Then, for* $r_1 > |x|$ *we have*

$$u(x) \geq v_{r_1}(x) \to m \quad \text{as } r_1 \to \infty,$$

so $u(x) \geq m$ *in* $\mathbb{R}^2 \setminus \overline{B}_{r_0}(0)$, *which contradicts* $u(x) \to 0$ *as* $|x| \to \infty$.

If $K = K_1 \cup K_2$ where K_1 has a finite number of components each of which is the closure of a C^2 domain and K_2 consists of a finite number of isolated points we have the following result.

Theorem 5.12 *Let* $K = K_1 \cup K_2 \subset \mathbb{R}^N$ $(N \geq 3)$ *where* K_1 *is a compact set having a finite number of components each of which is the closure of a* C^2 *domain and* $K_2 \subset \mathbb{R}^N$ *consists of a finite number of isolated points such that* $K_1 \cap K_2 = \emptyset$. *Then, inequality (5.1) has* C^2 *positive solutions if and only if (5.2) holds. Furthermore, if (5.2) is fulfilled, then there exists a minimal solution* $\tilde{u}$ *of (5.1) that satisfies*

$$\tilde{u} \in C^2(\mathbb{R}^N \setminus K) \cap C(\mathbb{R}^N \setminus \text{int}(K_1))$$

and

$$\begin{cases} -\Delta \tilde{u} = \varphi(\delta_K(x)) f(\tilde{u}),\ \tilde{u} > 0 & \text{in } \mathbb{R}^N \setminus K, \\ \tilde{u} = 0 & \text{on } \partial K_1, \\ \tilde{u} > 0 & \text{on } K_2, \\ \tilde{u}(x) \to 0 & \text{as } |x| \to \infty. \end{cases} \tag{5.35}$$

In addition, $\tilde{u}$ *has the same behavior around* K_1 *as described in (5.29).*

Proof. By Theorems 5.7 and 5.8, condition (5.2) is necessary in order to have C^2 positive solutions for (5.1). Assume now that (5.2) holds. Using, if necessary, a dilation argument, we can assume that $\mathrm{dist}(K_1, K_2) > 2$ and the distance between any two distinct points of K_2 is greater than 2. We fix $R > 0$ large enough such that

$$K_1 \cup \bigcup_{a \in K_2} \overline{B}_1(a) \subset B_R(0).$$

We now apply Lemma 5.5 for $L = \bigcup_{a \in K_2} \overline{B}_{1/n}(a)$ and $\Omega = B_{R+n}(a)$. Thus, there exists a unique solution u_n of

$$\begin{cases} -\Delta u_n = \varphi(\delta_K(x))f(u_n),\ u_n > 0 & \text{in } B_{R+n}(0) \setminus \Big(K_1 \cup \bigcup_{a \in K_2} \overline{B}_{1/n}(a)\Big), \\ u_n = 0 & \text{on } \partial B_{R+n}(0) \cup \partial K_1 \cup \bigcup_{a \in K_2} \partial B_{1/n}(a). \end{cases} \tag{5.36}$$

Extending $u_n = 0$ outside of $\overline{B}_{R+n}(0) \setminus \bigcup_{a \in K_2} \overline{B}_{1/n}(a)$, by Lemma 5.1 we obtain

$$0 \le u_1 \le u_2 \le \cdots \le u_n \le u_{n+1} \le \cdots \quad \text{in } \mathbb{R}^N \setminus K.$$

In order to pass to the limit in (5.36) we need to provide a barrier for $\{u_n\}$. Proceeding in the same manner as we did in the proof of Theorem 5.11 we can find a function

$$U \in C^2(\mathbb{R}^N \setminus K_1) \cap C(\mathbb{R}^N \setminus \mathrm{int}(K_1))$$

such that

$$\begin{cases} -\Delta U \ge \varphi(\delta_{K_1}(x))f(U),\ U > 0 & \text{in } \mathbb{R}^N \setminus K_1, \\ U(x) \to 0 & \text{as } |x| \to \infty. \end{cases} \tag{5.37}$$

We shall use a similar approach to construct a function $V \in C^2(\mathbb{R}^N \setminus K_2) \cap C(\mathbb{R}^N)$ such that

$$\begin{cases} -\Delta V \ge \varphi(\delta_{K_2}(x))f(V),\ V > 0 & \text{in } \mathbb{R}^N \setminus K_2, \\ V(x) \to 0 & \text{as } |x| \to \infty. \end{cases} \tag{5.38}$$

First, since (5.2) holds, by Lemma 5.6(iii), the function

$$D(r) := \int_r^\infty t^{1-N} \int_0^t s^{N-1} \varphi(s)\,ds\,dt \quad \text{for all } r \ge 0,$$

is well defined and $-\Delta D(|x|) = \varphi(|x|)$ in $\mathbb{R}^N \setminus \{0\}$. We next define $v : \mathbb{R}^N \to (0, \infty)$ by

$$\int_0^{v(x)} \frac{1}{f(t)} dt = D(|x|) \quad \text{for all } x \in \mathbb{R}^N.$$

Using the same arguments as in the proof of Theorem 5.11 we have $v \in C^2(\mathbb{R}^N \setminus \{0\}) \cap C(\mathbb{R}^N)$ and

$$\begin{cases} -\Delta v \geq \varphi(|x|) f(v),\ v > 0 & \text{in } \mathbb{R}^N \setminus \{0\}, \\ v(x) \to 0 & \text{as } |x| \to \infty. \end{cases} \tag{5.39}$$

Let now $V : \mathbb{R}^N \to (0,\infty)$ defined by

$$V(x) = \sum_{a \in K_2} v(x-a).$$

By (5.39) we have $V \in C^2(\mathbb{R}^N \setminus K_2) \cap C(\mathbb{R}^N)$, $V(x) \to 0$ as $|x| \to \infty$ and

$$\begin{aligned} -\Delta V(x) &= -\sum_{a \in K_2} \Delta v(x-a) \geq \sum_{a \in K_2} \varphi(|x-a|) f(v(x-a)) \\ &\geq \Big(\sum_{a \in K_2} \varphi(|x-a|) \Big) f(V(x)) \geq \varphi(\min_{a \in K_2} |x-a|) f(V(x)) \\ &= \varphi(\delta_{K_2}(x)) f(V(x)) \quad \text{for all } x \in \mathbb{R}^N \setminus K_2. \end{aligned}$$

Therefore, V fulfills (5.38). Now $W := U + V$ satisfies $W(x) \to 0$ as $|x| \to \infty$ and

$$\begin{aligned} -\Delta W(x) &\geq \varphi(\delta_{K_1}(x)) f(U) + \varphi(\delta_{K_2}(x)) f(V) \\ &\geq (\varphi(\delta_{K_1}(x)) + \varphi(\delta_{K_2}(x))) f(W) \\ &\geq \varphi\Big(\min\{\delta_{K_1}(x), \delta_{K_2}(x)\} \Big) f(W) \\ &= \varphi(\delta_K(x)) f(W) \quad \text{for all } x \in \mathbb{R}^N \setminus K. \end{aligned}$$

By Lemma 5.1 we obtain $u_n \leq W$ in $\mathbb{R}^N \setminus K$. Thus, passing to the limit in (5.36) and by elliptic arguments, we obtain that $\tilde{u} := \lim_{n\to\infty} u_n$ satisfies

$$-\Delta \tilde{u} = \varphi(\delta_K(x)) f(\tilde{u}) \quad \text{in } \mathbb{R}^N \setminus K.$$

The fact that $\tilde{u}$ is minimal, vanishes continuously on ∂K_1 and has the behavior near ∂K_1 as described by (5.29) follows exactly in the same way as in the proof of Theorem 5.11.

It remains to prove that $\tilde{u}$ can be continuously extended at any point of K_2 and $\tilde{u} > 0$ on K_2. To this aim, we state and prove the following auxiliary results.

Lemma 5.13 *Let $r > 0$ and $x \in \mathbb{R}^N \setminus \partial B_r(0)$, $N \geq 3$. Then*

$$\frac{1}{\sigma_N r^{N-1}} \int_{\partial B_r(0)} \frac{1}{|x-y|^{N-2}} d\sigma(y) = \begin{cases} \dfrac{1}{|x|^{N-2}} & \text{if } |x| > r, \\ \dfrac{1}{r^{N-2}} & \text{if } |x| < r. \end{cases}$$

Proof. Suppose first $|x| > r$. Then $u(y) = |y-x|^{2-N}$ is harmonic in $B_{r+\varepsilon}(0)$, for $\varepsilon > 0$ small. By the mean value theorem we have

$$\frac{1}{\sigma_N r^{N-1}} \int_{\partial B_r(0)} \frac{1}{|x-y|^{N-2}} d\sigma(y) = u(0) = \frac{1}{|x|^{N-2}}.$$

Assume now $|x| < r$. Since

$$v(x) := \frac{1}{\sigma_N r^{N-1}} \int_{\partial B_r(0)} \frac{1}{|x-y|^{N-2}} d\sigma(y)$$

is harmonic and radially symmetric, it follows that v is constant in $B_r(0)$. Thus $v(x) = v(0) = r^{2-N}$ for $x \in B_r(0)$. □

Lemma 5.14 *Let u be a C^2 positive solution of*

$$-\Delta u \geq 0 \quad \text{in } B_{2r_1}(0) \setminus \{0\}, N \geq 2.$$

Then

$$u(x) \geq m := \min_{|y|=r_1} u(y) \quad \text{for all } x \in \overline{B}_{r_1}(0) \setminus \{0\}.$$

Proof. For $0 < r_0 < r_1$ define $v_{r_0} : \mathbb{R}^N \setminus \{0\} \to \mathbb{R}$ by

$$v_{r_0}(x) = \frac{m(\Phi(r_0) - \Phi(|x|))}{\Phi(r_0) - \Phi(r_1)},$$

where

$$\Phi(r) = \begin{cases} \log \dfrac{1}{r} & \text{if } N = 2, \\ \dfrac{1}{r^{N-2}} & \text{if } N \geq 3. \end{cases}$$

Then v_{r_0} is harmonic in $\mathbb{R}^N \setminus \{0\}$ and $v_{r_0} \leq u$ on $\partial B_{r_1}(0) \cup \partial B_{r_0}(0)$. Thus, by the maximum principle, $v_{r_0} \leq u$ in $\overline{B}_{r_1}(0) \setminus B_{r_0}(0)$. Fix $x \in \overline{B}_{r_1}(0) \setminus \{0\}$. Then, for $0 < r_0 < |x|$ we have $u(x) \geq v_{r_0}(x) \to m$ as $r_0 \searrow 0$. This concludes the proof. □

Lemma 5.15 *Let $\varphi, f : (0,\infty) \to [0,\infty)$ be continuous functions such that $\int_0^1 r\varphi(r)dr < \infty$. Suppose that u is a C^2 positive bounded solution of $-\Delta u = \varphi(|x|)f(u)$ in a* *punctured neighborhood of the origin in* $\mathbb{R}^N$, $N \geq 3$. *Then, for some* $L > 0$ *we have* $u(x) \to L$ *as* $x \to 0$.

Proof. By Lemma 5.14 we can find $r_0 > 0$ small such that u is bounded away from zero in $\overline{B}_{r_0}(0) \setminus \{0\}$. Hence, for some $M > 0$ we have

$$f(u(x)) \leq M \quad \text{in } \overline{B}_{r_0}(0) \setminus \{0\}. \tag{5.40}$$

For $x \in \mathbb{R}^N$ let

$$I(x) := \frac{1}{\sigma_N} \int_{B_{r_0}(0)} \frac{\varphi(y) f(u(y))}{|x-y|^{N-2}} dy.$$

Then,

$$I(x) = \int_0^{r_0} F(x,r) dr, \quad \text{where} \quad F(x,r) = \frac{\varphi(r)}{\sigma_N} \int_{\partial B_r(0)} \frac{f(u(y))}{|x-y|^{N-2}} d\sigma(y).$$

Since, by (5.40) and Lemma 5.13 we have

(i) $F(x,r) \leq Mr\varphi(r)$ for $x \in \mathbb{R}^N$ and $0 < r < r_0$.
(ii) $\int_0^{r_0} r\varphi(r) dr < \infty$.
(iii) $F(x,r) \to F(0,r)$ as $x \to 0$ pointwise for $0 < r < r_0$,

it follows that I is bounded in $\mathbb{R}^N$ and by the dominated convergence theorem,

$$I(x) \to I(0) \quad \text{as } x \to 0. \tag{5.41}$$

Since $v := u - \frac{1}{N-2} I$ is harmonic and bounded in $B_{r_0}(0) \setminus \{0\}$, it is well known that $\lim_{x \to 0} v(x)$ exists. Thus, by (5.41), $\lim_{x \to 0} u(x)$ exists and is finite. □

Now, the fact that the minimal solution $\tilde{u}$ can be continuously extended on K_2 and $\tilde{u} > 0$ on K_2 follows by applying Lemma 5.15 for each point of K_2. This finishes the proof of Theorem 5.12. □

5.5 The Degenerate Case

In this section we study the inequality (5.1) in the case where $K \subset \mathbb{R}^N$ $(N \geq 3)$ reduces to a finite number of points. In this setting, the existence of a C^2 positive solution to (5.1) depends on both φ and f. To better emphasize this dependence we shall assume that $f(t) = t^{-p}$, $p > 0$. Therefore, we shall be concerned with the semilinear elliptic inequality

$$-\Delta u \geq \varphi(\delta_K(x))u^{-p} \quad \text{in } \mathbb{R}^N \setminus K. \tag{5.42}$$

Theorem 5.16 *Let $K \subset \mathbb{R}^N$ $(N \geq 3)$ be a finite set of points and $\varphi \in C^{0,\gamma}(0,\infty)$ $(0 < \gamma < 1)$ be a positive function such that $\varphi(r)$ is monotone for large values of r. Then, (5.42) has C^2 positive solutions if and only if*

$$\int_1^\infty r\varphi(r)dr < \infty \quad \text{and} \quad \int_0^1 r^{(1+p)(N-2)+1}\varphi(r)dr < \infty. \tag{5.43}$$

Furthermore, if (5.43) holds, then, there exists a minimal solution $\tilde{u}$ of (5.42) that satisfies

$$\begin{cases} -\Delta\tilde{u} = \varphi(\delta_K(x))\tilde{u}^{-p},\ \tilde{u} > 0 & \text{in } \mathbb{R}^N \setminus K, \\ \tilde{u}(x) \to 0 & \text{as } |x| \to \infty. \end{cases} \tag{5.44}$$

In addition, $\tilde{u}$ can be extended to a continuous positive function on the whole of $\mathbb{R}^N$ if and only if $\int_0^1 r\varphi(r)dr < \infty$.

Note that condition (5.43) above is weaker than (5.2) which is the optimal condition on φ in case when some components of K are the closure of C^2 domains.

Proof. Assume first that (5.42) has a C^2 positive solution u. From Theorem 5.7 it follows that $\int_1^\infty r\varphi(r)dr < \infty$. By translation one may assume that $0 \in K$ and fix $\rho > 0$ such that $\delta_K(x) = |x|$ in $B_\rho(0)$. Let now u_* be the image of u through the Kelvin transform, that is,

$$u_*(x) = |x|^{2-N}u\left(\frac{x}{|x|^2}\right), \quad x \in \mathbb{R}^N \setminus \{0\}. \tag{5.45}$$

Then u_* satisfies

$$\begin{aligned} -\Delta u_* &\geq |x|^{-2-N}\varphi\left(\frac{1}{|x|}\right)u^{-p}\left(\frac{x}{|x|^2}\right) \\ &= |x|^{-2-N-p(N-2)}\varphi\left(\frac{1}{|x|}\right)u_*^{-p}(x) \quad \text{in } \mathbb{R}^N \setminus B_{1/\rho}(0). \end{aligned}$$

With the same proof as in Theorem 5.7 (note that here we do not need $\varphi(r)$ to be monotone for small values of $r > 0$) we deduce

$$\int_1^\infty t^{-1-N-p(N-2)}\varphi\left(\frac{1}{t}\right)dt < \infty.$$

Now with the change of variable $r = t^{-1}$, $0 < r \le 1$ we derive the second condition in (5.43).

Conversely, assume now that (5.43) holds and let us construct a solution to (5.42). We first assume that $K = \{0\}$. The construction follows the general lines given in the proof of Theorem 5.11, the only difference is that one cannot use Lemmas 5.4 or 5.5 since K is degenerate. Instead, we shall use the Kelvin transform to reduce the construction of a solution to (5.42) near the origin to a solution of a similar inequality that holds in a neighborhood of infinity.

First, let

$$D(r) := \int_r^\infty t^{1-N} \int_1^t s^{N-1} \varphi(s) ds dt \quad \text{for all } r \ge 1.$$

Since (5.43) holds, by Lemma 5.6(ii) D is well defined and converges to zero at infinity. We now consider

$$u(x) := \Big[(p+1)D(|x|)\Big]^{1/(p+1)} \quad x \in \mathbb{R}^N \setminus \overline{B}_1(0).$$

Then $u \in C^2(\mathbb{R}^N \setminus \overline{B}_1(0))$ and in the same manner as in the proof of Theorem 5.11 we have

$$\begin{cases} -\Delta u \ge \varphi(|x|)u^{-p}(x),\ u > 0 & \text{in } \mathbb{R}^N \setminus \overline{B}_1(0),\\ u(x) \to 0 & \text{as } |x| \to \infty. \end{cases}$$

Proceeding similarly, with $r^{-2-N-p(N-2)}\varphi(1/r)$ instead of $\varphi(r)$ and then using the Kelvin transform, we obtain a function $v \in C^2(B_1(0) \setminus \{0\})$ such that

$$\begin{cases} -\Delta v \ge \varphi(|x|)v^{-p}(x),\ v > 0 & \text{in } \overline{B}_1(0) \setminus \{0\},\\ |x|^{N-2}v(x) \to 0 & \text{as } |x| \to 0. \end{cases}$$

From now on we proceed exactly as in the proof of Theorem 5.11. Let w be any C^2 extension of

$$x \longmapsto \begin{cases} u(x) & \text{if } x \in \mathbb{R}^N \setminus \overline{B}_1(0),\\ v(x) & \text{if } x \in B_{1/2}(0) \setminus \{0\}, \end{cases}$$

to the whole $\mathbb{R}^N \setminus \{0\}$. Now, one can find $M > 0$ large enough such that

$$U(x) := w(x) + M(1 + |x|^2)^{(2-N)/2}, \quad x \in \mathbb{R}^N \setminus \{0\}$$

fulfills

$$\begin{cases} -\Delta U \geq \varphi(|x|)U^{-p}(x),\ U > 0 & \text{in } \mathbb{R}^N \setminus \{0\}, \\ U(x) \to 0 & \text{as } |x| \to \infty, \\ |x|^{N-2}U(x) \to 0 & \text{as } |x| \to 0. \end{cases} \tag{5.46}$$

In particular U is a solution of (5.42) with $K = \{0\}$. In the general case, if K is a finite set of points, we consider $V(x) = \sum_{a \in K} U(x-a)$ for all $x \in \mathbb{R}^N \setminus K$. As in the proof of Theorem 5.12 we deduce that V satisfies (5.42).

Assume that condition (5.43) is satisfied. Then, the existence of the minimal solution $\tilde{u}$ of (5.42) is obtained with the same proof as in Theorem 5.12. Note that $\tilde{u}$ is obtained as a pointwise limit of the sequence $\{u_n\}$ where u_n satisfies (5.36) in which $K_1 = \emptyset$ and $K_2 = K$. It remains to prove that $\tilde{u}$ can be continuously extended to a positive continuous function in $\mathbb{R}^N$ if and only if $\int_0^1 r\varphi(r)dr < \infty$.

Assume first that the minimal solution $\tilde{u}$ of (5.42) is bounded. Using a translation argument, one can also assume that $0 \in K$. Fix $r_0 > 0$ such that $\delta_K(x) = |x|$ for all $x \in \overline{B}_{r_0}(0)$. Then averaging (5.42) we obtain

$$-(r^{N-1}\bar{u}'(r))' \geq cr^{N-1}\varphi(r) \quad \text{for all } 0 < r \leq r_0, \tag{5.47}$$

where $c > 0$. Hence $r^{N-1}\bar{u}'(r)$ is decreasing and its limit as $r \searrow 0$ must be zero for otherwise $\bar{u}$—and hence u—would be unbounded for small $r > 0$. Thus integrating (5.47) twice we obtain

$$\infty > \left(\limsup_{r \searrow 0} \bar{u}(r)\right) - \bar{u}(r_0) \geq c\int_0^{r_0} t^{1-N}\int_0^t s^{N-1}\varphi(s)dsdt,$$

which by Lemma 5.6(ii) yields $\int_0^1 r\varphi(r)dr < \infty$.

Assume now that $\int_0^1 r\varphi(r)dr < \infty$. The conclusion will follow by Lemma 5.15 once we prove that $\tilde{u}$ is bounded around each point of K. Again by translation and a scaling argument we may assume that $0 \in K$ and $\delta_K(x) = |x|$ for all $x \in B_1(0)$. Set

$$v(x) := M\int_{|x|}^2 t^{1-N}\int_0^t s^{N-1}\varphi(s)dsdt, \quad \text{for all } x \in B_2(0).$$

By Lemma 5.6(i), v is bounded and positive in $B_2(0)$ and

$$-\Delta v(x) = M\varphi(|x|) = M\varphi(\delta_K(x)) \quad \text{in } B_1(0) \setminus \{0\}. \tag{5.48}$$

Therefore, we can take $M > 1$ large enough such that

$$-\Delta v(x) \geq \varphi(\delta_K(x))v^{-p}(x) \quad \text{in } B_1(0)\setminus\{0\} \quad \text{and} \quad v \geq \tilde{u} \quad \text{on } \partial B_1(0). \quad (5.49)$$

Let u_n be the solution of (5.36) with $K_1 = \emptyset$ and $K_2 = K$. Recall that $\{u_n\}$ converges pointwise to $\tilde{u}$. Since $\tilde{u} \geq u_n$ in $\mathbb{R}^N \setminus K$, from (5.49) we have $v \geq u_n$ on $\partial B_1(0)$. According to the definition of u_n, this inequality also holds true on $\partial B_{1/n}(0)$. Now, by (5.49) and Lemma 5.1 it follows that $v \geq u_n$ in $B_1(0)\setminus B_{1/n}(0)$. Passing to the limit with $n \to \infty$ it follows that $v \geq \tilde{u}$ in $B_1(0)\setminus\{0\}$ and so, $\tilde{u}$ is bounded around zero. Proceeding similarly we derive that $\tilde{u}$ is bounded around every point of K. By Lemma 5.15 we now obtain that $\tilde{u}$ can be continuously extended on K. This finishes the proof of Theorem 5.16. □

In the remaining part of this section we shall be concerned with (5.3). The first result establishes the structure of the solution set of (5.3) when φ has a power-type growth near zero and near infinity. More precisely, we shall assume that φ satisfies

$$c_1 r^{\alpha} \leq \varphi(r) \leq c_2 r^{\alpha} \quad \text{for all } 0 < r < 1, \quad (5.50)$$

and

$$c_1 r^{\beta} \leq \varphi(r) \leq c_2 r^{\beta} \quad \text{for all } r > 1, \quad (5.51)$$

for some $c_1, c_2 > 0$. Our main result is the following.

Theorem 5.17 *Assume that $\varphi \in C^{0,\gamma}(0,\infty)$ $(0 < \gamma < 1)$ satisfies (5.50)–(5.51). Then, (5.3) has C^2 positive solutions if and only if*

$$N + \alpha + p(N-2) > 0 \quad \textit{and} \quad \beta < -2. \quad (5.52)$$

Furthermore, if (5.52) holds, then:

(i) For any $a, b \geq 0$ there exists a radially symmetric solution $u_{a,b}$ of (5.3) such that

$$\lim_{|x|\to 0} |x|^{N-2} u_{a,b}(x) = a \quad \textit{and} \quad \lim_{|x|\to\infty} u_{a,b}(x) = b. \quad (5.53)$$

(ii) The solution set of (5.3) consists only of $\{u_{a,b} : a, b \geq 0\}$. In particular, any C^2 positive solution of (5.3) is radially symmetric.

Proof. Condition (5.52) follows directly from Theorem 5.16. For the proof of (i), (ii) we divide our arguments into four steps.

Step 1: There exists a minimal solution $\xi : \mathbb{R}^N \setminus \{0\} \to (0,\infty)$ which in addition satisfies

$$\lim_{|x|\to 0} |x|^{N-2}\xi(x) = 0 \quad \text{and} \quad \lim_{|x|\to\infty} \xi(x) = 0. \tag{5.54}$$

Indeed, by Lemma 5.2 there exists a unique function ξ_n such that

$$\begin{cases} -\Delta\xi_n = \varphi(|x|)\xi_n^{-p},\ \xi_n > 0 & \text{in } B_n(0)\setminus\overline{B}_{1/n}(0), \\ \xi_n = 0 & \text{on } \partial B_n(0)\cup\partial B_{1/n}(0). \end{cases} \tag{5.55}$$

By uniqueness, it also follows that ξ_n is radially symmetric. We next extend $\xi_n = 0$ outside $B_n(0)\setminus\overline{B}_{1/n}(0)$. Now, by Lemma 5.1 we have that $\{\xi_n\}$ is nondecreasing. Since (5.52) holds, proceeding as in the proof of Theorem 5.16 we construct a function $U : \mathbb{R}^N\setminus\{0\} \to (0,\infty)$ that satisfies (5.46). By Lemma 5.1 it follows that $\xi_n \le U$ in $\mathbb{R}^N\setminus\{0\}$. Hence, there exists $\xi(x) := \lim_{n\to\infty}\xi_n(x)$, $x\in\mathbb{R}^N\setminus\{0\}$ and $\xi \le U$. Also ξ is radially symmetric and by standard elliptic arguments it follows that ξ is a solution of (5.3). From $\xi \le U$ it follows that ξ satisfies (5.54). Finally, if v is an arbitrary solution of (5.3), by Lemma 5.1 we deduce

$$\xi_n \le v \quad \text{in } B_n(0)\setminus\overline{B}_{1/n}(0).$$

Passing to the limit in the above inequality with $n\to\infty$, we obtain $\xi \le v$ in $\mathbb{R}^N\setminus\{0\}$. Therefore ξ is the minimal solution of (5.3).

Step 2: Proof of (i).

Fix $a,b \ge 0$. We shall construct a radially symmetric solution of (5.3) that satisfies (5.53) with the aid of the minimal solution ξ constructed at Step 1. By virtue of Lemma 5.2, for any $n\ge 2$ there exists a unique function

$$u_n \in C^2(B_n(0)\setminus\overline{B}_{1/n}(0))\cap C(\overline{B}_n(0)\setminus B_{1/n}(0))$$

such that

$$\begin{cases} -\Delta u_n = |x|^{\alpha}u_n^{-p},\ u_n > 0 & \text{in } B_n(0)\setminus\overline{B}_{1/n}(0), \\ u_n = a|x|^{2-N} + b + \xi(x) & \text{on } \partial B_n(0)\cup\partial B_{1/n}(0). \end{cases} \tag{5.56}$$

Since ξ is radially symmetric, so is u_n. Furthermore, $a|x|^{2-N}+b$ is a subsolution while $a|x|^{2-N}+b+\xi(x)$ is a supersolution of (5.56). Thus, in view of Lemma 5.1, we obtain

$$a|x|^{2-N}+b\le u_n(x)\le a|x|^{2-N}+b+\xi(x) \quad \text{in } B_n(0)\setminus B_{1/n}(0). \tag{5.57}$$

As usual we extend $u_n=0$ outside $B_n(0)\setminus\overline{B}_{1/n}(0)$. By standard elliptic regularity and a diagonal process, up to a subsequence there exists

$$u_{a,b}(x):=\lim_{n\to\infty}u_n(x), \quad x\in\mathbb{R}^N\setminus\{0\}$$

and $u_{a,b}$ is a solution of problem (5.3). Furthermore, from (5.57) we deduce that $u_{a,b}$ satisfies

$$a|x|^{2-N}+b\le u_{a,b}(x)\le a|x|^{2-N}+b+\xi(x) \quad \text{in } \mathbb{R}^N\setminus\{0\}. \tag{5.58}$$

Now, (5.54) and (5.58) imply (5.53).

Step 3: There exists $c>0$ depending on N,α,β,p such that any solution u of (5.3) satisfies

$$u(x)\ge c|x|^{(2+\alpha)/(1+p)} \quad \text{for all } x\in B_1(0)\setminus\{0\}, \tag{5.59}$$

$$u(x)\ge c|x|^{(2+\beta)/(1+p)} \quad \text{for all } x\in\mathbb{R}^N\setminus\overline{B}_1(0). \tag{5.60}$$

The proof uses an idea that goes back to Véron [199, Theorem 3.11] (see also [121, Section 1]). Let $R>1/2$ be fixed. By Harnack's inequality (see, e.g., [99, Theorem 8.18]) and a scaling argument, there exists a constant $C_0>0$ depending only on N such that

$$\int_{B_{3R}(0)\setminus B_{2R}(0)}u\,dx\le C_0R^N\inf_{B_{3R}(0)\setminus B_{2R}(0)}u. \tag{5.61}$$

Let $\psi\in C_0^\infty(\mathbb{R}^N)$ be such that $\operatorname{supp}\psi\subseteq B_{4R}(0)\setminus\overline{B}_R(0)$ and

$$0\le\psi\le1 \text{ in } \mathbb{R}^N, \quad \psi=1 \text{ on } B_{3R}(0)\setminus B_{2R}(0), \quad \text{and} \quad \|\nabla\psi\|_\infty\le 1/R. \tag{5.62}$$

Multiplying in (5.3) with ψ^2/u and then integrating we obtain

$$\int\nabla\left(\frac{\psi^2}{u}\right)\cdot\nabla u\,dx\ge\int|x|^\beta\psi^2u^{-1-p}dx. \tag{5.63}$$

In order to estimate the left-hand side in (5.63) we use (5.62). We have

$$
\begin{aligned}
\int \nabla\left(\frac{\psi^2}{u}\right)\cdot \nabla u\,dx &= 2\int \nabla\psi\cdot\nabla u\frac{\psi}{u}dx - \int |\nabla u|^2\left(\frac{\psi}{u}\right)^2 dx \\
&\le \int \left(|\nabla\psi|^2 + |\nabla u|^2\left(\frac{\psi}{u}\right)^2\right)dx - \int |\nabla u|^2\left(\frac{\psi}{u}\right)^2 dx \qquad (5.64)\\
&\le \int |\nabla\psi|^2 dx \le CR^{N-2}.
\end{aligned}
$$

To estimate the right-hand side in (5.63) we use Jensen's inequality together with (5.61). We have

$$
\begin{aligned}
\int |x|^\beta \psi^2 u^{-1-p}dx &\ge CR^\beta \int_{B_{3R}(0)\setminus B_{2R}(0)} u^{-1-p}dx \\
&\ge CR^{\beta+N(2+p)}\left(\int_{B_{3R}(0)\setminus B_{2R}(0)} u\,dx\right)^{-1-p} \qquad (5.65)\\
&\ge CR^{N+\beta}\left(\inf_{B_{3R}(0)\setminus B_{2R}(0)} u\right)^{-1-p}.
\end{aligned}
$$

Thus, combining (5.63), (5.64) and (5.65) we arrive at

$$
\inf_{B_{3R}(0)\setminus B_{2R}(0)} u \ge CR^{(2+\beta)/(1+p)},
$$

for some constant $C>0$ independent of u and R. This proves (5.60). For the proof of (5.59) we proceed in a similar way.

Step 4: Proof of (ii).

Let u be an arbitrary solution of (5.3). From (5.50) and (5.59) it follows that

$$
\varphi(|x|)u^{-p} \le c_0|x|^\alpha u^{-p} \le c|x|^{(\alpha-2p)/(1+p)} \quad \text{in } B_1(0)\setminus\{0\}, \qquad (5.66)
$$

for some $c_0, c>0$. Thus, by (5.52) we have

$$
-\Delta u = \varphi(|x|)u^{-p} \in L^1_{\mathrm{loc}}(\mathbb{R}^N). \qquad (5.67)
$$

Lemma 5.18 *There exists a real number $a\ge 0$ such that*

$$
\Delta u + \varphi(|x|)u^{-p} + a\delta(0) = 0 \quad \text{in } \mathscr{D}'(B_1(0)), \qquad (5.68)
$$

where $\delta(0)$ denotes the Dirac mass concentrated at zero.

Proof. Let $g(x)=\varphi(|x|)u^{-p}$, $x\in\bar{B}_1(0)\setminus\{0\}$ and denote by $\bar{u}$ and $\bar{g}$ the spherical averages of u and g respectively over $\partial B_r(0)$, $0<r\le 1$. From $-\Delta u = g(x)$ in

$\bar{B}_1(0)\setminus\{0\}$ we obtain $-\Delta\bar{u}(r)=\bar{g}(r)$, $0<r\le 1$, so that

$$-\left(r^{N-1}\bar{u}'(r)\right)'=r^{N-1}\bar{g}(r) \quad \text{for all } 0<r\le 1.$$

Integrating the above equation over $[r,1]$, $0<r\le 1$ we find

$$r^{N-1}\bar{u}'(r)=A+o(r) \quad \text{as } r\to 0, \tag{5.69}$$

where

$$A=\bar{u}'(1)+\int_0^1 t^{N-1}\bar{g}(t)dt.$$

We claim that $A\le 0$. Indeed, if $A>0$, then, from (5.69) there exists $r_0>0$ small such that

$$r^{n-1}\bar{u}'(r)>c>0 \quad \text{for all } 0<r\le r_0.$$

Integrating the above inequality over $[r,r_0]$ we obtain

$$\bar{u}(r_0)-\bar{u}(r)>\frac{C}{N-2}(r^{2-N}-r_0^{2-N}) \quad 0<r\le r_0,$$

which implies $\lim_{r\to 0}\bar{u}(r)=-\infty$. This is clearly a contradiction since $\bar{u}>0$. Therefore $A\le 0$.

Let now $\psi\in C_0^\infty(B_1(0))$. For $\varepsilon>0$ small enough we have

$$\begin{aligned}\int\limits_{B_1(0)\setminus B_\varepsilon(0)} u\Delta\psi dx &= \int\limits_{B_1(0)\setminus B_\varepsilon(0)} \psi\Delta u dx+\int\limits_{\partial B_\varepsilon(0)}\psi\frac{\partial u}{\partial n}d\sigma(x)-\int\limits_{\partial B_\varepsilon(0)} u\frac{\partial\psi}{\partial n}d\sigma(x)\\ &=-\int\limits_{B_1(0)} g(x)\psi(x)dx+\psi(0)\int\partial B_\varepsilon(0)\frac{\partial u}{\partial n}d\sigma(x)+o(\varepsilon),\end{aligned} \tag{5.70}$$

as $\varepsilon\to 0$, where n is the outer unit normal vector at $\partial B_\varepsilon(0)$. By (5.69) we have

$$\int\partial B_\varepsilon(0)\frac{\partial u}{\partial n}d\sigma(x)=\sigma_N\varepsilon^{n-1}\bar{u}'(\varepsilon)=\sigma_N A+o(\varepsilon) \quad \text{as } \varepsilon\to 0.$$

Combining this relation with (5.70) we find

$$\int\limits_{B_1(0)} u\Delta\psi dx=\lim_{\varepsilon ri0}\int\limits_{B_1(0)\setminus B_\varepsilon(0)} u\Delta\psi dx=-\int\limits_{B_1(0)} g(x)\psi(x)dx+\sigma_N A,$$

which proves (5.68) with $a=-\sigma_N A\ge 0$. □

Using Lemma 5.18 and standard potential theory arguments (see [134], [194]) it follows that

$$u(x) = a|x|^{2-N} + C\int_{B_1(0)} \frac{-\Delta u(y)}{|x-y|^{N-2}}dy + h(x) \quad \text{for all } x \in B_1(0)\setminus\{0\}, \tag{5.71}$$

where C is a positive constant depending on the dimension $N \geq 3$ and $h : B_1(0) \to \mathbb{R}$ is a harmonic function.

Lemma 5.19 $\lim_{|x|\to 0} |x|^{N-2}u(x) = a.$

Proof. In view of (5.71), we only need to show that

$$\int_{B_1(0)} \frac{-\Delta u(y)}{|x-y|^{N-2}}dy = o(|x|^{2-N}) \quad \text{as } |x| \to 0. \tag{5.72}$$

To this aim, we split the above integral into

$$\int_{B_1(0)} \frac{-\Delta u(y)}{|x-y|^{N-2}}dy = \int_{\substack{y\in B_1(0),\\ |y-x|\geq |x|/2}} \frac{-\Delta u(y)}{|x-y|^{N-2}}dy + \int_{\substack{y\in B_1(0),\\ |y-x|<|x|/2}} \frac{-\Delta u(y)}{|x-y|^{N-2}}dy. \tag{5.73}$$

Using (5.67), for $|x| > 0$ small we have

$$\begin{aligned}
\int_{\substack{y\in B_1(0),\\ |y-x|\geq |x|/2}} \frac{-\Delta u(y)}{|x-y|^{N-2}}dy &= \int_{\substack{y\in B_1(0),\\ |x|/2\leq|y-x|<\sqrt{|x|}}} \frac{-\Delta u(y)}{|x-y|^{N-2}}dy + \int_{\substack{y\in B_1(0),\\ |y-x|>\sqrt{|x|}}} \frac{-\Delta u(y)}{|x-y|^{N-2}}dy \\
&\leq \left(\frac{2}{|x|}\right)^{N-2} \int_{\substack{y\in B_1(0),\\ |y-x|<\sqrt{|x|}}} -\Delta u(y)dy + |x|^{(2-N)/2} \int_{\substack{y\in B_1(0),\\ |y-x|>\sqrt{|x|}}} -\Delta u(y)dy \\
&= o(|x|^{2-n}) \quad \text{as } |x| \to 0.
\end{aligned}$$

In order to evaluate the second integral on the right-hand side in (5.67) we first note that by (5.50) and (5.66) we have

$$-\Delta u = \varphi(|x|)u^{-p} = O(|x|^{-N}) \quad \text{as } |x| \to 0.$$

Thus, there exists $c > 0$ such that

$$-\Delta u \leq c|x|^{-N} \quad \text{for } |x| \text{ small and } |y-x| < \frac{|x|}{2}.$$

Let

$$r(x) = \left[\frac{1}{c\sigma_N |x|} \int_{|y-x|<|x|/2} -\Delta u(y)dy \right]^{1/N}.$$

Then $r(x) = o(|x|)$ as $|x| \to 0$. Since

$$\int_{|y-x|<r(x)} C|x|^{-N} dy = \int_{|y-x|<|x|/2} -\Delta u(y)dy,$$

it follows that $r(x) \geq |x|/2$ for small values of $|x|$. Therefore

$$\begin{aligned}
\int_{\substack{y\in B_1(0),\\ |y-x|<|x|/2}} \frac{-\Delta u(y)}{|x-y|^{N-2}} dy &\leq \int_{\substack{y\in B_1(0),\\ |y-x|<r(x)}} c|x|^{-N} \frac{dy}{|x-y|^{N-2}} \\
&= c|x|^{-N} \int_{|z|<r(x)} |z|^{2-N} dx \\
&= c'|x|^{-N} r(x)^2 = o(|x|^{2-N}) \quad \text{as } |x| \to 0.
\end{aligned}$$

This concludes the proof of (5.72).

By (5.71) we now deduce $\lim_{|x|\to 0} |x|^{N-2} u(x) = a$. □

Let u_* be the Kelvin transform of u given by (5.45). Then u_* satisfies

$$-\Delta u_* = |x|^{-2-N-p(N-2)} \varphi\left(\frac{1}{|x|}\right) u_*^{-p}(x) \quad \text{in } \mathbb{R}^N \setminus \{0\}.$$

Note that by (5.45), (5.51) and (5.59) we have

$$|x|^{-2-N-p(N-2)} \varphi\left(\frac{1}{|x|}\right) \leq c_2 |x|^{-\beta-2-N-p(N-2)} \quad \text{for all } x \in B_1(0) \setminus \{0\},$$

$$u_*(x) \geq c|x|^{2-N-(2+\beta)/(1+p)} \quad \text{for all } x \in B_1(0) \setminus \{0\}.$$

Combining the last estimates we arrive at

$$-\Delta u_*(x) = O(|x|^{-N}) \quad \text{as } |x| \to 0. \tag{5.74}$$

With the same arguments as above, there exists

$$b := \lim_{|x|\to 0} |x|^{N-2} u_*(x) \geq 0. \tag{5.75}$$

This yields

$$\lim_{|x|\to\infty} u(x) = b \geq 0. \tag{5.76}$$

Let $u_{a,b}$ be the solution of (5.3) that satisfies (5.53). We claim that $u \equiv u_{a,b}$. To this aim, for $\varepsilon > 0$ define

$$u_\varepsilon(x) := u(x) + \varepsilon(a|x|^{2-N} + b), \quad x \in \mathbb{R}^N \setminus \{0\}.$$

Also let u_n be the unique solution of (5.56). Taking into account Lemma 5.19, (5.76) and the definition of u_n in (5.56), we can find $n_0 = n_0(\varepsilon) \geq 2$ such that for all $n \geq n_0$ we have

$$u_\varepsilon \geq u_n \quad \text{in } \partial B_n(0) \cup \partial B_{1/n}(0),$$

for all $n \geq n_0$. We also have

$$\Delta u_\varepsilon + |x|^\alpha u_\varepsilon^{-p} \leq 0 \leq \Delta u_n + |x|^\alpha u_n^{-p} \quad \text{in } B_n(0) \setminus B_{1/n}(0).$$

Hence, by Lemma 5.1 we obtain

$$u_\varepsilon \geq u_n \quad \text{in } B_n(0) \setminus B_{1/n}(0),$$

for all $n \geq n_0$. Passing to the limit with $n \to \infty$ we obtain

$$u_\varepsilon \geq u_{a,b} \quad \text{in } \mathbb{R}^N \setminus \{0\}.$$

Now, letting $\varepsilon \searrow 0$ we deduce

$$u \geq u_{a,b} \quad \text{in } \mathbb{R}^N \setminus \{0\}. \tag{5.77}$$

Next, consider the spherical average $\bar{u}$ of u as defined in (5.22). Then

$$\lim_{r\searrow 0} r^{N-2}\bar{u}(r) = \lim_{r\searrow 0} r^{N-2} u_{a,b}(r) = a, \tag{5.78}$$

$$\lim_{r\to\infty} \bar{u}(r) = \lim_{r\to\infty} u_{a,b}(r) = b. \tag{5.79}$$

A straightforward computation shows that

$$\begin{aligned} r^{N-1}\bar{u}'(r) - \varepsilon^{N-1}\bar{u}'(\varepsilon) &= \frac{1}{\sigma_N}\int_{B_r(0)\setminus B_\varepsilon(0)} \Delta u(x)dx \\ &= -\frac{1}{\sigma_N}\int_{B_r(0)\setminus B_\varepsilon(0)} \varphi(|x|)u^{-p}(x)dx < 0, \end{aligned} \tag{5.80}$$

for all $0 < \varepsilon < r$. From (5.80) it follows that

$$(0,\infty) \ni r \longmapsto r^{N-1}\bar{u}'(r) \quad \text{is decreasing,}$$

so that there exists $\ell := \lim_{r\searrow 0} r^{N-1}\bar{u}'(r)$.

By l'Hospital's rule and (5.78) it follows that

$$\lim_{r\searrow 0} r^{N-1}\bar{u}'(r) = \ell = a(2-N).$$

On the other hand, from (5.67) we have $\varphi(|x|)u^{-p} \in L^1(B_r(0))$ for all $r > 0$. Hence, passing to the limit in (5.80) with $\varepsilon \searrow 0$ we obtain

$$r^{N-1}\bar{u}'(r) = a(2-N) - \frac{1}{\sigma_N}\int_{B_r(0)} \varphi(|x|)u^{-p}(x)dx \quad \text{for all } r > 0.$$

A similar relation holds for $u_{a,b}$ and using the fact that $u \geq u_{a,b}$ it follows that

$$\bar{u}'(r) \geq u'_{a,b}(r) \quad \text{for all } r > 0.$$

This means that $\bar{u} - u_{a,b}$ is increasing in $(0,\infty)$ and so, by (5.79),

$$\bar{u}(r) - u_{a,b}(r) \leq \lim_{t\to\infty}(\bar{u}(t) - u_{a,b}(t)) = 0 \quad \text{for all } r > 0.$$

Hence $\bar{u} \leq u_{a,b}$ in $(0,\infty)$ which by continuity and (5.77) implies $\bar{u} \equiv u \equiv u_{a,b}$. This completes the proof of Theorem 5.17. □

Remark 21 *If $N = 2$ then (5.3) has no C^2 positive solutions even for more general nonlinearities than u^{-p}. More precisely, if $u \in C^2(\mathbb{R}^2 \setminus \{0\})$ satisfies*

$$-\Delta u \geq 0, \;\; u \geq 0 \quad \text{in } \mathbb{R}^2 \setminus \{0\},$$

then u is constant (see [162, Theorem 29, page 130]).

Corollary 5.20 *Let $\alpha \in \mathbb{R}$, $p > 0$. Then, the equation*

$$-\Delta u = |x|^{\alpha}u^{-p} \quad \text{in } \mathbb{R}^N \setminus \{0\}, N \geq 3, \tag{5.81}$$

has solutions if and only if

$$N + \alpha + p(N-2) > 0 \quad \text{and} \quad \alpha < -2. \tag{5.82}$$

Furthermore, if (5.82) is fulfilled, then the same conclusion as in Theorem 5.17 holds true and the function

$$\xi(x) := \left[\frac{-(1+p)^2}{(\alpha+2)(p(N-2)+N-\alpha)}\right]^{1/(1+p)} |x|^{(2+\alpha)/(1+p)}, \quad x \in \mathbb{R}^N \setminus \{0\}, \tag{5.83}$$

is the minimal solution of (5.81).

Proof. The conclusion follows directly from Theorem 5.17. Also, if (5.82) holds, then the solution set of (5.81) consists of a two-parameter family of functions $\{u_{a,b} : a,b \geq 0\}$ that satisfy (5.53). It is easy to see that the function ξ defined by (5.83) satisfies (5.81) and (5.54). It follows that ξ is the minimal solution of (5.81). □

Using the approach in the proof of Theorem 5.17, one can obtain the same structure of the solution set for (5.3) for a large class of functions $\varphi(|x|)$ having not only a power-type behavior at zero or at infinity.

Corollary 5.21 *Let $\alpha \in \mathbb{R}$, $\beta, p > 0$ and $\varphi(r) = r^\alpha \log^\beta(1+r)$. Then, (5.3) has solutions if and only if*

$$N+\alpha+\beta+p(N-2) > 0 \quad \textit{and} \quad \alpha < -2. \tag{5.84}$$

Furthermore, if (5.84) holds, then the solution set of (5.3) consists of a two-parameter family of radially symmetric functions as described in Theorem 5.17.

Proof. Condition (5.84) follows directly from Theorem 5.16. For the remaining part, let ε be small enough such that $0 < \varepsilon < -\alpha-2$ and let $\phi : (0,\infty) \to (0,\infty)$ be a continuous function that satisfies

$$\phi(r) \geq \varphi(r) \quad \text{for all } r > 0,$$

$$\phi(r) \sim r^{\alpha+\beta} \quad \text{as } r \searrow 0 \quad \text{and} \quad \phi(r) \sim r^{\alpha+\varepsilon} \quad \text{as } r \to \infty.$$

The construction of the minimal solution ξ of (5.3) is obtained by considering the sequence $\{\xi_n\}$ where ξ_n satisfies (5.55). Since ϕ satisfies the condition (5.43), there exists a function $U : \mathbb{R}^N \setminus \{0\} \to (0,\infty)$ with the property

$$\begin{cases} -\Delta U \geq \phi(|x|)U^{-p}(x),\ U > 0 & \text{in } \mathbb{R}^N \setminus \{0\}, \\ U(x) \to 0 & \text{as } |x| \to \infty, \\ |x|^{N-2}U(x) \to 0 & \text{as } |x| \to 0. \end{cases}$$

Further, from $\phi \geq \varphi$ in $(0,\infty)$ we deduce that $\xi_n \leq U$ in $\mathbb{R}^N \setminus \{0\}$ which implies that $\xi(x) := \lim_{n\to\infty} \xi_n(x)$, $x \in \mathbb{R}^N \setminus \{0\}$ is well defined and it is the minimal solution of (5.3).

The construction of the two-parameter family of solutions to (5.3) is the same as in Step 2 of Theorem 5.17. We next show that this is the whole solution set of (5.3). Using the fact that $\varphi(|x|) \geq c|x|^{\alpha+\beta}$ in $B_1 \setminus \{0\}$ and $\varphi(|x|) > c|x|^{\alpha}$ in $\mathbb{R}^N \setminus \overline{B}_1(0)$, with the same argument as in the proof of Step 3 in Theorem 5.17 we obtain the existence of a positive constant $C > 0$ such that any solution u of (5.3) satisfies

$$u(x) \geq C|x|^{(2+\alpha+\beta)/(1+p)} \quad \text{for all } x \in B_1(0) \setminus \{0\} \tag{5.85}$$

and

$$u(x) \geq C|x|^{(2+\alpha)/(1+p)} \quad \text{for all } x \in \mathbb{R}^N \setminus \overline{B}_1(0). \tag{5.86}$$

Using (5.85) we have

$$-\Delta u(x) = \varphi(|x|)u^{-p} \leq |x|^{\alpha+\beta}u^{-p} \leq c|x|^{(\alpha+\beta-2p)/(1+p)} \leq |x|^{-N} \quad \text{as } |x| \to 0.$$

Now with the same method as in Lemmas 5.18 and 5.19 we find

$$\lim_{|x|\to 0} |x|^{N-2}u(x) = a.$$

Next, if u_* is the Kelvin transform of u as defined in (5.45), by (5.86) we have

$$\begin{aligned} -\Delta u_* &= |x|^{-2-N-p(N-2)}\varphi\left(\frac{1}{|x|}\right)u_*^{-p}(x) \\ &\leq C|x|^{-N-(2+\alpha)/(1+p)}\log^{\beta}\left(1+\frac{1}{|x|}\right) \\ &\leq C|x|^{-N} \quad \text{in } B_1(0) \setminus \{0\}. \end{aligned}$$

This yields (5.74) and then (5.75). From now on, we proceed exactly in the same way as we did in the proof of Theorem 5.17. □

With the same arguments we have.

Corollary 5.22 *Let $\alpha \in \mathbb{R}$, $\beta_1, \beta_2, \ldots, \beta_m, p > 0$ and*

$$\varphi(r) = r^{\alpha}\log^{\beta_1}(1+\log^{\beta_2}(1+\cdots+\log^{\beta_m}(1+r))\ldots), \quad t > 0.$$

Then, (5.3) has solutions if and only if

$$N+\alpha+\beta_1+\beta_2+\cdots+\beta_m+p(N-2)>0 \quad \textit{and} \quad \alpha<-2. \tag{5.87}$$

Furthermore, if (5.87) holds, then the solution set of (5.3) consists of a two-parameter family of radially symmetric functions as described in Theorem 5.17.

5.6 Application to Singular Elliptic Systems in Exterior Domains

Let Ω be an exterior domain (that is, $\mathbb{R}^N \setminus \Omega$ is bounded) in $\mathbb{R}^N$, $N \geq 3$, that does not contain the origin. In this section we study the existence of C^2 solutions to the elliptic system

$$\begin{cases} -\Delta u = f(|x|,v),\ u>0 & \text{in } \Omega, \\ -\Delta v = g(|x|,u),\ v>0 & \text{in } \Omega, \end{cases} \tag{5.88}$$

where $f,g \in C(0,\infty)$ are positive functions such that

(A) for all $r>0$ the mappings $f(r,\cdot)$ and $g(r,\cdot)$ are nonincreasing and convex.

Our first result concerns the case where Ω is nondegenerate with respect to the origin.

Theorem 5.23 *Assume $B_{r_0}(0) \subset \mathbb{R}^N \setminus \Omega$ for some $r_0>0$. Then system (5.88) has solutions if and only if there exists $c>0$ such that*

$$\int_1^\infty rf(r,c)dr<\infty \quad \textit{and} \quad \int_1^\infty rg(r,c)dr<\infty. \tag{5.89}$$

Proof. Assume that (5.88) has a solution (u,v) and let $w=u+v$. Adding the two equations in (5.88) we have

$$-\Delta w \geq f(|x|,w) \quad \text{in } \Omega, \tag{5.90}$$

$$-\Delta w \geq g(|x|,w) \quad \text{in } \Omega. \tag{5.91}$$

Fix $R>0$ such that $\mathbb{R}^N \setminus \Omega \subset B(0,R)$ and let $\bar{w}(r)$ be the average of w on $B(0,r)$, $r \geq R$, that is,

$$\bar{w}(r)=\frac{1}{\sigma_N r^{N-1}}\int_{\partial B_r(0)} w(y)d\sigma(y) \quad \text{for all } r \geq R, \tag{5.92}$$

where σ denotes the surface area measure in $\mathbb{R}^N$ and $\sigma_N=\sigma(\partial B_1(0))$. A straightforward computation using Green's formula yields

$$\bar{w}'(r) = \frac{1}{\sigma_N r^{N-1}} \int_{\partial B_r(0)} \frac{\partial w}{\partial \nu_r}(x)d\sigma(x) = \frac{1}{\sigma_N r^{N-1}} \int_{B_r(0)} \Delta u(x)dx, \tag{5.93}$$

where ν_r represents the outer unit normal at $\partial B_r(0)$.

Averaging (5.90) and using Jensen's inequality, we find

$$-\left[\bar{w}''(r) + \frac{N-1}{r}\bar{w}'(r)\right] \geq f(r, \bar{w}(r)) \quad \text{for all } r \geq R. \tag{5.94}$$

Let now

$$z(t) = \bar{w}(r), \quad t = r^{2-N}.$$

From (5.94) we find

$$-z''(t) \geq \frac{1}{(N-2)^2} t^{2(N-1)/(2-N)} f(t^{1/(2-N)}, z(t)),$$

for all $0 < t \leq T := R^{2-N}$. Since v is concave and positive, v is bounded from above by a constant $c > 0$ for $0 < t \leq T$. Hence

$$-z''(t) \geq Ct^{2(N-1)/(2-N)} f(t^{1/(2-N)}, c) \quad \text{for all } 0 < \rho \leq T.$$

Integrating this inequality twice we find

$$\begin{aligned}
\infty &> \int_0^T z'(t)\,dt - Tz'(T) \\
&\geq C \int_0^T \int_t^T s^{2(N-1)/(2-N)} f(s^{1/(2-N)}, c)\,dsdt \\
&= C \int_0^T s^{1+2(N-1)/(2-N)} f(s^{1/(2-N)}, c)\,ds \\
&= (N-2)C \int_R^{\infty} rf(r,c)\,dr.
\end{aligned}$$

Proceeding in the same way with v from (5.91) we deduce the second estimate in (5.89).

Assume now that (5.89) holds for some $c > 0$ and let us prove that (5.88) has solutions. More precisely, we shall construct a radially symmetric solution (u, v) of (5.88) in the larger exterior domain $\mathbb{R}^N \setminus B_{r_0}(0)$.

Let $A, B > 0$ be such that

$$A > c + \int_{r_0}^{\infty} rf(r,c)dr \quad \text{and} \quad B > c + \int_{r_0}^{\infty} rg(r,c)dr. \tag{5.95}$$

Consider $\tilde{f}, \tilde{g} : (0,\infty) \times \mathbb{R} \to (0,\infty)$ defined as

$$\tilde{f}(r,v) = \begin{cases} f(r,c) & \text{for } v \le c \\ f(r,v) & \text{for } v > c \end{cases} \quad \text{and } \tilde{g}(r,u) = \begin{cases} g(r,c) & \text{for } u \le c \\ g(r,u) & \text{for } u > c. \end{cases}$$

For all $n \ge 1$ let

$$u_n(r) = A - \int_{r_0}^{r} t^{1-N} \int_{r_0}^{t} s^{N-1} \tilde{f}(s, v_{n-1}(s)) ds dt, \quad r \ge r_0, \tag{5.96}$$

and

$$v_n(r) = B - \int_{r_0}^{r} t^{1-N} \int_{r_0}^{t} s^{N-1} \tilde{g}(s, u_{n-1}(s)) ds dt, \quad r \ge r_0, \tag{5.97}$$

where $u_0 \equiv A$, $v_0 \equiv B$. Remark first that $\{u_n\}$ and $\{v_n\}$ are well defined. Indeed, from Lemma 5.6(iii) we have

$$\begin{aligned} u_n(r) &\ge A - \int_{r_0}^{r} t^{1-N} \int_{r_0}^{t} s^{N-1} \tilde{f}(s,c) ds dt \\ &\ge A - \int_{r_0}^{\infty} t^{1-N} \int_{r_0}^{t} s^{N-1} f(s,c) ds dt \\ &= A - \int_{r_0}^{\infty} r f(r,c) dr \ge c, \end{aligned}$$

and similarly $v_n \ge c$ for all $n \ge 0$. It follows that u_n and v_n satisfy (5.96) and (5.97) with f and g instead of $\tilde{f}$ and $\tilde{g}$ respectively. Further, it is easy to see that $u_n'(r) \le 0$ and $v_n'(r) \le 0$ for all $r \ge r_0$. A straightforward induction argument yields $u_{n+1} \le u_n$ and $v_{n+1} \le v_n$ for all $n \ge 0$. Here, there exists $u(r) := \lim_{n\to\infty} u_n(r)$ and $v(r) := \lim_{n\to\infty} v_n(r)$, $r \ge r_0$. Passing to the limit in (5.96) and (5.97) we obtain

$$u(r) = A - \int_{r_0}^{r} t^{1-N} \int_{r_0}^{t} s^{N-1} f(s, v(s)) ds dt, \quad r \ge r_0,$$

$$v(r) = B - \int_{r_0}^{r} t^{1-N} \int_{r_0}^{t} s^{N-1} g(s, u(s)) ds dt, \quad r \ge r_0.$$

Hence $U(x) = u(|x|)$, $V(x) = v(|x|)$ is a solution of (5.88) in $\mathbb{R}^N \setminus B_{r_0}(0)$.

This finishes the proof of Theorem 5.23. □

Remark 22 *If f has separable variables, we do not require $f(r,\cdot)$ to be convex. Indeed, if $f(r,v) = a(r)b(v)$ with $a,b \in C(0,\infty)$ positive and b decreasing, we can always find $h \in C(0,\infty)$ convex and decreasing such that $b \ge h > 0$ in $(0,\infty)$. From (5.90) we deduce $-\Delta w \ge a(|x|)h(v)$ in Ω and we follow the same arguments as above (with $f(|x|,v)$ replaced by $(a|x|)h(v)$). In particular, if both f and g have separable variables we can remove the convexity assumption in hypothesis (A).*

We are next concerned with the case where $\Omega = \mathbb{R}^N \setminus \{0\}$. In this case we have:

Theorem 5.24 *Let* $\Omega = \mathbb{R}^N \setminus \{0\}$.
(i) The system (5.88) has solutions if and only if (5.89) together with

$$\int_0^1 r^{N-1} f(r, cr^{2-N})dr < \infty \quad \textit{and} \quad \int_0^1 r^{N-1} g(r, cr^{2-N})dr < \infty \tag{5.98}$$

hold.
(ii) The system (5.88) has solutions which are bounded in the neighborhood of the origin if and only if the following (stronger) condition holds

$$\int_0^\infty rf(r,c)dr < \infty \quad \textit{and} \quad \int_0^\infty rg(r,c)dr < \infty, \tag{5.99}$$

for some $c > 0$.

The proof of Theorem (5.24) relies on that of Theorem (5.23). To derive condition (5.89) we use the Kelvin transform.

Proof. (i) Assume first that (5.88) has a solution (u,v) in $\Omega = \mathbb{R}^N \setminus \{0\}$. Since (u,v) is also a solution of (5.88) in $\mathbb{R}^N \setminus B_2(0)$ by Theorem 5.23 it follows that (5.89) holds. Next, to deduce (5.98) let u_* and v_* be the Kelvin transforms of u and v respectively, that is

$$u_*(x) = |x|^{2-N} u\left(\frac{x}{|x|^2}\right), \quad x \in \mathbb{R}^N \setminus \{0\}, \tag{5.100}$$

$$v_*(x) = |x|^{2-N} v\left(\frac{x}{|x|^2}\right), \quad x \in \mathbb{R}^N \setminus \{0\}. \tag{5.101}$$

Then u_* satisfies

$$\begin{aligned}
-\Delta u_*(x) &= |x|^{-2-N}(-\Delta u)\left(\frac{x}{|x|^2}\right) \\
&= |x|^{-2-N} f\left(\frac{1}{|x|}, v\left(\frac{x}{|x|^2}\right)\right) \\
&= |x|^{-2-N} f\left(\frac{1}{|x|}, v_*(x)\right) \quad \text{in } \mathbb{R}^N \setminus B_1(0).
\end{aligned}$$

Similarly v_* satisfies

$$-\Delta v_*(x) = |x|^{-2-N} g\left(\frac{1}{|x|}, u_*(x)\right) \quad \text{in } \mathbb{R}^N \setminus B_1(0).$$

Hence, (u_*, v_*) are solutions of

$$\begin{cases} -\Delta u_* = |x|^{-2-N} f\left(\frac{1}{|x|}, v_*\right) & \text{in } \mathbb{R}^N \setminus B_1(0), \\ -\Delta v_* = |x|^{-2-N} g\left(\frac{1}{|x|}, u_*\right) & \text{in } \mathbb{R}^N \setminus B_1(0). \end{cases}$$

Since the mappings

$$\begin{aligned} \tilde{f}(r,v) &= r^{-2-N} f(r^{-1}, r^{N-2}v), \quad r,v>0 \\ \tilde{g}(r,u) &= r^{-2-N} g(r^{-1}, r^{N-2}u), \quad r,u>0 \end{aligned} \tag{5.102}$$

satisfy the assumption (A), by Theorem 5.23 we find

$$\int_1^\infty t^{-1-N} f(t^{-1}, ct^{N-2})dt < \infty \quad \text{and} \quad \int_1^\infty t^{-1-N} g(t^{-1}, ct^{N-2})dt < \infty,$$

for some $c>0$. Now the change of variable $t = r^{-1}$ leads us to (5.98).

Assume now that (5.89) and (5.98) hold. Note that by letting $c>1$ large enough, we can assume the same value for c in all the above integrals. Let

$$u_0(r) = v_0(r) = c + cr^{2-N}$$

and for all $n \geq 1$ define

$$u_n(r) = u_0(r) + \int_r^\infty t^{1-N} \int_{r_0}^t s^{N-1} f(s, v_{n-1}(s))dsdt, \quad r>0, \tag{5.103}$$

and

$$v_n(r) = v_0(r) + \int_r^r t^{1-N} \int_{r_0}^t s^{N-1} g(s, u_{n-1}(s))dsdt, \quad r>0. \tag{5.104}$$

Let us first remark that $\{u_n\}$ and $\{v_n\}$ are well defined. Indeed, since $u_n \geq u_0$ and $v_n \geq v_0$, by Lemma 5.6(ii) and (5.98) we have

$$\begin{aligned} &\int_r^\infty t^{1-N} \int_0^t s^{N-1} f(s, v_{n-1}(s))dsdt \\ &\qquad \leq \int_1^\infty t^{1-N} \int_0^1 s^{N-1} f(s, v_0(s))dsdt + \int_1^\infty t^{1-N} \int_1^t s^{N-1} f(s, v_0(s))dsdt \\ &\qquad \leq \frac{1}{N-2} \int_0^1 s^{N-1} f(s, cs^{2-N})ds + \int_1^\infty tf(t,c)dt < \infty. \end{aligned}$$

Using the fact that $u_1 \geq u_0$ and $v_1 \geq v_0$ we have

$$v_0 \le v_2 \le v_1 \quad \text{and} \quad u_0 \le u_2 \le u_1.$$

Further iterations imply

$$v_0 \le v_{2n} \le v_{2n+2} \le v_1 \quad \text{and} \quad u_0 \le u_{2n+1} \le u_{2n-1} \le u_1,$$

for all $n \ge 1$. Thus, there exists

$$u(r) = \lim_{n\to\infty} u_{2n-1}(r),\ v(r) = \lim_{n\to\infty} v_{2n}(r), r > 0.$$

Using (5.103) and (5.104) we find

$$u(r) = u_0(r) + \int_r^\infty t^{1-N} \int_{r_0}^t s^{N-1} f(s, v(s)) ds dt, \quad r > 0,$$

$$v(r) = v_0(r) + \int_r^r t^{1-N} \int_{r_0}^t s^{N-1} g(s, u(s)) ds dt, \quad r > 0.$$

It is now easy to check that $U(x) = u(|x|)$ and $V(x) = v(|x|)$, $x \in \mathbb{R}^N \setminus \{0\}$ is a solution of (5.88).

(ii) Let (u,v) be a solution of (5.88) such that $u, v \le M$ in $B_1(0) \setminus \{0\}$ and let $\bar{u}$, $\bar{v}$ be the spherical average of u and v respectively. Then $\bar{u}(r), \bar{v}(r) \le M$ for all $0 < r \le 1$. In the same manner as for (5.94) we find

$$-\left[\bar{u}''(r) + \frac{N-1}{r}\bar{u}'(r)\right] \ge f(r,M) \quad \text{for all } r > 0,$$

which yields

$$-(r^{N-1}\bar{u}')' \ge r^{N-1} f(r,M) > 0 \quad \text{for all } 0 < r \le 1. \tag{5.105}$$

In particular $(0,1] \ni r \longmapsto r^{N-1}\bar{u}'(r)$ is decreasing.

Claim: $\bar{u}'(r) \le 0$ for all $0 < r \le 1$.

Assume by contradiction that $\bar{u}'(R) > 0$ for some $0 < R \le 1$. Since $r^{N-1}\bar{u}'$ is decreasing, it follows that

$$\bar{u}'(r) \ge R^{N-1}\bar{u}'(R) r^{1-N} \quad \text{for all } 0 < r \le R.$$

Integrating the above inequality on $[r,R]$ we find

$$\bar{u}(r) \le \bar{u}(R) - \frac{C}{N-2} r^{2-N} \quad \text{for all } 0 < r \le R,$$

where $C = R^{N-1}\bar{u}'(R) > 0$. The above estimates yield $\bar{u}(r) \to -\infty$ as $r \to 0$, which is impossible, since $\bar{u} > 0$. Thus, the claim is proved.

Since $r^{N-1}\bar{u}'(r)$ is decreasing and negative in a neighborhood of the origin, there exists $\ell = \lim_{r\to 0^+} r^{N-1}\bar{u}'(r) \le 0$. Integrating in (5.105) we find

$$\begin{aligned} -\bar{u}'(r) &\ge -\ell r^{1-N} + r^{1-N}\int_0^r f(s,M)ds \\ &\ge r^{1-N}\int_0^r f(s,M)ds \quad \text{for all } 0 < r \le 1. \end{aligned}$$

A further integration over $[r,1]$ in the above estimate produces

$$M \ge \bar{u}(1) - \bar{u}(r) \ge \int_r^1 t^{1-N}\int_0^t f(s,M)dsdt\,,$$

for all $0 < r \le 1$. Hence, $\int_0^1 t^{1-N}\int_0^t f(s,M)dsdt$ is convergent which by Lemma 5.6(i) yields $\int_0^1 rf(r,M)dr < \infty$. This final estimate compounded with (5.89) yields (5.99).

For the converse part, assume now that (5.99) holds for some $c > 0$. We follow step by step the proof of the second part in Theorem 5.23 in which $r_0 = 0$. This concludes the proof. □

From Theorem 5.24 we deduce:

Corollary 5.25 *Let $p,q > 0$ and α,β be real numbers. Then the system*

$$\begin{cases} -\Delta u = |x|^{\alpha} v^{-p},\ u > 0 & \text{in } \mathbb{R}^N \setminus \{0\}, \\ -\Delta v = |x|^{\beta} u^{-q},\ v > 0 & \text{in } \mathbb{R}^N \setminus \{0\}, \end{cases}$$

has solutions if and only if $\alpha,\beta < -2$ and

$$N+\alpha+p(N-2) > 0, N+\beta+p(N-2) > 0.$$

Corollary 5.26 *Let $p,q > 0$ and α,β be real numbers. Then the system*

$$\begin{cases} -\Delta u = |x|^{\alpha} e^{v^{-p}},\ u > 0 & \text{in } \mathbb{R}^N \setminus \{0\}, \\ -\Delta v = |x|^{\beta} e^{u^{-q}},\ v > 0 & \text{in } \mathbb{R}^N \setminus \{0\}, \end{cases}$$

has solutions if and only if $-2 > \alpha,\beta > -N$.

Corollary 5.27 *Let $p,q>0$ and α,β be real numbers. Then the system*

$$\begin{cases} -\Delta u = (|x|^{\alpha}+v)^{-p},\ u>0 & \text{in } \mathbb{R}^N\setminus\{0\},\\ -\Delta v = (|x|^{\beta}+u)^{-q},\ v>0 & \text{in } \mathbb{R}^N\setminus\{0\}, \end{cases}$$

has solutions if and only if

$$\alpha>\frac{2}{p},\quad \beta>\frac{2}{q}.$$

Chapter 6
Two Quasilinear Elliptic Problems

> Science never solves a problem
> without creating ten more.
>
> George Bernard Shaw (1856–1950)

6.1 A Degenerate Elliptic Problem with Lack of Compactness

6.1.1 Introduction

In the last few decades, many researchers have been concerned with the study of degenerate elliptic problems. We start with the following example

$$\begin{cases} \operatorname{div}(a(x)\nabla u) + f(u) = 0 & \text{in } \Omega, \\ u = 0 & \text{on } \partial\Omega, \end{cases} \tag{6.1}$$

where Ω is an arbitrary domain in $\mathbb{R}^N$ ($N \geq 1$), and a is a nonnegative function that may have "essential" zeros at some points or even may be unbounded. The continuous function f satisfies $f(0) = 0$ and $tf(t)$ behaves like $|t|^p$ as $|t| \to \infty$, with $2 < p < 2^*$, where 2^* denotes the critical Sobolev exponent. Notice that equations of this type come from the consideration of standing waves in anisotropic Schrödinger equations (see [34, 117, 181, 185, 200]). Equations like (6.1) are also introduced as models for several physical phenomena related to equilibrium of anisotropic media which possibly are somewhere *"perfect" insulators* or *"perfect" conductors* (see [57], p. 79). Problem (6.1) has also some interest in the framework of optimization and G-convergence (see, e.g., [78] and the references therein).

M. Ghergu and V. Rădulescu, *Nonlinear PDEs*, Springer Monographs in Mathematics,
DOI 10.1007/978-3-642-22664-9_6, © Springer-Verlag Berlin Heidelberg 2012

Classical results (see [7, 168]) ensure the existence and the multiplicity of positive or nodal solutions for problem (6.1), provided that the differential operator $Tu := \operatorname{div}(a(x)\nabla u)$ is uniformly elliptic. Several difficulties occur both in the degenerate case (if $\inf_{\Omega} a = 0$) and in the singular case (if $\sup_{\Omega} a = +\infty$). In these situations the classical methods fail to be applied directly so that the existence and the multiplicity results (which hold in the nondegenerate case) may become a delicate matter that is closely related to some phenomena due to the degenerate character of the differential equation. These problems have been intensively studied starting with the pioneering paper by Murthy and Stampacchia [145] (see also [41, 71, 156], as well as the monograph [186]).

A natural question that arises in concrete applications is to see what happens if these elliptic (degenerate or nondegenerate) problems are affected by a certain perturbation. It is worth pointing out here that the idea of using perturbation methods in the treatment of nonlinear boundary value problems was introduced by Struwe [187].

Our aim in this chapter is to study the following degenerate perturbed problem

$$-\operatorname{div}(|x|^{-2a}\nabla u) = K(x)|x|^{-bp}|u|^{p-2}u + \lambda g(x) \quad \text{in } \mathbb{R}^N, \tag{6.2}$$

where

(i) if $N \geq 3: \ -\infty < a < \dfrac{N-2}{2}$, $a < b < a+1$ and $p = \dfrac{2N}{N-2+2(b-a)}$.

(ii) if $N = 2: \ -\infty < a < 0$, $a < b < a+1$ and $p = \dfrac{2}{b-a}$,

(iii) if $N = 1: \ -\infty < a < -\dfrac{1}{2}$, $a + \dfrac{1}{2} < b < a+1$ and $p = \dfrac{2}{-1+2(b-a)}$.

Equation (6.2) contains the critical Caffarelli–Kohn–Nirenberg exponent p (see Appendix A) defined as in (3.108). In this critical case, some concentration phenomena may occur, due to the action of the noncompact group of dilations in $\mathbb{R}^N$. The lack of compactness of problem (6.2) is also given by the fact that we are looking for entire solutions, that is, solutions defined on the whole space.

The reason for which we choose the parameters a, b, and p in the above range has to do with the Caffarelli–Kohn–Nirenberg inequality (see Appendix A):

$$\left(\int_{\mathbb{R}^N} |x|^{-bp}|u|^p dx\right)^{1/p} \leq C_{a,b}\left(\int_{\mathbb{R}^N} |x|^{-2a}|\nabla u|^2 dx\right)^{1/2}, \tag{6.3}$$

for all $u \in C_0^\infty(\mathbb{R}^N)$. We point out that the inequality (6.3) also holds true for $b = a+1$ (if $N \geq 1$) and $b = a$ (if $N \geq 3$) but, in these cases, the best embedding constant $C_{a,b}$ in (6.3) is never achieved (see [36] for details).

Throughout this chapter, the function K is assumed to fulfill:

$(K1)$ $K \in L^\infty(\mathbb{R}^N)$,

$(K2)$ $\text{esslim}_{|x|\to 0} K(x) = \text{esslim}_{|x|\to\infty} K(x) = K_0 \in (0,\infty)$ and $K(x) \geq K_0$ a.e. in $\mathbb{R}^N$,

$(K3)$ $\text{meas}\big(\{x \in \mathbb{R}^N : K(x) > K_0\}\big) > 0$.

The Palais–Smale condition (PS) plays a central role when variational methods are applied in the study of problem (6.2). In this chapter, we establish the existence and the multiplicity of nontrivial solutions of (6.2) with $\lambda > 0$ sufficiently small, in a case where the (PS) condition is not assumed even for $\lambda = 0$. More precisely, we will show that there exists at least two weak solutions of (6.2) for $g \not\equiv 0$ in an appropriate weighted Sobolev space and $\lambda > 0$ small enough. Our proof relies on Ekeland's variational principle and on the mountain pass theorem without the Palais–Smale condition (in the sense of Brezis and Nirenberg, see [29]), combined with a weighted variant of the Brezis–Lieb lemma.

The natural functional space to study problem (6.2) is $H_a^1(\mathbb{R}^N)$, defined as the completion of $C_0^\infty(\mathbb{R}^N)$ with respect to the norm

$$\|u\| = \left(\int_{\mathbb{R}^N} |x|^{-2a}|\nabla u|^2 dx\right)^{1/2}. \tag{6.4}$$

It turns out that $H_a^1(\mathbb{R}^N)$ is a Hilbert space with respect to the inner product

$$\langle u,v\rangle = \int_{\mathbb{R}^N} |x|^{-2a}\nabla u \cdot \nabla v dx, \quad \forall u,v \in H_a^1(\mathbb{R}^N).$$

It follows that (6.3) holds for all $u \in H_a^1(\mathbb{R}^N)$. Also we have

$$H_a^1(\mathbb{R}^N) = \overline{C_0^\infty(\mathbb{R}^N \setminus \{0\})}^{\|\cdot\|}, \tag{6.5}$$

where $\|\cdot\|$ is given by (6.4). Let $\|\cdot\|_{-1}$ denote the norm in the dual space $H_a^{-1}(\mathbb{R}^N)$ of $H_a^1(\mathbb{R}^N)$.

Throughout this chapter we suppose that $g \in H_a^{-1}(\mathbb{R}^N) \setminus \{0\}$.

For an arbitrary open set $\Omega \subset \mathbb{R}^N$, let $L_b^p(\Omega)$ be the space of all measurable real functions u defined on Ω such that $\int_\Omega |x|^{-bp}|u|^p dx$ is finite. By (6.3) it follows that the weighted Sobolev space $H_a^1(\Omega)$ is continuously embedded in $L_b^p(\Omega)$.

Definition 6.1 *We say that a function $u \in H_a^1(\mathbb{R}^N)$ is a weak solution of problem (6.2) if*

$$\int_{\mathbb{R}^N} |x|^{-2a}\nabla u \cdot \nabla v\,dx - \int_{\mathbb{R}^N} K(x)|x|^{-bp}|u|^{p-2}uv\,dx - \lambda \int_{\mathbb{R}^N} g(x)v\,dx = 0,$$

for all $u \in C_0^\infty(\mathbb{R}^N)$.

Obviously, the solutions of problem (6.2) correspond to critical points of the energy functional

$$J_\lambda(u) = \frac{1}{2}\int_{\mathbb{R}^N} |x|^{-2a}|\nabla u|^2 dx - \frac{1}{p}\int_{\mathbb{R}^N} K(x)|x|^{-bp}|u|^p dx - \lambda \int_{\mathbb{R}^N} g(x)u\,dx,$$

where $u \in H_a^1(\mathbb{R}^N)$.

Our main result is the following.

Theorem 6.2 *Suppose that assumptions $(K1)$, $(K2)$, $(K3)$ are fulfilled and fix $g \in H_a^{-1}(\mathbb{R}^N) \setminus \{0\}$. Then there exists $\lambda_0 > 0$ such that for all $\lambda \in (0, \lambda_0)$, problem (6.2) has at least two weak solutions.*

Since the embedding $H_a^1(\mathbb{R}^N) \hookrightarrow L_b^p(\mathbb{R}^N)$ is not compact, the energy functional J_λ fails to satisfy the (PS) condition. Such a failure brings about difficulty in applying a variational approach to (6.2). Furthermore, since $g \not\equiv 0$, then 0 is no longer a trivial solution of problem (6.2) and, therefore, the mountain pass theorem cannot be applied directly. We obtain the first solution by applying Ekeland's variational principle. Then, the mountain pass theorem without the Palais–Smale condition yields a bounded Palais–Smale sequence whose weak limit is a critical point of J_λ. The proof is concluded by showing that these two solutions are distinct because they realize different energy levels.

6.1.2 Auxiliary Results

Define the functionals $J_0, I : H_a^1(\mathbb{R}^N) \to \mathbb{R}$ by

$$J_0(u) = \frac{1}{2}\int_{\mathbb{R}^N} |x|^{-2a}|\nabla u|^2\,dx - \frac{1}{p}\int_{\mathbb{R}^N} K(x)|x|^{-bp}\,|u|^p\,dx,$$

$$I(u) = \frac{1}{2}\int_{\mathbb{R}^N} |x|^{-2a}|\nabla u|^2\,dx - \frac{1}{p}\int_{\mathbb{R}^N} K_0|x|^{-bp}\,|u|^p\,dx.$$

The Caffarelli–Kohn–Nirenberg inequality (6.3) and the conditions $(K1)$, $(K2)$ imply that the functionals J_λ, J_0, and I are well defined and $J_\lambda, J_0, I \in C^1(H_a^1(\mathbb{R}^N),\mathbb{R})$.

Remark 23 *If $\Omega \subset \mathbb{R}^N$ is a smooth bounded set such that $0 \notin \overline{\Omega}$ then, by the Sobolev inequality, we have*

$$\begin{aligned}\left(\int_\Omega |x|^{-bp}\,|u|^p dx\right)^{1/p} &\le C_1\left(\int_\Omega |u|^p dx\right)^{1/p}\\ &\le C_2\left(\int_\Omega |\nabla u|^2 dx\right)^{1/2}\\ &\le C_3\left(\int_\Omega |x|^{-2a}|\nabla u|^2 dx\right)^{1/2},\end{aligned}$$

for all $u \in H_a^1(\Omega)$. It follows that $H_a^1(\Omega)$ is compactly embedded in $L_b^p(\Omega)$.

Remark 23 implies that if $\{u_n\}$ is a sequence that converges weakly to some u_0 in $H_a^1(\mathbb{R}^N)$ then $\{u_n\}$ is bounded in $H_a^1(\mathbb{R}^N)$. Therefore, we can assume (up to a subsequence) that

$$u_n \to u_0 \text{ in } L^p_{b,\mathrm{loc}}(\mathbb{R}^N\setminus\{0\}) \quad \text{and} \quad u_n \to u_0 \text{ a.e. in } \mathbb{R}^N. \tag{6.6}$$

Definition 6.3 *Let X be a Banach space, $F : X \to \mathbb{R}$ be a C^1–functional and c be a real number. A sequence $\{u_n\} \subset X$ is called a $(PS)_c$ sequence of F if $F(u_n) \to c$ and $\|F'(u_n)\|_{X^*} \to 0$.*

Our first result shows that if a $(PS)_c$ sequence of J_λ is weakly convergent then its limit is a solution of problem (6.2).

Lemma 6.4 *Let $\{u_n\} \subset H_a^1(\mathbb{R}^N)$ be a $(PS)_c$ sequence of J_λ for some $c \in \mathbb{R}$. Suppose that $\{u_n\}$ converges weakly to some u_0 in $H_a^1(\mathbb{R}^N)$. Then u_0 is a solution of problem (6.2).*

Proof. Let $\phi \in C_0^\infty(\mathbb{R}^N\setminus\{0\})$ be an arbitrary function and set $\Omega := \operatorname{supp}\phi$. Since $J'_\lambda(u_n) \to 0$ in $H_a^{-1}(\mathbb{R}^N)$ we obtain $\langle J'_\lambda(u_n), \phi\rangle \to 0$ as $n \to \infty$, that is,

$$\lim_{n\to\infty}\left(\int_{\Omega}|x|^{-2a}\nabla u_n\cdot\nabla\phi\,dx-\int_{\Omega}K(x)|x|^{-bp}\,|u_n|^{p-2}u_n\phi\,dx-\lambda\int_{\Omega}g(x)\phi\,dx\right)=0. \tag{6.7}$$

Since $u_n\rightharpoonup u_0$ in $H_a^1(\mathbb{R}^N)$, it follows that

$$\lim_{n\to\infty}\int_{\Omega}|x|^{-2a}\nabla u_n\cdot\nabla\phi\,dx=\int_{\Omega}|x|^{-2a}\nabla u_0\cdot\nabla\phi\,dx. \tag{6.8}$$

The boundedness of $\{u_n\}$ in $H_a^1(\mathbb{R}^N)$ and the Caffarelli–Kohn–Nirenberg inequality imply that $\{|u_n|^{p-2}u_n\}$ is bounded in $L_b^{p/p-1}(\mathbb{R}^N)$. Since $|u_n|^{p-2}u_n\to|u_0|^{p-2}u_0$ a.e. in $\mathbb{R}^N$ (which is a consequence of (6.6)), we deduce that $|u_0|^{p-2}u_0$ is the weak limit in $L_b^{p/p-1}(\mathbb{R}^N)$ of the sequence $\{|u_n|^{p-2}u_n\}$. Therefore

$$\lim_{n\to\infty}\int_{\Omega}K(x)|x|^{-bp}\,|u_n|^{p-2}u_n\phi\,dx=\int_{\Omega}K(x)|x|^{-bp}\,|u_0|^{p-2}u_0\phi\,dx. \tag{6.9}$$

Consequently, relations (6.7), (6.8), and (6.9) yield

$$\int_{\Omega}|x|^{-2a}\nabla u_0\cdot\nabla\phi\,dx-\int_{\Omega}K(x)|x|^{-bp}\,|u_0|^{p-2}u_0\phi\,dx-\lambda\int_{\Omega}g(x)\phi\,dx=0.$$

By virtue of (6.5) we deduce that the above equality holds for all $\phi\in H_a^1(\mathbb{R}^N)$ which means that $J_\lambda'(u_0)=0$. The proof of our lemma is now complete. □

The next result is a weighted variant of the Brezis–Lieb lemma (see [27]).

Lemma 6.5 *Let $\{u_n\}$ be a sequence which is weakly convergent to u_0 in $H_a^1(\mathbb{R}^N)$. Then*

$$\lim_{n\to\infty}\int_{\mathbb{R}^N}K(x)|x|^{-bp}\left(|u_n|^p-|u_n-u_0|^p\right)dx=\int_{\mathbb{R}^N}K(x)|x|^{-bp}\,|u_0|^p\,dx.$$

Proof. Using the boundedness of $\{u_n\}$ in $H_a^1(\mathbb{R}^N)$ and the Caffarelli–Kohn–Nirenberg inequality, it follows that the sequence $\{u_n\}$ is bounded in $L_b^p(\mathbb{R}^N)$. Let $\varepsilon>0$ be a positive real number. By $(K1)$ and $(K2)$, we can choose $R_\varepsilon>r_\varepsilon>0$ such that

$$\int_{|x|<r_\varepsilon}K(x)|x|^{-bp}\,|u_0|^p\,dx<\varepsilon \tag{6.10}$$

and

$$\int_{|x|>R_\varepsilon}K(x)|x|^{-bp}\,|u_0|^p\,dx<\varepsilon. \tag{6.11}$$

Denote $\Omega_\varepsilon:=\overline{B(0,R_\varepsilon)}\setminus B(0,r_\varepsilon)$. We have

$$\begin{aligned}
&\left|\int_{\mathbb{R}^N} K(x)|x|^{-bp}\left(|u_n|^p-|u_0|^p-|u_n-u_0|^p\right)dx\right| \\
&\leq \left|\int_{\Omega_\varepsilon} K(x)|x|^{-bp}\left(|u_n|^p-|u_0|^p\right)dx\right| + \int_{\Omega_\varepsilon} K(x)|x|^{-bp}|u_n-u_0|^p\,dx \\
&\quad + \int_{|x|<r_\varepsilon} K(x)|x|^{-bp}|u_0|^p\,dx + \left|\int_{|x|<r_\varepsilon} K(x)|x|^{-bp}\left(|u_n|^p-|u_n-u_0|^p\right)dx\right| \\
&\quad + \int_{|x|>R_\varepsilon} K(x)|x|^{-bp}|u_0|^p\,dx + \left|\int_{|x|>R_\varepsilon} K(x)|x|^{-bp}\left(|u_n|^p-|u_n-u_0|^p\right)dx\right|.
\end{aligned}$$

By the Lagrange mean value theorem we have

$$\begin{aligned}
&\int_{|x|<r_\varepsilon} K(x)|x|^{-bp}\left(|u_n|^p-|u_n-u_0|^p\right)dx \\
&\qquad = p\int_{|x|<r_\varepsilon} K(x)|x|^{-bp}\,|\theta u_0+(u_n-u_0)|^{p-1}|u_0|\,dx,
\end{aligned} \tag{6.12}$$

where $0<\theta(x)<1$. Next, we employ the following elementary inequality: for all $s>0$ there exists a constant $c=c(s)$ such that

$$(x+y)^s \leq c(x^s+y^s), \quad \text{for any } x,y\in(0,\infty).$$

Then, by Hölder's inequality and relation (6.10) we deduce that

$$\begin{aligned}
&\int_{|x|<r_\varepsilon} K(x)|x|^{-bp}\,|\theta u_0+(u_n-u_0)|^{p-1}|u_0|\,dx \\
&\qquad \leq c\int_{|x|<r_\varepsilon} K(x)|x|^{-bp}\left(|u_0|^p+|u_n-u_0|^{p-1}|u_0|\right)dx \\
&\qquad = c\int_{|x|<r_\varepsilon} K(x)|x|^{-bp}|u_0|^p\,dx + c\int_{|x|<r_\varepsilon} K(x)|x|^{-bp}|u_n-u_0|^{p-1}|u_0|\,dx \\
&\qquad \leq c\,\varepsilon + c\left(\int_{|x|<r_\varepsilon} K(x)|x|^{-bp}|u_n-u_0|^p\,dx\right)^{(p-1)/p} \\
&\qquad\quad \times\left(\int_{|x|<r_\varepsilon} K(x)|x|^{-bp}|u_0|^p\,dx\right)^{1/p} \\
&\qquad \leq c_1\left(\varepsilon+\varepsilon^{1/p}\right),
\end{aligned}$$

where the constant c_1 is independent of n and ε. Using relation (6.12) we have

$$\begin{aligned}
&\int_{|x|<r_\varepsilon} K(x)|x|^{-bp}|u_0|^p\,dx + \left|\int_{|x|<r_\varepsilon} K(x)|x|^{-bp}\left(|u_n|^p-|u_n-u_0|^p\right)dx\right| \\
&\qquad \leq p\,\tilde{c}_1\left(\varepsilon+\varepsilon^{1/p}\right).
\end{aligned} \tag{6.13}$$

In a similar manner we obtain

$$\int_{|x|>R_\varepsilon} K(x)|x|^{-bp}|u_0|^p\,dx + \left|\int_{|x|>R_\varepsilon} K(x)|x|^{-bp}(|u_n|^p - |u_n-u_0|^p)\,dx\right| \le p\tilde{c}_2(\varepsilon+\varepsilon^{1/p}). \tag{6.14}$$

Since $u_n \rightharpoonup u_0$ in $H_a^1(\mathbb{R}^N)$, relation (6.6) yields

$$\lim_{n\to\infty}\int_{\Omega_\varepsilon} K(x)|x|^{-bp}(|u_n|^p - |u_0|^p)\,dx = 0, \quad \lim_{n\to\infty}\int_{\Omega_\varepsilon} K(x)|x|^{-bp}|u_n-u_0|^p\,dx = 0. \tag{6.15}$$

Now, by (6.13), (6.14), and (6.15) we find

$$\limsup_{n\to\infty}\left|\int_{\mathbb{R}^N} K(x)|x|^{-bp}(|u_n|^p - |u_0|^p - |u_n-u_0|^p)\,dx\right| \le (pC+1)(\varepsilon+\varepsilon^{1/p}).$$

Since $\varepsilon>0$ is arbitrary, it follows that

$$\lim_{n\to\infty}\int_{\mathbb{R}^N} K(x)|x|^{-bp}(|u_n|^p - |u_n-u_0|^p)\,dx = \int_{\mathbb{R}^N} K(x)|x|^{-bp}|u_0|^p\,dx.$$

This concludes the proof. □

Lemma 6.6 *Let $\{v_n\}$ be a sequence which converges weakly to 0 in $H_a^1(\mathbb{R}^N)$. Then the following properties hold true*

$$\lim_{n\to\infty}[J_\lambda(v_n) - I(v_n)] = 0,$$

$$\lim_{n\to\infty}[\langle J_\lambda'(v_n), v_n\rangle - \langle I'(v_n), v_n\rangle] = 0.$$

Proof. A straightforward computation yields

$$J_\lambda(v_n) = I(v_n) - \frac{1}{p}\int_{\mathbb{R}^N}(K(x)-K_0)|x|^{-bp}|v_n|^p\,dx - \lambda\int_{\mathbb{R}^N} g(x)v_n\,dx,$$

$$\langle J_\lambda'(v_n), v_n\rangle = \langle I'(v_n), v_n\rangle - \int_{\mathbb{R}^N}(K(x)-K_0)|x|^{-bp}|v_n|^p\,dx - \lambda\int_{\mathbb{R}^N} g(x)v_n\,dx.$$

Since $v_n \rightharpoonup 0$ in $H_a^1(\mathbb{R}^N)$, it follows from the above equalities that it suffices to prove that

$$\lim_{n\to\infty}\int_{\mathbb{R}^N}(K(x)-K_0)|x|^{-bp}|v_n|^p\,dx = 0. \tag{6.16}$$

Fix $\varepsilon>0$. By our assumptions $(K1)$ and $(K2)$, there exists $R_\varepsilon > r_\varepsilon > 0$ such that

$$|K(x)-K_0| = K(x)-K_0 < \varepsilon \quad \text{for a.e.} \quad x\in\mathbb{R}^N\setminus\Omega_\varepsilon,$$

where $\Omega_\varepsilon = \overline{B(0,R_\varepsilon)} \setminus B(0,r_\varepsilon)$. Next, we have

$$\begin{aligned}
&\int_{\mathbb{R}^N} (K(x)-K_0)|x|^{-bp}|v_n|^p\,dx \\
&\quad= \int_{\mathbb{R}^N\setminus\Omega_\varepsilon} (K(x)-K_0)|x|^{-bp}|v_n|^p\,dx + \int_{\Omega_\varepsilon} (K(x)-K_0)|x|^{-bp}|v_n|^p\,dx \\
&\quad\le \varepsilon \int_{\mathbb{R}^N\setminus\Omega_\varepsilon} |x|^{-bp}|v_n|^p\,dx + (\|K\|_\infty - K_0)\int_{\Omega_\varepsilon} |x|^{-bp}|v_n|^p\,dx \\
&\quad\le \varepsilon \int_{\mathbb{R}^N} |x|^{-bp}|v_n|^p\,dx + (\|K\|_\infty - K_0)\int_{\Omega_\varepsilon} |x|^{-bp}|v_n|^p\,dx.
\end{aligned}$$

Since $v_n \rightharpoonup 0$ in $H_a^1(\mathbb{R}^N)$, the Caffarelli–Kohn–Nirenberg inequality implies that $\{v_n\}$ is bounded in $L_b^p(\mathbb{R}^N)$. Moreover, by (6.6), it follows that $v_n \to 0$ in $L^p_{b,\mathrm{loc}}(\mathbb{R}^N \setminus \{0\})$. The above relations yield

$$\limsup_{n\to\infty} \int_{\mathbb{R}^N} (K(x)-K_0)|x|^{-bp}|v_n|^p\,dx \le C\varepsilon$$

for some constant $C>0$ independent of n and ε. Since $\varepsilon>0$ was arbitrarily chosen, we conclude that (6.16) holds and the proof of Lemma 6.6 is now complete. □

Lemma 6.7 *There exists $\lambda_1>0$ and $R=R(\lambda_1)>0$ such that for all $\lambda\in(0,\lambda_1)$, the functional J_λ admits a $(PS)_{c_{0,\lambda}}$ sequence with $c_{0,\lambda}=c_{0,\lambda}(R)=\inf_{u\in\overline{B}_R} J_\lambda(u)$. Moreover, $c_{0,\lambda}$ is achieved by some $u_0\in H_a^1(\mathbb{R}^N)$ with $J'_\lambda(u_0)=0$.*

Proof. Fix $\lambda\in(0,1)$. For all $u\in H_a^1(\mathbb{R}^N)$, the assumption (K_1) and the Caffarelli–Kohn–Nirenberg inequality imply

$$\begin{aligned}
J_\lambda(u) &= \frac{1}{2}\|u\|^2 - \frac{1}{p}\int_{\mathbb{R}^N} K(x)|x|^{-bp}|u|^p\,dx - \lambda\int_{\mathbb{R}^N} g(x)u\,dx \\
&\ge \frac{1}{2}\|u\|^2 - \frac{\|K\|_\infty}{p} C^p_{a,b}\|u\|^p - \lambda\|g\|_{-1}\|u\|.
\end{aligned}$$

We now apply the inequality $\alpha\beta \le \dfrac{\alpha^2+\beta^2}{2}$, for any $\alpha,\beta\ge 0$. Hence

$$J_\lambda(u) \ge \frac{1-\lambda}{2}\|u\|^2 - \frac{\|K\|_\infty}{p} C^p_{a,b}\|u\|^p - \frac{\lambda}{2}\|g\|^2_{-1}. \tag{6.17}$$

Since $p>2$ and the right-hand side of (6.17) is a decreasing function on λ, we find $\lambda_1>0$ and $R=R(\lambda_1)>0$, $\delta=\delta(\lambda_1)>0$ such that

$$J_\lambda(u) \ge -\frac{\lambda}{2}\|g\|^2_{-1}, \quad \text{for all } u\in\overline{B}_R \text{ and } \lambda\in(0,\lambda_1) \tag{6.18}$$

and

$$J_\lambda(u) \geq \delta > 0, \quad \text{for all } u \in \partial B_R \text{ and } \lambda \in (0, \lambda_1). \tag{6.19}$$

For instance, we can take

$$\lambda_1 := \min\left\{\frac{1}{2}, \frac{1}{2\|g\|_{-1}^2}\left(\frac{1}{2} - \frac{1}{p}\right) r_0^2\right\}, \qquad r_0 := \left[\frac{1}{2\|K\|_\infty C_{a,b}^p}\right]^{1/(p-2)}$$

and

$$R := \left[\frac{1-\lambda_1}{\|K\|_\infty C_{a,b}^p}\right]^{1/(p-2)}, \qquad \delta(\lambda_1) := \frac{\lambda_1}{2}\|g\|_{-1}^2.$$

Using now the estimate (6.17), we easily deduce (6.18) and (6.19).

Next, we define $c_{0,\lambda} := c_{0,\lambda}(R) = \inf\{J_\lambda(u) \,:\, u \in \overline{B}_R\}$. We first note that $c_{0,\lambda} \leq J_\lambda(0) = 0$. The set $\overline{B}_R$ becomes a complete metric space with respect to the distance

$$\operatorname{dist}(u,v) := \|u - v\|, \qquad \text{for any } u, v \in \overline{B}_R.$$

The functional J_λ is lower semicontinuous and bounded from below on $\overline{B}_R$. Then, by Ekeland's variational principle, for any positive integer n there exists u_n such that

$$c_{0,\lambda} \leq J_\lambda(u_n) \leq c_{0,\lambda} + \frac{1}{n} \tag{6.20}$$

and

$$J_\lambda(w) \geq J_\lambda(u_n) - \frac{1}{n}\|u_n - w\|, \quad \text{for all } w \in \overline{B}_R. \tag{6.21}$$

We first show that $\|u_n\| < R$ for n large enough. Indeed, if not, then $\|u_n\| = R$ for infinitely many n, and so (up to a subsequence) we can assume that $\|u_n\| = R$ for all $n \geq 1$. It follows that $J_\lambda(u_n) \geq \delta > 0$. Using (6.20) and letting $n \to \infty$, we have $0 \geq c_{0,\lambda} \geq \delta > 0$, which is a contradiction.

We now claim that $J_\lambda'(u_n) \to 0$ in $H_a^{-1}(\mathbb{R}^N)$. Fix $u \in H_a^1(\mathbb{R}^N)$ with $\|u\| = 1$ and let $w_n = u_n + tu$. For some fixed n, we have $\|w_n\| \leq \|u_n\| + t < R$ if $t > 0$ is small enough. Then relation (6.21) yields

$$J_\lambda(u_n + tu) \geq J_\lambda(u_n) - \frac{t}{n}\|u\|,$$

that is,

$$\frac{J_\lambda(u_n + tu) - J_\lambda(u_n)}{t} \geq -\frac{1}{n}\|u\| = -\frac{1}{n}.$$

Letting $t \searrow 0$ it follows that $\langle J'_\lambda(u_n), u\rangle \geq -\frac{1}{n}$. Arguing in a similar way for $t \nearrow 0$, we obtain $\langle J'_\lambda(u_n), u\rangle \leq \frac{1}{n}$. Since $u \in H^1_a(\mathbb{R}^N)$ with $\|u\| = 1$ has been arbitrarily chosen, we have

$$\|J'_\lambda(u_n)\| = \sup_{u\in H^1_a(\mathbb{R}^N),\|u\|=1} |\langle J'_\lambda(u_n), u\rangle| \leq \frac{1}{n} \to 0 \quad \text{as } n \to \infty.$$

We have proved the existence of a $(PS)_{c_{0,\lambda}}$ sequence, that is, a sequence $\{u_n\} \subset H^1_a(\mathbb{R}^N)$ with

$$J_\lambda(u_n) \to c_{0,\lambda} \quad \text{and} \quad J'_\lambda(u_n) \to 0 \text{ in } H^1_a(\mathbb{R}^N). \tag{6.22}$$

Since $\|u_n\| \leq R$, it follows that $\{u_n\}$ converges weakly (up to a subsequence) in $H^1_a(\mathbb{R}^N)$ to some u_0. Moreover, relations (6.6) and (6.22) yield

$$u_n \rightharpoonup u_0 \quad \text{in } H^1_a(\mathbb{R}^N), \qquad u_n \to u_0 \quad \text{a.e. in } \mathbb{R}^N \tag{6.23}$$

and

$$J'_\lambda(u_0) = 0. \tag{6.24}$$

Next, we prove that $J_\lambda(u_0) = c_{0,\lambda}$. Indeed, using relations (6.22) and (6.23) we have

$$\begin{aligned} o(1) &= \langle J'_\lambda(u_n), u_n\rangle \\ &= \int_{\mathbb{R}^N} |x|^{-2a}|\nabla u_n|^2\,dx - \int_{\mathbb{R}^N} K(x)|x|^{-bp}\,|u_n|^p\,dx - \lambda\int_{\mathbb{R}^N} g(x)u_n\,dx. \end{aligned}$$

Therefore

$$J_\lambda(u_n) = \left(\frac{1}{2} - \frac{1}{p}\right)\int_{\mathbb{R}^N} K(x)|x|^{-bp}|u_n|^p\,dx - \frac{\lambda}{2}\int_{\mathbb{R}^N} g(x)u_n\,dx + o(1).$$

Hence

$$J_\lambda(u_0) = \left(\frac{1}{2} - \frac{1}{p}\right)\int_{\mathbb{R}^N} K(x)|x|^{-bp}|u_0|^p\,dx - \frac{\lambda}{2}\int_{\mathbb{R}^N} g(x)u_0\,dx + o(1).$$

Fatou's lemma and relations (6.22), (6.23), (6.24) imply

$$
\begin{aligned}
c_{0,\lambda} &= \liminf_{n\to\infty} J_\lambda(u_n) \\
&\geq \left(\frac{1}{2}-\frac{1}{p}\right)\int_{\mathbb{R}^N} K(x)|x|^{-bp}|u_0|^p\,dx - \frac{\lambda}{2}\int_{\mathbb{R}^N} g(x)u_0\,dx \\
&= J_\lambda(u_0).
\end{aligned}
$$

Thus, $c_{0,\lambda} \geq J_\lambda(u_0)$. On the other hand, since $u_0 \in \overline{B}_R$, we deduce that $J_\lambda(u_0) \geq c_{0,\lambda}$, so $J_\lambda(u_0) = c_{0,\lambda}$. This concludes the proof of Lemma 6.7. □

6.1.3 Proof of the Main Result

Define

$$
\mathscr{S} := \{u \in H_a^1(\mathbb{R}^N)\setminus\{0\} : \langle I'(u), u\rangle = 0\}.
$$

We claim that $\mathscr{S} \neq \emptyset$. For this purpose we fix $u \in H_a^1(\mathbb{R}^N)\setminus\{0\}$ and set, for any $\lambda > 0$,

$$
\Psi(\lambda) = \langle I'(\lambda u), \lambda u\rangle = \lambda^2 \int_{\mathbb{R}^N} |x|^{-2a}|\nabla u|^2\,dx - \lambda^p \int_{\mathbb{R}^N} K_0|x|^{-bp}\,|u|^p\,dx.
$$

Since $p > 2$, it follows that $\Psi(\lambda) < 0$ for λ large enough and $\Psi(\lambda) > 0$ for λ sufficiently close to the origin. So, there exists $\lambda > 0$ such that $\Psi(\lambda) = 0$, that is, $\lambda u \in \mathscr{S}$.

Proposition 6.8 *Let $I_\infty := \inf\{I(u) : u \in \mathscr{S}\}$. Then there exists $\bar{u} \in H_a^1(\mathbb{R}^N)$ such that*

$$
I_\infty = I(\bar{u}) = \sup_{t\geq 0} I(t\bar{u}). \tag{6.25}
$$

Proof. For some fixed $\phi \in H_a^1(\mathbb{R}^N)\setminus\{0\}$ denote

$$
f(t) = I(t\phi) = \frac{t^2}{2}\int_{\mathbb{R}^N} |x|^{-2a}|\nabla\phi|^2\,dx - \frac{K_0}{p}t^p \int_{\mathbb{R}^N} |x|^{-bp}\,|\phi|^p\,dx.
$$

We have

$$
f'(t) = t\int_{\mathbb{R}^N} |x|^{-2a}|\nabla\phi|^2\,dx - K_0 t^{p-1}\int_{\mathbb{R}^N} |x|^{-bp}\,|\phi|^p\,dx.
$$

Then f attains its maximum at

$$
t_0 = t_0(\phi) := \left\{\frac{\displaystyle\int_{\mathbb{R}^N} |x|^{-2a}|\nabla\phi|^2\,dx}{\displaystyle\int_{\mathbb{R}^N} K_0|x|^{-bp}|\phi|^p\,dx}\right\}^{1/(p-2)}.
$$

Hence

$$f(t_0) = I(t_0\phi) = \sup_{t\geq 0} I(t\phi) = \left(\frac{1}{2} - \frac{1}{p}\right) \left\{ \frac{\displaystyle\int_{\mathbb{R}^N} |x|^{-2a} |\nabla \phi|^2 \, dx}{\left(\displaystyle\int_{\mathbb{R}^N} K_0 |x|^{-bp} |\phi|^p \, dx\right)^{2/p}} \right\}^{p/(p-2)}.$$

It follows that

$$\inf_{\phi \in H_0^1(\mathbb{R}^N)\setminus\{0\}} \sup_{t\geq 0} I(t\phi) = \left(\frac{1}{2} - \frac{1}{p}\right) [S(a,b)]^{p/(p-2)}, \tag{6.26}$$

where

$$S(a,b) = \inf_{\phi \in H_0^1(\mathbb{R}^N)\setminus\{0\}} \left\{ \frac{\displaystyle\int_{\mathbb{R}^N} |x|^{-2a} |\nabla \phi|^2 \, dx}{\left(\displaystyle\int_{\mathbb{R}^N} K_0 |x|^{-bp} |\phi|^p \, dx\right)^{2/p}} \right\}. \tag{6.27}$$

We now easily observe that for every $u \in \mathscr{S}$ we have $t_0(u) = 1$. So, by (6.26), it follows that

$$I(u) = \sup_{t\geq 0} I(tu), \quad \text{for all } u \in \mathscr{S}. \tag{6.28}$$

By Remark A.1 in Appendix A, the infimum in (6.27) is achieved by a function $U \in H_a^1(\mathbb{R}^N)$ such that $\int_{\mathbb{R}^N} K_0 |x|^{-bp} |U|^p dx = 1$. Letting $\bar{u} = [S(a,b)]^{1/(p-2)} U$, we see that $\bar{u} \in \mathscr{S}$ and

$$I(\bar{u}) = \left(\frac{1}{2} - \frac{1}{p}\right) [S(a,b)]^{p/(p-2)}. \tag{6.29}$$

Relations (6.28) and (6.29) yield

$$\begin{aligned} I_\infty &= \inf_{u\in\mathscr{S}} I(u) \\ &= \inf_{u\in\mathscr{S}} \sup_{t\geq 0} I(tu) \\ &\geq \inf_{u \in H_0^1(\mathbb{R}^N)\setminus\{0\}} \sup_{t\geq 0} I(tu) \\ &= \left(\frac{1}{2} - \frac{1}{p}\right) [S(a,b)]^{p/(p-2)} = I(\bar{u}), \end{aligned}$$

which concludes our proof. □

Proposition 6.9 *Assume that* $\{u_n\}$ *is a* $(PS)_c$ *sequence of* J_λ *which is weakly convergent in* $H_a^1(\mathbb{R}^N)$ *to some* u_0. *Then the following alternative holds: either* $\{u_n\}$ *converges strongly in* $H_a^1(\mathbb{R}^N)$, *or* $c \geq J_\lambda(u_0) + I_\infty$.

Proof. Since $\{u_n\}$ is a $(PS)_c$ sequence and $u_n \rightharpoonup u_0$ in $H_a^1(\mathbb{R}^N)$ we have

$$J_\lambda(u_n) = c + o(1) \quad \text{and} \quad \langle J'_\lambda(u_n), u_n\rangle = o(1). \tag{6.30}$$

Denote $v_n = u_n - u_0$. It follows that $v_n \rightharpoonup 0$ in $H_a^1(\mathbb{R}^N)$ which implies

$$\lim_{n\to\infty} \int_{\mathbb{R}^N} |x|^{-2a} \nabla v_n \cdot \nabla u_0 \, dx = 0,$$

$$\lim_{n\to\infty} \int_{\mathbb{R}^N} g(x) v_n \, dx = 0.$$

The above relations imply

$$\begin{aligned} \|u_n\|^2 &= \|u_0\|^2 + \|v_n\|^2 + o(1), \\ J_\lambda(v_n) &= J_0(v_n) + o(1). \end{aligned} \tag{6.31}$$

Using Lemmas 6.4 to 6.6 and relations (6.30), (6.31) we deduce that

$$o(1) + c = J_\lambda(u_n) = J_\lambda(u_0) + J_\lambda(v_n) + o(1) = J_\lambda(u_0) + I(v_n) + o(1), \tag{6.32}$$

$$\begin{aligned} o(1) &= \langle J'_\lambda(u_n), u_n\rangle \\ &= \langle J'_\lambda(u_0), u_0\rangle + \langle J'_\lambda(v_n), v_n\rangle + o(1) \\ &= \langle I'(v_n), v_n\rangle + o(1). \end{aligned} \tag{6.33}$$

If $v_n \to 0$ in $H_a^1(\mathbb{R}^N)$ then $u_n \to u_0$ in $H_a^1(\mathbb{R}^N)$. It follows that $J_\lambda(u_0) = \lim_{n\to\infty} J_\lambda(u_n)$. If the sequence $\{v_n\}$ does not converge strongly to 0 in $H_a^1(\mathbb{R}^N)$, then, since $v_n \rightharpoonup 0$ in $H_a^1(\mathbb{R}^N)$, we can assume (up to a subsequence) that $\|v_n\| \to l > 0$.

By virtue of (6.32), it remains only to show that $I(v_n) \geq I_\infty + o(1)$. Taking $t > 0$ we have

$$\langle I'(tv_n), tv_n\rangle = t^2 \int_{\mathbb{R}^N} |x|^{-2a} |\nabla v_n|^2 \, dx - t^p K_0 \int_{\mathbb{R}^N} |x|^{-bp} |v_n|^p \, dx.$$

If we prove the existence of a sequence $\{t_n\} \subset (0,\infty)$ with $t_n \to 1$ and $\langle I'(t_n v_n), t_n v_n\rangle = 0$, then $t_n v_n \in \mathscr{S}$. This implies that

$$\begin{aligned} I(v_n) &= I(t_n v_n) + \frac{1-t_n^2}{2}\|v_n\|^2 - \frac{1-t_n^p}{p} K_0 \int_{\mathbb{R}^N} |x|^{-bp} |v_n|^p \, dx \\ &= I(t_n v_n) + o(1) \geq I_\infty + o(1), \end{aligned}$$

and the conclusion follows. For this purpose, we denote

$$\alpha_n = \int_{\mathbb{R}^N} |x|^{-2a} |\nabla v_n|^2 \, dx = \|v_n\|^2 \geq 0,$$

$$\beta_n = K_0 \int_{\mathbb{R}^N} |x|^{-bp} |v_n|^p \, dx \geq 0,$$

$$\mu_n = \alpha_n - \beta_n.$$

From (6.33) it follows that $\mu_n = \langle I'(v_n), v_n \rangle \to 0$ as $n \to \infty$. If $\mu_n = 0$, then we take $t_n = 1$. Next, we assume that $\mu_n \neq 0$. Let $\delta \in \mathbb{R}$ with $|\delta| > 0$ sufficiently small and $t = 1 + \delta$. Then

$$\begin{aligned} \langle I'(tv_n), tv_n \rangle &= (1+\delta)^2 \alpha_n - (1+\delta)^p \beta_n \\ &= (1+\delta)^2 \alpha_n - (1+\delta)^p (\alpha_n - \mu_n) \\ &= \alpha_n (2\delta - p\delta + o(\delta)) + (1+\delta)^p \mu_n \\ &= \alpha_n (2-p)\delta + \alpha_n o(\delta) + (1+\delta)^p \mu_n. \end{aligned}$$

Since $p > 2$, $\alpha_n \to l^2 > 0$ and $\mu_n \to 0$, for n large enough we can define $\delta_n^+ = \dfrac{2|\mu_n|}{\alpha_n(p-2)}$ and $\delta_n^- = -\dfrac{2|\mu_n|}{\alpha_n(p-2)}$. It follows that

$$\delta_n^+ \searrow 0 \quad \text{and} \quad \langle I'((1+\delta_n^+)v_n), (1+\delta_n^+)v_n \rangle < 0,$$

$$\delta_n^- \nearrow 0 \quad \text{and} \quad \langle I'((1+\delta_n^-)v_n), (1+\delta_n^-)v_n \rangle < 0.$$

From the above relations we deduce the existence of some $t_n \in (1+\delta_n^-, 1+\delta_n^+)$ such that $t_n \to 1$ and $\langle I'(t_n v_n), t_n v_n \rangle = 0$. This concludes the proof. □

We now fix $\bar{u} \in H_a^1(\mathbb{R}^N)$ such that (6.25) holds. Since $p > 2$, there exists $\bar{t}$ such that

$$\begin{aligned} I(t\bar{u}) < 0, &\quad \text{for all} \quad t > \bar{t}, \\ J_\lambda(t\bar{u}) < 0, &\quad \text{for all} \quad t > \bar{t} \text{ and } \lambda > 0. \end{aligned}$$

Set

$$\mathscr{P} := \{\gamma \in C([0,1], H_a^1(\mathbb{R}^N)) \, : \, \gamma(0) = 0, \, \gamma(1) = \bar{t}\bar{u}\}, \tag{6.34}$$

$$c_g := \inf_{\gamma \in \mathscr{P}} \sup_{u \in \gamma} J_\lambda(u). \tag{6.35}$$

Proposition 6.10 *There exists $\lambda_0 > 0$, $R_0 = R_0(\lambda_0) > 0$, $\delta_0 = \delta_0(\lambda_0) > 0$ such that for all $\lambda \in (0, \lambda_0)$, $J_\lambda \geq \delta_0$ on ∂B_{R_0} and $c_g < c_{0,\lambda} + I_\infty$, where $c_{0,\lambda} := \inf_{u \in \overline{B}_{R_0}} J_\lambda(u)$.*

Proof. By our hypothesis $(K3)$ and the definition of I we can assume that

$$J_0(t\bar{u}) < I(t\bar{u}), \qquad \text{for all } t > 0.$$

An elementary computation implies the existence of some $t_0 \in (0, \bar{t})$ such that

$$\sup_{t \geq 0} J_0(t\bar{u}) = J_0(t_0\bar{u}) < I(t_0\bar{u}) \leq \sup_{t \geq 0} I(t\bar{u}) = I_\infty.$$

So, we can choose $\varepsilon_0 \in (0,1)$ such that

$$\sup_{t\geq 0} J_0(t\bar{u}) < I_\infty - \varepsilon_0. \tag{6.36}$$

Set

$$\lambda_0 := \min\left\{\lambda_1, \frac{\varepsilon_0}{2\bar{t}\,\|\bar{u}\|\,\|g\|_{-1}}, \frac{\varepsilon_0}{2\|g\|_{-1}^2}\right\}. \tag{6.37}$$

By the above definition of λ_0 and applying Lemma 6.7, it follows that there exists $R_0 = R_0(\lambda_0) > 0$ such that for all $\lambda \in (0,\lambda_0)$ the conclusion of Lemma 6.7 holds. Moreover, by virtue of its proof, there exists $\delta_0 = \delta(\lambda_0) > 0$ such that $J_\lambda \geq \delta_0$ on $\partial\overline{B}_{R_0}$. Then relations (6.18) and (6.37) yield

$$c_{0,\lambda} = \inf_{u\in\overline{B}_{R_0}} J_\lambda(u) \geq -\frac{\lambda}{2}\|g\|_{-1}^2 > -\frac{\varepsilon_0}{2}, \quad \text{for all } \lambda \in (0,\lambda_0). \tag{6.38}$$

Fix $u \in \gamma_0 := \{t\bar{t}\bar{u} : 0 \leq t \leq 1\} \in \mathscr{P}$. Then

$$|J_\lambda(u) - J_0(u)| = \lambda\left|\int_{\mathbb{R}^N} g(x)u\,dx\right| \leq \lambda\,\bar{t}\,\|\bar{u}\|\,\|g\|_{-1} \leq \frac{\varepsilon_0}{2}, \quad \text{for all } \lambda \in (0,\lambda_0).$$

Therefore

$$J_\lambda(u) \leq J_0(u) + \frac{\varepsilon_0}{2}, \quad \text{for all } \lambda \in (0,\lambda_0). \tag{6.39}$$

Using relations (6.36), (6.38) and (6.39) we obtain

$$\begin{aligned} c_g &= \inf_{\gamma\in\mathscr{P}} \sup_{u\in\gamma} J_\lambda(u) \leq \sup_{u\in\gamma_0} J_\lambda(u) \\ &\leq \sup_{u\in\gamma_0} J_0(u) + \frac{\varepsilon_0}{2} \leq \sup_{t\geq 0} J_0(t\bar{u}) + \frac{\varepsilon_0}{2} < I_\infty - \frac{\varepsilon_0}{2} < I_\infty + c_{0,\lambda}. \end{aligned}$$

This completes the proof. □

Proof of Theorem 6.2 continued. Consider $R_0 > 0$ and $\delta_0 > 0$ given by Proposition 6.10. In view of its proof, we deduce that for all $\lambda \in (0,\lambda_0)$ the conclusion of Lemma 6.7 holds. Therefore, we obtain the existence of a solution u_0 of problem (6.2) such that $J_\lambda(u_0) = c_{0,\lambda}$.

On the other hand, applying the mountain pass theorem without the Palais–Smale condition, there exists a $(PS)_{c_g}$ sequence $\{u_n\}$ of J_λ, that is,

$$J_\lambda(u_n) = c_g + o(1) \quad \text{and} \quad J'_\lambda(u_n) \to 0 \text{ in } H_a^{-1}(\mathbb{R}^N).$$

Therefore

$$c_g + o(1) + \frac{1}{p}\|J'_\lambda(u_n)\|_{-1}\|u_n\| \geq J_\lambda(u_n) - \tfrac{1}{p}\langle J'_\lambda(u_n), u_n\rangle$$
$$\geq \left(\frac{1}{2} - \frac{1}{p}\right)\|u_n\|^2 - \lambda\left(1 - \frac{1}{p}\right)\|g\|_{-1}\|u_n\|.$$

The above inequality shows that $\{u_n\}$ is bounded in $H_a^1(\mathbb{R}^N)$. Thus, we can assume (up to a subsequence) that $u_n \rightharpoonup u_1$ in $H_a^1(\mathbb{R}^N)$. By Lemma 6.4 it follows that u_1 is a weak solution of problem (6.2).

We claim that $u_0 \neq u_1$. Indeed, by Proposition 6.9, the following alternative holds: either $u_n \to u_1$ in $H_a^1(\mathbb{R}^N)$, which gives

$$J_\lambda(u_1) = \lim_{n\to\infty} J_\lambda(u_n) = c_g > 0 \geq c_{0,\lambda} = J_\lambda(u_0)$$

and the conclusion follows; or

$$c_g = \lim_{n\to\infty} J_\lambda(u_n) \geq J_\lambda(u_1) + I_\infty.$$

In the last case, if we suppose that $u_1 = u_0$ then $J_\lambda(u_1) = J_\lambda(u_0) = c_{0,\lambda}$ and so $c_g \geq c_{0,\lambda} + I_\infty$, which contradicts Proposition 6.10. The proof of Theorem 6.2 is now complete. □

6.2 A Quasilinear Elliptic Problem for p-Laplace Operator

Nonlinear elliptic equations with convex–concave nonlinearities in bounded domains have been studied starting with the seminal paper by Ambrosetti, Brezis and Cerami [6]. They considered the Dirichlet problem

$$\begin{cases} -\Delta u = \lambda u^{q-1} + u^{p-1} & \text{in } \Omega \\ u > 0 & \text{in } \Omega \\ u = 0 & \text{on } \partial\Omega, \end{cases} \tag{6.40}$$

where λ is a positive parameter, $\Omega \subset \mathbb{R}^N$ is a bounded domain with smooth boundary, and $1 < q < 2 < p < 2^\star$ ($2^\star = 2N/(N-2)$ if $N \geq 3$, $2^\star = \infty$ if $N = 1, 2$). Ambrosetti, Brezis and Cerami proved that there exists $\lambda_0 > 0$ such that problem (6.40) admits at least two solutions for all $\lambda \in (0, \lambda_0)$, has one solution for $\lambda = \lambda_0$, and no solution exists provided that $\lambda > \lambda_0$.

Further, Alama and Tarantello [5] studied the related Dirichlet problem with indefinite weights

$$\begin{cases} -\Delta u - \lambda u = k(x)u^q - h(x)u^p & \text{in } \Omega, \\ u > 0 & \text{in } \Omega, \\ u = 0 & \text{on } \partial\Omega, \end{cases} \tag{6.41}$$

where $\lambda \in \mathbb{R}$, $\Omega \subset \mathbb{R}^N$, $N \geq 3$, is a bounded open set with smooth boundary, the functions $h, k \in L^1(\Omega)$ are nonnegative, and $1 < p < q$. For $\lambda \in \mathbb{R}$ in a neighborhood of λ_1 (the first eigenvalue of the Laplace operator in $H_0^1(\Omega)$), the solvability of (6.41) (and corresponding multiplicities) is obtained under various assumptions on h and k. So far, existence, nonexistence and multiplicity results depending on λ and according to the integrability properties of the ratio k^{p-1}/h^{q-1} are already known.

Motivated by these results, we are concerned in this section with the existence and multiplicity of solutions in the quasilinear case. More precisely, we consider the problem

$$\begin{cases} -\text{div}(|\nabla u|^{m-2}\nabla u) + |u|^{m-2}u = \lambda |u|^{q-2}u - h(x)|u|^{p-2}u & \text{in } \mathbb{R}^N \\ u \geq 0, & \text{in } \mathbb{R}^N, \end{cases} \tag{6.42}$$

where $\Delta_m u := \text{div}(|\nabla u|^{m-2}\nabla u)$ is the standard m-Laplace operator, $h \in C(\mathbb{R}^N) \cap L^{q/(q-p)}(\mathbb{R}^N)$ is positive, $\lambda > 0$ and

$$2 \leq m < q < p < m^\star,$$

with

$$m^\star = \begin{cases} \dfrac{Nm}{N-m} & \text{if } N > m \\ \infty & \text{if } N \leq m. \end{cases}$$

Without altering the proof arguments below, the coefficient 1 of the dominating term $|u|^{m-2}u$ can be replaced by any function $f \in L^\infty(\mathbb{R}^N)$ with $\inf\text{ess}\, \|f\|_{L^\infty} > 0$. Hence (6.42) is the renormalized form.

In the sequel we denote by $W^{1,m}(\mathbb{R}^N)$ the Sobolev space equipped with the norm

$$\|u\|_{W^{1,m}} = \left(\int_{\mathbb{R}^N} (|\nabla u|^m + |u|^m)\, dx \right)^{1/m}.$$

For simplicity we often denote the above norm by $\|u\|$.

By $L_r^p(\mathbb{R}^N)$, $1 \leq p < \infty$, we denote the weighted Lebesgue space

$$L_r^p(\mathbb{R}^N) = \left\{ u \in L^1_{\text{loc}}(\mathbb{R}^N) \, : \, \int_{\mathbb{R}^N} r(x)|u|^p dx < \infty \right\}, \quad r \in C(\mathbb{R}^N), r > 0,$$

where $r(x)$ is a positive continuous function on $\mathbb{R}^N$, equipped with the norm

$$\|u\|_{r,p} = \left(\int_{\mathbb{R}^N} r(x)|u|^p dx \right)^{1/p}.$$

If $r(x) \equiv 1$ on $\mathbb{R}^N$, the norm is denoted by $\| \cdot \|_p$.

We are concerned with the existence and multiplicity of weak solutions of problem (6.42) in a subspace E of $W^{1,m}(\mathbb{R}^N)$, which is defined by

$$E = \left\{ u \in W^{1,m}(\mathbb{R}^N) \, : \, \int_{\mathbb{R}^N} h(x)|u|^p dx < \infty \right\}.$$

Then E is a Banach space if equipped with the norm

$$\|u\|_E = \left(\|u\|^m_{W^{1,m}} + \|u\|^m_{r,p} \right)^{1/m}.$$

We define a *weak solution* of problem (6.42) as a function $u \in E$ with $u(x) \geq 0$ a.e. in $\mathbb{R}^N$, satisfying

$$\int_{\mathbb{R}^N} |\nabla u|^{m-2} \nabla u \nabla v dx + \int_{\mathbb{R}^N} |u|^{m-2} uv dx - \lambda \int_{\mathbb{R}^N} |u|^{q-2} uv dx + \int_{\mathbb{R}^N} h(x)|u|^{p-2} uv dx = 0,$$

for all $v \in E$.

The main result in the present section establishes the following properties: the nonexistence of nontrivial solutions to problem (6.42) if λ is small enough; the existence of at least two nontrivial solutions for problem (6.42) if λ is large enough.

Theorem 6.11 *Under the above hypotheses there exists $\lambda^\star > 0$ such that*

(i) *if $0 < \lambda < \lambda^\star$, then problem (6.42) does not possess any nontrivial weak solution.*

(ii) *if $\lambda > \lambda^\star$, then problem (6.42) admits at least two nontrivial weak solutions.*

(iii) *if $\lambda = \lambda^\star$, then problem (6.42) has at least one nontrivial weak solution.*

Before proceeding with the proof, let us outline the main ideas:

(a) There exists $\lambda^\star > 0$ such that problem (6.42) does not have any solution for any $\lambda < \lambda^\star$. This means that if a solution exists then λ must be sufficiently large. One of the key arguments in this proof is based on the assumption $p > q$. In particular, this proof yields an energy lower bound of solutions in term of λ that will be useful to conclude that problem (6.42) has a nontrivial solution if $\lambda = \lambda^\star$.

(b) There exists $\lambda^{\star\star} > 0$ such that problem (6.42) admits at least two solutions for all $\lambda > \lambda^{\star\star}$. Next, by the properties of $\lambda^\star$ and $\lambda^{\star\star}$ we deduce that $\lambda^\star = \lambda^{\star\star}$.

Proof. We shall perform the proof in several steps.

Step 1: Nonexistence for $\lambda > 0$ small.

Let $\Phi_\lambda : E \to \mathbb{R}$ be the energy functional given by

$$\Phi_\lambda(u) = \frac{1}{m}\|u\|^m - \frac{\lambda}{q}\|u\|_q^q + \frac{1}{p}\|u\|_{h,p}^p.$$

Then $\Phi_\lambda \in C^1(E,\mathbb{R})$ and for all $u, v \in E$

$$\langle \Phi_\lambda'(u), v\rangle = \int_{\mathbb{R}^N}(|\nabla u|^{m-2}\nabla u \nabla v + |u|^{m-2}uv)dx - \lambda \int_{\mathbb{R}^N}|u|^{q-2}uvdx + \int_{\mathbb{R}^N} h(x)|u|^{p-2}uvdx.$$

Weak solutions of problem (6.42) are found as the critical points of the functional Φ_λ in E.

Let us now assume that $u \in E$ is a weak solution of problem (6.42). Then

$$\|u\|^m + \|u\|_{h,p}^p = \lambda\|u\|_q^q. \tag{6.43}$$

Define

$$H := \int_{\mathbb{R}^N} h(x)^{q/(q-p)}dx \in \mathbb{R}^+. \tag{6.44}$$

To proceed further, we need Young's inequality

$$ab \le \frac{a^\alpha}{\alpha} + \frac{b^\beta}{\beta} \quad \text{for all } a,b > 0,$$

where α, $\beta > 1$ satisfy $1/\alpha + 1/\beta = 1$.

Taking $a = h(x)^{q/p}|u|^q$, $b = \lambda/[h(x)]^{q/p}$, $\alpha = p/q$ and $\beta = p/(p-q)$, we obtain

$$h(x)^{q/p}|u|^q \frac{\lambda}{h(x)^{q/p}} \le \frac{q}{p}(h(x)^{q/p}|u|^q)^{p/q} + \frac{p-q}{p}\left(\frac{\lambda}{h(x)^{q/p}}\right)^{p/(p-q)}.$$

Integrating over $\mathbb{R}^N$ we have

$$\lambda\|u\|_q^q \le \frac{q}{p}\|u\|_{h,p}^p + \frac{p-q}{p}\lambda^{p/(p-q)}\int_{\mathbb{R}^N} h(x)^{q/(q-p)}dx.$$

The above inequality and relation (6.43) imply

$$\begin{aligned}\|u\|^m &\leq \frac{p-q}{p}\lambda^{p/(p-q)}\int_{\mathbb{R}^N} h(x)^{q/(q-p)}dx + \frac{q-p}{p}\|u\|_{h,p}^p \\ &\leq \frac{p-q}{p}\lambda^{p/(p-q)}H,\end{aligned} \tag{6.45}$$

for $q < p$.

Since $m < q < m^\star$, the Sobolev embedding $W^{1,m}(\mathbb{R}^N) \subset L^q(\mathbb{R}^N)$ is continuous, so that there exists a positive constant C_q such that

$$C_q\|v\|_q^m \leq \|v\|^m \quad \text{for all } v \in W^{1,m}(\mathbb{R}^N).$$

On the other hand, for $\|u\|_{h,p} \geq 0$, it follows from (6.43) that

$$\|u\|^m \leq \lambda\|u\|_q^q.$$

Combining the last two inequalities we obtain

$$C_q\|u\|_q^m \leq \|u\|^m \leq \lambda\|u\|_q^q. \tag{6.46}$$

Retaining the first and the last terms of (6.46) we get

$$(C_q\lambda^{-1})^{q/(q-m)} \leq \|u\|_q^q.$$

That inequality combined with (6.46) leads to

$$C_q[(C_q\lambda^{-1})^{q/(q-m)}]^{m/q} \leq \|u\|^m.$$

By relation (6.45) and the above inequality we have

$$C_q(C_q\lambda^{-1})^{m/(q-m)} \leq \|u\|^m \leq \frac{p-q}{p}\lambda^{p/(p-q)}H. \tag{6.47}$$

Retaining the first and the last term it follows that

$$\lambda > \left(C_q^{q/(m-q)}\frac{p-q}{p}H\right)^{(q-p)(q-m)/q(p-m)},$$

for $H > 0$ by (6.44). Denoting the term in the right-hand side of the above inequality by μ, we conclude that Theorem 6.11–(i) holds true, by putting

$$\lambda^\star := \sup\{\lambda > 0 : \text{(6.42) does not admit any nontrivial weak solution}\}. \tag{6.48}$$

Clearly $\lambda^\star \geq \mu > 0$.

Step 2: Existence if λ is large

We start with several auxiliary results.

Lemma 6.12 *The functional Φ_λ is coercive.*

Proof. We need the following elementary inequality: for every $k_1 > 0$, $k_2 > 0$ and $0 < s < r$ we have

$$k_1|t|^s - k_2|t|^r \leq C_{rs}k_1\left(\frac{k_1}{k_2}\right)^{s/(r-s)} \quad \text{for all } t \in \mathbb{R}, \tag{6.49}$$

where $C_{rs} > 0$ is a constant depending on r and s.

Taking $k_1 = \lambda/q$, $k_2 = (m-1)h(x)/mp$, $s = q$ and $r = p$ (so that $s < r$ is verified, for $q < p$), in (6.49) for all $x \in \mathbb{R}^N$ we obtain

$$\begin{aligned}\frac{\lambda}{q}|u(x)|^q - \frac{(m-1)h(x)}{mp}|u(x)|^p &\leq C_{pq}\frac{\lambda}{q}\left(\frac{\lambda/q}{(m-1)h(x)/mp}\right)^{q/(p-q)} \\ &= \frac{C_{pq}}{q}\left(\frac{mp}{q(m-1)}\right)^{q/(p-q)}\lambda^{p/(p-q)}h(x)^{q/(q-p)},\end{aligned}$$

where $C_{pq} > 0$ is a constant depending on p and q. Integrating the above inequality over $\mathbb{R}^N$, we find

$$\int_{\mathbb{R}^N}\left(\frac{\lambda}{q}|u|^q - \frac{(m-1)h(x)}{mp}|u|^p\right)dx \leq K\lambda^{p/(p-q)}\int_{\mathbb{R}^N} h(x)^{q/(q-p)}dx.$$

By assumption (6.44) there exists a constant $C_\lambda > 0$ such that

$$\frac{\lambda}{q}\|u\|_q^q - \frac{m-1}{mp}\|u\|_{h,p}^p \leq C_\lambda.$$

Therefore

$$\begin{aligned}\Phi_\lambda(u) &= \frac{1}{m}\|u\|^m - \left[\frac{\lambda}{q}\|u\|_q^q - \frac{m-1}{mp}\|u\|_{h,p}^p\right] - \frac{m-1}{mp}\|u\|_{h,p}^p + \frac{1}{p}\|u\|_{h,p}^p \\ &\geq \frac{1}{m}\|u\|^m + \frac{1}{mp}\|u\|_{h,p}^p - C_\lambda,\end{aligned}$$

and so Φ_λ is coercive in E. □

Lemma 6.13 *If $\{u_n\}$ is a sequence in E such that $\{\Phi_\lambda(u_n)\}$ is bounded in $\mathbb{R}$, then there exists a subsequence of $\{u_n\}$, still relabeled $\{u_n\}$, which converges weakly in E to some $u_0 \in E$ and*

$$\Phi_\lambda(u_0) \leq \liminf_{n\to\infty} \Phi_\lambda(u_n).$$

Proof. By the fact that $\{\Phi_\lambda(u_n)\}$ is bounded, it follows that both sequences $\{\|u_n\|\}$ and $\{\|u_n\|_{h,p}\}$ are bounded. Therefore, $\{\|u_n\|_E\}$ is bounded and there exists $u_0 \in E$ such that

$$\begin{aligned} u_n &\rightharpoonup u_0 \text{ in } W^{1,m}(\mathbb{R}^N), \\ u_n &\to u_0 \text{ in } L^p_h(\mathbb{R}^N), \\ u_n &\to u_0 \text{ in } L^s_{\text{loc}}(\mathbb{R}^N) \quad \text{for all } s \in [1, m^\star). \end{aligned}$$

Let us define

$$F(x,u) = \frac{\lambda}{q}|u|^q - h(x)\frac{|u|^p}{p}$$

and

$$f(x,u) = F_u(x,u) = \lambda|u|^{q-2}u - h(x)|u|^{p-2}u,$$

so that

$$f_u(x,u) = \lambda(q-1)|u|^{q-2} - h(x)(p-1)|u|^{p-2}.$$

Using again inequality (6.49) for $k_1 = \lambda(q-1)$, $k_2 = h(x)(p-1)$, $s = q-2$, $r = p-2$, we obtain

$$\begin{aligned} f_u(x,u) &= \lambda(q-1)|u|^{q-2} - h(x)(p-1)|u|^{p-2} \\ &\leq C \cdot \lambda \cdot (q-1) \cdot \left(\frac{\lambda(q-1)}{h(x)(p-1)}\right)^{(q-2)/(p-q)}, \end{aligned}$$

where C is a positive constant depending only on p and q.

This yields,

$$f_u(x,u) \leq C_{pq} \cdot \lambda \cdot \left(\frac{\lambda}{h(x)}\right)^{(q-2)/(p-q)}, \tag{6.50}$$

where C_{pq} is a positive constant depending only on p and q. According to the definition of Φ_λ and F we obtain the following estimate for $\Phi_\lambda(u_0) - \Phi_\lambda(u_n)$

$$\begin{aligned} \Phi_\lambda(u_0) - \Phi_\lambda(u_n) = \frac{1}{m}(\|u_0\|^m - \|u_n\|^m) \\ + \int_{\mathbb{R}^N} [F(x,u_n) - F(x,u_0)]dx. \end{aligned} \tag{6.51}$$

By position

$$\int_0^s f_u(x,u_0+t(u_n-u_0))dt = \frac{1}{u_n-u_0}[f(x,u_0+s(u_n-u_0))-f(x,u_0)]$$
$$= \frac{1}{u_n-u_0}[F_u(x,u_0+s(u_n-u_0))-F_u(x,u_0)].$$

Integrating the above relation over $[0,1]$, we obtain

$$\int_0^1 \left(\int_0^s f_u(x,u_0+t(u_n-u_0))dt\right) ds$$
$$= \frac{1}{u_n-u_0}\int_0^1 [F_u(x,u_0+s(u_n-u_0))-F_u(x,u_0)]ds$$
$$= \frac{1}{(u_n-u_0)^2}[F(x,u_n)-F(x,u_0)] - \frac{f(x,u_0)}{u_n-u_0}.$$

The above equality can be rewritten in the following way

$$F(x,u_n)-F(x,u_0) = (u_n-u_0)^2\int_0^1\left(\int_0^s f_u(x,u_0+t(u_n-u_0))dt\right)ds \tag{6.52}$$
$$+(u_n-u_0)f(x,u_0).$$

Introducing relation (6.52) in (6.51) we get

$$\begin{aligned}\Phi_\lambda(u_0)-\Phi_\lambda(u_n) &= \frac{1}{m}(\|u_0\|^m-\|u_n\|^m)+\int_{\mathbb{R}^N}(u_n-u_0)f(x,u_0)dx\\ &+\int_{\mathbb{R}^N}(u_n-u_0)^2\int_0^1\int_0^s f_u(x,u_0+t(u_n-u_0))dtdsdx\\ &\le \frac{1}{m}(\|u_0\|^m-\|u_n\|^m)+\int_{\mathbb{R}^N}(u_n-u_0)f(x,u_0)dx\\ &+C_1\int_{\mathbb{R}^N}(u_n-u_0)^2h(x)^{(q-2)/(q-p)}dx,\end{aligned} \tag{6.53}$$

where the last inequality follows from (6.50) and $C_1 = C_{pq}\lambda^{(p-2)/(p-q)}$. It remains to show that the last two integrals converge to 0 as $n\to\infty$.

We define $J: E\to\mathbb{R}$ by

$$J(v) = \int_{\mathbb{R}^N} f(x,u_0)\,v dx.$$

Obviously, J is linear. We shall show that J is also continuous. Indeed,

$$\begin{aligned}|J(v)| &\le \int_{\mathbb{R}^N}|f(x,u_0)|\cdot|v|\,dx\\ &\le \lambda\int_{\mathbb{R}^N}|u_0|^{q-1}|v|\,dx+\int_{\mathbb{R}^N}h(x)|u_0|^{p-1}|v|\,dx.\end{aligned} \tag{6.54}$$

On the other hand, using Hölder's inequality, it follows that

$$\int_{\mathbb{R}^N} |u_0|^{q-1}|v|dx \le \|u_0\|_q^{q-1}\|v\|_q.$$

Since $W^{1,m}(\mathbb{R}^N)$ is continuously embedded in $L^q(\mathbb{R}^N)$ we deduce that there exists a constant $C>0$ such that

$$\|v\|_q \le C\|v\|_{W^{1,m}(\mathbb{R}^N)} \quad \text{for all } v\in W^{1,m}(\mathbb{R}^N).$$

Combining the last two inequalities with the fact that

$$\|v\|_{W^{1,m}(\mathbb{R}^N)} \le \|v\|_E\,,$$

we deduce that there exists a positive constant $c_q>0$ such that

$$\int_{\mathbb{R}^N} |u_0|^{q-1}|v|\,dx \le c_q\|v\|_E\,. \tag{6.55}$$

Applying again Hölder's inequality we obtain

$$\begin{aligned}\int_{\mathbb{R}^N} h(x)|u_0|^{p-1}|v|dx &= \int_{\mathbb{R}^N} (h(x)^{(p-1)/p}|u_0|^{p-1})(h(x)^{1/p}|v|)dx\\ &\le \|u_0\|_{h,p}^{p-1}\|v\|_{h,p} \le C_0\|v\|_{h,p}\\ &\le C_0\|v\|_E\,,\end{aligned} \tag{6.56}$$

where C_0 is a positive constant.

By (6.54), (6.55) and (6.56) we conclude that there exists a positive constant κ such that

$$|J(v)| \le \kappa\|v\|_E \quad \text{for all } v\in E,$$

and so J is continuous.

Since $\{u_n\}$ converges weakly to u_0 in E and J is linear and continuous we deduce that

$$J(u_n)\to J(u_0),$$

in other words

$$\lim_{n\to\infty}\int_{\mathbb{R}^N} f(x,u_0)(u_n-u_0)\,dx = 0. \tag{6.57}$$

In order to show that

$$\lim_{n\to\infty}\int_{\mathbb{R}^N} (u_n-u_0)^2 h(x)^{(q-2)/(q-p)}dx = 0, \tag{6.58}$$

we first note that for all $R > 0$

$$\begin{aligned}\int_{\mathbb{R}^N}(u_n-u_0)^2h(x)^{(q-2)/(q-p)}dx &= \int_{\{|x|<R\}}(u_n-u_0)^2h(x)^{(q-2)/(q-p)}dx\\ &+\int_{\{|x|\geq R\}}(u_n-u_0)^2h(x)^{(q-2)/(q-p)}dx\\ &\leq \left(\int_{\{|x|<R\}}h(x)^{q/(q-p)}dx\right)^{(q-2)/q}\cdot\left(\int_{\{|x|<R\}}|u_n-u_0|^q dx\right)^{2/q}\\ &+\left(\int_{\{|x|\geq R\}}h(x)^{q/(q-p)}\right)^{(q-2)/q}\cdot\left(\int_{\{|x|\geq R\}}|u_n-u_0|^q dx\right)^{2/q}.\end{aligned} \tag{6.59}$$

By hypothesis (6.44) we have

$$\int_{\{|x|<R\}}h(x)^{q/(q-p)}dx \leq \int_{\mathbb{R}^N}h(x)^{q/(q-p)}dx = H < \infty \quad \text{for all } R > 0.$$

On the other hand, for all $\varepsilon > 0$ there exists $R_\varepsilon > 0$ such that

$$\int_{\{|x|\geq R_\varepsilon\}}h(x)^{q/(q-p)}dx < \varepsilon.$$

Using the fact that $m < q < m^\star$ we deduce that $W^{1,m}(B_{R_\varepsilon}(0))$ is compactly embedded in $L^q(B_{R_\varepsilon}(0))$ and thus

$$\lim_{n\to\infty}\left(\int_{|x|<R_\varepsilon}|u_n-u_0|^q\,dx\right)^{2/q} = 0.$$

Since $\{u_n-u_0\}$ is bounded in E, it is also bounded in $L^q(\mathbb{R}^N)$ and so there exists a positive constant $M > 0$ such that

$$\left(\int_{|x|\geq R_\varepsilon}|u_n-u_0|^q dx\right)^{2/q} \leq \|u_n-u_0\|_q^2 \leq M.$$

Combining the above information with relation (6.59), we conclude that for any $\varepsilon > 0$ there exists $N_\varepsilon > 0$ such that for all $n \geq N_\varepsilon$ we have

$$\int_{\mathbb{R}^N}(u_n-u_0)^2h(x)^{(q-2)/(q-p)}dx \leq H\varepsilon + M\varepsilon^{(q-2)/q}.$$

Therefore, (6.58) holds true.

Since $(u_n)_n$ converges weakly to u_0 in $W^{1,m}(\mathbb{R}^N)$ we have

$$\liminf_{n\to\infty}\|u_n\|^m_{W^{1,m}(\mathbb{R}^N)} \geq \|u_0\|^m_{W^{1,m}(\mathbb{R}^N)}.$$

Passing to the limit in (6.53) and taking into account that (6.57) and (6.58) hold true, we obtain

$$\Phi_\lambda(u_0) \leq \liminf_{n\to\infty} \Phi_\lambda(u_n).$$

Thus, Φ_λ is weakly lower semicontinuous.

The proof of Lemma 6.13 is now complete. □

Proof of Theorem 6.11 continued. Using Lemmas 6.12, 6.13 and Theorem 1.2 in [188] we deduce that there exists a global minimizer $u \in E$ of Φ_λ, that is,

$$\Phi_\lambda(u) = \inf_{v\in E} \Phi_\lambda(v).$$

It is obvious that u is a weak solution of problem (6.42). We prove that $u \not\equiv 0$ in E. To do that we show that $\inf_E \Phi_\lambda < 0$, provided that the parameter λ is sufficiently large.

Let us set

$$\overline{\lambda} = \inf_{u\in E} \left\{ \frac{q}{m}\|u\|^m + \frac{q}{p}\|u\|^p_{h,p} \,:\, \|u\|_q = 1 \right\}.$$

We point out that $\overline{\lambda} > 0$. Indeed, for any $u \in E$ with $\|u\|_q = 1$ by Hölder's inequality and by (6.44) we have

$$\begin{aligned} 1 = \|u\|_q^q &\leq \left(\int_{\mathbb{R}^N} h(x)^{q/(q-p)} dx \right)^{(p-q)/p} \cdot \left(\int_{\mathbb{R}^N} h(x)|u|^p \, dx \right)^{q/p} \\ &= H^{(p-q)/p} \|u\|^q_{h,p}, \end{aligned}$$

so that

$$\overline{\lambda} \geq \frac{q}{p} H^{(q-p)/q} > 0,$$

for $H > 0$ by assumption. Let $\lambda > \overline{\lambda}$. Then there exists a function $u_1 \in E$, with $\|u_1\|_q = 1$, such that

$$\lambda \|u_1\|_q^q = \lambda > \frac{q}{m}\|u_1\|^m + \frac{q}{p}\|u_1\|^p_{h,p}.$$

This can be rewritten as

$$\Phi_\lambda(u_1) = \frac{1}{m}\|u_1\|^m - \frac{\lambda}{q}\|u_1\|_q^q + \frac{1}{p}\|u_1\|^p_{h,p} < 0$$

and consequently $\inf_{u\in E}\Phi_\lambda(u) < 0$. Therefore, there exists $\lambda_0 = \overline{\lambda} > 0$ such that problem (6.42) has a nontrivial weak solution $u_1 \in E$ for any $\lambda > \lambda_0$, and $\Phi_\lambda(u_1) < 0$. Since $\Phi_\lambda(u_1) = \Phi_\lambda(|u_1|)$ and $|u_1| \in E$, we may assume that $u_1 \ge 0$ a.e. in $\mathbb{R}^N$. □

In the following we are looking for the second nontrivial weak solution for problem (6.42).

Fix $\lambda \ge \lambda_0$. Set

$$g(x,t) = \begin{cases} 0 & \text{if } t < 0, \\ \lambda t^{q-1} - h(x)t^{p-1} & \text{if } 0 \le t \le u_1(x), \\ \lambda u_1(x)^{q-1} - h(x)u_1(x)^{p-1} & \text{if } t > u_1(x), \end{cases}$$

and

$$G(x,t) = \int_0^t g(x,s)ds.$$

Define the functional $\Psi : E \to \mathbb{R}$ by

$$\Psi(u) = \frac{1}{m}\|u\|^m - \int_{\mathbb{R}^N} G(x,u)dx.$$

Clearly, $\Psi \in C^1(E,\mathbb{R})$ and

$$\langle \Psi'(u), v\rangle = \int_{\mathbb{R}^N} (|\nabla u|^{m-2}\nabla u \nabla v + |u|^{m-2}uv)\,dx - \int_{\mathbb{R}^N} g(x,u)\,v\,dx,$$

for all u, $v \in E$. Moreover, *if u is a critical point of Ψ, then $u \ge 0$ a.e. in $\mathbb{R}^N$.*

Next, we are concerned with the location of critical points of the energy functional Ψ.

Lemma 6.14 *If u is a critical point of Ψ, then $u \le u_1$.*

Proof. For a function v we define the positive part $v^+(x) = \max\{v(x), 0\}$. Then $v^+ \in E$ whenever $v \in E$. We have

$$\begin{aligned} 0 &= \langle \Psi'(u) - \Phi'_\lambda(u_1), (u-u_1)^+\rangle \\ &= \int_{\mathbb{R}^N} (|\nabla u|^{m-2}\nabla u - |\nabla u_1|^{m-2}\nabla u_1)\nabla(u-u_1)^+ dx \\ &\quad + \int_{\mathbb{R}^N} (|u|^{m-2}u - |u_1|^{m-2}u_1)(u-u_1)^+ dx \\ &\quad - \int_{\mathbb{R}^N} [g(x,u) - \lambda u_1^{q-1} + h(x)u_1^{p-1}](u-u_1)^+ dx, \end{aligned}$$

which yields

$$
\begin{aligned}
0 = & \int_{\{u>u_1\}} (|\nabla u|^{m-2}\nabla u - |\nabla u_1|^{m-2}\nabla u_1)(\nabla u - \nabla u_1)\,dx \\
& + \int_{\{u>u_1\}} (|u|^{m-2}u - |u_1|^{m-2}u_1)(u-u_1)\,dx \\
\geq & \int_{\{u>u_1\}} (|\nabla u|^{m-1} - |\nabla u_1|^{m-1})(|\nabla u| - |\nabla u_1|)\,dx \\
& + \int_{\{u>u_1\}} (|u|^{m-1} - |u_1|^{m-1})(|u| - |u_1|)\,dx \geq 0.
\end{aligned}
$$

Thus, we obtain $u \leq u_1$ and the proof of Lemma 6.14 is complete. □

In the following, via the mountain pass theorem, we determine a critical point $u_2 \in E$ of Ψ such that $\Psi(u_2) > 0$. By the above lemma we shall deduce that $0 \leq u_2 \leq u_1$ in Ω. Therefore,

$$
g(x,u_2) = \lambda u_2^{q-1} - h(x)u_2^{p-1} \quad \text{and} \quad G(x,u_2) = \frac{\lambda}{q}u_2^q - \frac{h(x)}{p}u_2^p,
$$

so that

$$
\Psi(u_2) = \Phi_\lambda(u_2) \quad \text{and} \quad \Psi'(u_2) = \Phi'_\lambda(u_2).
$$

More precisely, we find

$$
\Phi_\lambda(u_2) > 0 = \Phi_\lambda(0) > \Phi_\lambda(u_1) \qquad \text{and} \qquad \Phi'_\lambda(u_2) = 0.
$$

This shows that u_2 is a weak solution of problem (6.42) such that $0 \leq u_2 \leq u_1$, $u_2 \neq 0$ and $u_2 \neq u_1$.

In order to find u_2 described above we prove the following result.

Lemma 6.15 *There exists $\rho \in (0, \|u_1\|)$ and $a > 0$ such that*

$$
\Psi(u) \geq a \quad \textit{for all } u \in E \ \textit{with} \ \|u\| = \rho.
$$

Proof. We have

$$
\begin{aligned}
\Psi(u) &= \frac{1}{m}\|u\|^m - \int_{\{u>u_1\}} G(x,u)\,dx - \int_{\{u\leq u_1\}} G(x,u)\,dx \\
&= \frac{1}{m}\|u\|^m - \frac{\lambda}{q}\int_{\{u>u_1\}} u_1^q dx + \frac{1}{p}\int_{\{u>u_1\}} h(x)u_1^p dx \\
&\quad - \frac{\lambda}{q}\int_{\{0\leq u\leq u_1\}} u^q dx + \frac{1}{p}\int_{\{0\leq u\leq u_1\}} h(x)u^p dx \\
&\geq \frac{1}{m}\|u\|^m - \frac{\lambda}{q}\|u\|_q^q.
\end{aligned}
$$

On the other hand, the continuous Sobolev embedding of E into $L^q(\mathbb{R}^N)$ implies that there exists a positive constant $L > 0$ such that

$$\|v\|_q \le L\|v\| \quad \text{for all } v \in E.$$

The above inequalities yield

$$\Psi(u) \ge \frac{1}{m}\|u\|^m - L_1\|u\|^q = \|u\|^m\left(\frac{1}{m} - L_1\|u\|^{q-m}\right),$$

where $L_1 = \lambda L^q/q$ is a positive constant. Since $q > m$ it is clear that Lemma 6.15 holds true. □

Lemma 6.16 *The functional Ψ is coercive.*

Proof. For each $u \in E$ we have

$$\begin{aligned}
\Psi(u) &= \frac{1}{m}\|u\|^m - \frac{\lambda}{q}\int_{\{u>u_1\}} u_1^q dx + \frac{1}{p}\int_{\{u>u_1\}} h(x)u_1^p dx \\
&\quad - \frac{\lambda}{q}\int_{\{0\le u\le u_1\}} u^q dx + \frac{1}{p}\int_{\{0\le u\le u_1\}} h(x)u^p dx \\
&\ge \frac{1}{m}\|u\|^m - \frac{\lambda}{q}\int_{\mathbb{R}^N} u_1^q dx \\
&= \frac{1}{m}\|u\|^m - L_2,
\end{aligned}$$

where L_2 is a positive constant, for $u_1 \neq 0$. The above inequality implies that $\Psi(u) \to \infty$ as $\|u\| \to \infty$, that is, Ψ is coercive, as required. □

Proof of Theorem 6.11 continued. Using Lemma 6.15 and the mountain pass theorem, there exists a sequence $\{v_n\} \subset E$ such that

$$\Psi(v_n) \to c > 0 \quad \text{and} \quad \Psi'(v_n) \to 0, \tag{6.60}$$

where

$$c = \inf_{\gamma\in\Gamma}\max_{t\in[0,1]} \Psi(\gamma(t))$$

and

$$\Gamma = \{\gamma \in C([0,1],E) : \gamma(0) = 0,\ \gamma(1) = u_1\}.$$

By relation (6.60) and Lemma 6.16 we obtain that $\{v_n\}$ is bounded and thus passing eventually to a subsequence, still denoted by $\{v_n\}$, we may assume that there exists $u_2 \in E$ such that v_n converges weakly to u_2. Standard arguments based on the Sobolev embeddings will show that

$$\lim_{n\to\infty} \langle \Psi'(v_n), \varphi \rangle = \langle \Psi'(u_2), \varphi \rangle$$

for any $\varphi \in C_0^\infty(\mathbb{R}^N)$. Taking into account that $E \subset W^{1,m}(\mathbb{R}^N)$ and $C_0^\infty(\mathbb{R}^N)$ is dense in $W^{1,m}(\mathbb{R}^N)$, the above information implies that u_2 is a weak solution of problem (6.42).

We conclude that problem (6.42) admits at least two nontrivial weak solutions for all $\lambda > \lambda_0$.

Set

$$\lambda^{**} := \inf\{\lambda > 0 : \text{problem (6.42) admits a nontrivial weak solution}\}.$$

Then $\lambda^{**} \geq \lambda^\star > 0$, where $\lambda^\star$ is the parameter defined in (6.48).

Let us consider the constrained minimization problem

$$\Lambda := \inf_{v\in E} \left\{ \frac{1}{m}\|v\|^m + \frac{1}{p}\|v\|_{h,p}^p \, : \, \|v\|_q^q = q \right\}. \tag{6.61}$$

Let $\{v_n\} \subset E$ be a minimizing sequence of (6.61). Then $\{v_n\}$ is bounded in E, hence we can assume, without loss of generality, that it converges weekly to some $v \in E$ with $\|v\|_q^q = q$. Moreover, by lower semicontinuity arguments we have

$$\Lambda = \frac{1}{m}\|v\|^m + \frac{1}{p}\|v\|_{h,p}^p.$$

Thus, $\Phi_\lambda(v) = \Lambda - \lambda$ for all $\lambda > \Lambda$.

To complete the proof of Theorem 6.11 it is enough to show the following crucial facts:

(a) problem (6.42) has at least two distinct solutions for any $\lambda > \lambda^{**}$.
(b) $\lambda^{**} = \lambda^\star$ and problem (6.42) admits a nontrivial weak solution if $\lambda = \lambda^\star$.

Claim (a) follows by standard monotonicity techniques; as for claim (b) we shall use some arguments from Filippucci, Pucci and Rădulescu [76].

Fix $\lambda > \lambda^{**}$. By the definition of λ^{**}, there exists $\mu \in (\lambda^{**}, \lambda)$ such that Φ_μ has a nontrivial critical point $u_\mu \in E$. Clearly, u_μ is a subsolution of (6.42). In order to find a supersolution of (6.42) which dominates u_μ, we consider the constrained minimization problem

$$\inf\left\{\frac{1}{m}\|w\|^m + \frac{1}{p}\|w\|_p^p - \frac{\lambda}{q}\|w\|_q^q \,:\, w \in E \text{ and } w \geq u_\mu\right\}.$$

As before, one can show that the above minimization problem has a solution $u_\lambda \geq u_\mu$ which is also a weak solution of problem (6.42), provided $\lambda > \lambda^{**}$. Thus, problem (6.42) has a weak solution for all $\lambda > \lambda^{**}$ so $\lambda^{**} = \lambda^*$.

It remains to show that (6.42) has solutions for $\lambda = \lambda^*$. To this aim, let $\{\lambda_n\}$ be a decreasing sequence converging to λ^* and let $\{u_n\}$ be a corresponding sequence of nonnegative weak solutions of (6.42). By the properties of Ψ, the sequence $\{u_n\}$ is bounded in E, so that, without loss of generality, we may assume that it converges weakly in E, strongly in $L_h^p(\Omega)$, and pointwise to some $u^* \in E$, with $u^* \geq 0$.

Replacing u by u_n in the definition of a weak solution and then passing to the limit as $n \to \infty$ we find that U^* is a weak solution of (6.42) for $\lambda = \lambda^*$.

It remains to show that $u^* \neq 0$.

A key ingredient in this argument is the lower bound energy given in (6.46). Hence, since u_n is a nontrivial weak solution of problem (7.2) corresponding to λ_n, we have $\|u_n\|_{a,b}^p \geq (C^r/\lambda^p)^{1/(r-p)}$ by (6.46), where $C > 0$ is a positive constant independent of λ_n. Next, since $\lambda_n \searrow \lambda^*$ as $n \to \infty$ and $\lambda^* > 0$, it is enough to show that

$$\|u_n - u^*\|_{a,b} \to 0 \quad \text{as } n \to \infty. \tag{6.62}$$

Since u_n and u^* are weak solutions of (6.42) corresponding to λ_n and λ^*, we have

$$\begin{aligned}
&\int_{\mathbb{R}^N} \left(|\nabla u_n|^{m-2}\nabla u_n - |\nabla u^*|^{p-2}\nabla u^*\right) \cdot \nabla(u_n - u^*)dx \\
&\quad + \int_{\mathbb{R}^N} \left(|u_n|^{m-2}u_n - |u^*|^{m-2}u^*\right)(u_n - u^*)dx \\
&\quad + \int_{\mathbb{R}^N} h(x)(|u_n|^{p-2}u_n - |u^*|^{p-2}u^*)(u_n - u^*)dx \\
&= \int_{\mathbb{R}^N} \left(\lambda_n |u_n|^{q-2}u_n - \lambda^* |u^*|^{q-2}u^*\right)(u_n - u^*)dx.
\end{aligned}$$

Elementary monotonicity properties imply that

$$\int_{\mathbb{R}^N} h(x)\left(|u_n|^{p-2}u_n - |u^*|^{p-2}u^*\right)(u_n - u^*)dx \geq 0.$$

Hence

$$\begin{aligned}&\int_{\mathbb{R}^N} \left(|\nabla u_n|^{m-2}\nabla u_n - |\nabla u^*|^{p-2}\nabla u^*\right)\cdot\nabla(u_n-u^*)dx\\&+\int_{\mathbb{R}^N}\left(|u_n|^{m-2}u_n-|u^*|^{m-2}u^*\right)(u_n-u^*)dx\\&\leq\int_{\mathbb{R}^N}\left(\lambda_n|u_n|^{q-2}u_n-\lambda^*|u^*|^{q-2}u^*\right)(u_n-u^*)dx\to 0 \text{ as } n\to\infty.\end{aligned}\tag{6.63}$$

On the other hand, since $m\geq 2$, there exists a positive constant $c=c(m)>0$ such that

$$|\xi-\zeta|^m\leq c(|\xi|^{m-2}\xi-|\zeta|^{m-2}\zeta)(\xi-\zeta)\quad\text{for all }\xi,\zeta\in\mathbb{R}^N.$$

Combining this fact with (6.63) we find

$$\|u_n-u^*\|\leq c\int_{\mathbb{R}}^{N}\left(\lambda_n|u_n|^{q-2}u_n-\lambda^*|u^*|^{q-2}u^*\right)(u_n-u^*)dx\to 0\text{ as }n\to\infty.$$

Therefore, $u_n\to U^*$ in E. Finally, from (6.47) we have

$$\|u_n\|^m\geq C\lambda^{m/(m-q)}\qquad\text{for all }n\geq 1.$$

Passing to the limit in the above estimate we find $\|u^*\|^m\geq C(\lambda^*)^{m/(m-q)}$ so $u^*\not\equiv 0$. This completes the proof of Theorem 6.11. □

We point out that (6.42) can be studied also in the case when p is supercritical using similar arguments, since the $|u|^p$ term in the energy continues to be coercive. In these cases standard regularity results will lead to stronger results in what concerns the smoothness of solutions, since $W^{1,m}$ is embedded into C^1.

Chapter 7
Some Classes of Polyharmonic Problems

As for everything else, so for a mathematical theory: beauty can be perceived but not explained.

Arthur Cayley (1821–1881)

In this chapter we present several results concerning elliptic problems involving the polyharmonic operator. The first section is devoted to the study of an eigenvalue problem that exhibits a continuous spectrum. The second section of the chapter deals with a boundary value problem with infinitely many solutions while the last section is concerned with a biharmonic problem involving a singular nonlinearity. By taking a different approach in each of these situations we emphasize the complex structure of elliptic problems involving the polyharmonic operator.

7.1 An Eigenvalue Problem with Continuous Spectrum

Let $B = B_R(0)$ be the ball in $\mathbb{R}^N$, $N \geq 1$, centered at the origin and having radius $R > 0$. Consider the linear eigenvalue problem

$$\begin{cases} (-\Delta)^K u = \lambda u & \text{in } B \\ u = Du = \cdots = D^{K-1}u = 0 & \text{on } \partial B, \end{cases} \tag{7.1}$$

where K is a positive integer. Then the lowest eigenvalue λ_1 of problem (7.1) is *simple*, that is, the associated eigenfunctions are merely multiples of each other. Moreover they are radial, strictly monotone in $r = |x|$ and never change sign in B. We

M. Ghergu and V. Rădulescu, *Nonlinear PDEs*, Springer Monographs in Mathematics,
DOI 10.1007/978-3-642-22664-9_7, © Springer-Verlag Berlin Heidelberg 2012

refer to Pucci and Serrin [163] for further properties of eigenvalues of polyharmonic operators.

In this section we are concerned with the nonlinear eigenvalue problem

$$\begin{cases} (-\Delta)^K u = \lambda f(x,u) & \text{in } B \\ u = Du = \cdots = D^{K-1}u = 0 & \text{on } \partial B, \end{cases} \tag{7.2}$$

where λ is a positive parameter and the nonlinear function f is given by

$$f(x,t) = \begin{cases} t, & \text{if } t < 0 \\ h(x,t), & \text{if } t \geq 0, \end{cases} \tag{7.3}$$

where $h : B \times \mathbb{R}_0^+ \to \mathbb{R}$ is a Carathéodory function, $H(x,t) := \int_0^t h(x,s)ds$, and the following conditions are fulfilled:

(H_1) *There exists* $c \in (0,1)$ *such that* $|h(x,t)| \leq ct$ *for all* $t \in \mathbb{R}$ *and a.a.* $x \in B$.
(H_2) *There exists* $t_0 > 0$ *such that* $H(x,t_0) > 0$ *for a.a.* $x \in B$.
(H_3) $\lim_{t\to\infty} \frac{h(x,t)}{t} = 0$ *uniformly in* $B \setminus \mathscr{O}$, *where* $\mathscr{O}$ *is a set of zero Lebesgue measure.*

Some examples of functions h verifying (H_1)–(H_3) in $B \times \mathbb{R}_0^+$ are as follows:

(i) $h(x,t) = \sin(ct)$;
(ii) $h(x,t) = c\log(1+t)$;
(iii) $h(x,t) = g(x)[t^{q(x)-1} - t^{p(x)-1}]$, where $c \in (0,1)$, p, $q : \overline{B} \to (1,2)$ continuous in $\overline{B}$, $\max_{\overline{B}} p(x) < \min_{\overline{B}} q(x)$, $g \in L^\infty(B)$, $\|g\|_\infty = c$.

The main result of this section is the following.

Theorem 7.1 *Suppose that f is of type (7.3) and that (H_1)–(H_3) are fulfilled. Then the first eigenvalue λ_1 of (7.1) is an isolated eigenvalue of problem (7.2) and the corresponding set of eigenfunctions is a cone. Moreover, any $\lambda \in (0,\lambda_1)$ is not an eigenvalue of (7.2), while there exists $\mu_1 > \lambda_1$ such that any $\lambda \in (\mu_1,\infty)$ is an eigenvalue of (7.2).*

Proof. Consider the standard higher order Hilbertian Sobolev space $H_0^K(B)$, endowed with the scalar product

$$\langle u,v\rangle_K = \begin{cases} \displaystyle\int_B (\Delta^m u)(\Delta^m v)dx, & \text{if } K = 2m, \\ \displaystyle\int_B (D\Delta^m u)(D\Delta^m v)dx, & \text{if } K = 2m+1, \end{cases} \tag{7.4}$$

and denote by $\|\cdot\|_K$ the corresponding norm. In case of higher order Sobolev space $H_0^K(B)$, the decomposition in the positive and negative part of $u \in H_0^K(B)$ is no longer admissible in $H_0^K(B)$. However, we have the following result.

Lemma 7.2 *For any $u \in H_0^K(B)$ there exists a unique couple $(u_1, u_2) \in \mathcal{K} \times \mathcal{K}'$ such that $u = u_1 + u_2$ and $\langle u_1, u_2 \rangle_K = 0$, where $\mathcal{K}$ is the convex closed cone of positive functions*

$$\mathcal{K} = \{v \in H_0^K : v(x) \geq 0 \text{ a.e. in } B\},$$

while $\mathcal{K}'$ is the dual cone of $\mathcal{K}$, that is

$$\mathcal{K}' = \{w \in H_0^K : \langle w, v \rangle_K \leq 0 \text{ for all } v \in \mathcal{K}\}.$$

Moreover, $\mathcal{K}'$ is contained in the cone of negative functions, that is,

$$\mathcal{K}' \subseteq \{w \in H_0^K : w(x) \leq 0 \text{ a.e. in } B\}.$$

Proof. Let u_1 be the projection of u onto $\mathcal{K}$ defined by

$$\|u - u_1\|_K = \inf_{v \in \mathcal{K}} \|u - v\|_K,$$

and let $u_2 = u - u_1$. Then, for any $v \in \mathcal{K}$ we have

$$\begin{aligned} \|u_2\|_K = \|u - u_1\|_K &\leq \|u - (u_1 + \varepsilon v)\|_K \\ &= \|u - u_1\|_K - 2\varepsilon \langle u - u_1, v \rangle_K + \varepsilon^2 \|v\|_K^2. \end{aligned}$$

This implies

$$2\langle u_2, v \rangle_K \leq \varepsilon \|v\|_K^2 \quad \text{for all } v \in \mathcal{K}.$$

By letting $\varepsilon \to 0$, it follows that $u_2 \in \mathcal{K}'$. Replacing v by $\pm\varepsilon u_1$ (note that $u + \varepsilon v = u_1 - \varepsilon u_1 \in \mathcal{K}$) we deduce $\langle u_1, u_2 \rangle_K = 0$. In order to prove the uniqueness, let

$$u = u_1 + u_2 = v_1 + v_2$$

be two decompositions of u such that $u_1, v_1 \in \mathcal{K}$ and $u_2, v_2 \in \mathcal{K}'$ and $\langle u_1, u_2 \rangle_K = \langle v_1, v_2 \rangle_K = 0$. Then

$$
\begin{aligned}
0 &= \langle u_1 + u_2 - v_1 - v_2, u_1 + u_2 - v_1 - v_2 \rangle_K \\
&= \langle (u_1 - v_1) + (u_2 - v_2), (u_1 - v_1) + (u_2 - v_2) \rangle_K \\
&= \|u_1 - v_1\|_K + \|u_2 - v_2\|_K - 2\langle u_1 - v_1, u_2 - v_2 \rangle_K \\
&= \|u_1 - v_1\|_K + \|u_2 - v_2\|_K - 2\langle u_1, v_2 \rangle_K - \langle v_1, u_2 \rangle_K \\
&\geq \|u_1 - v_1\|_K + \|u_2 - v_2\|_K.
\end{aligned}
$$

Thus, $u_1 = v_1$ and $u_2 = v_2$. It remains to prove the fact that $\|' \subset -\mathscr{K}$. To this aim, let $w \in \mathscr{K}'$ and $\psi \in C_0^\infty(B) \cap \mathscr{K}$. Consider $v \in H_0^K(B)$ that satisfies

$$
\begin{cases} (-\Delta)^K v = \psi & \text{in } B \\ u = Du = \cdots = D^{K-1} u = 0 & \text{on } \partial B, \end{cases}
$$

Using the fact that the Green function in any ball is positive (so that the maximum principle holds) it follows that $v \in \mathscr{K}$. Thus

$$
0 \geq \langle w, v \rangle_K = \int_B w \cdot (-\Delta)^K v dx = \int_B w \psi dx
$$

for all $\psi \in C_0^\infty(B) \cap \mathscr{K}$. By density, the above inequality holds for all $\psi \in L^2(B)$, hence $w \leq 0$ a.e. in B. This concludes the proof. □

The number $\lambda > 0$ is an eigenvalue of problem (7.2), with f of the type (7.3), if there exists $u \in H_0^K \setminus \{0\}$ such that

$$
\langle u, v \rangle_K = \lambda \int_B f(x, u) v dx \tag{7.5}
$$

for any $v \in H_0^K$.

Lemma 7.3 *If $\lambda > 0$ is an eigenvalue of (7.2), then $\lambda \geq \lambda_1$.*

Proof. Assume that $\lambda > 0$ is and eigenvalue of (7.2), with corresponding eigenfunction $u \in H_0^K \setminus \{0\}$. Letting $v = u$ in (7.5), and putting $B_- = \{x \in B : u(x) \leq 0\}$ and $B_+ = \{x \in B : u(x) \geq 0\}$, we get by (H_1)

$$
\|u\|_K^2 = \lambda \left[\int_{B_+} h(x,u) u dx + \int_{B_-} u^2 dx \right] \leq \lambda \left[c \int_{B_+} u^2 dx + \int_{B_-} u^2 dx \right] \leq \lambda |u|_2^2,
$$

for $c \in (0,1)$. By the definition of λ_1

$$
\lambda_1 |u|_2^2 \leq \|u\|_K^2 \leq \lambda |u|_2^2.
$$

Since $u \neq 0$, then the above inequality shows that $\lambda \geq \lambda_1$. □

Lemma 7.4 *The first eigenvalue λ_1 of (7.1) is also an eigenvalue of (7.2) and the set of the corresponding eigenfunctions is a cone of H_0^K.*

Proof. As already noted in the introduction the lowest eigenvalue λ_1 of (7.1) is simple, so that there exists a first eigenfunction $\varphi \in H_0^K \setminus \{0\}$, with $\varphi < 0$ in B. Hence φ is an eigenfunction also of (7.2), since clearly satisfies (7.5) with $\lambda = \lambda_1$ as

$$\langle \varphi, v\rangle_K = \lambda_1 \int_B \varphi v dx = \lambda_1 \int_B f(x,\varphi) v dx,$$

by (7.3). Moreover the set of the corresponding eigenfunctions lies in a cone of H_0^K. □

Lemma 7.5 *The first eigenvalue λ_1 of (7.1) is isolated in the set of eigenvalues of (7.2).*

Proof. Let $\lambda > 0$ be an eigenvalue of (7.2) whose corresponding eigenfunction u has Moreau's decomposition with $u_1 \not\equiv 0$. Then, for $u_1 \in H_0^K$, we take $v = u_1$ in (7.5), and by the definition of λ_1 and (H_1) we get

$$\lambda_1 |u_1|_2^2 \le \|u_1\|_K^2 = \lambda \left[\int_{B_+} h(x,u) u_1 dx + \int_{B_-} u u_1 dx \right] \le \lambda c |u_1|_2^2.$$

Hence $\lambda \ge \lambda_1/c > \lambda_1$, for $c \in (0,1)$. In particular, any eigenfunction u corresponding to an eigenvalue $\lambda \in (0, \lambda_1/c)$ has decomposition $u = u_2$, so that u is also an eigenfunction of (7.1), since $u = u_2 \le 0$ a.e. in B. It is known, as noted in the introduction, that $\lambda_1 < \lambda_2$, where λ_2 is the second eigenvalue of (7.1). Hence any $\lambda \in (\lambda_1, \delta)$, with $\delta = \min\{\lambda_1/c, \lambda_2\}$, cannot be an eigenvalue of (7.1) and in turn is not an eigenvalue of (7.2), by the argument above. This completes the proof. □

As already noted, $\lambda > 0$ is an eigenvalue of the problem

$$\begin{cases} (-\Delta)^K u = \lambda h(x,u_+) & \text{in } B \\ u = Du = \cdots = D^{K-1} u = 0 & \text{on } \partial B, \end{cases} \tag{7.6}$$

if there exists $u \in H_0^K \setminus \{0\}$ such that $\langle u, v\rangle_K = \lambda \int_B h(x,u_+) v dx$ for all $v \in H_0^K$, that is if and only if u is a nontrivial critical point of the C^1 functional $I_\lambda : H_0^K \to \mathbb{R}$ defined by

$$I_\lambda(u) = \frac{1}{2}\|u\|_K^2 - \lambda \int_B H(x,u_+) dx.$$

If $\lambda > 0$ is an eigenvalue of (7.6), with corresponding eigenfunction $u = u_1 + u_2$, then taking as test function $v = u_2$ by (H_1) we get, for $\langle u_1, u_2\rangle_K = 0$ and $h(x,0) = 0$

a.e. in B,

$$\|u_2\|_K^2 = \langle u, u_2\rangle_K = \lambda \int_B h(x,u_+)u_2 dx = \lambda \int_{B_+} h(x,u)u_2 dx \le 0,$$

for $u_2 \le 0$ a.e. in B, that is $u = u_1 \ge 0$ in B and $u \ne 0$. In particular, *any eigenvalue* λ *of* (7.6) *is also an eigenvalue of* (7.2). Assumption (H_3) implies that for every $\lambda > 0$ there exists $C_\lambda > 0$ such that

$$\lambda H(x,t) \le C_\lambda + \lambda_1 t^2/4 \quad \text{for a.a. } x \in B \text{ and all } t \in \mathbb{R},$$

where λ_1 is the first eigenvalue of (7.1). Hence, by the definition of λ_1, we have that for all $u \in H_0^K$

$$I_\lambda(u) \ge \frac{1}{2}\|u\|_K^2 - \frac{\lambda_1}{4}|u|_2^2 - C_\lambda|B| \ge \frac{1}{4}\|u\|_K^2 - C_\lambda|B|,$$

in other words I_λ *is bounded from below, weakly lower semicontinuous and coercive on* H_0^K. □

Lemma 7.6 *There exists* $\lambda^* > 0$ *such that* $\inf_{H_0^K} I_\lambda(u) < 0$ *for all* $\lambda \ge \lambda^*$.

Proof. By (H_2) there exists $t_0 > 0$ such that $H(x,t_0) > 0$ a.e. in B. Let $\Omega \subset B$ be a compact subset, sufficiently large, such that

$$|B \setminus \Omega| < \frac{1}{ct_0^2}\int_\Omega H(x,t_0)dx,$$

where $c \in (0,1)$ is given in (H_1). Take $u_0 \in C_0^\infty(B)$, with $u_0(x) = t_0$ if $x \in \Omega$ and $0 \le u_0(x) \le t_0$ if $x \in B \setminus \Omega$. Hence, by (H_1),

$$\int_B H(x,u_0(x))dx \ge \int_\Omega H(x,t_0)dx - ct_0^2|B \setminus \Omega| > 0,$$

and so $I_\lambda(u_0) < 0$ for $\lambda > 0$ sufficiently large. The lemma follows at once. □

Now, we return to the proof of Theorem 7.1. Since I_λ is bounded from below, weakly lower semicontinuous and coercive on H_0^K, then Lemma 7.5 and [188, Theorem 1.2] show that I_λ has a negative global minimum for $\lambda > 0$ sufficiently large. This means that all such λ are eigenvalues of problem (7.6) and, consequently, of (7.2). This fact and Lemmas 7.3–7.5 complete the proof of Theorem 7.1. □

7.2 Infinitely Many Solutions for Perturbed Nonlinearities

Let $\Omega \subset \mathbb{R}^N$, $N \geq 1$, be a smooth and bounded domain, $K \geq 1$ be a positive integer such that $N > 2K$. In this section we study the problem

$$\begin{cases} (-\Delta)^K u = |u|^{p-2}u + \phi & \text{in } \Omega, \\ u = \partial_\nu u = \cdots = \partial_\nu^{K-1} u = 0 & \text{on } \partial\Omega, \end{cases} \tag{7.7}$$

where $\phi \in L^2(\Omega)$, ν is the exterior unit normal at $\partial\Omega$ and

$$\partial_\nu^j u = \frac{\partial^j u}{\partial \nu^j}, \quad 0 \leq j \leq K-1.$$

The exponent p satisfies

$$2 < p < \frac{2(N-K)}{N-2K}. \tag{7.8}$$

We say that $u \in H_0^K(\Omega)$ is a solution of (7.7) if

$$\langle u, v \rangle_K = \int_\Omega |u|^{p-2} uv dx + \int_\Omega \phi v dx \quad \text{for all } v \in H_0^K(\Omega),$$

where $\langle \cdot, \cdot \rangle_K$ is the scalar product defined by (7.4).

Before stating the main result concerning (7.7), let us recall what is known regarding the case $K = 1$. If $\phi \equiv 0$ the problem (7.7) is symmetric and multiplicity results can be obtained from the equivalent Lusternik–Schnirelmann theory. In turn, if $\phi \not\equiv 0$, the symmetry of the problem fails to hold and different techniques are needed. The classical critical point theory still applies provided p is close to 2. The problem of whether (7.7) has infinitely many solutions for all exponents p in the range $2 < p < 2N/(N-2)$ is still open. For a dense subset of function ϕ in $L^2(\Omega)$ a positive answer was given by Bahri and Lions [10].

In the study of (7.7) we shall employ a method devised by Bolle [22] as described in Sect. 1.3.4.

The key is to exhibit a continuous path of functionals $\{J_\theta\}_{0 \leq \theta \leq 1}$ such that J_0 is symmetric and J_1 is the functional associated to (7.7), that is,

$$J_1(u) = \frac{1}{2}\|u\|_K^2 - \frac{1}{p}\int_\Omega |u|^p dx - \int_\Omega \phi u dx.$$

Further, as θ varies, we wish to control the min-max critical levels of J_θ, thus getting estimates for critical points of J_1. Due to the compact embedding of $H_0^K(\Omega)$ into

$L^p(\Omega)$ for all

$$2 < p < K_* := \frac{2N}{N-2K},$$

we would expect to formulate our result for all exponents p in the above range. However, the method we employ requires a further restriction on p as stated in (7.8). Our main result regarding (7.7) is the following.

Theorem 7.7 *Assume p satisfies (7.8). Then, for all $\phi \in L^2(\Omega)$ the problem (7.7) has infinitely many solutions.*

Proof. For all $0 \le \theta \le 1$ consider the functional

$$J_\theta : H_0^K(\Omega) \to \mathbb{R}, \; J_\theta(u) = \frac{1}{2}\|u\|_K^2 - \frac{1}{p}\int_\Omega |u|^p dx - \theta \int_\Omega \phi u dx,$$

and let $J(\theta,u) = J_\theta(u)$. Remark that Φ_0 is even and any solution u of (7.7) corresponds to a critical point of Φ_1.

We first wish to apply Theorem 1.15 to J_0. It is clear that J_0 satisfies the hypotheses $(A_1)-(A_2)$ in Sect. 1.3.3. Since

$$J_0'(u)v = \langle u,v\rangle_K - \int_\Omega |u|^{p-2}uvdx \quad \text{for all } v \in H_0^K(\Omega),$$

and the mapping

$$H_0^K(\Omega) \xrightarrow{T} L^{K_*/(p-1)}(\Omega) \xrightarrow{((-\Delta)^K)^{-1}} H_0^K(\Omega), \text{ where } T(u) = |u|^{p-2}u, \tag{7.9}$$

is compact, it follows that J_0 satisfies (A_3).

For any $k \ge 1$, let φ_k be the k-th eigenfunction of $(-\Delta)^K$ with homogeneous Dirichlet boundary conditions. Let

$$X_k = \operatorname{span}\{\varphi_1, \varphi_2, \ldots, \varphi_k\} \quad \text{so that } X = \overline{\bigcup_{k=1}^{\infty} X_k}. \tag{7.10}$$

For any $k \ge 1$, let $R_k > 0$ be such that

$$J_0(u) < 0 \quad \text{for all } u \in X_k \text{ with } \|u\|_K \ge R.$$

Define

$$\mathscr{C}_k = \left\{ \gamma \in C(X_k \cap \overline{B(0,R_k)}, H_0^K(\Omega)) \; : \gamma \text{ is odd and } \gamma\Big|_{X_k \cap \partial B(0,R_k)} = Id \right\}$$

and

$$b_k = \inf_{\gamma \in \mathscr{C}_k} \sup_{u \in X_k \cap \overline{B(0,R_k)}} J_0(\gamma(u)). \tag{7.11}$$

By Theorem 1.15 there exists a sequence $\{u_k\}$ of critical points of J_0 in $H_0^K(\Omega)$ such that

$$J_0(u_k) \leq b_k \quad \text{and} \quad m^*(J_0, u_k) \geq k. \tag{7.12}$$

Our next result provides a lower bound for b_k.

Proposition 7.8 *There exists a positive constant $A > 0$ such that*

$$b_k \geq A k^{2Kp/(N(p-2))} \quad \textit{for all } k \geq 1. \tag{7.13}$$

Proof. Fix $k \geq 1$ and let $\{\mu_j\}$ be the sequence of eigenvalues (repeated according to their multiplicity) of the operator

$$(-\Delta)^K - (p-1)|u_k|^{p-2}, k \geq 1.$$

Since

$$J_0''(u_k)(v,v) = \langle ((-\Delta)^K - (p-1)|u_k|^{p-2})v, v \rangle_K,$$

the definition of the large Morse index (see Definition 1.14) together with the second relation in (7.12) imply

$$|\{j \geq 1 : \mu_j \leq 0\}| = m^*(J_0, u_k) \geq k. \tag{7.14}$$

To derive an upper bound for the set $|\{j \geq 0 \ : \ \mu_j \leq 0\}|$, we state without proof a more general result regarding spectral properties of higher order Schrödinger operators.

Lemma 7.9 *Let $N > 2K$, $V \in L^{N/(2K)}(\Omega)$ and let $\{\mu_j\}$ be the sequence of eigenvalues (repeated according to their multiplicity) of the operator $(-\Delta)^K + V(x)$. Then, there exists $M = M(N,K) > 0$ such that*

$$|\{j \geq 1 : \mu_j \leq 0\}| \leq M \int_{\mathbb{R}^N} V^-(x)^{N/(2K)} dx.$$

For the proof, we refer the reader to Rozenbljum [173, Theorem 3]. We apply the above lemma for

$$V = \begin{cases} -(p-1)|u_k|^{p-2} & \text{in } \Omega, \\ 0 & \text{in } \mathbb{R}^N \setminus \Omega. \end{cases}$$

Using Lemma 7.9 together with (7.14) we find

$$k \leq |\{j \geq 1 : \mu_j \leq 0\}| \leq M\|u_k\|_{N(p-2)/(2K)}^{N(p-2)/(2K)}$$

so

$$\|u_k\|_{N(p-2)/(2K)}^{N(p-2)/(2K)} \geq Ck,$$

for some $C > 0$. On the other hand, from $\langle J_0'(u_k), u_k\rangle_K = 0$ we find

$$b_k \geq J_0(u_k) = \left(\frac{1}{2} - \frac{1}{p}\right) \|u_k\|_p^p.$$

Combining the last two estimates we find

$$b_k \geq \left(\frac{1}{2} - \frac{1}{p}\right) \|u_k\|_p^p \geq C\|u_k\|_{N(p-2)/(2K)}^p \geq Ak^{2kp/(N(p-2))}$$

which proves (7.13). □

We next prove that J_θ satisfies conditions $(B_1) - (B_4)$ in Sect. 1.3.4.

Lemma 7.10 *Let* $\{(\theta_n, u_n)\} \subset [0,1] \times H_0^K(\Omega)$ *be such that*

$$\{J_{\theta_n}(u_n)\} \textit{ is bounded and } \lim_{n\to\infty} J_{\theta_n}'(u_n) = 0.$$

Then, up to a subsequence, $\{(\theta_n, u_n\}$ *converges in* $[0,1] \times H_0^K(\Omega)$.

Proof. We have

$$(J_{\theta_n}'(u_n), u_n) = \|u\|_K^2 - \int_\Omega |u|^p dx - \theta \int_\Omega \phi u dx.$$

Since $(J_{\theta_n}'(u_n), u_n) = 0(\|u_n\|_K)$ as $n \to \infty$, for $M > 1$ large enough and $1/p < \rho < 1/2$ we find

$$\begin{aligned} M + \rho\|u\|_K &\geq J_{\theta_n}(u_n) - \rho(J_{\theta_n}'(u_n), u_n) \\ &= \left(\frac{1}{2} - \rho\right) \|u\|_K^2 + \left(\rho - \frac{1}{p}\right) \int_\Omega |u|^p dx - \theta(1-\rho) \int_\Omega \phi u_n dx. \end{aligned} \tag{7.15}$$

By Young's inequality we obtain

$$(1-\rho) \int_\Omega \phi u_n dx \leq \left(\rho - \frac{1}{p}\right) \int_\Omega |u|^p dx + C\|\phi\|_{L^{p'}(\Omega)}^{p'}, \tag{7.16}$$

where $p' = p/(p-1)$ and $C > 0$ is a positive constant depending on ρ and p only. Combining (7.15) and (7.16) we derive

$$M + \rho\|u\|_K \geq \left(\frac{1}{2} - \rho\right)\|u\|_K^2 - C(p, \|\phi\|_{L^{p'}(\Omega)}),$$

which shows that $\{u_n\}$ is bounded in $H_0^K(\Omega)$. Since the mapping in (7.9) is compact, it follows that $\{u_n\}$ converges strongly in $H_0^K(\Omega)$. This finishes the proof of lemma. □

Lemma 7.11 *For any $b > 0$ there exists $C > 0$ such that*

$$\left|\frac{\partial J}{\partial \theta}(\theta, u)\right| \leq C(1 + \|J'_\theta(u)\|)(1 + \|u\|_K),$$

for all $(\theta, u) \in [0,1] \times H_0^K(\Omega)$ satisfying $|J_\theta(u)| \leq b$.

Proof. From $|J_\theta(u)| \leq b$ it follows that

$$\theta \int_\Omega \phi u dx \geq \frac{p}{2}\|u\|_K^2 - \int_\Omega |u|^p dx - (p-1)\theta \int_\Omega \phi u dx - pb.$$

Hence

$$\begin{aligned} -(J'_\theta(u), u) &= -\|u\|_K^2 + \int_\Omega |u|^p dx + \theta \int_\Omega \phi u dx \\ &\geq \left(\frac{p}{2} - 1\right)\|u\|_K^2 - (p-1)\theta \int_\Omega \phi u dx - pb. \end{aligned}$$

In particular

$$-(J'_\theta(u), u) \geq c_1\|u\|_K^2 - c_2,$$

where c_1, c_2 are two positive constants. On the other hand,

$$\left|\frac{\partial J}{\partial \theta}(\theta, u)\right| \leq \int_\Omega |\phi||u| dx \leq \|u\|_{L^p(\Omega)}\|\phi\|_{L^{p'}(\Omega)} \leq \varepsilon\|u\|_K^2 + C(\varepsilon, p, \|\phi\|_{L^{p'}(\Omega)}).$$

Now the conclusion follows from the last two inequalities from above. □

Lemma 7.12 *There exist two flows $\eta_i : [0,1] \times \mathbb{R} \to \mathbb{R}$ such that $\eta_i(\theta, \cdot)$ are Lispchitz continuous for all $\theta \in [0,1]$ and*

$$\eta_1(\theta, J_\theta(u)) \leq \frac{\partial J}{\partial \theta}(\theta, u) \leq \eta_2(\theta, J_\theta(u)) \tag{7.17}$$

at each critical point u of J_θ.

Proof. Let $u \in H_0^K(\Omega)$ be a critical point of J_θ. By Hölder's inequality we have

$$\left|\frac{\partial J}{\partial \theta}(\theta,u)\right| \leq \|u\|_{L^p(\Omega)}\|\phi\|_{L^{p'}(\Omega)}. \tag{7.18}$$

On the other hand,

$$0 = (J'_\theta(u),u) = \|u\|_K^2 - \int_\Omega |u|^p dx - \theta \int_\Omega \phi u dx = 0,$$

which yields

$$\begin{aligned}\|u\|^p_{L^p(\Omega)} &= \|u\|_K^2 - \theta \int_\Omega \phi u dx \\ &= 2J_\theta(u) + \frac{2}{p}\int_\Omega |u|^p dx + \theta \int_\Omega \phi u dx \\ &\leq 2J_\theta(u) + \left(\varepsilon + \frac{2}{p}\right)\int_\Omega |u|^p dx + C(\varepsilon, p, \|\phi\|_{L^{p'}(\Omega)}).\end{aligned}$$

By taking $\varepsilon > 0$ small, we deduce

$$\|u\|^p_{L^p(\Omega)} \leq C(J_\theta(u)+1). \tag{7.19}$$

Combining (7.18) with (7.19) we obtain

$$\left|\frac{\partial J}{\partial \theta}(\theta,u)\right| \leq C\|u\|_{L^p(\Omega)} \leq C(J_\theta(u)+1)^{1/p} \leq 2C(J^2_\theta(u)+1)^{2/p}.$$

It is now enough to consider

$$\eta_2(\theta,t) = 2C(t^2+1)^{2/p}, \quad \eta_1(\theta,t) = -\eta_2(\theta,t).$$

□

Lemma 7.13 *For any finite dimensional subspace W of $H_0^K(\Omega)$ we have*

$$\lim_{u\in W, \|u\|_K\to\infty} \sup_{\theta\in[0,1]} J_\theta(u) = -\infty.$$

Proof. By Hölder's inequality and the continuous embedding of $H_0^K(\Omega)$ into $L^p(\Omega)$ we can find $C > 0$ such that

$$J_\theta(u) \leq C(\|u\|_K^2 - \|u\|_K^p - 1) \quad \text{for all } (\theta,u) \in [0,1]\times W.$$

This proves our lemma. □

We are now able to complete the argument and finish the proof of Theorem 7.7. Let

$$\mathscr{C} = \left\{ \zeta \in C(H_0^K(\Omega), H_0^K(\Omega)) : \zeta \text{ is odd and } \zeta\Big|_{H_0^K(\Omega)\setminus\overline{B(0,R)}} = Id \right\},$$

and

$$c_k = \inf_{\zeta\in\mathscr{C}} \sup_{u\in X_k} J_0(\zeta(u)),$$

where X_k are defined in (7.10). In order to apply Theorem 1.16 we only have to check that the sequence

$$\left\{ \frac{c_{k+1}-c_k}{\bar{\eta}_1(c_{k+1})+\bar{\eta}_2(c_k)+1} \right\} \quad \text{is unbounded},$$

where η_i are given by Lemma 7.12. Assuming the contrary, it follows that

$$\left\{ \frac{c_{k+1}-c_k}{c_{k+1}^{1/p}-c_k^{1/p}+1} \right\} \quad \text{is bounded},$$

so that

$$\left\{ \left| c_{k+1}^{(p-1)/p} - c_k^{(p-1)/p} \right| \right\}$$

is bounded. Therefore, there exists a positive constant $B>0$ such that

$$c_k \le Bk^{p/(p-1)} \qquad \text{for all } k \ge 1.$$

Combining this estimate with Proposition 7.8 and $b_k \le c_k$ we reach a contradiction in view of (7.8). Thus, one can apply Theorem 1.16, which implies the existence of infinitely many solutions of problem (7.7). This concludes the proof of Theorem 7.7. □

Remark 24 *The approach in this section follows an idea from Lancelotti et al. [127]. In fact, in [127] is obtained the existence of infinitely many solutions for the nonhomogeneous problem*

$$\begin{cases} (-\Delta)^K u = |u|^{p-2}u+\phi & \text{in } \Omega, \\ \partial_\nu^j u = \phi_j,\, j=0,1,\dots,K-1 & \text{on } \partial\Omega, \end{cases}$$

where $\phi \in L^2(\Omega)$, $\phi_j \in H^{K-j-\frac{1}{2}}(\partial\Omega)$, $j=0,1,\dots,K-1$ *and* $2<p<2(N+K)/N$.

7.3 A Biharmonic Problem with Singular Nonlinearity

In this section we study the biharmonic elliptic problem

$$\begin{cases} \Delta^2 u = u^{-\alpha}\,,\ u>0 & \text{in } \Omega, \\ u = \partial_\nu u = 0 & \text{on } \partial\Omega, \end{cases} \tag{7.20}$$

where $0<\alpha<1$, $\Omega\subset\mathbb{R}^N$ $(N\geq 2)$ is a smooth bounded domain, ν is the exterior unit normal at $\partial\Omega$ and $\partial_\nu = \frac{\partial}{\partial\nu}$ is the outer normal derivative at $\partial\Omega$.

We denote by $G(\cdot,\cdot)$ the Green function associated with the biharmonic operator Δ^2 subject to Dirichlet boundary conditions.

Throughout this section we assume that $\Omega\subset\mathbb{R}^N$, $N\geq 2$ satisfies:

$(A1)$ The boundary $\partial\Omega$ is of class C^{16} if $N=2$ and of class C^{12} if $n\geq 3$.
$(A2)$ The Green function $G(\cdot,\cdot)$ is positive.

The assumption $(A1)$ on the regularity of $\partial\Omega$ goes back to Krasovskiĭ [122] (see also Dell'Acqua and Sweers [55]) and allows us to employ some sharp estimates for the biharmonic Green function (see Appendix B). The need for condition $(A2)$ will become more clear once we specify what it is understood by a solution of (7.20).

Definition 7.14 *We say that u is a solution of (7.20) if*

$$u\in C(\overline{\Omega}),\quad u>0\quad \textit{in } \Omega,$$

and u satisfies the integral equation

$$u(x) = \int_\Omega G(x,y)u^{-\alpha}(y)dy\quad \textit{for all } x\in\Omega. \tag{7.21}$$

The restriction $0<\alpha<1$ is needed in order to make the integral in (7.21) finite. It will appear several times in the proofs in the following sections. Note also that condition $(A2)$ above implies the standard maximum principle for the biharmonic operator in Ω.

Let φ_1 be the first eigenfunction of $(-\Delta)$ in $H_0^1(\Omega)$. It is well known that φ_1 has constant sign in Ω, so by a suitable normalization we may assume $\varphi_1>0$ in Ω. Therefore, φ_1 satisfies

$$\begin{cases} -\Delta\varphi_1 = \lambda_1\varphi_1\,,\ \varphi_1>0 & \text{in } \Omega, \\ \varphi_1 = 0 & \text{on } \partial\Omega, \end{cases} \tag{7.22}$$

where $\lambda_1 > 0$ is the first eigenvalue of $(-\Delta)$. By the Hopf maximum principle [162] we have $\partial_\nu \varphi_1 < 0$ on $\partial\Omega$. Also, by the regularity of Ω we have $\varphi_1 \in C^4(\overline{\Omega})$ and

$$c\delta(x) \le \varphi_1(x) \le \frac{1}{c}\delta(x) \quad \text{in } \Omega, \tag{7.23}$$

for some $0 < c < 1$, where $\delta(x) = \text{dist}(x, \partial\Omega)$.

Proposition 7.15 *Let u be a solution of problem (7.20). Then, there exist $c_1, c_2 > 0$ such that u satisfies*

$$c_1\delta^2(x) \le u(x) \le c_2\delta^2(x) \quad \textit{in } \Omega. \tag{7.24}$$

Proof. Let $a(x) = \varphi_1^2(x), x \in \overline{\Omega}$. It is easy to see that since $\varphi_1 \in C^4(\overline{\Omega})$ then $f := \Delta^2 a$ is bounded in $\overline{\Omega}$, so, by the continuity of u there exists $m > 0$ small enough such that

$$u(x) - ma(x) = \int_\Omega G(x,y)\Big[u^{-\alpha}(y) - mf(y)\Big]dy \ge 0 \quad \text{for all } x \in \Omega.$$

Therefore,

$$u(x) \ge ma(x) \ge c_0\delta^2(x) \quad \text{in } \Omega, \tag{7.25}$$

for some $c_0 > 0$. This proves the first part of the inequality in (7.31). For the second part, assume first $N > 4$ and let $x \in \Omega$. Using Proposition B.1(ii1), for all $y \in \Omega$ we have

$$\begin{aligned} G(x,y) &\le c|x-y|^{2-N}\delta^2(x)\min\left\{1, \frac{\delta(y)}{|x-y|}\right\}^2 \\ &\le c|x-y|^{2-N}\delta^2(x)\min\left\{1, \frac{\delta(y)}{|x-y|}\right\}^{2\alpha} \\ &= c|x-y|^{2-2\alpha-N}\delta^2(x)\delta^{2\alpha}(y). \end{aligned} \tag{7.26}$$

Now, from (7.25) and (7.26) we have

$$\begin{aligned} u(x) &= \int_\Omega G(x,y)u^{-\alpha}(y)dy \\ &\le c_1\int_\Omega G(x,y)\delta^{-2\alpha}(y)dy \le c_2\delta^2(x)\int_\Omega |x-y|^{2-2\alpha-N}dy \\ &\le c_2\delta^2(x)\int_{0\le|x-y|\le \text{diam}(\Omega)} |x-y|^{2-2\alpha-N}dy \\ &= c_2\delta^2(x)\int_0^{\text{diam}(\Omega)} t^{1-2\alpha}dt \\ &\le c_3\delta^2(x). \end{aligned} \tag{7.27}$$

Let now $N = 4$. We use Proposition B.1(ii2) to derive a similar inequality to (7.26). More precisely, for all $y \in \Omega$ we have

$$\begin{aligned} G(x,y) &\le c\log\left(2+\frac{\delta(y)}{|x-y|}\right)\min\left\{1,\frac{\delta(x)}{|x-y|}\right\}^2 \min\left\{1,\frac{\delta(y)}{|x-y|}\right\}^{2\alpha} \\ &\le c|x-y|^{-2-2\alpha}\delta^2(x)\delta^{2\alpha}(y)\log\left(2+\frac{\operatorname{diam}(\Omega)}{|x-y|}\right). \end{aligned} \tag{7.28}$$

If $N = 3$, let $\beta = \max\{0, 2\alpha - 1/2\} < 3/2$ and by Proposition B.1(ii4) we have

$$\begin{aligned} G(x,y) &\le c\delta^{1/2}(x)\delta^{1/2}(y)\min\left\{1,\frac{\delta(x)}{|x-y|}\right\}^{3/2} \min\left\{1,\frac{\delta(y)}{|x-y|}\right\}^{3/2} \\ &\le c|x-y|^{-3/2-\beta}\delta^2(x)\delta^{\beta+1/2}(y) \\ &\le C|x-y|^{-3/2-\beta}\delta^2(x)\delta^{2\alpha}(y). \end{aligned} \tag{7.29}$$

Finally, if $N = 2$, let $\beta = \max\{0, 2\alpha - 1\} < 1$ and by Proposition B.1(ii3) we have

$$\begin{aligned} G(x,y) &\le c\delta(x)\delta(y)\min\left\{1,\frac{\delta(x)}{|x-y|}\right\}\min\left\{1,\frac{\delta(y)}{|x-y|}\right\} \\ &\le c|x-y|^{-1}\delta^2(x)\delta(y)\min\left\{1,\frac{\delta(y)}{|x-y|}\right\}^{\beta} \\ &\le c|x-y|^{-1-\beta}\delta^2(x)\delta^{1+\beta}(y) \\ &\le C|x-y|^{-1-\beta}\delta^2(x)\delta^{2\alpha}(y). \end{aligned} \tag{7.30}$$

We now use the estimates (7.28)–(7.30) to derive a similar inequality to that in (7.27).

This completes the proof of Proposition 7.15. □

Proposition 7.16 *Let* $0 < \alpha < 1$ *and* $u \in C(\overline{\Omega})$ *be such that* $u(x) \ge c_0\delta^2(x)$ *in* Ω *for some* $c_0 > 0$. *Consider*

$$w(x) = \int_\Omega G(x,y)u^{-\alpha}(y)dy \quad \text{for all } x \in \overline{\Omega}.$$

Then

(i) $w \in C^2(\overline{\Omega})$.
(ii) $w \in C^3(\overline{\Omega})$ *for any* $0 < \alpha < 1/2$.

Proof. With the same proof as in Proposition 7.15 it is easy to see that v is well defined. For $0 < \varepsilon < 1$ small, define $\Omega_\varepsilon = \{x \in \overline{\Omega} : \delta(x) < \varepsilon\}$. Set $u_\varepsilon = \max\{u, c_0\varepsilon^2\}$ and

$$w_\varepsilon(x) = \int_\Omega G(x,y)u_\varepsilon^{-\alpha}(y)dy \quad \text{for all } x \in \overline{\Omega}.$$

It is easy to see that $w_\varepsilon = w$ on $\Omega \setminus \Omega_\varepsilon$. Since $u_\varepsilon^{-\alpha}$ is bounded in $\overline{\Omega}$, by the estimates in Proposition B.1 it follows that $w_\varepsilon \in C^3(\overline{\Omega})$ and

$$D_x^k w_\varepsilon(x) = \int_\Omega D_x^k G(x,y)u_\varepsilon^{-\alpha}(y)dy \quad \text{for all } x \in \overline{\Omega},$$

for any $N-$dimensional multi-index k with $|k| \le 3$. The proof of this fact is similar to that of Lemma 4.1 in [99]. We employ in the following the same approach as in [99] to show that $w \in C^2(\overline{\Omega})$ (resp. $w \in C^3(\overline{\Omega})$ if $0 < \alpha < 1/2$).

Assume first $N > 4$ and let k be an $N-$dimensional multi-index with $|k| \le 2$. Fix $\beta > 0$ such that $2\alpha < \beta < 2$.

By Proposition B.1(i1) (if $|k| = 2$) and (ii1) (if $|k| \le 1$) we have

$$\begin{aligned}
&\left| D_x^k w_\varepsilon(x) - \int_\Omega D_x^k G(x,y)u^{-\alpha}(y)dy \right| \\
&\le \int_{\Omega_\varepsilon} |D_x^k G(x,y)|(u^{-\alpha}(y) + (c_0\varepsilon^2)^{-\alpha})dy \\
&\le c_1\varepsilon^{-2\alpha} \int_{\Omega_\varepsilon} |x-y|^{4-|k|-N} \min\left\{1, \frac{\delta(y)}{|x-y|}\right\}^2 dy \\
&\le c_1\varepsilon^{-2\alpha} \int_{\Omega_\varepsilon} |x-y|^{4-|k|-N} \min\left\{1, \frac{\delta(y)}{|x-y|}\right\}^\beta dy \\
&\le c_1\varepsilon^{-2\alpha} \int_{\Omega_\varepsilon} |x-y|^{4-|k|-\beta-N} \delta^\beta(y)dy \\
&\le c_1\varepsilon^{\beta-2\alpha} \int_\Omega |x-y|^{4-|k|-\beta-N} dy \\
&\le c_1\varepsilon^{\beta-2\alpha} \int_{0 \le |x-y| \le \operatorname{diam}(\Omega)} |x-y|^{4-|k|-\beta-N} dy \\
&\le c_1\varepsilon^{\beta-2\alpha} \int_0^{\operatorname{diam}(\Omega)} t^{3-|k|-\beta} dt \\
&\le c_2\varepsilon^{\beta-2\alpha} \int_0^{\operatorname{diam}(\Omega)} t^{1-\beta} dt \\
&\le c_3\varepsilon^{\beta-2\alpha} \to 0 \quad \text{as } \varepsilon \to 0.
\end{aligned}$$

The case $2 \le N \le 4$ can be analyzed in the same way. For instance, if $N = 3$ and $|k| = 1$, we use Proposition B.1(ii2) to derive

$$\begin{aligned}
&\left|D_x^k w_\varepsilon(x) - \int_\Omega D_x^k G(x,y)u^{-\alpha}(y)dy\right| \\
&\quad \le c_1\varepsilon^{-2\alpha}\int_{\Omega_\varepsilon}\log\left(2+\frac{\delta(y)}{|x-y|}\right)\min\left\{1,\frac{\delta(y)}{|x-y|}\right\}^2 dy \\
&\quad \le c_1\varepsilon^{-2\alpha}\int_{\Omega_\varepsilon}|x-y|^{-\beta}\delta^\beta(y)\log\left(2+\frac{\delta(y)}{|x-y|}\right)dy \\
&\quad \le c_1\varepsilon^{\beta-2\alpha}\int_{\Omega_\varepsilon}|x-y|^{-\beta}\log\left(2+\frac{\operatorname{diam}(\Omega)}{|x-y|}\right)dy \\
&\quad \le c_2\varepsilon^{\beta-2\alpha}\int_0^{\operatorname{diam}(\Omega)} t^{2-\beta}\log\left(2+\frac{\operatorname{diam}(\Omega)}{t}\right)dt \\
&\quad \le c_3\varepsilon^{\beta-2\alpha}\to 0 \quad \text{as } \varepsilon\to 0.
\end{aligned}$$

We have obtained that

$$D_x^k w_\varepsilon \to \int_\Omega D_x^k G(\cdot,y)u^{-\alpha}(y)dy \quad \text{uniformly as } \varepsilon\to 0,$$

for any $N-$dimensional multi-index k with $0\le|k|\le 2$. It follows that $w\in C^2(\overline{\Omega})$ and

$$D_x^k w(x) = \int_\Omega D_x^k G(x,y)u^{-\alpha}(y)dy \quad \text{for all } x\in\overline{\Omega},$$

for any multi-index k with $0\le|k|\le 2$.

(ii) Let k be a multi-index with $|k|=3$ and $2\alpha<\beta<1$. From Proposition B.1(i1) we have

$$\begin{aligned}
\left|D_x^k w_\varepsilon(x) - \int_\Omega D_x^k G(x,y)u^{-\alpha}(y)dy\right| &\le 2(c_0\varepsilon^2)^{-\alpha}\int_{\Omega_\varepsilon}|D_x^k G(x,y)|dy \\
&\le c_1\varepsilon^{-2\alpha}\int_{\Omega_\varepsilon}|x-y|^{1-N}\min\left\{1,\frac{\delta(y)}{|x-y|}\right\}^\beta dy \\
&\le c_1\varepsilon^{\beta-2\alpha}\int_\Omega |x-y|^{1-N-\beta}dy \\
&\le c_1\varepsilon^{\beta-2\alpha}\int_0^{\operatorname{diam}(\Omega)} t^{-\beta}dt \\
&\le c_2\varepsilon^{\beta-2\alpha}\to 0 \quad \text{as } \varepsilon\to 0,
\end{aligned}$$

since $\beta<1$. With the same arguments as above we find $w\in C^3(\overline{\Omega})$. This ends the proof. □

Our main result concerning (7.20) is the following.

Theorem 7.17 *Assume* $0 < \alpha < 1$ *and conditions* (A1), (A2) *hold. Then, the problem (7.20) has a unique solution u and there exist* $c_1, c_2 > 0$ *such that*

$$c_1\delta^2(x) \le u(x) \le c_2\delta^2(x) \quad \text{in } \Omega, \tag{7.31}$$

where $\delta(x) = \operatorname{dist}(x, \partial\Omega)$. *Moreover,* $u \in C^2(\overline{\Omega})$ *and if* $0 < \alpha < 1/2$ *then* $u \in C^3(\overline{\Omega})$.

The existence of a solution will be obtained by means of the Schauder fixed point theorem. To this aim, we employ the estimates for the biharmonic Green function stated in Appendix B. The uniqueness relies heavily on the boundary estimate (7.31) which is obtained by using the behavior of the Green function (see Proposition 7.15).

Let $a(x) = \varphi_1^2(x)$, $x \in \overline{\Omega}$. Motivated by Proposition 7.15 we will be looking for solutions u of (7.20) in the form

$$u(x) = a(x)v(x)$$

where $v \in C(\overline{\Omega})$, $v > 0$ in $\overline{\Omega}$. This leads us to the following integral equation for v:

$$v(x) = \frac{1}{a(x)}\int_\Omega \frac{G(x,y)}{a^\alpha(y)}v^{-\alpha}(y)dy \quad \text{for all } x \in \overline{\Omega}. \tag{7.32}$$

We can now regard (7.32) as the fixed point problem

$$\mathscr{F}(v) = v,$$

where

$$\mathscr{F}(v) = \frac{1}{a(x)}\int_\Omega \frac{G(x,y)}{a^\alpha(y)}v^{-\alpha}(y)dy.$$

Remark that $\mathscr{F}$ is an integral operator of the form

$$\mathscr{F}(v) = \int_\Omega K(x,y)v^{-\alpha}(y)dy,$$

where the kernel K is given by

$$K : \overline{\Omega}\times\Omega \to [0,\infty], \quad K(x,y) = \begin{cases} \dfrac{G(x,y)}{a(x)a^\alpha(y)} & \text{if } x, y \in \Omega, \\[2ex] \dfrac{\partial_\nu^2 G(x,y)}{\partial_\nu^2 a(x)a^\alpha(y)} & \text{if } x \in \partial\Omega, y \in \Omega. \end{cases}$$

Note that K is well defined since $\partial_\nu^2 a(x) = 2(\partial_\nu\varphi_1(x))^2 > 0$ on $\partial\Omega$.

We first need the following result.

Lemma 7.18 *(i) For any $y \in \Omega$, the function $K(\cdot, y) : \overline{\Omega} \to [0,\infty]$ is continuous;*

(ii) The mapping

$$\overline{\Omega} \ni x \mapsto \int_\Omega K(x,y)dy$$

is continuous and there exists $M > 1$ such that

$$\frac{1}{M} \le \int_\Omega K(x,y)dy \le M \quad \textit{for all } x \in \overline{\Omega}. \tag{7.33}$$

Proof. (i) By the results in the Appendix B, $G : \Omega \times \Omega \to (0,\infty]$ is continuous (in the extended sense). Therefore $K(\cdot,y)$ is continuous (in the extended sense) in Ω. It remains to prove the continuity of $K(\cdot,y)$ on $\partial\Omega$. Let $\varepsilon > 0$. Since $G(\cdot,y) \in C^4(\overline{\Omega}\setminus\{y\})$ and $a \in C^4(\overline{\Omega})$, for any $z \in \partial\Omega$ we have

$$G(z+t\nu,y) = t^2\left(\frac{1}{2}\partial_\nu^2 G(z,y) + G_1(z,t)\right) \quad \text{as } t \nearrow 0,$$
$$a(z+t\nu,y) = t^2\left(\frac{1}{2}\partial_\nu^2 a(z,y) + a_1(z,t)\right) \quad \text{as } t \nearrow 0,$$

where

$$\lim_{t\nearrow 0} G_1(z,t) = \lim_{t\nearrow 0} a_1(z,t) = 0 \quad \text{uniformly for } z \in \partial\Omega.$$

Hence, as $t \nearrow 0$ we have

$$\begin{aligned}|K(z+t\nu,y) - K(z,y)| &= \left|\frac{\frac{1}{2}\partial_\nu^2 G(z,y) + G_1(z,t)}{\frac{1}{2}\partial_\nu^2 a(z,y) + a_1(z,t)} - \frac{\partial_\nu^2 G(z,y)}{\partial_\nu^2 a(z,y)}\right| \\ &\le \frac{|G_1(z,y)|\partial_\nu^2 a(z,y) + |a_1(z,t)||\partial_\nu^2 G(z,y)|}{\partial_\nu^2 a(z,y)|\frac{1}{2}\partial_\nu^2 a(z,y) + a_1(z,t)|}.\end{aligned}$$

Thus, there exists $\eta_1 > 0$ such that

$$|K(z+t\nu,y) - K(z,y)| < \frac{\varepsilon}{2} \quad \text{for all } z \in \partial\Omega \text{ and } -\eta_1 < t < 0. \tag{7.34}$$

Also, by the smoothness of the boundary $\partial\Omega$ there exists $\eta_2 > 0$ such that

$$|K(z,y) - K(\bar{z},y)| < \frac{\varepsilon}{2} \quad \text{for all } z,\bar{z} \in \partial\Omega, |z-\bar{z}| < \eta_2. \tag{7.35}$$

Define $\eta = \min\{\eta_1,\eta_2\}/2$ and fix $z \in \partial\Omega$. Let now $x \in \overline{\Omega}$ be such that $|x-z| < \eta$. Also, let $\bar{x} \in \partial\Omega$ be such that $|x-\bar{x}| = \delta(x) = \operatorname{dist}(x,\partial\Omega)$. Then $|x-\bar{x}| \le |x-z| < \eta$ and $|\bar{x}-z| \le |x-\bar{x}| + |z-x| < 2\eta < \eta_2$ so by (7.35) we have

$$|K(\bar{x},y) - K(z,y)| < \frac{\varepsilon}{2}. \tag{7.36}$$

Now, from (7.34) and (7.36) we obtain

$$|K(x,y)-K(z,y)| \le |K(x,y)-K(\bar{x},y)| + |K(\bar{x},y)-K(z,y)| < \varepsilon$$

so $K(\cdot,y)$ is continuous at $z \in \partial\Omega$. This completes the proof of (i).

(ii) Assume first $N > 4$. Using (7.23) and Proposition B.1(ii1) we have

$$\begin{aligned}
K(x,y) &\le c_1\delta^{-2}(x)\delta^{-2\alpha}(y)G(x,y)\\
&\le c_2|x-y|^{2-N}\delta^{-2\alpha}(y)\min\left\{1,\frac{\delta(y)}{|x-y|}\right\}^2\\
&\le c_2|x-y|^{2-N}\delta^{-2\alpha}(y)\min\left\{1,\frac{\delta(y)}{|x-y|}\right\}^{2\alpha}\\
&\le c_2|x-y|^{2-2\alpha-N} \quad \text{for all } x,y \in \Omega.
\end{aligned}$$

Since $0 < \alpha < 1$, the mapping $x \mapsto |x-y|^{2-2\alpha-N}$ is integrable on Ω, so by means of Lebesgue's dominated convergence theorem we deduce that $\overline{\Omega} \ni x \mapsto \int_\Omega K(x,y)dy$ is continuous. This fact combined with $K > 0$ in Ω proves the existence of a number $M > 1$ that satisfies (7.33).

For $2 \le N \le 4$ we proceed similarly with different estimates (as in the proof of Proposition 7.15) to derive the same conclusion. This finishes the proof of Lemma 7.18. □

We are now ready to prove Theorem 7.17. Let $M > 1$ satisfy (7.33) and fix $0 < \varepsilon < 1$ such that

$$\varepsilon^{1-\alpha^2} \le M^{-1-\alpha}. \tag{7.37}$$

Define

$$g_\varepsilon : \mathbb{R} \to \mathbb{R}, \quad g_\varepsilon(t) = \begin{cases} \varepsilon^{-\alpha} & \text{if } t < \varepsilon,\\ t^{-\alpha} & \text{if } t \ge \varepsilon,\end{cases}$$

and for any $v \in C(\overline{\Omega})$, $v > 0$ in $\overline{\Omega}$ consider the operator

$$T_\varepsilon(v)(x) = \int_\Omega K(x,y)g_\varepsilon(v(y))dy \quad \text{for all } x \in \overline{\Omega}.$$

If $v \in C(\overline{\Omega})$ satisfies $v > 0$ in $\overline{\Omega}$, then $g_\varepsilon(v) \le \varepsilon^{-\alpha}$ in $\overline{\Omega}$ so by (7.33) we find $T_\varepsilon(v) \le M\varepsilon^{-\alpha}$ in $\overline{\Omega}$. Let now

$$v_1 \equiv M^{-1-\alpha}\varepsilon^{\alpha^2}, \quad v_2 \equiv M\varepsilon^{-\alpha}.$$

and

$$[v_1,v_2] = \{v \in C(\overline{\Omega}) : v_1 \le v \le v_2\}.$$

By Lemma 7.18 it is easy to see that $T_\varepsilon([v_1,v_2]) \subseteq [v_1,v_2]$. Further, by Lemma 7.18 and the Arzela–Ascoli theorem, it follows that

$$T_\varepsilon : [v_1,v_2] \to [v_1,v_2] \quad \text{is compact.}$$

Hence, by Schauder's fixed point theorem, there exists $v \in C(\overline{\Omega})$, $v_1 \le v \le v_2$ in $\overline{\Omega}$ such that $T_\varepsilon(v) = v$. By (7.37) it follows that $v \ge v_1 \ge \varepsilon$ in $\overline{\Omega}$, so $g_\varepsilon(v) = v^{-\alpha}$. Therefore, v satisfies (7.32), that is, $u = av$ is a solution of (7.20). Now, the boundary estimate (7.31) and the regularity of solution u follows from Proposition 7.15 and Proposition 7.16 respectively. In the following we derive the uniqueness of the solution to (7.20).

Let u_1, u_2 be two solutions of (7.20). Using Proposition 7.15 there exists $0 < c < 1$ such that

$$c\delta^2(x) \le u_i(x) \le \frac{1}{c}\delta^2(x) \quad \text{in } \Omega,\ i = 1,2. \tag{7.38}$$

This means that we can find a constant $C > 1$ such that $Cu_1 \ge u_2$ and $Cu_2 \ge u_1$ in Ω.

We claim that $u_1 \ge u_2$ in Ω. Supposing the contrary, let

$$M = \inf\{A > 1 : Au_1 \ge u_2 \ \text{in } \Omega\}.$$

By our assumption, we have $M > 1$. From $Mu_1 \ge u_2$ in Ω, it follows that

$$M^\alpha u_2(x) - u_1(x) = \int_\Omega G(x,y)\left[M^\alpha u_2^{-\alpha}(y) - u_1^{-\alpha}(y)\right]dy \ge 0 \quad \text{for all } x \in \Omega,$$

and then

$$M^{\alpha^2} u_1(x) - u_2(x) = \int_\Omega G(x,y)\left[M^{\alpha^2} u_1^{-\alpha}(y) - u_2^{-\alpha}(y)\right]dy \ge 0 \quad \text{for all } x \in \Omega.$$

We have thus obtained $M^{\alpha^2} u_1 \ge u_2$ in Ω. Since $M > 1$ and $\alpha^2 < 1$, this last inequality contradicts the minimality of M. Hence, $u_1 \ge u_2$ in Ω. Similarly we deduce $u_1 \le u_2$ in Ω, so $u_1 \equiv u_2$ and the uniqueness is proved. This finishes the proof of Theorem 7.17. □

Chapter 8
Large Time Behavior of Solutions for Degenerate Parabolic Equations

> The saddest aspect of life right now is that science gathers knowledge faster than society gathers wisdom.
>
> Isaac Asimov (1902–1992)

8.1 Introduction

We are concerned in this chapter with degenerate parabolic problems of type

$$\begin{cases} \partial_t u = a(\delta(x))u^p \Delta u + g(x,u) & \text{in } \Omega \times (0,\infty), \\ u = 0 & \text{in } \partial\Omega \times (0,\infty), \\ u(x,0) = u_0(x) & \text{in } \Omega, \end{cases} \tag{8.1}$$

where $\Omega \subset \mathbb{R}^N$ $(N \geq 1)$ is a smooth bounded domain, $\delta(x) = \text{dist}(x, \partial\Omega)$ and $p \geq 1$. The initial data u_0 verifies $u_0 \in C^{\alpha}(\overline{\Omega})$ and $u_0 > 0$ in Ω.

If $a \equiv 1$, $p = 2$ and $g(x,u) = u$, problem (8.1) arises in a model for the resistive diffusion of a forced free magnetic field in a plasma confined between two walls (we refer to Friedman and McLeod [79] and the references therein for further details). We mention here the works of Winkler [208–212] that deal with the case $a \equiv 1$. However, the case where a is not constant have been less investigated.

Our aim in this chapter is to provide conditions such that (8.1) admits solutions which are global in time. If such solutions exist we also investigate their behavior as $t \to \infty$. It turns out that both existence and behavior of global solutions is

M. Ghergu and V. Rădulescu, *Nonlinear PDEs*, Springer Monographs in Mathematics,
DOI 10.1007/978-3-642-22664-9_8, © Springer-Verlag Berlin Heidelberg 2012

strongly related to the behavior of g (resp. a) at infinity (resp. around the origin). These features are observed in our study by a close analysis of the associated elliptic equations whose solutions are the stabilizers for the time dependent problem (8.1). In our subsequent analysis we shall distinguish between the superlinear and sublinear case. The linear case will be considered in the last section of this chapter.

8.2 Superlinear Case

We are concerned in this section with the following parabolic problem

$$\begin{cases} \partial_t u = a(\delta(x))u^p \Delta u + g(u) & \text{in } \Omega \times (0,\infty), \\ u = 0 & \text{in } \partial\Omega \times (0,\infty), \\ u(x,0) = u_0(x) & \text{in } \Omega. \end{cases} \tag{8.2}$$

We assume that $g \in C^1(0,\infty) \cap C[0,\infty)$ satisfies $g > 0$ in $(0,\infty)$ and

$(g1)$ the mapping $(0,\infty) \ni s \longmapsto \frac{g(s)}{s^{p+1}}$ is nondecreasing.

As a consequence, g is increasing and $g(0) = 0$. Moreover, there exists

$$\ell := \lim_{s \searrow 0} \frac{g(s)}{s^{p+1}} \in [0,\infty). \tag{8.3}$$

The potential $a : [0,\infty) \to [0,\infty)$ is Hölder continuous and nondecreasing such that $a(0) = 0$ and $a > 0$ in $(0,\infty)$.

The main result in this section is the following.

Theorem 8.1 *Assume that the potential a fulfills the condition*

$$\int_0^1 \frac{s}{a(s)} ds < \infty. \tag{8.4}$$

Then, problem (8.2) has a unique solution u which, in addition, satisfies

$$\lim_{t \to \infty} \|(1+pt)^{1/p} u(\cdot,t) - W\|_{L^\infty(\Omega)} = 0,$$

where $W \in C^2(\Omega) \cap C(\overline{\Omega})$ is the unique solution of

$$\begin{cases} -a(\delta(x))\Delta W + \ell W = W^{1-p} & \text{in } \Omega, \\ W > 0 & \text{in } \Omega, \\ W = 0 & \text{on } \partial\Omega. \end{cases} \tag{8.5}$$

Condition (8.4) appears as a naturally one in the context of stationary problems associated to (8.2). More precisely we have

Proposition 8.2 *Assume that a satisfies (8.4). Then, for all $\ell \geq 0$ problem (8.5) has a unique classical solution $W \in C^2(\Omega) \cap C(\overline{\Omega})$.*

In our setting we prove that condition (8.4) is also sufficient in order to ensure the existence of global solutions to the parabolic problem (8.2). This requirement enables us to determine the asymptotic profile of the unique solution to (8.2) as $t \to \infty$.

Proof of Theorem 8.1. We divide the proof into several steps.

Step 1: Existence. Let $(u_{n,0})_{n\geq 1} \subset C^2(\overline{\Omega})$ be a positive smooth sequence such that $(u_{n,0})_{n\geq 1}$ is decreasing and $\|u_{n,0} - u_0\|_{L^\infty(\Omega)} \to 0$ as $n \to \infty$.

For $0 < \varepsilon < 1$ consider the approximated problem

$$\begin{cases} \partial_t u = a_\varepsilon(\delta(x)) f_\varepsilon(u)\Delta u - g(u) & \text{in } \Omega \times (0,T), \\ u = u_{n,0} + \varepsilon & \text{on } \Sigma_T, \end{cases}$$

where $a_\varepsilon(\delta(x)) = a(\delta(x)) + \varepsilon$ and $f_\varepsilon : \mathbb{R} \to [0,\infty)$ is a C^1 function defined by

$$f_\varepsilon(s) = \begin{cases} \varepsilon^p & \text{if } s \leq \varepsilon/2, \\ s^p & \text{if } s > \varepsilon. \end{cases}$$

By standard parabolic arguments, for all $0 < \varepsilon < 1$ there exists a local solution $u_{n,\varepsilon} \in C^{2,1}(\overline{\Omega} \times [0, T_{n,\varepsilon}))$ such that $u_{n,\varepsilon} \geq \varepsilon$. Hence, $u_{n,\varepsilon}$ satisfies

$$\begin{cases} \partial_t u_{n,\varepsilon} = a_\varepsilon(\delta(x)) u_{n,\varepsilon}^p \Delta u_{n,\varepsilon} - g(u_{n,\varepsilon}) & \text{in } \Omega \times (0, T_{n,\varepsilon}), \\ u_{n,\varepsilon} = u_{n,0} + \varepsilon & \text{on } \Sigma_{T_{n,\varepsilon}}. \end{cases} \tag{8.6}$$

We claim that $T_{n,\varepsilon} = \infty$. To this aim, we provide uniform bounds for $u_{n,\varepsilon}$. Let ζ be a positive superharmonic function in Ω such that $\zeta \geq u_{n,0}$ in Ω; for instance we may consider $\zeta = M - |x|^2$ with $M > 0$ large enough. Then $\overline{u}_{n,\varepsilon} := \zeta + \varepsilon$ satisfies $\overline{u}_{n,\varepsilon} \geq u_{n,0} + \varepsilon$ on $\Sigma_{T_{n,\varepsilon}}$ and

$$\partial_t \overline{u}_{n,\varepsilon} - a_\varepsilon(\delta(x)) \overline{u}_{n,\varepsilon}^p \Delta \overline{u}_{n,\varepsilon} + g(\overline{u}_{n,\varepsilon}) \geq 0 \quad \text{in } \Omega \times (0, T_{n,\varepsilon}).$$

By Theorem 1.5 we have

$$u_{n,\varepsilon} \le \overline{u}_{n,\varepsilon} \le \zeta + 1 \quad \text{in } \Omega \times (0, T_{n,\varepsilon}). \tag{8.7}$$

Let now $\eta > 0$ be small enough and $\Omega_\eta = \{x \in \Omega : \text{dist}(x, \partial\Omega) > \eta\}$. We denote by $\lambda_{1,\eta}$, $\varphi_{1,\eta}$ the first eigenvalue and the first eigenfunction of $(-\Delta)$ in Ω_η. By (8.7) we also may define

$$\beta := \sup_{0<\varepsilon<1} \max_{\Omega \times (0,T_{n,\varepsilon})} \left[f_\varepsilon(u_\varepsilon)(a(\delta)+1)\lambda_{1,\eta} + \frac{g(u_\varepsilon)}{u_\varepsilon} \right] < \infty.$$

Let now $\underline{u}_{n,\varepsilon} = Ce^{-\beta t}\varphi_{1,\eta}$, where $C = C(n,\eta) > 0$ is small enough such that $C\varphi_{1,\eta} < u_{n,0}$ in Ω_δ. Then $\underline{u}_{n,\varepsilon}$ satisfies

$$\underline{u}_{n,\varepsilon} \le u_{n,0} + \varepsilon \quad \text{on } (\partial\Omega_\eta \times (0, T_{n,\varepsilon})) \cup (\Omega_\eta \times \{0\})$$

and

$$\begin{aligned} \partial_t \underline{u}_{n,\varepsilon} - a_\varepsilon(\delta(x)) \underline{u}_{n,\varepsilon}^p \Delta \underline{u}_{n,\varepsilon} + \frac{g(u_{n,\varepsilon})}{u_{n,\varepsilon}} \underline{u}_{n,\varepsilon} & \\ = \underline{u}_{n,\varepsilon} & \left[-\beta + f_\varepsilon(u_{n,\varepsilon})(a(\delta)+1)\lambda_{1,\eta} + \frac{g(u_{n,\varepsilon})}{u_{n,\varepsilon}} \right] \\ \ge 0 & \quad \text{in } \Omega_\eta \times (0, T_{n,\varepsilon}). \end{aligned}$$

By Theorem 1.5 we now derive

$$u_{n,\varepsilon} \ge \underline{u}_{n,\varepsilon} = Ce^{-\beta t}\varphi_{1,\eta} \quad \text{in } \Omega_\eta \times (0, T_{n,\varepsilon}).$$

This last estimate combined with (8.7) leads us to $T_{n,\varepsilon} = \infty$. Let $v_{n,\varepsilon} = \partial_t u_{n,\varepsilon}$. By (8.6) we obtain

$$\begin{aligned} \partial_t v_{n,\varepsilon} &= a_\varepsilon(\delta(x)) p u_{n,\varepsilon}^{p-1} v_{n,\varepsilon} \Delta u_{n,\varepsilon} + a_\varepsilon(\delta(x)) u_{n,\varepsilon}^p \Delta v_{n,\varepsilon} - g'(u_{n,\varepsilon}) v_{n,\varepsilon} \\ &= a_\varepsilon(\delta(x)) u_{n,\varepsilon}^p \Delta v_{n,\varepsilon} + v_{n,\varepsilon} \frac{p v_{n,\varepsilon} + p g(u_{n,\varepsilon}) - u_{n,\varepsilon} g'(u_{n,\varepsilon})}{u_{n,\varepsilon}} \quad \text{in } \Omega \times (0,\infty), \end{aligned}$$

and $v_{n,\varepsilon} = 0$ on Σ_∞. Again by Theorem 1.5 and taking into account that $\nabla^2 v_{n,\varepsilon} \in L^\infty(\Omega \times (0,\infty))$ we deduce $\partial_t u_{n,\varepsilon} = v_{n,\varepsilon} \ge 0$ in $\Omega \times (0,\infty)$. In particular this implies that $\Delta u_{n,\varepsilon} \ge 0$ in $\Omega \times (0,\infty)$. This allows us to apply Theorem 1.5 once more in order to derive that $\{u_{n,\varepsilon}\}_{n\ge 1}$ is nonincreasing in $\Omega \times (0,\infty)$.

For all $(x,t) \in \Omega \times (0,\infty)$, let $u_n(x,t) = \lim_{\varepsilon \searrow 0} u_{n,\varepsilon}(x,t)$. By standard parabolic arguments $u_n \in C^{2,1}(\Omega \times (0,\infty)) \cap C(\overline{\Omega} \times [0,\infty))$ and u_n satisfies

$$\begin{cases} \partial_t u_n = a(\delta(x))u_n^p \Delta u_n - g(u_n) & \text{in } \Omega \times (0,\infty), \\ u_n = u_{n,0} & \text{on } \Sigma_\infty. \end{cases} \tag{8.8}$$

Since $u_{n,\varepsilon} \ge u_{n+1,\varepsilon}$ on Σ_∞ we may apply Theorem 1.5 in order to deduce that $\{u_n\}$ is decreasing. Let now $u(x,t) = \lim_{n\to\infty} u_n(x,t)$, for all $(x,t) \in \Omega \times (0,\infty)$. Then, by standard estimates we deduce from (8.8) that u is a solution of (8.2). This completes the proof of the existence part.

Step 2: Uniqueness. We first consider the case $p > 1$. Let $v \in C^{2,1}(\Omega \times [0,T)) \cap C(\overline{\Omega} \times [0,T])$ be another solution of problem (8.2). By Theorem 1.5 we deduce $v \le u_\varepsilon$ in $\Omega \times [0,T]$ so that, passing to the limit with $\varepsilon \to 0$ we obtain

$$v \le u \quad \text{in } \Omega \times [0,T]. \tag{8.9}$$

Let K be a compact subset of $\Omega \times (0,T)$ and fix $\Omega_0 \subset\subset \Omega$ such that $K \subset\subset \Omega_0$. Also denote by μ_1, ϕ_1 the first eigenvalue resp. the first eigenfunction corresponding to the Laplace operator $(-\Delta)$ in $H_0^1(\Omega_0)$. Subtracting the two equations corresponding to u and v we find

$$\frac{d}{dt}\left[\frac{v^{1-p}(x,t) - u^{1-p}(x,t)}{(p-1)a(\delta(x))}\right] = \Delta(u-v) + \frac{h(u)-h(v)}{a(\delta(x))} \quad \text{in } \Omega \times (0,T),$$

where $h(s) = g(s)/s^p$, $s > 0$. We now integrate over $[\eta,t]$, $0 < \eta < t < T$ in the above equality and then we let $\eta \to 0$. We find

$$\frac{v^{1-p}(x,t) - u^{1-p}(x,t)}{(p-1)a(\delta(x))} = \int_0^t \left[\Delta(u-v) + \frac{h(u)-h(v)}{a(\delta(x))}\right] ds \quad \text{in } \Omega \times (0,T).$$

Next we multiply by ϕ_1 and we integrate over Ω_0. We obtain

$$\int_{\Omega_0} \frac{v^{1-p}(x,t) - u^{1-p}(x,t)}{(p-1)a(\delta(x))}\phi_1 dx = \int_0^t \int_{\Omega_0} \left[\Delta(u-v) + \frac{h(u)-h(v)}{a(\delta(x))}\right]\phi_1 dx ds. \tag{8.10}$$

Setting

$$y(t) := \int_{\Omega_0} \left[\frac{v^{1-p}(x,t) - u^{1-p}(x,t)}{a(\delta(x))}\right]\phi_1 dx, \quad 0 < t < T, \tag{8.11}$$

from (8.10) we have

$$y(t) \le C \int_0^t \int_{\Omega_0} \left[\Delta(u_\lambda - v_\lambda) + \frac{h(u)-h(v)}{a(\delta(x))}\right]\phi_1 dx ds, \quad 0 < t < T.$$

By Green's identity and (8.9) we further obtain

$$\begin{aligned} y(t) \le& C\mu_1 \int_0^t \int_{\Omega_0} (v-u)\phi_1 dxds + C\int_0^t \int_{\partial\Omega_0} (v-u)\frac{\partial\phi_1}{\partial\nu} d\sigma(x)ds \\ &+ C\lambda \int_0^t \int_{\Omega_0} \frac{h(u)-h(v)}{a(\delta(x))}\phi_1 dxds \\ \le& C\int_0^t \int_{\partial\Omega_0} (v-u)\frac{\partial\phi_1}{\partial\nu} d\sigma(x)ds + C\lambda \int_0^t \int_{\Omega_0} \frac{h(u)-h(v)}{a(\delta(x))}\phi_1 dxds \\ \le& c_1 t \max_{\partial\Omega_0} w + C\lambda \int_0^t \int_{\Omega_0} \frac{h(u)-h(v)}{a(\delta(x))}\phi_1 dxds, \end{aligned} \tag{8.12}$$

where $w = u - v$. Using the hypothesis $(g1)$ we have that the mapping $(0,\infty) \ni s \longmapsto h(s)/s$ is nondecreasing so that $(h(s)/s)' \le 0$ for all $s > 0$. Thus,

$$sh'(s) \le h(s) \quad \text{for all } s > 0. \tag{8.13}$$

By the mean value theorem we deduce

$$\frac{h(u)-h(v)}{v^{1-p}-u^{1-p}} = \frac{h'(\theta)\theta^p}{p-1} \le \frac{\sup\{h'(s)s^p : 0 \le s \le \|u\|_{L^\infty(\Omega\times[0,T])}\}}{p-1} < \infty.$$

Hence, there exists a positive constant $c > 0$ depending only on $\|u\|_{L^\infty(\Omega\times[0,T])}$ such that

$$h(u) - h(v) \le c(v^{1-p} - u^{1-p}) \quad \text{in } \Omega \times (0,T). \tag{8.14}$$

By (8.11)–(8.14) we finally obtain

$$y(t) \le c_1 T \max_{\partial\Omega_0} w_\lambda(x) + c_2 \int_0^t y(s)ds \quad \text{for all } 0 < t < T.$$

Now, Gronwall's inequality leads us to

$$\int_K (v^{1-p}(x,t) - u^{1-p}(x,t))\phi_1 dx \le y(t) \le C \max_{\partial\Omega_0} w_\lambda(x).$$

Since the right-hand side tends to zero as $\Omega_0 \to \Omega$, it follows that $u = v$ in K. Hence, $u \equiv v$ and the problem (8.2) has a unique solution.

If $p = 1$ we proceed in the same manner. We have only to replace the definition of $y(t)$ in (8.11) by

$$y(t) := \int_{\Omega_0} (\ln u(x,t) - \ln v(x,t))\phi_1 dx, \quad 0 < t < T,$$

and then use (8.13) to derive the uniqueness of the global solution.

Step 3: Large time behavior.

Fix $0<\varepsilon<1$ and let

$$v_\varepsilon(x,t)=(C_\varepsilon+pt)^{-1/p}(W(x)+\varepsilon),\quad (x,t)\in\Omega\times(0,\infty),$$

where W is the unique solution of (8.5) and $C_\varepsilon>0$ is small enough such that

$$\varepsilon\ge C_\varepsilon^{1/p}\|u_0\|_{L^\infty(\Omega)}. \tag{8.15}$$

Then $v_\varepsilon\ge u$ on Σ_∞ and

$$\begin{aligned}
\partial_t v_\varepsilon - &a(\delta(x))v_\varepsilon\Delta v_\varepsilon+g(v_\varepsilon)\\
&\ge \partial_t v_\varepsilon - a(\delta(x))v_\varepsilon\Delta v_\varepsilon+\ell v_\varepsilon^{p+1}\\
&=(C_\varepsilon+pt)^{-\frac{1+p}{p}}\left[-(W+\varepsilon)-a(\delta(x))(W+\varepsilon)^p\Delta W+\ell(W+\varepsilon)^{p+1}\right]\\
&=(C_\varepsilon+pt)^{-\frac{1+p}{p}}\left[-(W+\varepsilon)+(W+\varepsilon)^p(W^{1-p}-\ell W)+\ell(W+\varepsilon)^{p+1}\right]\\
&\ge (C_\varepsilon+pt)^{-\frac{1+p}{p}}\left[-(W+\varepsilon)+(W+\varepsilon)\left(\frac{W+\varepsilon}{W}\right)^{p-1}\right]\\
&\ge 0 \quad \text{in } \Omega\times(0,\infty).
\end{aligned}$$

By Theorem 1.5 we obtain $u\le v_\varepsilon$ in $\Omega\times(0,\infty)$. This yields

$$\begin{aligned}
(1+pt)^{1/p}u(x,t)-W &\le \left(\frac{C_\varepsilon+pt}{1+pt}\right)^{1/p}(W+\varepsilon)-W\\
&=\left[1+\mathcal{O}\left(\frac{1}{t}\right)\right](W+\varepsilon)-W \quad \text{as } t\to\infty\\
&=\mathcal{O}\left(\frac{1}{t}\right)(W+\varepsilon)+\varepsilon \quad \text{as } t\to\infty.
\end{aligned}$$

Hence, we can find $t_1=t_1(\varepsilon)>0$ such that

$$\sup_{x\in\overline{\Omega}}\left[(1+pt)^{1/p}u(x,t)-W\right]\le 2\varepsilon \quad \text{for all } t\ge t_1. \tag{8.16}$$

According to Proposition 8.2, there exists $W_\eta\in C^2(\Omega)\cap C(\overline{\Omega})$ a unique solution of

$$\begin{cases}
-a(\delta(x))\Delta W+(\ell+\eta)W=W^{1-p} & \text{in } \Omega_\eta,\\
W>0 & \text{in } \Omega_\eta,\\
W=0 & \text{on } \partial\Omega_\eta.
\end{cases} \tag{8.17}$$

We claim that $\|W_\eta-W\|_{L^\infty(\Omega_\eta)}\to 0$ as $\eta\to\infty$.

Indeed, extending W_η by zero on $\overline{\Omega}\setminus\Omega_\eta$, by Theorem 1.2 we obtain that $\{W_\eta\}$ is decreasing as $\eta \searrow 0$ and $W_\eta \le W$ in Ω. By standard elliptic arguments the sequence $\{W_\eta\}_{0<\eta<1}$ converges in $C^2_{\mathrm{loc}}(\Omega)$ to a $C^2(\Omega)\cap C(\overline{\Omega})$ function which is a solution of (8.5). Since (8.5) has a unique solution we find $\|W_\eta - W\|_{L^\infty(\Omega_\eta)} \to 0$ as $\eta \to \infty$. This proves our claim.

In view of the above arguments, there exists $\eta > 0$ such that

$$\|W - W_\eta\|_{L^\infty(\Omega_\eta)} < \varepsilon \quad \text{and} \quad \|W\|_{L^\infty(\Omega\setminus\Omega_\eta)} < \varepsilon, \tag{8.18}$$

where W_η is the unique solution of (8.17).

Define

$$w_\eta(x,t) = (C_\eta + pt)^{-1/p} W(x), \quad (x,t) \in \Omega \times (0,\infty),$$

where $C_\eta > 0$ is such that

$$w_\eta(x,0) = C_\eta^{-1/p} W(x) < u_0(x) \quad \text{in } \Omega_\eta. \tag{8.19}$$

Furthermore, taking into account the hypothesis $(g1)$ and (8.3) we may choose $C_\eta > 0$ such that

$$g(w_\eta) < (\ell+\eta) w_\eta^{p+1} \quad \text{in } \Omega_\eta. \tag{8.20}$$

Then $w_\eta \le u$ on $(\partial\Omega_\eta \times (0,\infty)) \cup (\Omega_\delta \times \{0\})$ and

$$\begin{aligned}
\partial_t w_\eta - a(\delta(x)) w_\eta \Delta w_\eta + g(w_\eta) &\\
\le \partial_t w_\eta - a(\delta(x)) w_\eta \Delta w_\eta + (\ell+\eta) v_\varepsilon^{p+1} &\\
= (C_\varepsilon + pt)^{-\frac{1+p}{p}} \left[-W_\eta - a(\delta(x)) W_\eta^p \Delta W_\eta + (\ell+\eta) W^{p+1}\right] &\\
= (C_\varepsilon + pt)^{-\frac{1+p}{p}} W_\eta^p \left[-W_\eta^{1-p} + a(\delta(x)) W_\eta^p \Delta W_\eta + \ell (W+\varepsilon)^{p+1}\right] &\\
= 0 \quad \text{in } \Omega_\eta \times (0,\infty). &
\end{aligned}$$

Again by Theorem 1.5 we derive that $w_\eta \le u$ in $\Omega_\eta \times (0,\infty)$ which yields

$$W_\eta - (1+pt)^{1/p} u(x,t) \le \left[(C_\eta + pt)^{1/p} u(x,t) - (1+pt)^{1/p}\right] u(x,t) \to 0 \quad \text{as } t \to 0.$$

Hence, we may choose $t_2 = t_2(\varepsilon,\eta) > 0$ such that

$$\sup_{x\in\overline{\Omega}_\eta} \left[W_\eta(x) - (1+pt)^{1/p} u(x,t)\right] \le \varepsilon \quad \text{for all } t \ge t_2. \tag{8.21}$$

By (8.18) and (8.21) we then obtain

$$\begin{aligned}
&\sup_{x\in\overline{\Omega}_\eta}\Big[W(x)-(1+pt)^{1/p}u(x,t)\Big]\\
&\le \|W-W_\eta\|_{L^\infty(\Omega_\delta)}+\sup_{x\in\overline{\Omega}_\eta}\Big[W_\eta(x)-(1+pt)^{1/p}u(x,t)\Big]\\
&\le 2\varepsilon \quad \text{for all } t\ge t_2.
\end{aligned}$$

Also by (8.18) we have

$$\sup_{x\in\Omega\setminus\Omega_\eta}\Big[W(x)-(1+pt)^{1/p}u(x,t)\Big]\le \|W\|_{L^\infty(\Omega\setminus\Omega_\eta)}<\varepsilon \quad \text{for all } t\ge 0.$$

Hence

$$\sup_{x\in\Omega}\Big[W(x)-(1+pt)^{1/p}u(x,t)\Big]\le 2\varepsilon \quad \text{for all } t\ge 0. \tag{8.22}$$

Combining now (8.16) and (8.22) we obtain the conclusion. This completes the proof. □

8.3 Sublinear Case

In this section we are concerned with the following parabolic problem

$$\begin{cases}
\partial_t u = a(\delta(x))u^p(\Delta u+\lambda g(u)) & \text{in } \Omega\times(0,T),\\
u=0 & \text{in } \partial\Omega\times(0,T),\\
u(x,0)=u_0(x) & \text{in } \Omega,
\end{cases} \tag{8.23}$$

where $\Omega\subset\mathbb{R}^N$ ($N\ge 1$) is a smooth bounded domain, $\delta(x)=\mathrm{dist}(x,\partial\Omega)$, $T>0$, $\lambda>0$, and $p\ge 1$. Here we assume that $g\in C^1(0,\infty)\cap C[0,\infty)$ is a nondecreasing function such that $g>0$ in $(0,\infty)$ and g has a sublinear growth, that is,

$(g1)$ The mapping $(0,\infty)\ni s\longmapsto \frac{g(s)}{s}$ is nonincreasing.
$(g2)$ $\lim_{s\searrow 0}\frac{g(s)}{s}=\infty$ and $\lim_{s\to+\infty}\frac{g(s)}{s}=0$.

The standard example of function g that satisfies $(g1)-(g2)$ is $g(s)=s^q$ with $0<q<1$. We also may consider:

$$g(s)=s^q\ln(1+s^r),\quad g(s)=s^q\ln^r(1+s),\quad g(s)=s^q\arctan s^r,$$

where $q,r\ge 0$ with $0<q+r<1$.

The initial data u_0 satisfies $u_0 \in C^1(\overline{\Omega})$, $u_0 > 0$ in Ω and

$$c_0\delta(x) \le u_0(x) \le c_1\delta(x) \quad \text{in } \Omega, \tag{8.24}$$

for some positive constants $c_0, c_1 > 0$.

The main result in this section is the following.

Theorem 8.3 *Assume that conditions $(g1)-(g2)$ hold. Then, for all $\lambda > 0$, problem (8.23) has a unique global solution u_λ such that:*

(i) *The mapping $(0,\infty) \ni \lambda \longmapsto u_\lambda(x,t)$ is increasing for all $(x,s) \in \Omega \times (0,\infty)$.*
(ii) *We have:*

$$\lim_{t\to\infty} \|u_\lambda(\cdot,t) - w_\lambda\|_{L^\infty(\Omega)} = 0, \tag{8.25}$$

where $w_\lambda \in C^2(\Omega)\cap C(\overline{\Omega})$ is the unique solution of

$$\begin{cases} -\Delta w = \lambda g(w) & \text{in } \Omega,\\ w > 0 & \text{in } \Omega,\\ w = 0 & \text{on } \partial\Omega. \end{cases} \tag{8.26}$$

The first step in proving Theorem 8.3 is the following.

Proposition 8.4 *Assume that $g \in C^1(0,\infty)\cap C[0,\infty)$ is positive on $(0,\infty)$ and satisfies $(g1)-(g2)$. Then, for all $\lambda > 0$, problem (8.26) has a unique classical solution $w_\lambda \in C^2(\Omega)\cap C(\overline{\Omega})$ such that*

$$c_1\delta(x) \le w_\lambda \le c_2\delta(x) \quad \text{in } \Omega, \tag{8.27}$$

for some positive constants c_1, c_2.

Moreover, if for fixed $\lambda > 0$ we denote by $w_\varepsilon \in C^2(\Omega)\cap C(\overline{\Omega})$, $0 < \varepsilon < 1$, the unique solution of

$$\begin{cases} -\Delta w_\varepsilon = \lambda g(w_\varepsilon) & \text{in } \Omega,\\ w_\varepsilon > 0 & \text{in } \Omega,\\ w_\varepsilon = \varepsilon & \text{on } \partial\Omega, \end{cases} \tag{8.28}$$

then the sequence $\{w_\varepsilon\}_{0<\varepsilon<1}$ converges to w_λ as $\varepsilon \searrow 0$ uniformly in $\overline{\Omega}$.

Proof. We divide the proof of Theorem 8.3 into several steps.
Step 1: Existence. Fix $\lambda > 0$ and for $0 < \varepsilon < 1$ consider the approximated problem

$$\begin{cases} \partial_t u_\varepsilon = a_\varepsilon(\delta(x)) f_\varepsilon(u_\varepsilon)(\Delta u_\varepsilon + \lambda g(u_\varepsilon)) & \text{in } \Omega \times (0,T), \\ u_\varepsilon = \varepsilon & \text{in } \partial\Omega \times (0,T), \\ u_\varepsilon(x,0) = u_0(x) + \varepsilon & \text{in } \Omega, \end{cases}$$

where $a_\varepsilon(\delta(x)) = a(\delta(x)) + \varepsilon$ and $f_\varepsilon : \mathbb{R} \to [0,\infty)$ is defined by

$$f_\varepsilon(s) = \begin{cases} \varepsilon^p & \text{if } s \le \varepsilon, \\ s^p & \text{if } s > \varepsilon. \end{cases}$$

By standard parabolic arguments, for all $0 < \varepsilon < 1$ there exists a local solution $u_\varepsilon \in C^{2,1}(\overline{\Omega} \times [0,T_\varepsilon))$ and $u_\varepsilon \ge \varepsilon$. Hence, u_ε satisfies

$$\begin{cases} \partial_t u_\varepsilon = a_\varepsilon(\delta(x)) u_\varepsilon^p(\Delta u_\varepsilon + \lambda g(u_\varepsilon)) & \text{in } \Omega \times (0,T_\varepsilon), \\ u_\varepsilon = \varepsilon & \text{in } \partial\Omega \times (0,T_\varepsilon), \\ u_\varepsilon(x,0) = u_0(x) + \varepsilon & \text{in } \Omega. \end{cases} \tag{8.29}$$

We claim that $T_\varepsilon = \infty$. To this aim, we provide a uniform upper bound for u_ε. Let w_λ and w_ε be the unique solutions of (8.26) and (8.28) respectively. Set $\overline{u}_\varepsilon := Mw_\varepsilon$ where $M > 1$ is large. In view of $(g1)$ we derive that $g(\overline{u}_\varepsilon) \le Mg(w_\varepsilon)$ in Ω. Furthermore we have

$$\begin{aligned} \partial_t \overline{u}_\varepsilon - a_\varepsilon(\delta(x)) \overline{u}_\varepsilon^p \Big(\Delta \overline{u}_\varepsilon + \lambda g(\overline{u}_\varepsilon)\Big) & \\ &= -a_\varepsilon(\delta(x)) \overline{u}_\varepsilon^p \Big(\Delta \overline{u}_\varepsilon + \lambda g(\overline{u}_\varepsilon)\Big) \\ &\ge -a_\varepsilon(\delta(x)) M^{p+1} w_\varepsilon^p \Big(\Delta w_\varepsilon + \lambda g(w_\varepsilon)\Big) \\ &= 0 \quad \text{in } \Omega \times (0,T_\varepsilon). \end{aligned} \tag{8.30}$$

In view of (8.27) and Proposition 8.4, for all $x \in \Omega$ we also have

$$\overline{u}_\varepsilon = Mw_\varepsilon \ge \varepsilon + (M-1)w_\varepsilon \ge \varepsilon + (M-1)w_\lambda \ge \varepsilon + (M-1)c_1\delta(x).$$

Letting $M > 1$ sufficiently large, by (8.24) we deduce

$$\overline{u}_\varepsilon(x,0) \ge u_0(x) + \varepsilon \quad \text{in } \Omega. \tag{8.31}$$

Combining (8.30) and (8.31), by virtue of Lemma 4.5 we find $\varepsilon \le u_\varepsilon \le Mw_\varepsilon$ in $\Omega \times (0,T_\varepsilon)$. Hence, we may continue the local solution u_ε of (8.29) for all times. This means that $T_\varepsilon = \infty$ for all $0 < \varepsilon < 1$.

Next, let $\underline{u}_\varepsilon := m\varphi_1 + \varepsilon$. Using (8.24), we may find $0 < m < 1$ small enough such that $m\varphi_1 \le u_0$ in Ω. In $\Omega \times (0,\infty)$ we also have

$$\begin{aligned} \partial_t \underline{u}_\varepsilon - a_\varepsilon(\delta(x))\underline{u}_\varepsilon^p\Big(\Delta \underline{u}_\varepsilon + \lambda g(\underline{u}_\varepsilon)\Big) & \\ &= -a_\varepsilon(\delta(x))\underline{u}_\varepsilon^p\Big(\Delta \underline{u}_\varepsilon + \lambda g(\underline{u}_\varepsilon)\Big) \\ &\le a_\varepsilon(\delta(x))\underline{u}_\varepsilon^p\Big(m\lambda_1\varphi_1 - \lambda g(m\varphi_1+\varepsilon)\Big) \\ &\le a_\varepsilon(\delta(x))\underline{u}_\varepsilon^p\Big(\lambda_1(m\varphi_1+\varepsilon) - \lambda g(m\varphi_1+\varepsilon)\Big). \end{aligned} \tag{8.32}$$

Since $\lim_{s\searrow 0} g(s)/s = \infty$, we may choose $0 < m, \varepsilon_0 < 1$ sufficiently small such that

$$\frac{g(m\varphi_1+\varepsilon)}{m\varphi_1+\varepsilon} > \frac{\lambda_1}{\lambda} \quad \text{in } \Omega, \text{ for all } 0 < \varepsilon < \varepsilon_0. \tag{8.33}$$

By (8.32) and (8.33) we deduce

$$\partial_t \underline{u}_\varepsilon - a_\varepsilon(\delta(x))\underline{u}_\varepsilon^p\Big(\Delta \underline{u}_\varepsilon + \lambda g(\underline{u}_\varepsilon)\Big) \le 0 \quad \text{in } \Omega \times (0,\infty).$$

Since $\underline{u}_\varepsilon = \varepsilon$ on $\partial\Omega$ and $\underline{u}_\varepsilon \le \varepsilon + u_0$ in Ω, by Lemma 4.5 we obtain $m\varphi_1 + \varepsilon \le u_\varepsilon$ in $\Omega \times (0,\infty)$. Hence, for all $0 < \varepsilon < \varepsilon_0$ we have

$$m\varphi_1 + \varepsilon \le u_\varepsilon \le Mw_\varepsilon \quad \text{in } \Omega \times (0,\infty). \tag{8.34}$$

Now, interior Hölder and Schauder estimates can be deduced in order to show that $(u_\varepsilon)_{0<\varepsilon<1}$ converges monotonically in any compact subset of $\Omega \times (0,\infty)$ to a function $u_\lambda \in C^{2,1}(\Omega \times (0,\infty)) \cap C(\overline{\Omega} \times [0,\infty))$ which is a solution of (8.23). Furthermore, from (8.34) we derive

$$m\varphi_1 \le u_\lambda \le Mw_\lambda \quad \text{in } \Omega \times (0,\infty). \tag{8.35}$$

Step 2: Uniqueness and dependence on λ. We first consider the case $p > 1$. Let $v_\lambda \in C^{2,1}(\Omega \times [0,T)) \cap C(\overline{\Omega} \times [0,T])$ be another solution of problem (8.23). By Lemma 4.5 we deduce $v_\lambda \le u_\varepsilon$ in $\Omega \times [0,T]$ so that, passing to the limit and by (8.35) we obtain

$$v_\lambda \le u_\lambda \le w_\lambda \quad \text{in } \Omega \times [0,T]. \tag{8.36}$$

Let K be a compact subset of Ω and fix $\Omega_0 \subset \Omega$ such that $K \subset\subset \Omega_0 \subset\subset \Omega$. Also denote by μ_1, ϕ_1 the first eigenvalue resp. the first eigenfunction corresponding to the Laplace operator $(-\Delta)$ in $H_0^1(\Omega_0)$. Subtracting the two equations corresponding to u and v we find

$$\frac{\partial_t (v_\lambda^{1-p} - u_\lambda^{1-p})}{(p-1)a(\delta(x))} = \Delta(u_\lambda - v_\lambda) + \lambda g(u_\lambda) - \lambda g(v_\lambda) \quad \text{in } \Omega \times (0,T).$$

We now integrate over $[\eta, t]$, $0 < \eta < t < T$ in the above equality and then we let $\eta \to 0$. We find

$$\frac{v_\lambda^{1-p}(x,t) - u_\lambda^{1-p}(x,t)}{(p-1)a(\delta(x))} = \int_0^t \Big(\Delta(u_\lambda - v_\lambda) + \lambda g(u_\lambda) - \lambda g(v_\lambda) \Big) ds \quad \text{in } \Omega \times (0,T).$$

Next we multiply the last equality by ϕ_1 and we integrate over Ω_0. We obtain

$$\int_{\Omega_0} \frac{v_\lambda^{1-p}(x,t) - u_\lambda^{1-p}(x,t)}{(p-1)a(\delta(x))} \phi_1 dx = \int_0^t \int_{\Omega_0} \Big(\Delta(u_\lambda - v_\lambda) + \lambda g(u_\lambda) - \lambda g(v_\lambda) \Big) \phi_1 dx ds. \tag{8.37}$$

Since $v_\lambda \le u_\lambda$ in $\Omega \times (0,T)$ and $p > 1$, we deduce

$$\frac{v_\lambda^{1-p}(x,t) - u_\lambda^{1-p}(x,t)}{a(\delta(x))} \ge \frac{v_\lambda^{1-p}(x,t) - u_\lambda^{1-p}(x,t)}{\|a \circ \delta\|_{L^\infty(\Omega)}} \quad \text{in } \Omega \times (0,T). \tag{8.38}$$

Setting

$$y(t) := \int_{\Omega_0} (v_\lambda^{1-p}(x,t) - u_\lambda^{1-p}(x,t)) \phi_1(x) dx, \quad 0 < t < T, \tag{8.39}$$

from (8.37) and (8.38) we have

$$y(t) \le C \int_0^t \int_{\Omega_0} \Big(\Delta(u_\lambda - v_\lambda) + \lambda g(u_\lambda) - \lambda g(v_\lambda) \Big) \phi_1 dx ds, \quad 0 < t < T,$$

where $C = (p-1)\|a \circ \delta\|_{L^\infty(\Omega)}$. By Green's identity and (8.36) we further obtain

$$\begin{aligned} y(t) \le & C\mu_1 \int_0^t \int_{\Omega_0} (v_\lambda - u_\lambda)\phi_1 dx ds + C \int_0^t \int_{\partial\Omega_0} (v_\lambda - u_\lambda) \frac{\partial \phi_1}{\partial \nu} d\sigma(x) ds \\ & + C\lambda \int_0^t \int_{\Omega_0} (g(u_\lambda) - g(v_\lambda))\phi_1 dx ds \\ \le & C \int_0^t \int_{\partial\Omega_0} (v_\lambda - u_\lambda) \frac{\partial \phi_1}{\partial \nu} d\sigma(x) ds + C\lambda \int_0^t \int_{\Omega_0} (g(u_\lambda) - g(v_\lambda))\phi_1 dx ds \\ \le & c_1 t \max_{\partial\Omega_0} w_\lambda + C\lambda \int_0^t \int_{\Omega_0} (g(u_\lambda) - g(v_\lambda))\phi_1 dx ds. \end{aligned} \tag{8.40}$$

Using the hypothesis $(g1)$ we have $(g(s)/s)' \le 0$ for all $s > 0$. Thus,

$$sg'(s) \le g(s) \quad \text{for all } s > 0, \tag{8.41}$$

and by the mean value theorem we deduce

$$\frac{g(u_\lambda)-g(v_\lambda)}{v_\lambda^{1-p}-u_\lambda^{1-p}}=\frac{g'(\theta)\theta^p}{p-1}\le\frac{\sup\{g'(s)s^p:0\le s\le\|u_\lambda\|_{L^\infty(\Omega\times[0,T])}\}}{p-1}<\infty.$$

Hence, there exists a positive constant $c>0$ depending only on $\|u_\lambda\|_{L^\infty(\Omega\times[0,T])}$ such that

$$g(u_\lambda)-g(v_\lambda)\le c(v_\lambda^{1-p}-u_\lambda^{1-p})\quad\text{in } \Omega\times(0,T). \tag{8.42}$$

By (8.39)–(8.42) we finally obtain

$$y(t)\le c_1 T\max_{\partial\Omega_0} w_\lambda(x)+c_2\int_0^t y(s)ds\quad\text{for all } 0<t<T. \tag{8.43}$$

Now, Gronwall's inequality leads us to

$$\int_K (v_\lambda^{1-p}(x,t)-u_\lambda^{1-p}(x,t))\phi_1(x)dx\le y(t)\le C\max_{\partial\Omega_0} w_\lambda(x).$$

Since the right-hand side tends to zero as $\Omega_0\to\Omega$, it follows that $u_\lambda=v_\lambda$ in K. Hence, $u_\lambda\equiv v_\lambda$ and so problem (8.23) has a unique solution.

If $p=1$ we proceed in the same manner. We have only to replace the definition of $y(t)$ in (8.39) by

$$y(t):=\int_{\Omega_0}(\ln u_\lambda(x,t)-\ln v_\lambda(x,t))\phi_1(x)dx,\quad 0<t<T.$$

Then, estimate (8.40) holds in our case. Also by (8.41) and the mean value property we have

$$\frac{g(u_\lambda)-g(v_\lambda)}{\ln u_\lambda-\ln v_\lambda}=g'(\theta)\theta\le\sup\{g'(s)s:0\le s\le\|u_\lambda\|_{L^\infty(\Omega\times[0,T])}\}<\infty.$$

Hence, there exists a positive constant $c>0$ depending only on $\|u_\lambda\|_{L^\infty(\Omega\times[0,T])}$ such that

$$g(u_\lambda)-g(v_\lambda)\le c(\ln u_\lambda-\ln v_\lambda)\quad\text{in } \Omega\times(0,T).$$

Therefore, we again obtain (8.43) and use Gronwall's inequality which finally produces $u_\lambda\equiv v_\lambda$.

In order to prove the dependence on λ let us fix $0<\lambda_1<\lambda_2$ and let $u_{\lambda_1,\varepsilon}$ and $u_{\lambda_2,\varepsilon}$ be the solutions of (8.29) corresponding to λ_1 and λ_2 respectively. By Lemma 4.5 we obtain $u_{\lambda_1,\varepsilon}\le u_{\lambda_2,\varepsilon}$ in $\overline{\Omega}\times[0,\infty)$. Then, passing to the limit with $\varepsilon\searrow 0$ we find $u_{\lambda_1} \le u_{\lambda_2}$ in $\overline{\Omega} \times [0,\infty)$. By the strong maximum principle we further deduce $u_{\lambda_1} < u_{\lambda_2}$ in $\Omega \times (0,\infty)$.

Step 3: Asymptotic behavior. Fix $\lambda > 0$. For all $0 < \varepsilon < \varepsilon_0$ (where ε_0 is defined by (8.33)) let u_ε^- and u_ε^+ be the unique solutions of

$$\begin{cases} \partial_t u_\varepsilon^- = a_\varepsilon(\delta(x))(u_\varepsilon^-)^p(\Delta u_\varepsilon^- + \lambda g(u_\varepsilon^-)) & \text{in } \Omega \times (0,\infty), \\ u_\varepsilon^- = m\varphi_1 + \psi_\varepsilon^-(t) & \text{on } \Sigma_\infty, \end{cases} \tag{8.44}$$

and

$$\begin{cases} \partial_t u_\varepsilon^+ = a_\varepsilon(\delta(x))(u_\varepsilon^+)^p(\Delta u_\varepsilon^+ + \lambda g(u_\varepsilon^+)) & \text{in } \Omega \times (0,\infty), \\ u_\varepsilon^+ = Mw_\varepsilon + \psi_\varepsilon^+(t) & \text{on } \Sigma_\infty, \end{cases} \tag{8.45}$$

where w_ε is the unique solution of (8.28) and $\psi_\varepsilon^\pm : [0,\infty) \to \mathbb{R}$ are continuous differentiable functions such that

$$\psi_\varepsilon^-(0) = \frac{\varepsilon}{2}, \; (\psi_\varepsilon^-)' \ge 0 \text{ in } [0,\infty), \text{ and } \lim_{t\to\infty} \psi_\varepsilon^-(t) = \varepsilon,$$

$$\psi_\varepsilon^+(0) = 0, \; (\psi_\varepsilon^+)' \le 0 \text{ in } [0,\infty), \text{ and } \lim_{t\to\infty} \psi_\varepsilon^+(t) = \varepsilon(1-M) < 0.$$

It is easy to see that $u_\varepsilon^\pm \in C^{2,1}(\overline{\Omega} \times [0,\infty))$ and for all $0 < \varepsilon < \varepsilon_0$ there holds

$$m\varphi_1 + \frac{\varepsilon}{2} \le u_\varepsilon^- \le u_\varepsilon \le u_\varepsilon^+ \le Mw_\varepsilon \quad \text{in } \overline{\Omega} \times [0,\infty).$$

Furthermore, letting $v_\varepsilon^\pm := \partial_t u_\varepsilon^\pm$, by (8.44) and (8.45) we derive

$$\partial_t v_\varepsilon^\pm = a_\varepsilon(\delta(x))(u_\varepsilon^p)^\pm(\Delta v_\varepsilon + \lambda g'(u_\varepsilon^\pm)v_\varepsilon^\pm) + p\frac{(v_\varepsilon^\pm)^2}{u_\varepsilon^\pm} \quad \text{in } \Omega \times (0,\infty). \tag{8.46}$$

Also notice that the coefficients in (8.46) are bounded. Since $\pm v_\varepsilon^\pm \le 0$ on Σ_∞ by Lemma 4.5 we deduce $\pm v_\varepsilon^\pm \le 0$ in $\overline{\Omega} \times [0,\infty)$. Thus, there exists $w_\varepsilon^\pm : \overline{\Omega} \to [0,\infty)$ such that

$$u_\varepsilon^-(x,t) \nearrow w_\varepsilon^-(x) \quad \text{and} \quad u_\varepsilon^+(x,t) \searrow w_\varepsilon^+(x) \quad \text{as } t \to \infty.$$

Letting $u_{n,\varepsilon}^\pm(x,t) := u_\varepsilon^\pm(x,t+n)$, $(x,t) \in \overline{\Omega} \times [0,2]$, with the same arguments as in [207, Theorem 3.2] we obtain

$$u_{n,\varepsilon}^\pm \to w_\varepsilon^\pm \quad \text{in } C^{2,1}(\overline{\Omega} \times [0,2]) \text{ as } n \to \infty.$$

This also yields

$$u_\varepsilon^\pm(\cdot,t) \to w_\varepsilon^\pm \quad \text{uniformly in } \overline{\Omega} \text{ as } t \to \infty. \tag{8.47}$$

Furthermore, $w_\varepsilon^\pm$ satisfies

$$\begin{cases} -\Delta w_\varepsilon^\pm = \lambda g(w_\varepsilon^\pm) & \text{in } \Omega, \\ w_\varepsilon^\pm > 0 & \text{in } \Omega, \\ w_\varepsilon^\pm = \varepsilon & \text{in } \partial\Omega. \end{cases}$$

By the uniqueness of (8.28) it follows that $w_\varepsilon^- \equiv w_\varepsilon^+ \equiv w_\varepsilon$.

Let $\eta > 0$. Since $\|w_\varepsilon - w_\lambda\|_{L^\infty(\Omega)} \to 0$ as $\varepsilon \searrow 0$, we can find $\varepsilon > 0$ such that

$$|w_\varepsilon - w_\lambda| \le \eta \quad \text{in } \overline{\Omega}.$$

On the other hand, by (8.47) we can find $T_1 > 0$ such that $u_\varepsilon^+ \le w_\varepsilon + \eta$ in $\overline{\Omega} \times [T_1, \infty)$. Hence,

$$u \le u_\varepsilon \le u_\varepsilon^+ \le w_\varepsilon + \eta \le w_\lambda + 2\eta \quad \text{in } \overline{\Omega} \times [T_1, \infty). \tag{8.48}$$

Let now $\widetilde{\Omega} \subset\subset \Omega$ be such that

$$|w_\lambda - \widetilde{w}_\lambda| < \eta \quad \text{in } \widetilde{\Omega}, \tag{8.49}$$

where $\widetilde{w}_\lambda$ is the unique solution of (8.26) in $\widetilde{\Omega}$. This yields $w_\lambda \le \eta$ on $\partial\widetilde{\Omega}$. Using the asymptotic behavior of w_λ described in (8.27) we obtain $\delta(x) \le \eta/c_1$ on $\partial\widetilde{\Omega}$ and

$$w_\lambda \le \frac{c_2}{c_1}\eta \quad \text{in } \overline{\Omega} \setminus \widetilde{\Omega}. \tag{8.50}$$

Let $\widetilde{\Sigma}_\infty = (\partial\widetilde{\Omega} \times (0,\infty)) \cup (\widetilde{\Omega} \times \{0\})$ and $0 < \widetilde{\varepsilon} < \inf_{\widetilde{\Sigma}_\infty} u$ be such that for all $0 < \varepsilon < \widetilde{\varepsilon}$ the solution $\widetilde{u}_\varepsilon^-$ of (8.44) in $\widetilde{\Omega} \times [0,\infty)$ converges uniformly to some $\widetilde{w}_\varepsilon$ as $t \to \infty$. As we have already proved, $\widetilde{w}_\varepsilon$ is the unique solution of (8.28) in $\widetilde{\Omega}$ and $\widetilde{w}_\varepsilon \to \widetilde{w}_\lambda$ uniformly in $\widetilde{\Omega}$ as $\varepsilon \searrow 0$.

Fix $0 < \varepsilon < \widetilde{\varepsilon}$ and $T_2 > 0$ such that

$$\widetilde{w}_\varepsilon \ge \widetilde{w}_\lambda - \eta \ge w_\lambda - 2\eta \quad \text{in } \widetilde{\Omega} \quad \text{and} \quad \widetilde{u}_\varepsilon^- \ge \widetilde{w}_\varepsilon - \eta \quad \text{in } \widetilde{\Omega} \times [T_2, \infty). \tag{8.51}$$

Furthermore, since $u_\lambda \ge \widetilde{u}_\varepsilon^-$ on $\widetilde{\Sigma}_\infty$, by Lemma 4.5 we deduce $u_\lambda \ge \widetilde{u}_\varepsilon^-$ in $\widetilde{\Omega} \times [0,\infty)$.

Combining (8.49)–(8.51) we obtain

$$u_\lambda \geq \widetilde{u_\varepsilon} \geq \widetilde{w}_\varepsilon - \eta \geq w_\lambda - 3\eta \quad \text{in } \widetilde{\Omega} \times [T_2, \infty),$$

$$u_\lambda \geq 0 \geq w_\lambda - \frac{c_2}{c_1}\eta \quad \text{in } (\overline{\Omega} \setminus \widetilde{\Omega}) \times [0, \infty).$$

Therefore,

$$u_\lambda - w_\lambda \geq -c\eta \quad \text{in } \Omega \times [T_2, \infty), \tag{8.52}$$

for some $c > 0$ independent of ε and η. Now, by (8.48) and (8.52) we finally obtain

$$\lim_{t \to \infty} \|u_\lambda(\cdot, t) - w_\lambda\|_{L^\infty(\Omega)} = 0.$$

This finishes the proof. □

8.4 Linear Case

In this section we study the parabolic problem (8.23) for $g(u) = u$. We also assume that the potential a satisfies

$$\int_0^1 \frac{s}{a(s)} ds < \infty. \tag{8.53}$$

Theorem 8.5 *Assume that $g(u) = u$ and condition (8.53) holds. Then, for all $0 < \lambda < \lambda_1$, problem (8.23) has a unique global solution u_λ which is increasing with respect to λ.*

Moreover, if $U_\lambda(x,t) := u_\lambda(x,t)(1+pt)^{1/p}$, $(x,t) \in \overline{\Omega} \times [0, \infty)$ then:

(i) *For all $0 < \lambda < \lambda_1$ we have*

$$\lim_{t \to \infty} \|U_\lambda(\cdot, t) - W_\lambda\|_{L^\infty(\Omega)} = 0,$$

where $W_\lambda \in C^2(\Omega) \cap C(\overline{\Omega})$ is the unique solution of

$$\begin{cases} -\Delta W = \dfrac{1}{a(\delta(x))} W^{1-p} + \lambda W & \text{in } \Omega, \\ W > 0 & \text{in } \Omega, \\ W = 0 & \text{on } \partial\Omega. \end{cases} \tag{8.54}$$

(ii) *If $1 \leq p < 2$ then $\lim_{\lambda \nearrow \lambda_1} \|U_\lambda(x, \cdot)\|_{L^\infty(0,\infty)} = \infty$ uniformly on compact subsets of Ω.*

Before we start the proof of Theorem 8.5 we present some qualitative results regarding problem (8.54). More precisely, let us consider

$$\begin{cases} -\Delta W = \dfrac{1}{a(\delta(x))} W^{-\alpha} + \lambda W & \text{in } \Omega, \\ W > 0 & \text{in } \Omega, \\ W = 0 & \text{on } \partial\Omega, \end{cases} \tag{8.55}$$

where a satisfies (8.53) and $\alpha \geq 0$. We have:

Proposition 8.6 *Let $\alpha \geq 0$, $\lambda > 0$, and a satisfy (8.53). We have:*

(i) *Problem (8.55) has classical solutions if and only if $\lambda < \lambda_1$.*

(ii) *For all $0 < \lambda < \lambda_1$ problem (8.55) has a unique classical solution W_λ. In addition, W_λ has the following properties:*

(ii1) *The mapping $(0, \lambda_1) \ni \lambda \longmapsto W_\lambda(x)$ is increasing for all $x \in \Omega$.*

(ii2) *There exists a nondecreasing function $H : [0, \infty) \to [0, \infty)$ such that for all $0 < \lambda < \lambda_1$ we can find $0 < m < M$ (depending on λ) with the property*

$$m\delta(x) \leq W_\lambda \leq MH(\delta(x)) \quad \text{in } \Omega. \tag{8.56}$$

(ii3) *If $0 \leq \alpha < 1$ then $\lim_{\lambda \nearrow \lambda_1} W_\lambda = \infty$ uniformly on compact subsets of Ω.*

Proof of Theorem 8.5. Let us remark first that u is a global solution of (8.23) if and only if

$$U(x,t) := u(x,t)(1+pt)^{1/p}, \quad (x,t) \in \overline{\Omega} \times [0, \infty)$$

satisfies

$$\begin{cases} (1+pt)\partial_t U = a(\delta(x))U^p(\Delta U + \lambda U) + U & \text{in } \Omega \times (0,\infty), \\ U = 0 & \text{on } \partial\Omega \times (0,\infty), \\ U(x,0) = u_0(x) & \text{on } \Omega. \end{cases}$$

Next, with the change of variable

$$t = \frac{1}{p}(e^{ps} - 1), \quad s \geq 0,$$

we find that $V(x,s) := U(x, \frac{1}{p}(e^{ps} - 1))$ verifies

$$\begin{cases} \partial_s V = a(\delta(x))V^p(\Delta V + \lambda V) + V & \text{in } \Omega \times (0,\infty), \\ V = 0 & \text{on } \partial\Omega \times (0,\infty), \\ V(x,0) = u_0(x) & \text{on } \Omega. \end{cases} \tag{8.57}$$

As in the proof of Theorem 8.3, for all $0 < \varepsilon < 1$ there exists a unique solution V_ε of the approximated problem

$$\begin{cases} \partial_s V = a_\varepsilon(\delta(x))V^p(\Delta V + \lambda V) + V & \text{in } \Omega \times (0,s_\varepsilon), \\ V = \varepsilon & \text{on } \partial\Omega \times (0,s_\varepsilon), \\ V(x,0) = u_0(x) + \varepsilon & \text{on } \Omega, \end{cases} \tag{8.58}$$

in a maximal interval $[0, s_\varepsilon)$ with respect to the time variable. In order to prove that $s_\varepsilon = \infty$, let us consider $\underline{V}_\varepsilon := m\varphi_1 + \varepsilon$. By (8.24) we can find $m > 0$ small enough such that $\underline{V}_\varepsilon \le u_0 + \varepsilon$ in Ω. We have

$$\partial_s \underline{V}_\varepsilon - a_\varepsilon(\delta(x))\underline{V}_\varepsilon^p(\Delta \underline{V}_\varepsilon + \lambda \underline{V}_\varepsilon) - \underline{V}_\varepsilon \le \Big[(a(\delta(x))+1)(m\varphi_1+\varepsilon)^p(\lambda_1 - \lambda) - 1\Big] m\varphi_1,$$

in $\Omega \times (0, s_\varepsilon)$. Thus, we can chose $m, \varepsilon_0 > 0$ small enough such that

$$\partial_s \underline{V}_\varepsilon - a_\varepsilon(\delta(x))\underline{V}_\varepsilon^p(\Delta \underline{V}_\varepsilon + \lambda \underline{V}_\varepsilon) - \underline{V}_\varepsilon \le 0 \quad \text{in } \Omega \times (0, s_\varepsilon),$$

for all $0 < \varepsilon < \varepsilon_0$. By Lemma 4.5 we deduce $V_\varepsilon \ge \underline{V}_\varepsilon$ in $\overline{\Omega} \times [0, s_\varepsilon)$.

Let $W_\varepsilon \in C^2(\Omega) \cap C(\overline{\Omega})$ be the unique solution of

$$\begin{cases} -\Delta W = \dfrac{1}{a(\delta(x)) + \varepsilon} W^{-\alpha} + \lambda W & \text{in } \Omega, \\ W > 0 & \text{in } \Omega, \\ W = \varepsilon & \text{on } \partial\Omega. \end{cases} \tag{8.59}$$

Due to the presence of the potential $a(\delta(x)) + \varepsilon$ in (8.59) we are *not* able to show that $\{W_\varepsilon\}_{0<\varepsilon<1}$ is monotone with respect to ε. However, we still have the following result which is essential in our further analysis.

Proposition 8.7 *The sequence* $(W_\varepsilon)_{0<\varepsilon<1}$ *converges uniformly as* $\varepsilon \searrow 0$ *to the unique solution* W_λ *of problem (8.55).*

It is now easy to see that $\overline{V}_\varepsilon := MW_\varepsilon$ satisfies

$$\partial_s \overline{V}_\varepsilon - a_\varepsilon(\delta(x))\overline{V}_\varepsilon^p(\Delta \overline{V}_\varepsilon + \lambda \overline{V}_\varepsilon) - \overline{V}_\varepsilon \ge 0 \quad \text{in } \Omega \times (0, s_\varepsilon). \tag{8.60}$$

On the other hand, $c\varphi_1+\varepsilon$ is a subsolution of (8.59) for $c>0$ sufficiently small and by Theorem 1.2 we deduce $W_\varepsilon \geq c\varphi_1+\varepsilon$ in Ω. Hence, taking $M>1$ large enough, in view of (8.24) we have

$$\overline{V}_\varepsilon = MW_\varepsilon \geq M(c\varphi_1+\varepsilon) \geq u_0+\varepsilon \quad \text{in } \Omega. \tag{8.61}$$

By (8.60), (8.61) and Theorem 1.5 we obtain

$$\underline{V}_\varepsilon \leq V_\varepsilon \leq \overline{V}_\varepsilon \quad \text{in } \overline{\Omega}\times[0,s_\varepsilon).$$

This shows that V_ε is actually a global solution of (8.58). From now on we can employ the same technique as in the proof of Theorem 8.3 in order to show that for all $0<\lambda<\lambda_1$ problem (8.57) has a unique global solution $V_\lambda \in C^{2,1}(\Omega\times(0,\infty))\cap C(\overline{\Omega}\times[0,\infty))$ such that the mapping $(0,\lambda_1)\ni\lambda\longmapsto V_\lambda(x,s)$ is increasing for all $(x,s)\in\Omega\times(0,\infty)$.

Next, the estimate (ii2) in Proposition 8.6 allows us to employ the same arguments as in Step 3 of Theorem 8.3 in order to obtain

$$\lim_{s\to\infty} \|V_\lambda(\cdot,s)-W_\lambda\|_{L^\infty(\Omega)}=0, \tag{8.62}$$

where $W_\lambda \in C^2(\Omega)\cap C(\overline{\Omega})$ is the unique solution of (8.55). The only difference is that due to the estimate (8.56), relation (8.50) should be replaced by $W_\lambda \leq MH(\eta/m)$ in $\overline{\Omega}\setminus\widetilde{\Omega}$, where $m,M>0$ are the constants from (8.56).

We now prove that asymptotic behavior in (ii) holds for V_λ. To this aim, assume that $1\leq p<2$ and let $\omega\subset\subset\Omega$ be fixed. By Proposition 8.6 (ii3), for any $M>0$ there exists $0<\lambda_0<\lambda_1$ such that $W_{\lambda_0}>M+1$ in ω. In view of (8.62) we can find $s_0>0$ such that

$$|V_{\lambda_0}-W_{\lambda_0}|\leq 1 \quad \text{in } \omega\times[s_0,\infty),$$

which yields $V_{\lambda_0}\geq W_{\lambda_0}-1\geq M$ in $\omega\times[s_0,\infty)$. Since V_λ is increasing with respect to λ we find that

$$V_\lambda \geq M \quad \text{in } \omega\times[s_0,\infty), \text{ for all } \lambda_0\leq\lambda<\lambda_1.$$

Hence $\lim_{\lambda\nearrow\lambda_1}\|V_\lambda(x,\cdot)\|_{L^\infty(0,\infty)}=\infty$ uniformly on compact subsets of Ω.

Finally, since U_λ and V_λ are related by $U_\lambda(x,t)=V_\lambda(x,\frac{1}{p}\ln(1+pt))$, the conclusion in Theorem 8.5 follows immediately. This completes the proof. □

Chapter 9
Reaction-Diffusion Systems Arising in Chemistry

To accomplish great things, we must not only act, but also dream; not only plan, but also believe.

Anatole France (1844–1924)

9.1 Introduction

Many physical, chemical, biological, environmental and even sociological processes are driven by reaction-diffusion systems. These are multi-component models involving two different mechanisms: on the one hand there is diffusion, a random particle movement, and on the other hand there are chemical, biological or sociological reactions representing instantaneous interactions, which depend on the state variables themselves and possibly also explicitly on the particles' position.

In the early 1950s, the British mathematician Alan M. Turing [197] proposed a model that accounts for pattern formation in morphogenesis. Turing [197] suggested that under certain conditions, chemicals can react and diffuse in such a way to produce steady-state heterogeneous spatial patterns of chemical or morphogen concentrations. He showed that a system of two reacting and diffusing chemicals could give rise to spatial patterns from initial near-homogeneity. The idea behind Turing's model is the existence of a low-range diffusing activator and a wide-range diffusing inhibitor. The activator production is inhibited by the presence of inhibitors and enhanced by the presence of the activator. In contrast, the inhibitor

M. Ghergu and V. Rădulescu, *Nonlinear PDEs*, Springer Monographs in Mathematics,
DOI 10.1007/978-3-642-22664-9_9, © Springer-Verlag Berlin Heidelberg 2012

is not self-enhancing, that is, its production is not linked to the presence of other inhibitors, but to the presence of activators. Turing systems show a very rich behavior from the pattern formation point of view, varying from spots to stripes and from lamellar to chaotic structures.

Lately, many Turing-type models described by coupled systems of reaction-diffusion equations have been used for generating patterns in both organic and inorganic systems.

9.2 Brusselator Model

In this section we shall be concerned with Turing patterns in a general Brusselator model for autocatalytic oscillating chemical reactions. An autocatalytic reaction is one in which a species acts to increase the rate of its producing reaction. In many autocatalytic systems complex dynamics are seen, including multiple steady-states and periodic orbits. The study of oscillating reactions has only been the subject of interest for the last fifty years, starting with the Belousov–Zhabotinsky chemical reactions.

There is now a large number of real and hypothetical systems that provide insight into the complex behavior of autocatalytic systems. Among them we mention the Brusselator model [161], Gray–Scott model [100], Lengyel–Epstein model [133], Oregonator model [73], Schnakenberg model [174], Sel'klov model [175].

The Brusselator model was introduced by Prigogine and Lefever [161] in 1968. It consists of the following four intermediate reaction steps

$$A \to X, \qquad B+X \to Y+D, \qquad 2X+Y \to 3X, \qquad X \to E.$$

The global reaction is $A+B \to D+E$ and corresponds to the transformation of input products A and B into output products D and E.

Using the law of mass action, we can derive the reaction-diffusion system associated to the above reactions as

$$\begin{cases} u_t - d_1 \Delta u = a - (b+1)u + f(u)v & \text{in } \Omega \times (0,T), \\ v_t - d_2 \Delta v = bu - f(u)v & \text{in } \Omega \times (0,T). \end{cases}$$

The unknowns u, v in the above system represent the concentrations of two intermediary reactants having the diffusion rates $d_1, d_2 > 0$ while a, $b > 0$ are fixed concentrations. The Brusselator system has been extensively investigated in recent decades from both analytical and numerical points of view (see for instance [9, 32, 69, 84, 113, 124, 157, 159, 198, 217]).

In the following we shall be concerned with the more general system

$$\begin{cases} u_t - d_1\Delta u = a - (b+1)u + f(u)v & \text{in } \Omega \times (0,T), \\ v_t - d_2\Delta v = bu - f(u)v & \text{in } \Omega \times (0,T), \\ u(x,0) = u_0(x), v(x,0) = v_0(x) & \text{on } \Omega, \\ \dfrac{\partial u}{\partial \nu}(x,t) = \dfrac{\partial u}{\partial \nu}(x,t) = 0 & \text{on } \partial\Omega \times (0,T), \end{cases} \tag{9.1}$$

where $\Omega \subset \mathbb{R}^N$ $(N \geq 1)$ is a smooth and bounded domain, a, b, d_1, d_2 are positive constants and $f \in C^1(0,\infty) \cap C[0,\infty)$ is a nonnegative and nondecreasing function such that $f > 0$ in $(0,\infty)$. The initial data u_0, v_0 are nonnegative continuous functions in $\overline{\Omega}$.

Our further analysis will reveal the fact that the dynamics of (9.1) and its associated steady-state is strongly related to the behavior of the nonlinearity f. We shall assume that f satisfies one of the following hypotheses:

either f is sublinear, that is,

$(f1)$ the mapping $(0,\infty) \ni s \longrightarrow \frac{f(s)}{s}$ is nonincreasing;

of f has a superlinear character, namely,

$(f2)$ the mapping $(0,\infty) \ni s \longrightarrow \frac{f(s)}{s}$ is nondecreasing.

Particular attention will be paid to the steady states of (9.1), that is, solutions of

$$\begin{cases} -d_1\Delta u = a - (b+1)u + f(u)v & \text{in } \Omega, \\ -d_2\Delta v = bu - f(u)v & \text{in } \Omega, \\ \dfrac{\partial u}{\partial \nu} = \dfrac{\partial v}{\partial \nu} = 0 & \text{on } \partial\Omega. \end{cases} \tag{9.2}$$

It is easily seen that there exists a unique uniform steady state of (9.2), namely

$$(u,v) = \left(a, \frac{ab}{f(a)}\right). \tag{9.3}$$

We shall also investigate the asymptotic stability of the above constant solution. In particular, we shall see that if f has a sublinear growth, then the constant solution

(u,v) defined by (9.3) is uniformly asymptotically stable. Moreover, in the sublinear case on f we prove that (9.3) is the unique solution of system (9.2), so there are no Turing patterns in this case. In turn, if f satisfies $(f2)$, the analysis of (9.2) is more involved. The existence of Turing patterns (and implicitly of nonconstant solutions to (9.2)) is strongly dependent on the diffusion coefficients d_1, d_2 and on the parameters a, b. The most important issue in the study of steady-state solutions are the *a priori* estimates. Using a similar approach to that in [84], we are able to find precise upper and lower bounds for solutions to (9.2) in terms of a, b, d_1, d_2 for any dimension $N \geq 1$. This allows us to extend the study of the standard Brusselator system started in [32, 84, 159]. As a consequence, we are able to provide existence results in terms of a, b, d_1 and d_2 in case where f has a superlinear growth.

9.2.1 Existence of Global Solutions

In this section we establish the existence of global solutions to (9.1). Our first result concerns the case where f is sublinear.

Theorem 9.1 *Assume that f satisfies $(f1)$ and* $\lim_{s\to\infty} f(s)/s = 0$. *Then, for any $a, b, d_1, d_2 > 0$ and any nonnegative continuous functions u_0, v_0, the system (9.1) has at least one global solution.*

Proof. The proof relies on the invariant region method (see, e.g., [182, 215]). To this aim, we rewrite the system (9.1) in the vectorial form

$$\mathbf{w}_t = \begin{pmatrix} d_1 & 0 \\ 0 & d_2 \end{pmatrix} \Delta \mathbf{w} + F(\mathbf{w}) \quad \text{in } \Omega \times (0,\infty), \tag{9.4}$$

where $\mathbf{w} = (u,v)^T$ and

$$F(\mathbf{w}) = \begin{pmatrix} a-(b+1)u+f(u)v \\ bu-f(u)v \end{pmatrix}.$$

We claim that the rectangle $\Sigma := [0,c_1] \times [0,c_2]$ is an invariant region for (9.4) provided $c_1, c_2 > 0$ are large enough. In view of $(f1)$ we can choose $c_1 > \max\{2a, \|u_0\|_{L^\infty}\}$ such that

$$\frac{(b+1/2)c_1}{f(c_1)} > \|v_0\|_{L^\infty}$$

and define

$$c_2 := \frac{(b+1/2)c_1}{f(c_1)}.$$

We also write Σ as

$$\Sigma = \{\mathbf{w} = (u,v)^T \in C(\overline{\Omega}) \cap C(\overline{\Omega}) : G_i(\mathbf{w}) \leq 0, 1 \leq i \leq 4\},$$

where

$$G_1(\mathbf{w}) = -u, \;\; G_2(\mathbf{w}) = u - c_1, \;\; G_3(\mathbf{w}) = -v, \;\; G_4(\mathbf{w}) = v - c_2.$$

It is obviously that the initial data (u_0, v_0) belongs to the interior of Σ. If $\mathbf{w} = (u,v)^T \in \partial\Sigma$, by the definition of c_1 and c_2 we have

$$\nabla G_1 \cdot F_{|_{u=0}} = -a - f(0)v < 0,$$

$$\nabla G_2 \cdot F_{|_{u=c_1}} = a - (b+1)c_1 + f(c_1)v \leq a - (b+1)c_1 + f(c_1)c_2 = a - \frac{c_1}{2} \leq 0,$$

$$\nabla G_3 \cdot F_{|_{v=0}} = -bu \leq 0,$$

$$\nabla G_4 \cdot F_{|_{v=c_2}} = bu - f(u)v = u\left(b - \frac{f(u)}{u}\right) \leq u\left(b - \frac{f(c_1)}{c_1}c_2\right) < 0.$$

By Theorem 14.13 in [182] it follows that Σ is invariant for (9.4). Thus, there exists a global solution of (9.4). □

Next, we turn our attention to the case where f is superlinear. For the standard Brusselator model, that is, $f(u) = u^2$, the existence of a global solution to (9.1) was obtained by Rothe [172]. Here, the existence of global solution to (9.1) is derived for more general nonlinearities f under the restriction $d_1 = d_2$ and the initial data u_0 is strictly positive in $\overline{\Omega}$ (see [96]).

Theorem 9.2 *Assume that $d_1 = d_2 > 0$, the initial data u_0, v_0 are continuous functions in $\overline{\Omega}$ such that $u_0 > 0$, $v_0 \geq 0$ in $\overline{\Omega}$ and the nonlinearity f satisfies $(f2)$ and $\lim_{s\to 0} f(s)/s = 0$. Then, for any $a, b > 0$, the system (9.1) has a global solution.*

Proof. With the change of variable we can assume $d_1 = d_2 = 1$. For $\varepsilon > 0$ small enough we consider the related problem

$$\begin{cases} u_t - \Delta u = a - (b+1)u + f(u)v & \text{in } \Omega \times (0,\infty), \\ v_t - \Delta v = bu - f(u)v & \text{in } \Omega \times (0,\infty), \\ u(x,0) = u_0(x), v(x,0) = v_0(x) + \varepsilon & \text{on } \Omega, \\ \frac{\partial u}{\partial \nu}(x,t) = \frac{\partial v}{\partial \nu}(x,t) = 0 & \text{on } \partial\Omega \times (0,\infty), \end{cases} \tag{9.5}$$

By standard parabolic arguments, there exists a classical solution $(u^\varepsilon, u^\varepsilon)$ of (9.5) in a maximal interval $(0, T^\varepsilon_{\max})$. We claim that $T^\varepsilon_{\max} = \infty$. First, by (9.5) we have that U^ε satisfies

$$u^\varepsilon_t - \Delta u^\varepsilon + (b+1)u^\varepsilon \geq a > 0 \quad \text{in } \Omega \times (0, T^\varepsilon_{\max}).$$

Since $u_0 > 0$ in $\overline{\Omega}$, there exists a constant $k > 0$ independent of ε such that

$$u^\varepsilon \geq k \quad \text{in } \Omega \times (0, T^\varepsilon_{\max}). \tag{9.6}$$

Since $\lim_{s\to 0} f(s)/s = 0$, we can choose $k > 0$ small enough such that

$$v_0 + 1 \leq \frac{bk}{f(k)} \quad \text{in } \overline{\Omega}. \tag{9.7}$$

The function v satisfies

$$\begin{cases} v^\varepsilon_t - \Delta v^\varepsilon = bu^\varepsilon - f(u^\varepsilon)v^\varepsilon & \text{in } \Omega \times (0, T^\varepsilon_{\max}), \\ v^\varepsilon(x,0) = v_0(x) + \varepsilon & \text{on } \Omega, \\ \frac{\partial v^\varepsilon}{\partial \nu}(x,t) = 0 & \text{on } \partial\Omega \times (0, T^\varepsilon_{\max}). \end{cases} \tag{9.8}$$

Using (9.6) and that fact that f satisfies $(f2)$ we can easily deduce that the interval

$$\Sigma := [0, bk/f(k)]$$

is an invariant region for (9.8). This means that

$$v(x,t) \leq \frac{bk}{f(k)} = \text{const.} \quad \text{in } \overline{\Omega} \times (0, T^\varepsilon_{\max}). \tag{9.9}$$

Adding the first two equations in (9.5) we have

$$(u^\varepsilon + u^\varepsilon)_t - \Delta(u^\varepsilon + u^\varepsilon) + \frac{1}{d_1}(u^\varepsilon + u^\varepsilon) \leq a + \frac{bk}{d_1 f(k)} \quad \text{in } \Omega \times (0, T^\varepsilon_{\max}).$$

By maximum principle we deduce that $u^\varepsilon + v^\varepsilon \leq M$ in $\overline{\Omega} \times (0, T^\varepsilon_{\max})$, for some constant $M > 0$ independent of ε. Therefore, for $\varepsilon > 0$ small enough, u^ε, u^ε satisfy

$$\varepsilon \le u^\varepsilon, v^\varepsilon \le M \quad \text{in } \overline{\Omega} \times (0, T^\varepsilon_{\max}).$$

This yields $T^\varepsilon_{\max} = \infty$, so u^ε and v^ε exist globally. Also by standard parabolic arguments and up to a subsequence, u^ε and v^ε converge to some functions u and v which are global solutions to (9.1). This finishes the proof of Theorem 9.2. □

9.2.2 Stability of the Uniform Steady State

The linearization of (9.4) at $\mathbf{w}_0 = (a, ab/f(a))^T$ is

$$\mathbf{w}_t = \begin{pmatrix} d_1 & 0 \\ 0 & d_2 \end{pmatrix} \Delta \mathbf{w} + \nabla F(\mathbf{w}_0) \cdot \mathbf{w} + \mathscr{O}(\|\mathbf{w} - \mathbf{w}_0\|^2). \tag{9.10}$$

Denote by

$$0 = \mu_0 < \mu_1 < \mu_2 < \cdots < \mu_n < \cdots$$

the eigenvalues of $-\Delta$ with homogeneous Neumann boundary condition. For any $k \ge 0$ we also denote by $e(\mu_k)$ the multiplicity of μ_k. Consider

$$\mathbf{X} = \left\{ \mathbf{w} = (u, v) \in C^1(\overline{\Omega}) \times C^1(\overline{\Omega}) : \frac{\partial u}{\partial \nu} = \frac{\partial v}{\partial \nu} = 0 \text{ on } \partial\Omega \right\} \tag{9.11}$$

and decompose

$$\mathbf{X} = \bigoplus_{k \ge 0} \mathbf{X}_k, \tag{9.12}$$

where $\mathbf{X}_k$ denotes the eigenspace corresponding to μ_k, $k \ge 0$.

Theorem 9.3 *Assume that*

$$f(a) > \frac{baf'(a)}{f(a)} - b - 1 \tag{9.13}$$

and the first eigenvalue μ_1 *of the Dirichlet operator subject to homogeneous Neumann condition satisfies*

$$\mu_1 > \frac{1}{d_1} \left(\frac{baf'(a)}{f(a)} - b - 1 \right) - \frac{f(a)}{d_2}. \tag{9.14}$$

Then the steady-state $\mathbf{w}_0$ *is uniformly asymptotically stable.*

Proof. Define $\mathscr{L} : \mathbf{X} \to C(\overline{\Omega}) \times C(\overline{\Omega})$ by

$$\mathscr{L} = \begin{pmatrix} d_1\Delta + \frac{baf'(a)}{f(a)} - b - 1 & f(a) \\ b - \frac{baf'(a)}{f(a)} & d_2\Delta - f(a) \end{pmatrix}.$$

Then $\mathbf{X}_k$ is invariant for $\mathscr{L}$ and ξ_k is an eigenvalue of $\mathscr{L}$ on $\mathbf{X}_k$ if and only if ξ is an eigenvalue of the matrix

$$A_k = \begin{pmatrix} -d_1\mu_k + \frac{baf'(a)}{f(a)} - b - 1 & f(a) \\ b - \frac{baf'(a)}{f(a)} & -d_2\mu_k - f(a) \end{pmatrix}.$$

The determinant and trace of A_k are

$$\begin{aligned} \det(A_k) &= \mu_k \left[d_1 d_2 \mu_k + d_1 f(a) - d_2 \left(\frac{baf'(a)}{f(a)} - b - 1 \right) \right] + f(a), \\ \mathrm{Tr}(A_k) &= \frac{baf'(a)}{f(a)} - b - 1 - f(a) - (d_1 + d_2)\mu_k. \end{aligned} \tag{9.15}$$

Remark that for any $k \geq 0$ we have

$$\det(A_k) > 0 > \mathrm{Tr}(A_k).$$

Denote by ξ_k^+ and ξ_k^- the two eigenvalues of A_k, $k \geq 0$.

If ξ_k^+, ξ_k^- are complex numbers, then by (9.14) we have

$$\mathrm{Re}(\xi_k^+) = \mathrm{Re}(\xi_k^-) = \frac{1}{2}\mathrm{Tr}(A_k) \leq \frac{1}{2}\left(\frac{baf'(a)}{f(a)} - b - 1 - f(a) \right) < 0.$$

If ξ_k^+, ξ_k^- are real numbers, then by (9.14) we have

$$\begin{aligned} \xi_k^- \leq \xi_k^+ &= \frac{\mathrm{Tr}(A_k) + \sqrt{\mathrm{Tr}^2(A_k) - 4\det(A_k)}}{2} \\ &= \frac{2\det(A_k)}{\mathrm{Tr}(A_k) - \sqrt{\mathrm{Tr}^2(A_k) - 4\det(A_k)}} \\ &\leq \frac{\det(A_k)}{\mathrm{Tr}(A_k)} \\ &< 0. \end{aligned}$$

Since $\mu_k \to \infty$ as $k \to \infty$, from the above estimate we deduce $\xi_k^+ \to -\infty$ as $k \to \infty$.

Hence, in both the above cases we can find $\delta > 0$ such that the spectrum of $\mathscr{L}$ lies in the region $\{z \in \mathbb{C} : \mathrm{Re}(z) < -\delta\}$. By Theorem 5.1.1. in [110] we obtain that $\mathbf{w}_0$ is asymptotically uniformly stable for (9.4). This ends the proof. □

If f satisfies $(f1)$ then $\frac{baf'(a)}{f(a)} - b - 1 < 0$ so that both conditions (9.13) and (9.14) are satisfied. In this case we obtain

Corollary 9.4 *If f satisfies $(f1)$ then $\mathbf{w}_0$ is uniformly asymptotically stable.*

9.2.3 Diffusion-Driven Instability

In this section we point out that under certain conditions on the parameters a and b, the uniform steady state (u_0, v_0) defined by (9.3) can be linearly stable in the absence of diffusion but unstable in the presence of diffusion. This is the well-known phenomenon of *diffusion-driven instability* emphasized by Turing in his pioneering work [197].

Let us consider the spatially homogeneous system corresponding to (9.1):

$$\begin{cases} \dfrac{du}{dt} = a - (b+1)u + f(u)v, & t > 0, \\ \dfrac{dv}{dt} = bu - f(u)v, & t > 0. \end{cases} \tag{9.16}$$

We have the following result.

Theorem 9.5 *Assume that*

$$f(a) > \frac{baf'(a)}{f(a)} - b - 1 > 0. \tag{9.17}$$

Then, there exist $d^*, D^* > 0$ such that for all

$$0 < d_1 < d^* \quad \textit{and} \quad d_2 > D^*,$$

the steady-state $\mathbf{w}_0 = (a, ba/f(a))^T$ is uniformly asymptotically stable for the system (9.16) and instable for the system (9.1), that is, Turing instabilities occur.

Remark that (9.17) does not hold if f satisfies $(f1)$.

Proof. Using the same approach as in Theorem 9.3 we have that $\mathbf{w}_0$ is uniformly asymptotically stable for (9.16) provided (9.17) holds. Also by (9.17) we can choose $D^* > 0$ large enough such that

$$\mu_1 D^* \left(\frac{baf'(a)}{f(a)} - b - 1 \right) > f(a).$$

Using (9.15), for all $d_2 > D^*$ we have

$$\lim_{d_1 \searrow 0} \det(A_1) \le f(a) - \mu_1 D^* \left(\frac{baf'(a)}{f(a)} - b - 1 \right) < 0.$$

Therefore we can find $d^* > 0$ such that

$$\det(A_1) < 0 \quad \text{for all } 0 < d_1 < d^*, d_2 > D^*.$$

This implies that A_1, and so the operator $\mathscr{L}$, has at least one positive eigenvalue. By [110, Corollary 5.1.1] it follows that $\mathbf{w}_0$ is uniformly asymptotically unstable. This finishes the proof. □

9.2.4 A Priori Estimates

Using Theorem 1.1 we first derive that if f satisfies $(f1)$ then (9.2) has no nonconstant solutions. More precisely we have.

Theorem 9.6 *Assume that f satisfies $(f1)$. Then, $(u,v) = \left(a, \frac{ab}{f(a)}\right)$ is the unique solution of system (9.2).*

Proof. Let (u,v) be a classical solution of (9.2). Let also x_1 (resp. x_2) be a maximum point of u (resp. v) and x_3 (resp. x_4) be a minimum point of u (resp. v) in $\overline{\Omega}$. Using Theorem 1.1(i) in the first equation of (9.2) we have

$$(b+1)u(x_1) \le a + f(u(x_1))v(x_1). \tag{9.18}$$

Now, Theorem 1.1(i) applied to the second equation in (9.2) yields

$$bu(x_2) \ge f(u(x_2))v(x_2),$$

that is, $v(x_2) \le b\frac{u(x_2)}{f(u(x_2))}$. By virtue of $(f1)$ we next derive

$$v(x_1) \le v(x_2) \le b\frac{u(x_2)}{f(u(x_2))} \le b\frac{u(x_1)}{f(u(x_1))}. \tag{9.19}$$

Therefore (9.18) and (9.19) imply $(b+1)u(x_1) \le a + bu(x_1)$, that is,

$$u \le u(x_1) \le a \quad \text{in } \Omega. \tag{9.20}$$

On the other hand, Theorem 1.1(ii) applied to the second equation of (9.2) leads us to $v(x_4) \geq b\frac{u(x_4)}{f(u(x_4))}$. Again by $(f1)$ it follows that

$$v(x_3) \geq v(x_4) \geq b\frac{u(x_4)}{f(u(x_4))} \geq b\frac{u(x_3)}{f(u(x_3))}. \tag{9.21}$$

Next, Theorem 1.1(ii) applied to the first equation in (9.2) yields

$$(b+1)u(x_3) \geq a + f(u(x_3))v(x_3) \geq a + bu(x_3),$$

which implies

$$u \geq u(x_3) \geq a \qquad \text{in } \Omega. \tag{9.22}$$

Now (9.20) and (9.22) produce $u \equiv a$ in Ω and by (9.2) we also have $v \equiv ab/f(a)$. This ends the proof. □

When f satisfies $(f2)$ the analysis of the steady state system (9.2) is more delicate. In some cases, depending on the parameters a, b, d_1, d_2 we obtain the existence of nonconstant solutions to (9.1). We start this study with the following crucial result that provides *a priori* estimates for solutions to (9.2).

Theorem 9.7 *Assume that f satisfies $(f2)$. Then, any solution (u,v) of (9.2) satisfies*

$$\frac{a}{b+1} \leq u \leq a + \frac{d_2}{d_1} \cdot \frac{ab}{(b+1)f(\frac{a}{b+1})} \qquad \text{in } \Omega, \tag{9.23}$$

and

$$\frac{ab}{(b+1)f\left(a + \frac{d_2}{d_1} \cdot \frac{ab}{(b+1)f(\frac{a}{b+1})}\right)} \leq v \leq \frac{ab}{(b+1)f(\frac{a}{b+1})} \qquad \text{in } \Omega. \tag{9.24}$$

Proof. Consider first a minimum point $x_0 \in \overline{\Omega}$ of u. By Theorem 1.1(ii) it follows

$$a - (b+1)u(x_0) + f(u(x_0))v(x_0) \leq 0$$

which implies $u(x_0) \geq a/(b+1)$. Hence

$$u \geq \frac{a}{b+1} \qquad \text{in } \Omega. \tag{9.25}$$

At maximum point of v we have $bu - f(u)v \geq 0$, that is, $v \leq bu/f(u)$. By virtue of $(f2)$ and (9.25) we deduce

$$v \leq \frac{ab}{(b+1)f(\frac{a}{b+1})} \qquad \text{in } \Omega. \tag{9.26}$$

Let $w = d_1 u + d_2 v$. Adding the first two relations in (9.2) we have

$$-\Delta w = a - u \text{ in } \Omega, \qquad \frac{\partial w}{\partial \nu} = 0 \text{ on } \partial\Omega.$$

Let now $x_1 \in \overline{\Omega}$ be a maximum point of w. According to Theorem 1.1(i) we have $a - u(x_1) \geq 0$, that is, $u(x_1) \leq a$. By virtue of (9.26), for all $x \in \overline{\Omega}$ we have

$$d_1 u(x) \leq w(x) \leq w(x_1) \leq d_1 a + d_2 \cdot \frac{ab}{(b+1)f(\frac{a}{b+1})} \qquad \text{in } \Omega.$$

This yields

$$u \leq a + \frac{d_2}{d_1} \cdot \frac{ab}{(b+1)f(\frac{a}{b+1})} \qquad \text{in } \Omega. \tag{9.27}$$

We have proved that u satisfies (9.23). Again by Theorem 1.1(ii), at minimum points of v we have $bu - f(u)v \leq 0$, which yields $v \geq bu/f(u)$. Combining this inequality with (9.27) we obtain the first estimate in (9.24). This concludes our proof. □

From the estimates (9.23)–(9.24) in Theorem 9.7 we derive the following.

Proposition 9.8 *Assume that f satisfies $(f2)$ and let $a, b, D_1, D_2 > 0$ be fixed. Then, there exist two positive constants $C_1, C_2 > 0$ depending on a, b, D_1, D_2 such that for all*

$$d_1 \geq D_1, \quad 0 < d_2 \leq D_2,$$

any solution (u, v) of (9.2) satisfies

$$C_1 < u, v < C_2 \qquad \text{in } \overline{\Omega}.$$

Furthermore, by standard elliptic arguments and Theorem 9.7 we now obtain:

Proposition 9.9 *Assume that f satisfies $(f2)$ and let $a, b, D_1, D_2 > 0$ be fixed. Then, for any positive integer $k \geq 1$ there exists a constant*

$$C = C(a, b, D_1, D_2, k, N, \Omega) > 0$$

such that for all

$$d_1 \geq D_1, \quad 0 < d_2 \leq D_2,$$

any solution (u, v) of (9.2) satisfies

$$\|u\|_{C^k(\overline{\Omega})} + \|v\|_{C^k(\overline{\Omega})} \leq C.$$

In particular, any solution of (9.2) belongs to $C^\infty(\overline{\Omega}) \times C^\infty(\overline{\Omega})$.

9.2.5 Nonexistence Results

Theorem 9.10 *(i) Let $a, b, d_2 > 0$ be fixed. There exists $D = D(a, b, d_2) > 0$ such that system (9.2) has no nonconstant solutions for all $d_1 > D$.*
(ii) Let $a, d_1, d_2 > 0$ be fixed. There exists $B = B(a, d_1, d_2) > 0$ such that system (9.2) has no nonconstant solutions for all $0 < b < B$.

Proof. (i) Remark first that if (u, v) is a solution of (9.2), then, integrating the two equations in (9.2) over Ω and adding up we have

$$\int_\Omega u(x)dx = a|\Omega|. \tag{9.28}$$

Lemma 9.11 *Let $a, b, d_2 > 0$ be fixed and let $\{\delta_n\} \subset (0, \infty)$ be such that $\delta_n \to \infty$ as $n \to \infty$. If (u_n, v_n) is a solution of (9.2) with $d_1 = \delta_n$ then*

$$(u_n, v_n) \to \left(a, \frac{ab}{f(a)}\right) \qquad \text{in } C^2(\overline{\Omega}) \times C^2(\overline{\Omega}) \text{ as } n \to \infty.$$

Proof. By Proposition 9.9 the sequence $\{(u_n, v_n)\}$ is bounded in $C^3(\overline{\Omega}) \times C^3(\overline{\Omega})$. Hence, passing to a subsequence if necessary, $\{(u_n, v_n)\}$ converges in $C^2(\overline{\Omega}) \times C^2(\overline{\Omega})$ to some $(u, v) \in C^2(\overline{\Omega}) \times C^2(\overline{\Omega})$. We divide by δ_n in the corresponding equation to u_n and then we pass to the limit with $n \to \infty$. We obtain that (u, v) satisfies

$$\begin{cases} -\Delta u = 0 & \text{in } \Omega, \\ -d_2 \Delta v = bu - f(u)v & \text{in } \Omega, \\ \dfrac{\partial u}{\partial \nu} = \dfrac{\partial v}{\partial \nu} = 0 & \text{on } \partial\Omega. \end{cases} \tag{9.29}$$

Also, u_n and u satisfy (9.28). Now, the first equation in (9.29) together with $\partial u / \partial \nu = 0$ on $\partial\Omega$ implies that u is constant. Combining this fact with (9.28) it follows that $u \equiv a$. Thus, from (9.29), v satisfies

$$-d_2 \Delta v = ab - f(a)v \text{ in } \Omega, \quad \frac{\partial v}{\partial \nu} = 0 \text{ on } \partial\Omega.$$

Multiplying the above equality with $ab - f(a)v$ and then integrating over Ω we obtain

$$0 \le \frac{d_2}{f(a)} \int_\Omega |\nabla(ab - f(a)v)|^2 dx = - \int_\Omega (ab - f(a)v)^2 dx \le 0.$$

Hence $v \equiv \frac{ab}{f(a)}$ and the proof follows. □

We first introduce the function spaces

$$H^2_{\rm n}(\Omega) = \left\{ w \in W^{2,2}(\Omega) : \frac{\partial w}{\partial \nu} = 0 \right\}, \qquad L^2_0(\Omega) = \left\{ w \in L^2(\Omega) : \int_\Omega w = 0 \right\}.$$

Thus, letting $w = u - a$, by (9.28) and the standard elliptic regularity, system (9.2) is equivalent to

$$\begin{cases} -\Delta w = \delta(a - (b+1)(w+a) + f(w+a)v) & \text{in } \Omega, \\ -d_2 \Delta v = b(w+a) - f(w+a)v & \text{in } \Omega, \\ w \in H^2_{\rm n}(\Omega) \cap L^2_0(\Omega),\ v \in H^2_{\rm n}(\Omega), \end{cases} \tag{9.30}$$

where $\delta = 1/d_1$. Define

$$\mathscr{F} : \mathbb{R} \times (H^2_{\rm n}(\Omega) \cap L^2_0(\Omega)) \times H^2_{\rm n}(\Omega) \to L^2_0(\Omega) \times L^2(\Omega),$$

by

$$\mathscr{F}(\delta, w, v) = \begin{pmatrix} \Delta w + \delta \mathscr{P}(a - (b+1)(w+a) + f(w+a)v) \\ d_2 \Delta v + b(w+a) - f(w+a)v \end{pmatrix},$$

where $\mathscr{P} : L^2(\Omega) \to L^2_0(\Omega)$ is the projection operator from $L^2(\Omega)$ onto $L^2_0(\Omega)$, namely,

$$\mathscr{P}(z) = z - \frac{1}{|\Omega|} \int_\Omega z(x) dx, \qquad \text{for all } z \in L^2(\Omega).$$

Now (9.30) is equivalent to

$$\mathscr{F}(\delta, w, v) = \mathbf{0}.$$

Indeed, if $\mathscr{F}(\delta, w, v) = \mathbf{0}$, then

$$d_2 \Delta v + b(w+a) - f(w+a)v = 0 \text{ in } \Omega, \qquad v \in H^2_{\rm n}(\Omega).$$

It is easy to see that the above relations imply $b(w+a) - f(w+a)v \in L^2_0(\Omega)$. Since $w \in L^2_0(\Omega)$, this yields

$$a-(b+1)(w+a)+f(w+a)v\in L_0^2(\Omega),$$

so that

$$\mathscr{P}(a-(b+1)(w+a)+f(w+a)v)=a-(b+1)(w+a)+f(w+a)v.$$

Therefore (9.30) is satisfied.

We have that the equation $\mathscr{F}(0,w,v)=\mathbf{0}$ has the unique solution $(w,v)=(0,ab/f(a))$. Next it is easy to see that

$$D_{(w,v)}\mathscr{F}(0,0,ab/f(a)):(H_{\mathrm{n}}^2(\Omega)\cap L_0^2(\Omega))\times H_{\mathrm{n}}^2(\Omega)\to L_0^2(\Omega)\times L^2(\Omega),$$

is given by

$$D_{(w,v)}\mathscr{F}(0,0,ab/f(a))=\begin{pmatrix}\Delta & 0\\ b\frac{f(a)-af'(a)}{f(a)} & d_2\Delta-f(a)\end{pmatrix}.$$

Thus $D_{(w,v)}\mathscr{F}(0,0,ab/f(a))$ is invertible and we are in the frame of the implicit function theorem. It follows that there exists $\delta_0,r>0$ such that $(0,0,ab/f(a))$ is the unique solution of

$$\mathscr{F}(\delta,w,v)=\mathbf{0}\qquad \text{in } [0,\delta_0]\times B_r\left(0,\frac{ab}{f(a)}\right),$$

where $B_r(0,\frac{ab}{f(a)})$ denotes the open ball in $(H_{\mathrm{n}}^2(\Omega)\cap L_0^2(\Omega))\times H_{\mathrm{n}}^2(\Omega)$ centered at $(0,ab/f(a))$ and having the radius $r>0$.

Let now $\{\delta_n\}$ be a sequence of positive real numbers such that $\delta_n\to\infty$ as $n\to\infty$ and let (u_n,v_n) be an arbitrary solution of (9.2) for a,b,d_2 fixed and $d_1=\delta_n$. Letting $w_n=u_n-a$, it follows that

$$\mathscr{F}\left(\frac{1}{\delta_n},w_n,v_n\right)=\mathbf{0}.$$

According to Lemma 9.11 we have

$$(w_n,v_n)\to\left(0,\frac{ab}{f(a)}\right)\qquad \text{in } C^2(\overline{\Omega})\times C^2(\overline{\Omega}) \text{ as } n\to\infty.$$

This means that for $n\geq 1$ large enough there holds $(1/\delta_n,w_n,v_n)\in(0,\delta_0)\times B_r(0,\frac{ab}{f(a)})$ which yields $(w_n,v_n)=(0,\frac{ab}{f(a)})$. Hence, for $d_1=1/\delta_n$ small enough, system (9.2) has only the constant solution $(a,\frac{ab}{f(a)})$. The proof of (ii) is similar. □

9.2.6 Existence Results

Let $\mathbf{X}$ be the space defined in (9.11) and let

$$\mathbf{X}^+ = \{(u,v) \in X : u,v > 0 \text{ in } C(\overline{\Omega})\}.$$

We write the system (9.2) in the form

$$-\Delta \mathbf{w} = \mathscr{G}(\mathbf{w}), \qquad \mathbf{w} \in \mathbf{X}^+, \tag{9.31}$$

where

$$\mathscr{G}(\mathbf{w}) = \begin{pmatrix} \frac{1}{d_1}(a-(b+1)u+f(u)v) \\ \\ \frac{1}{d_2}(bu-f(u)v) \end{pmatrix}.$$

It is more convenient to write (9.31) in the form

$$\mathscr{F}(\mathbf{w}) = \mathbf{0}, \qquad \mathbf{w} \in \mathbf{X}^+, \tag{9.32}$$

where

$$\mathscr{F}(\mathbf{w}) = \mathbf{w} - (\mathbf{I}-\Delta)^{-1}(\mathscr{G}(\mathbf{w})+\mathbf{w}), \qquad \mathbf{w} \in \mathbf{X}^+. \tag{9.33}$$

Let $\mathbf{w}_0 = (a, ab/f(a))^T$ be the uniform steady state solution of (9.2). Then

$$\nabla\mathscr{F}(\mathbf{w}_0) = \mathbf{I} - (\mathbf{I}-\Delta)^{-1}(\mathbf{I}+A),$$

where

$$A := \nabla\mathscr{G}(\mathbf{w}_0) = \begin{pmatrix} \frac{1}{d_1}\left(b\frac{af'(a)-f(a)}{f(a)} - 1\right) & \frac{f(a)}{d_1} \\ \\ -\frac{b}{d_2}\frac{af'(a)-f(a)}{f(a)} & -\frac{f(a)}{d_2} \end{pmatrix}.$$

If $\nabla\mathscr{F}(\mathbf{w}_0)$ is invertible, by [154, Theorem 2.8.1] the index of $\mathscr{F}$ at $\mathbf{w}_0$ is given by

$$\text{index}(\mathscr{F}, \mathbf{w}_0) = (-1)^\gamma, \tag{9.34}$$

where γ denotes the number of negative eigenvalues of $\nabla\mathscr{F}(\mathbf{w}_0)$. On the other hand, using the decomposition (9.12) we have that $\mathbf{X}_i$ is an invariant space under $\nabla\mathscr{F}(\mathbf{w}_0)$ and $\xi \in \mathbb{R}$ is an eigenvalue of $\nabla\mathscr{F}(\mathbf{w}_0)$ in $\mathbf{X}_i$ if and only if ξ is an eigenvalue of $(\mu_i+1)^{-1}(\mu_i\mathbf{I}-A)$. Therefore, $\nabla\mathscr{F}(\mathbf{w}_0)$ is invertible if and only if for any $i \geq 0$ the matrix $(\mu_i\mathbf{I}-A)$ is invertible.

Let us define

$$H(a,b,d_1,d_2,\mu) = \det(\mu \mathbf{I} - A). \tag{9.35}$$

Then, if $(\mu_i \mathbf{I} - A)$ is invertible for any $i \geq 0$, with the same arguments as in [158] we have

$$\gamma = \sum_{\substack{i \geq 0, \\ H(a,b,d_1,d_2,\mu_i)<0}} e(\mu_i). \tag{9.36}$$

A straightforward computation yields

$$H(a,b,d_1,d_2,\mu) = \mu^2 - \Big(\frac{abf'(a)-(b+1)f(a)}{d_1 f(a)} - \frac{f(a)}{d_2}\Big)\mu + \frac{f(a)}{d_1 d_2}.$$

If

$$b\frac{af'(a)-f(a)}{f(a)} > \Big(1+\sqrt{\frac{d_1}{d_2}f(a)}\Big)^2, \tag{9.37}$$

then the equation $H(\mu) = 0$ has two positive solutions $\mu^{\pm}(a,b,d_1,d_2)$ given by

$$\mu^{\pm}(a,b,d_1,d_2) = \frac{1}{2}\Big(\theta(a,b,d_1,d_2) \pm \sqrt{\theta(a,b,d_1,d_2)^2 - 4f(a)/(d_1 d_2)}\Big),$$

where

$$\theta(a,b,d_1,d_2) = \frac{abf'(a)-(b+1)f(a)}{d_1 f(a)} - \frac{f(a)}{d_2}.$$

We have the following result.

Theorem 9.12 *Assume that condition (9.37) holds and there exist $i > j \geq 0$ such that*

(i) $\mu_i < \mu^+(a,b,d_1,d_2) < \mu_{i+1}$ *and* $\mu_j < \mu^-(a,b,d_1,d_2) < \mu_{j+1}$.
(ii) $\sum_{k=j+1}^{i} e_k$ *is odd.*

Then (9.2) has at least one nonconstant solution.

Proof. The proof uses some topological degree arguments (see [23, 24]). By Theorem 9.10(i) we can fix $D > d_1$ such that

(a) System (9.2) with diffusion coefficients D and d_2 has no nonconstant solutions,
(b) $H(a,b,D,d_2,\mu) > 0$ for all $\mu \geq 0$.

Further, by Proposition 9.8 one can find $C_1, C_2 > 0$ depending on a,b,d_1,d_2 such that for any $d \geq d_1$, any solution (u,v) of (9.2) with diffusion coefficients d and d_2 satisfies

$$C_1 < u,v < C_2 \quad \text{in } \overline{\Omega}.$$

Set

$$\mathscr{M} = \{(u,v) \in C(\overline{\Omega}) \times C(\overline{\Omega}) : C_1 < u, v < C_2 \text{ in } \overline{\Omega}\},$$

and define

$$\Psi : [0,1] \times \mathscr{M} \to C(\overline{\Omega}) \times C(\overline{\Omega}),$$

by

$$\Psi(t,\mathbf{w}) = (-\Delta + \mathbf{I})^{-1} \begin{pmatrix} u + \left(\frac{1-t}{D} + \frac{t}{d_1}\right)(a - (b+1)u + f(u)v) \\ v + \frac{1}{d_2}(bu - f(u)v) \end{pmatrix}.$$

It is easy to see that solving (9.2) is equivalent to find a fixed point of $\Psi(1,\cdot)$ in $\mathscr{M}$. Further, from the definition of $\mathscr{M}$ and Proposition 9.8, we have that $\Psi(t,\cdot)$ has no fixed points in $\partial\mathscr{M}$ for all $0 \le t \le 1$. Therefore, the Leray–Schauder topological degree $\deg(\mathbf{I} - \Psi(t,\cdot), \mathscr{M}, 0)$ is well defined.

Using (9.33) we have $\mathbf{I} - \Psi(1,\cdot) = \mathscr{F}$. Thus, if (9.2) has no other solutions except the constant one $\mathbf{w}_0$, then by (9.34) and (9.36) we have

$$\deg(\mathbf{I} - \Psi(1,\cdot), \mathscr{M}, 0) = \operatorname{index}(\mathscr{F}, \mathbf{w}_0) = (-1)^{\sum_{k=j+1}^{i} e(\mu_k)} = -1. \tag{9.38}$$

On the other hand, from the invariance of the Leray–Schauder degree at the homotopy we deduce

$$\deg(\mathbf{I} - \Psi(1,\cdot), \mathscr{M}, 0) = \deg(\mathbf{I} - \Psi(0,\cdot), \mathscr{M}, 0). \tag{9.39}$$

Remark that by our choice of D, we have that $\mathbf{w}_0$ is the only fixed point of $\Psi(0,\cdot)$. Furthermore by (b) above we have

$$\deg(\mathbf{I} - \Psi(0,\cdot), \mathscr{M}, 0) = \operatorname{index}(\mathbf{I} - \Psi(\cdot,0), \mathbf{w}_0) = 1. \tag{9.40}$$

Now, from (9.38) to (9.40) we reach a contradiction. Therefore, there exists a nonconstant solution of (9.2). This ends the proof. □

Corollary 9.13 *Let $a, b, d_2 > 0$ be fixed. Assume that*

$$abf'(a) > (b+1)f(a) \tag{9.41}$$

and all the eigenvalues μ_i have odd multiplicity. Then, there exists a sequence of intervals $\{(k_n, K_n)\}$ with $0 < k_n < K_n < k_{n-1} \to 0$ (as $n \to \infty$) such that the steady-state system (9.2) has at least one nonconstant solution for all $d_1 \in \bigcup_{n \ge 1}(k_n, K_n)$.

Proof. In view of (9.41), condition (9.37) holds for small values of $d_1 > 0$. Also for $a, b, d_2 > 0$ fixed we have

$$\mu^-(a,b,d_1,d_2) \to \frac{f(a)^2}{d_2(abf'(a)-(b+1)f(a))} \qquad \text{as } d_1 \to 0,$$

$$\mu^+(a,b,d_1,d_2) \to \infty \qquad \text{as } d_1 \to 0.$$

Therefore we can find a sequence of intervals $\{(k_n, K_n)\}_n$ such that

$$\sum_{\substack{i\geq 0,\\ \mu^-(a,b,d_1,d_2)<\mu_i<\mu^+(a,b,d_1,d_2)}} e(\mu_i) \ \ \text{is odd} \tag{9.42}$$

for all $d_1 \in \bigcup_{n\geq 1}(k_n, K_n)$. Therefore, conditions (i)–(ii) in Theorem 9.12 are fulfilled. □

Corollary 9.14 *Let $a, b, d_1 > 0$ be fixed. Assume that (9.41) holds and*

$$\sum_{\substack{i\geq 0,\\ 0<\mu_i<\frac{abf'(a)-(b+1)f(a)}{d_1 f(a)}}} e(\mu_i) \ \ \textit{is odd.} \tag{9.43}$$

Then there exists $D > 0$ such that the steady-state system (9.2) has at least one nonconstant solution for any $d_2 > D$.

Proof. By virtue of (9.41), for any $d_2 > 0$ large enough condition (9.37) holds. Also for any a, b, d_1 fixed we have

$$0 < \mu^-(a,b,d_1,d_2) < \mu^+(a,b,d_1,d_2) < \frac{abf'(a)-(b+1)f(a)}{d_1 f(a)}$$

and

$$\mu^-(a,b,d_1,d_2) \to 0\,, \ \ \mu^+(a,b,d_1,d_2) \to \frac{abf'(a)-(b+1)f(a)}{d_1 f(a)} \qquad \text{as } d_2 \to \infty.$$

Therefore, for $d_2 > 0$ large, condition (9.43) implies (i)–(ii) in Theorem 9.12. This concludes the proof. □

Corollary 9.15 *Let $a, d_1, d_2 > 0$ be fixed. Assume that $af'(a) > f(a)$ and all the eigenvalues μ_i have odd multiplicity. Then, there exists a sequence of intervals*

$\{(b_n, B_n)\}$ with $0 < b_n < B_n < b_{n+1} \to \infty$ (as $n \to \infty$) such that the steady-state system (9.2) has at least one nonconstant solution for all $b \in \cup_{n\geq 1}(b_n, B_n)$.

Proof. We proceed similarly. Since $af'(a) > f(a)$, for large values of b condition (9.37) is fulfilled. Also for $a, d_1, d_2 > 0$ fixed we have

$$\mu^-(a,b,d_1,d_2) \to 0\,, \quad \mu^+(a,b,d_1,d_2) \to \infty \qquad \text{as } b \to \infty.$$

Hence, we can find a sequence of nonoverlapping intervals $\{(b_n, B_n)\}$ such that $b_n \to \infty$ as $n \to \infty$ and (9.42) holds for all $b \in \bigcup_{n\geq 1}(b_n, B_n)$. □

If $f(s) = s^m$, $m > 1$, then condition (9.41) is independent of a. We obtain

Corollary 9.16 *Let $f(s) = s^m$, $m > 1$. Assume that $b(m-1) > 1$ and*

$$\sum_{\substack{i\geq 0,\\ 0<\mu_i<(b(m-1)-1)/d_1}} e(\mu_i) \quad \text{is odd.} \tag{9.44}$$

Then there exists $A > 0$ such that the steady-state system (9.2) has at least one nonconstant solution for any $0 < a < A$.

Proof. It is easy to see that (9.37) holds for small values of $a > 0$. As before

$$0 < \mu^-(a,b,d_1,d_2) < \mu^+(a,b,d_1,d_2) < \frac{b(m-1)-1}{d_1}$$

and

$$\mu^-(a,b,d_1,d_2) \to 0\,, \quad \mu^+(a,b,d_1,d_2) \to \frac{b(m-1)-1}{d_1} \qquad \text{as } a \to 0.$$

Therefore, for $a > 0$ small, condition (9.44) implies (i)–(ii) in Theorem 9.12. This ends the proof. □

9.3 Schnackenberg Model

The Schnakenberg model [174] was introduced in 1979 to describe chemical reactions with limit cycle behavior. This is a two-species model for trimolecular reactions that reads as

$$A \rightleftharpoons X, \qquad B \to Y, \qquad 2X + Y \to 3X.$$

Using the law of mass action, one can derive the nondimensional equations for the concentrations u and v of the chemical products X and Y as follows

$$u_t - d_1\Delta u = a - u + u^2 v \quad \text{and} \quad v_t - d_2\Delta v = b - u^2 v.$$

In the above equations d_1, d_2 are the diffusion coefficients of the chemicals X, Y and a, b are the concentrations of A and B. It is also assumed that A and B are in abundance so a and b are approximately constant. The Schnackenberg model has received considerable attention in recent decades from both qualitative and quantitative points of view, see [15, 147, 184, 202, 203] and the references therein.

In this section we shall be concerned with the following general reaction-diffusion system

$$\begin{cases} u_t - d_1\Delta u = a - u + u^p v & \text{in } \Omega\times(0,\infty),\\ v_t - d_2\Delta v = b - u^p v & \text{in } \Omega\times(0,\infty),\\ u(x,0) = u_0(x), v(x,0) = v_0(x) & \text{on } \Omega,\\ \dfrac{\partial u}{\partial \nu}(x,t) = \dfrac{\partial u}{\partial \nu}(x,t) = 0 & \text{on } \partial\Omega\times(0,\infty), \end{cases} \tag{9.45}$$

where $\Omega\subset\mathbb{R}^N$ $(N\geq 1)$, $a, b, d_1, d_2 > 0$ and $p > 1$. For $p = 2$ we obtain the standard model derived by Schnackenberg [174]. We will also investigate existence and nonexistence of nonconstant steady-states, that is, solutions of

$$\begin{cases} -d_1\Delta u = a - u + u^p v & \text{in } \Omega,\\ -d_2\Delta v = b - u^p v & \text{in } \Omega,\\ \dfrac{\partial u}{\partial \nu} = \dfrac{\partial v}{\partial \nu} = 0 & \text{on } \partial\Omega. \end{cases} \tag{9.46}$$

9.3.1 The Evolution System and Global Solutions

We are concerned in this section with the parabolic system (9.45). Our main result is Theorem 9.17 below.

Theorem 9.17 *Assume that* $1 < p < (N+4)/N$ *and the initial data* u_0, v_0 *satisfy*

$$u_0, v_0 \in C^1(\overline{\Omega}), \quad u_0 > 0, v_0 \geq 0 \quad \text{in } \overline{\Omega}.$$

Then, the system (9.45) has at least one solution which is bounded and global in time.

Proof. We first recall the following useful result, the proof of which can be found in [215].

Proposition 9.18 (see [215, Proposition 2.2]) *Let w be a classical solution of*

$$\begin{cases} w_t - \Delta w = f(x,w) & \text{in } \Omega \times (0,T), \\ w(x,0) = w_0(x) & \text{on } \Omega, \\ \dfrac{\partial w}{\partial \nu}(x,t) = 0 & \text{on } \partial\Omega \times (0,T), \end{cases}$$

where $w_0 \in C(\overline{\Omega})$ and $f : \overline{\Omega} \times [0,\infty) \to \mathbb{R}$ satisfies

$$f(x,s) \le C(1+s^\gamma) \quad \text{for all } (x,s) \in \Omega \times [0,\infty),$$

for some $C > 0$ and $1 \le \gamma < 1 + 2q/N$, $q > 1$. If

$$\sup_t \|w(\cdot,t)\|_q < \infty,$$

then there exist $C, \sigma > 0$ independent of T such that

$$\sup_t \|w(\cdot,t)\|_\infty \le C \max\{1, \|w\|_q^\sigma\}.$$

By [172, Theorem 1, pag. 111] there exists a solution (u,v) of (9.45) defined in a maximal time interval $[0,T_{\max})$. Furthermore, if $T_{\max} < \infty$, then

$$\lim_{t \nearrow T_{\max}} \Big[\|u(\cdot,t)\|_\infty + \|v(\cdot,t)\|_\infty \Big] = \infty. \tag{9.47}$$

We divide our arguments into three steps.

Step 1: There exists $m = m(u_0,a) > 0$ such that $u \ge m$ in $\Omega \times [0,T_{\max})$.

Indeed, let $m = \min\{\min_{\overline{\Omega}} u_0, a\}$. Since $u_0 > 0$ in Ω, we have $m > 0$. We multiply by $(m-u)^+$ in the first equation of (9.45) and integrate over Ω. We obtain

$$\begin{aligned} -\frac{1}{2}\frac{d}{dt}\int_\Omega |(m-u)^+|^2 dx =& d_1 \int_\Omega |\nabla (m-u)^+|^2 dx + \int_\Omega (a-u)(m-u)^+ dx \\ &+ \int_\Omega u^p (m-u)^+ dx \ge 0. \end{aligned}$$

Hence,

$$\int_\Omega |(m-u(x,t))^+|^2 dx \le \int_\Omega |(m-u(x,0))^+|^2 dx = 0 \qquad \text{for all } 0 \le t < T_{\max},$$

that is, $u \ge m$ in $\Omega \times [0, T_{\max})$.

Step 2: There exists $M = M(u_0, v_0, a, b) > 0$ such that $v \le M$ in $\Omega \times [0, T_{\max})$.

Using the fact that $u \ge m > 0$ in $\Omega \times [0, T_{\max})$, from the second equation of (9.45) we have

$$v_t - d_2 \Delta v + m^p v \le b \quad \text{ in } \Omega \times [0, T_{\max}).$$

By standard comparison results we find $v \le M = M(u_0, v_0, a, b)$ in $\Omega \times [0, T_{\max})$.

Step 3: $T_{\max} = \infty$.

Adding the two equations in (9.45) we find

$$(u+v)_t - \Delta(d_1 u + d_2 v) = a + b - u \quad \text{ in } \Omega \times [0, T_{\max}).$$

We multiply the above equality by $(u+v)$ and then integrate over Ω. We obtain

$$\frac{1}{2}\frac{d}{dt}\int_\Omega (u+v)^2 dx + \int_\Omega \nabla(d_1 u + d_2 v) \cdot \nabla(u+v) dx = \int_\Omega (a+b-u)(u+v) dx,$$

that is,

$$\begin{aligned} \frac{d}{dt}\int_\Omega (u+v)^2 dx + 2\int_\Omega \Big[d_1 |\nabla u|^2 + (d_1+d_2)\nabla u \cdot \nabla v + d_2 |\nabla^2\Big] dx \\ = 2\int_\Omega (a+b-u)(u+v) dx. \end{aligned} \tag{9.48}$$

Using the inequality

$$d_1 |\nabla u|^2 + (d_1+d_2)\nabla u \cdot \nabla v + d_2 |\nabla v|^2 \ge \frac{d_1}{2}|\nabla u|^2 - \frac{d_1^2 + d_2^2}{2d_1}|\nabla v|^2 \quad \text{ in } \Omega,$$

from (9.48) we find

$$\frac{d}{dt}\int_\Omega (u+v)^2 dx + \frac{d_1}{2}\int_\Omega |\nabla u|^2 - \frac{d_1^2+d_2^2}{d_1}\int_\Omega |\nabla v|^2 dx \le 2\int_\Omega (a+b-u)(u+v) dx. \tag{9.49}$$

Next, we consider the second equation in (9.45) and we integrate it over Ω. We find

$$\frac{d}{dt}\int_\Omega v^2 dx + 2d_2 \int_\Omega |\nabla v|^2 = 2\int_\Omega (b - u^p v) v dx \le 2\int_\Omega (b - m^p v) v dx \le \frac{b^2}{2m^p}|\Omega|. \tag{9.50}$$

We now multiply (9.50) with $(d_1^2+d_2^2)/(2d_1d_2)$ and add it to (9.49). We deduce

$$\frac{d}{dt}\int_\Omega \left[(u+v)^2+\frac{d_1^2+d_2^2}{2d_1d_2}v^2\right]dx \le C_1+2\int_\Omega (a+b-u)(u+v)dx,$$

where C_1 depends on a,b,d_1,d_2,u_0,v_0 and $|\Omega|$. Hence

$$\begin{aligned}\frac{d}{dt}\int_\Omega \left[(u+v)^2+\frac{d_1^2+d_2^2}{2d_1d_2}v^2\right]dx+2\int_\Omega (u+v)^2dx &\le C_1+2\int_\Omega (a+b+v)(u+v)dx.\\ &\le C_2+2\|a+b+v\|_2\|u+v\|_2\\ &\le C_2+\|a+b+v\|_2^2+\|u+v\|_2^2.\end{aligned}$$

Using now the fact that v is bounded from above by M, we find

$$\left(1+\frac{d}{dt}\right)\int_\Omega \left[(u+v)^2+\frac{d_1^2+d_2^2}{2d_1d_2}v^2\right]dx \le C_3,$$

for some constant $C_3>0$ independent of $T_{\max}$. Integrating the above inequality we now obtain $\|u+v\|_2$ is bounded from above by a positive constant independent of $T_{\max}$. Since $v\le M$ in $\Omega\times[0,T_{\max})$ it follows that $\|u\|_2\le C$ for some constant C independent of $T_{\max}$.

It remains now to apply Proposition 9.18 to deduce that $\sup_t \|u(\cdot,t)\|_\infty<\infty$ so by (9.47) it follows that $T_{\max}=\infty$. This concludes the proof. □

9.3.2 A Priori Estimates

Using Theorem 1.1 we can establish various *a priori* estimates for solutions to (9.46).

Lemma 9.19 *Any solution* (u,v) *of (9.46) satisfies*

$$a\le u\le a+b+\frac{d_2b}{d_1a^p} \qquad in\ \Omega, \tag{9.51}$$

and

$$b\left(a+b+\frac{d_2b}{d_1a^p}\right)^{-p}\le v\le ba^{-p} \qquad in\ \Omega. \tag{9.52}$$

Proof. Let $x_0\in\overline{\Omega}$ be a minimum point of u. By Theorem 1.1(ii) it follows that $a-u(x_0)+u^p(x_0)v(x_0)\le 0$ which implies $u(x_0)\ge a$. Hence

$$u\ge a \qquad \text{in } \Omega. \tag{9.53}$$

At any maximum point of v we have $b-u^p v \geq 0$, that is, $v \leq b/u^{-p}$. By (9.53) we deduce

$$v \leq \frac{b}{a^p} \qquad \text{in } \Omega. \tag{9.54}$$

Let $w = d_1 u + d_2 v$. Adding the first two relations in (9.46) we have

$$-\Delta w = a + b - u \text{ in } \Omega, \qquad \frac{\partial w}{\partial \nu} = 0 \text{ on } \partial\Omega.$$

Let now $x_1 \in \overline{\Omega}$ be a maximum point of w. According to Theorem 1.1(i) we deduce $a+b-u(x_1) \geq 0$, that is, $u(x_1) \leq a+b$. By virtue of (9.54), for all $x \in \overline{\Omega}$ we have

$$d_1 u(x) \leq w(x) \leq w(x_1) \leq d_1(a+b) + \frac{d_2 b}{a^p} \qquad \text{in } \Omega.$$

This yields

$$u \leq a+b+\frac{d_2 b}{d_1 a^p} \qquad \text{in } \Omega. \tag{9.55}$$

We have proved that u satisfies (9.51). Again by Theorem 1.1(ii), at minimum points of v we have $b - u^p v \leq 0$, which yields $v \geq b/u^p$. Combining this inequality with (9.55), we obtain the first estimate in (9.52). This concludes our proof. □

From the estimates in Lemma 9.19 we derive the following.

Proposition 9.20 *Let $a, B, D_1, D_2 > 0$ be fixed. Then, there exist two positive constants $C_1, C_2 > 0$ depending on a, B, D_1, D_2 such that for all*

$$0 < b < B, \quad d_1 > D_1, \quad 0 < d_2 < D_2,$$

any solution (u, v) of (9.46) satisfies

$$C_1 \leq u, v \leq C_2 \qquad \text{in } \Omega.$$

Furthermore, by standard elliptic arguments and Lemma 9.19 we now obtain:

Proposition 9.21 *Let $a, B, D_1, D_2 > 0$ be fixed. Then, for any positive integer $k \geq 1$ there exists a constant*

$$C = C(a, B, D_1, D_2, k, N, \Omega) > 0$$

such that for all

$$0 < b < B, \quad d_1 > D_1, \quad 0 < d_2 < D_2,$$

any solution (u, v) of (9.46) satisfies

$$\|u\|_{C^k(\overline{\Omega})} + \|v\|_{C^k(\overline{\Omega})} \le C.$$

In particular, any solution of (9.46) belongs to $C^\infty(\overline{\Omega}) \times C^\infty(\overline{\Omega})$.

For any solution (u,v) of (9.46) we denote by $\bar{u}$ and $\bar{v}$ the average over Ω of u respectively v, that is

$$\bar{u} = \frac{1}{|\Omega|}\int_\Omega u dx, \quad \bar{v} = \frac{1}{|\Omega|}\int_\Omega v dx.$$

Integrating in (9.46) we deduce

$$\bar{u} = a + b \quad \text{and} \quad \int_\Omega u^p v dx = b|\Omega|. \tag{9.56}$$

Let now

$$\phi = u - \bar{u}, \quad \psi = v - \bar{v}.$$

Our next result provides energy estimates for ϕ and ψ.

Proposition 9.22 *Let* (u,v) *be a nonconstant solution of (9.46). Then:*

(i) $\displaystyle \frac{\mu_1^2}{(2\mu_1^2+2\mu_1+1)}\left(\frac{d_2}{d_1}\right)^2 \le \frac{\|\nabla\phi\|_2^2}{\|\nabla\psi\|_2^2} \le \left(\frac{d_2}{d_1}\right)^2,$

(ii) $\displaystyle \frac{\mu_1^3}{(\mu_1+1)(2\mu_1^2+2\mu_1+1)}\left(\frac{d_2}{d_1}\right)^2 \le \frac{\|\nabla\phi\|_2^2+\|\phi\|_2^2}{\|\nabla\psi\|_2^2+\|\psi\|_2^2} \le \frac{\mu_1+1}{\mu_1}\left(\frac{d_2}{d_1}\right)^2.$

Proof. (i) Adding the first two equations in (9.46) we obtain $-\Delta(d_1 u + d_2 v) = a + b - u$ in Ω, that is,

$$\Delta w = \phi \qquad \text{in } \Omega, \tag{9.57}$$

where $w = d_1\phi + d_2\psi$. Multiplying by ϕ in (9.57) and integrating over Ω we have

$$\int_\Omega \nabla w \nabla\phi dx = -\int_\Omega \phi^2 dx,$$

which yields

$$d_2\int_\Omega \nabla\phi\nabla\psi dx = -\int_\Omega \phi^2 dx - d_1\int_\Omega |\nabla\phi|^2 dx. \tag{9.58}$$

Now from (9.58) we have

$$\begin{aligned} 0 \le \int_\Omega |\nabla w|^2 dx &= \int_\Omega |\nabla(d_1\phi + d_2\psi)|^2 \\ &= d_1^2\int_\Omega |\nabla\phi|^2 dx + 2d_1 d_2\int_\Omega \nabla\phi\nabla\psi dx + d_2^2\int_\Omega |\nabla\psi|^2 dx \\ &= -d_1^2\int |\nabla\phi|^2 dx - 2d_2\int_\Omega \phi^2 dx + d_2\int_\Omega |\nabla\psi|^2 dx. \end{aligned}$$

This implies

$$d_2^2\int_{\Omega}|\nabla\psi|^2dx-d_1^2\int|\nabla\phi|^2dx\geq 0,$$

which proves the last estimate in (i). Next we multiply (9.57) by w and integrate over Ω. We obtain

$$\int_{\Omega}|\nabla w|^2dx=-\int_{\Omega}w\phi dx,$$

that is,

$$d_1^2\int|\nabla\phi|^2dx+2d_1d_2\int_{\Omega}\nabla\phi\nabla\psi dx+d_2^2\int_{\Omega}|\nabla\psi|^2dx=-d_1^2\int_{\Omega}\phi^2dx-d_1d_2\int_{\Omega}\phi\psi dx.$$

From (9.58) we have

$$d_2^2\int_{\Omega}|\nabla\psi|^2dx=d_1^2\int_{\Omega}|\nabla\phi|^2dx+d_1^2\int_{\Omega}\phi^2dx-d_1d_2\int_{\Omega}\phi\psi dx.$$

On the other hand

$$-d_1d_2\phi\psi\leq\frac{d_1^2}{2\mu_1}\phi^2+\frac{\mu_1d_2^2}{2}\psi^2,$$

where μ_1 is the first positive eigenvalue of the negative Laplace operator in Ω subject to homogeneous Neumann boundary condition. Combining the last two estimates we find

$$d_2^2\int_{\Omega}|\nabla\psi|^2dx\leq d_1^2\left(1+\frac{1}{2\mu_1}\right)\int_{\Omega}\phi^2dx+d_1^2\int_{\Omega}|\nabla\phi|^2dx+\frac{\mu_1d_2^2}{2}\int_{\Omega}\psi^2dx. \tag{9.59}$$

By Poincaré's inequality we have

$$\int_{\Omega}\phi^2dx\leq\frac{1}{\mu_1}\int_{\Omega}|\nabla\phi|^2dx,\qquad\int_{\Omega}\psi^2dx\leq\frac{1}{\mu_1}\int_{\Omega}|\nabla\psi|^2dx. \tag{9.60}$$

Therefore, from (9.59) to (9.60) we now obtain

$$d_2^2\int_{\Omega}|\nabla\psi|^2dx\leq\frac{d_1^2(2\mu_1^2+2\mu_1+1)}{\mu_1^2}\int_{\Omega}|\nabla\phi|^2dx,$$

which completes the proof of (i).

(ii) The proof in this part follows directly from (i) together with the following estimate which is a direct consequence of Poincaré's inequality

$$\frac{\mu_1}{\mu_1+1}\frac{\|\nabla\phi\|_2^2}{\|\nabla\psi\|_2^2}\leq\frac{\|\nabla\phi\|_2^2+\|\phi\|_2^2}{\|\nabla\psi\|_2^2+\|\psi\|_2^2}\leq\frac{\mu_1+1}{\mu_1}\frac{\|\nabla\phi\|_2^2}{\|\nabla\psi\|_2^2}.$$

This finishes the proof. □

9.3.3 Nonexistence of Nonconstant Steady States

Our first result in this part establishes that a nonconstant steady state may exist only if the first eigenvalue μ_1 of $-\Delta$ subject to Neumann boundary condition is sufficiently small. We have:

Theorem 9.23 *Let $a,b,d_1,d_2>0$ be fixed. Then, there exists $L=L(a,b,d_1,d_2)>0$ such that system (9.2) has no nonconstant solutions if $\mu_1>L$.*

Proof. By Lemma 9.19, there exist two positive constants C_1,C_2 depending only on a,b,d_1,d_2 such that any solution (u,v) of (9.2) satisfies

$$C_1\le u,v\le C_2 \quad \text{in } \Omega. \tag{9.61}$$

We first multiply by $\phi=u-\bar u$ in the first equation of (9.2) and integrate over Ω. By (9.61) and Poincaré's inequality we obtain

$$\begin{aligned}
d_1\int_\Omega|\nabla\phi|^2dx&=-\int_\Omega\phi^2dx+\int_\Omega u^pv\phi dx\le\int_\Omega u^pv\phi dx\\
&=\int_\Omega u^p\phi\psi dx+\bar v\int_\Omega(u^p-\bar u^p)\phi dx\\
&\le c\int_\Omega\phi\psi dx+p\bar v\int_\Omega\xi^{p-1}\phi^2dx \text{ (with } \xi \text{ between } u \text{ and } \bar u)\\
&\le c_1\int_\Omega(|\phi||\psi|+\int_\Omega\phi^2)dx\le 2c_1\int_\Omega(\phi^2+\psi^2)dx\\
&\le\frac{2c_1}{\mu_1}\int_\Omega(|\nabla\phi|^2+|\nabla\psi|^2)dx,
\end{aligned}$$

where c_1,c_2 depend only on C_1,C_2 from (9.61). Similarly we have

$$d_2\int_\Omega|\nabla\psi|^2dx\le\frac{c_2}{\mu_1}\int_\Omega(|\nabla\phi|^2+|\nabla\psi|^2)dx.$$

Adding the above two relations we find

$$\min\{d_1,d_2\}\int_\Omega(|\nabla\phi|^2+|\nabla\psi|^2)dx\le\frac{C}{\mu_1}\int_\Omega(|\nabla\phi|^2+|\nabla\psi|^2)dx, \tag{9.62}$$

where C depends only on a,b,d_1 and d_2. From (9.62) it follows that if μ_1 is large enough then $\int_\Omega|\nabla\phi|^2dx=\int_\Omega|\nabla\psi|^2dx=0$, that is, u and v are constant functions. This ends the proof. □

Theorem 9.24 *(i) Let $a,b,d_2>0$ be fixed. Then, there exists $D=D(a,b,d_2)>0$ such that system (9.2) has no nonconstant solutions for all $d_1>D$.*
(ii) Let $a,d_1,d_2>0$ be fixed. Then, there exists $B=B(a,d_1,d_2)>0$ such that system (9.2) has no nonconstant solutions for all $0<b<B$.

Proof. We first prove the following result.

Lemma 9.25 *Let $a,b,d_2>0$ be fixed and let $\{\delta_n\}\subset(0,\infty)$ be such that $\delta_n\to\infty$ as $n\to\infty$. If (u_n,v_n) is a solution of (9.2) with $d_1=\delta_n$ then*

$$u_n\to a+b,\quad v_n\to\frac{b}{(a+b)^p}\qquad \text{in } C^2(\overline{\Omega}) \text{ as } n\to\infty.$$

Proof. By Proposition 9.21 the sequence $\{(u_n,v_n)\}$ is bounded in $C^3(\overline{\Omega})\times C^3(\overline{\Omega})$. Hence, passing to a subsequence if necessary, $\{(u_n,v_n)\}$ converges in $C^2(\overline{\Omega})\times C^2(\overline{\Omega})$ to some $(u,v)\in C^2(\overline{\Omega})\times C^2(\overline{\Omega})$. We divide by δ_n in the corresponding equation to u_n and then we pass to the limit with $n\to\infty$. We obtain that (u,v) satisfies

$$\begin{cases}-\Delta u=0 & \text{in } \Omega,\\ -d_2\Delta v=b-u^pv & \text{in } \Omega,\\ \dfrac{\partial u}{\partial\nu}=\dfrac{\partial v}{\partial\nu}=0 & \text{on } \partial\Omega.\end{cases}\tag{9.63}$$

Also, u_n and u satisfy (9.56). Now, the first equation in (9.63) together with $\partial u/\partial\nu=0$ on $\partial\Omega$ implies that u is a constant. Combining this fact with (9.56) it follows that $u\equiv a+b$. Thus, from (9.63), v satisfies

$$-d_2\Delta v=b-(a+b)^pv \text{ in } \Omega,\quad \frac{\partial v}{\partial\nu}=0 \text{ on } \partial\Omega.$$

Multiplying the above equality with $b-(a+b)^pv$ and then integrating over Ω we obtain

$$0\le\frac{d_1}{(a+b)^p}\int_\Omega|\nabla(b-(a+b)^pv)|^2dx=-\int_\Omega(b-(a+b)^pv)^2dx\le 0.$$

Hence $v\equiv b(a+b)^{-p}$ and the proof of Lemma 9.25 is now complete. □

We first introduce the function spaces

$$H^2_{\mathrm{n}}(\Omega)=\left\{w\in W^{2,2}(\Omega):\frac{\partial w}{\partial\nu}=0\right\},\qquad L^2_0(\Omega)=\left\{w\in L^2(\Omega):\int_\Omega wdx=0\right\}.$$

Thus, letting $w = u - (a+b)$ and $\delta = 1/d_1$, by (9.56) and the standard elliptic regularity theory, the system (9.2) is equivalent to

$$\begin{cases} -\Delta u = \delta(-b - w + (w+a+b)^p v) & \text{in } \Omega, \\ -d_2 \Delta v = b - (w+a+b)^p v & \text{in } \Omega, \\ w \in H^2_{\mathrm{n}}(\Omega) \cap L^2_0(\Omega),\ v \in H^2_{\mathrm{n}}(\Omega). \end{cases} \tag{9.64}$$

Define

$$\mathscr{F} : \mathbb{R} \times (H^2_{\mathrm{n}}(\Omega) \cap L^2_0(\Omega)) \times H^2_{\mathrm{n}}(\Omega) \to L^2_0(\Omega) \times L^2(\Omega),$$

by

$$\mathscr{F}(\delta, w, v) = \begin{pmatrix} \Delta w + \delta \mathscr{P}(-w + (w+a+b)^p v) \\ \\ d_2 \Delta v + b - (w+a+b)^p v \end{pmatrix},$$

where $\mathscr{P} : L^2(\Omega) \to L^2_0(\Omega)$ is the projection operator from $L^2(\Omega)$ onto $L^2_0(\Omega)$, namely,

$$\mathscr{P}(z) = z - \frac{1}{|\Omega|} \int z(x) dx, \qquad \text{for all } z \in L^2(\Omega).$$

Now (9.64) is equivalent to

$$\mathscr{F}(\delta, w, v) = \mathbf{0}.$$

Indeed, if $\mathscr{F}(\delta, w, v) = \mathbf{0}$, then

$$d_2 \Delta v + b - (w+a+b)^p v = 0 \ \text{ in } \Omega, \qquad v \in H^2_{\mathrm{n}}(\Omega).$$

By integration, it is easy to see that the above relations imply $b - (w+a+b)^p v \in L^2_0(\Omega)$. Since $w \in L^2_0(\Omega)$, this yields

$$-b - w + (w+a+b)^p v \in L^2_0(\Omega),$$

so that

$$\mathscr{P}(-w + (w+a+b)^p v) = -b - w + f(w+a+b)v.$$

Therefore (9.64) is satisfied.

With the same method as in the proof of Lemma 9.25 we have that the equation

$$\mathscr{F}(0, w, v) = \mathbf{0}$$

has the unique solution $(w, v) = (0, b(a+b)^{-p})$. Also remark that

$$D_{(w,v)}\mathscr{F}(0,0,b(a+b)^{-p}) : (H^2_{\mathrm{n}}(\Omega)\cap L^2_0(\Omega)) \times H^2_{\mathrm{n}}(\Omega) \to L^2_0(\Omega) \times L^2(\Omega),$$

is given by

$$D_{(w,v)}\mathscr{F}(0,0,b(a+b)^{-p}) = \begin{pmatrix} \Delta & 0 \\ \\ -\frac{bp}{(a+b)^p} & d_2\Delta - (a+b)^p \end{pmatrix}.$$

Thus $D_{(w,v)}\mathscr{F}(0,0,b(a+b)^{-p})$ is invertible and we are in the frame of the implicit function theorem. It follows that there exist $\delta_0, r > 0$ such that $(0,0,b(a+b)^{-p})$ is the unique solution of

$$\mathscr{F}(\delta,w,v) = \mathbf{0} \qquad \text{in } [0,\delta_0] \times B_r(0,b(a+b)^{-p}),$$

where $B_r(0,b(a+b)^{-p})$ denotes the open ball in $(H^2_{\mathrm{n}}(\Omega)\cap L^2_0(\Omega)) \times H^2_{\mathrm{n}}(\Omega)$ centered at $(0,b(a+b)^{-p})$ and having the radius $r > 0$.

Let now $\{\delta_n\}$ be a sequence of positive real numbers such that $\delta_n \to \infty$ as $n \to \infty$ and let (u_n, v_n) be an arbitrary solution of (9.2) for a, b, d_2 fixed and $d_1 = \delta_n$. Letting $w_n = u_n - (a+b)$, it follows that

$$\mathscr{F}\left(\frac{1}{\delta_n}, w_n, v_n\right) = \mathbf{0}.$$

According to Lemma 9.25 we have

$$(w_n, v_n) \to (0, b(a+b)^{-p}) \text{ in } C^2(\overline{\Omega}) \times C^2(\overline{\Omega}) \text{ as } n \to \infty.$$

This means that for $n \geq 1$ large enough there holds $(1/\delta_n, w_n, v_n) \in (0,\delta_0) \times B_r(0,b(a+b)^{-p})$ which yields $(w_n, v_n) = (0, b(a+b)^{-p})$. Hence, for $d_1 = 1/\delta_n > 0$ small enough, the system (9.2) has only the constant solution $(a+b, b(a+b)^{-p})$. The proof of (ii) is similar. □

9.3.4 Existence Results

Let

$$\mathbf{X} = \left\{\mathbf{w} = (u,v) \in C^1(\overline{\Omega}) \times C^1(\overline{\Omega}) : \frac{\partial u}{\partial \nu} = \frac{\partial v}{\partial \nu} = 0 \text{ on } \partial\Omega\right\} \tag{9.65}$$

and decompose

$$\mathbf{X} = \bigoplus_{k\geq 1} \mathbf{X}_k, \tag{9.66}$$

where $\mathbf{X}_k$ denotes the eigenspace corresponding to μ_k, $k \geq 1$. Also consider

$$\mathbf{X}^+ = \{(u,v) \in X : u,v > 0 \text{ in } \overline{\Omega}\}.$$

We write the system (9.2) in the form

$$-\Delta \mathbf{w} = \mathscr{G}(\mathbf{w}), \qquad \mathbf{w} \in \mathbf{X}^+, \tag{9.67}$$

where

$$\mathscr{G}(\mathbf{w}) = \begin{pmatrix} \frac{1}{d_1}(a - u + u^p v) \\ \\ \frac{1}{d_2}(b - u^p v) \end{pmatrix}.$$

It is more convenient to write (9.67) in the form

$$\mathscr{F}(\mathbf{w}) = \mathbf{0}, \qquad \mathbf{w} \in \mathbf{X}^+,$$

where

$$\mathscr{F}(\mathbf{w}) = \mathbf{w} - (\mathbf{I} - \Delta)^{-1}(\mathscr{G}(\mathbf{w}) + \mathbf{w}), \qquad \mathbf{w} \in \mathbf{X}^+. \tag{9.68}$$

Furthermore, at the uniform stationary steady state $\mathbf{w}_0$ we have

$$\nabla \mathscr{F}(\mathbf{w}_0) = \mathbf{I} - (\mathbf{I} - \Delta)^{-1}(\mathbf{I} + A),$$

where

$$A := \nabla \mathscr{G}(\mathbf{w}_0) = \begin{pmatrix} \frac{1}{d_1}\left(\frac{bp}{a+b} - 1\right) & \frac{(a+b)^p}{d_1} \\ \\ -\frac{bp}{d_2(a+b)} & -\frac{(a+b)^p}{d_2} \end{pmatrix}.$$

If $\nabla \mathscr{F}(\mathbf{w}_0)$ is invertible, by [154, Theorem 2.8.1] the index of $\mathscr{F}$ at $\mathbf{w}_0$ is given by

$$\text{index}(\mathscr{F}, \mathbf{w}_0) = (-1)^\gamma, \tag{9.69}$$

where γ denotes the number of the negative eigenvalues of $\nabla \mathscr{F}(\mathbf{w}_0)$. On the other hand, using the decomposition (9.66) we have that $\mathbf{X}_i$ is an invariant space under $\nabla \mathscr{F}(\mathbf{w}_0)$ and $\xi \in \mathbb{R}$ is an eigenvalue of $\nabla \mathscr{F}(\mathbf{w}_0)$ in $\mathbf{X}_i$ if and only if ξ is an eigenvalue of $(\mu_i + 1)^{-1}(\mu_i \mathbf{I} - A)$. Therefore, $\nabla \mathscr{F}(\mathbf{w}_0)$ is invertible if and only if for any $i \geq 0$ the matrix $(\mu_i \mathbf{I} - A)$ is invertible.

Let us define

$$S(a,b,d_1,d_2,\mu) = \det(\mu\mathbf{I} - A). \tag{9.70}$$

Then, if $\mu_i\mathbf{I} - A$ is invertible for any $i \geq 0$, we have

$$\gamma = \sum_{i\geq 0,\, H(\mu_i)<0} e(\mu_i). \tag{9.71}$$

A straightforward computation yields

$$S(a,b,d_1,d_2,\mu) = \mu^2 - \left[\frac{1}{d_1}\left(\frac{bp}{a+b} - 1\right) - \frac{(a+b)^p}{d_2}\right]\mu + \frac{(a+b)^p}{d_1 d_2}. \tag{9.72}$$

If

$$\frac{bp}{a+b} > \left(1 + \sqrt{\frac{d_1}{d_2}(a+b)^p}\right)^2, \tag{9.73}$$

then the equation $S(a,b,d_1,d_2,\mu) = 0$ has two positive solutions $\mu^{\pm}(a,b,d_1,d_2)$ given by

$$\mu^{\pm}(a,b,d_1,d_2) = \frac{1}{2}\left(\theta(a,b,d_1,d_2) \pm \sqrt{\theta(a,b,d_1,d_2)^2 - \frac{4(a+b)^p}{d_1 d_2}}\right),$$

where

$$\theta(a,b,d_1,d_2) = \frac{1}{d_1}\left(\frac{bp}{a+b} - 1\right) - \frac{(a+b)^p}{d_2}.$$

Our main existence result is the following.

Theorem 9.26 *Assume that condition (9.73) holds and there exist $i > j \geq 0$ such that*

(i) $\mu_i < \mu^+(a,b,d_1,d_2) < \mu_{i+1}$ and $\mu_j < \mu^-(a,b,d_1,d_2) < \mu_{j+1}$.
(ii) $\sum_{k=j+1}^{i} e_k$ is odd.

Then (9.2) has at least one nonconstant solution.

Proof. We follow a similar approach to that in the previous section. According to Theorem 9.24 and the definition of S in (9.70) and (9.72), we can choose $D > 0$ large enough such that system (9.2) with diffusion coefficients D and d_2 has no nonconstant solutions and

$$S(a,b,D,d_2,\mu) > 0 \quad \text{for all } \mu \geq 0. \tag{9.74}$$

Further, by Proposition 9.20 there exist C_1, C_2 depending on a, b, d_1, d_2 such that any solution (u, v) of (9.2) with diffusion coefficients d and d_2 $(d \geq d_1)$ satisfies

$$C_1 < u, v < C_2 \quad \text{in } \overline{\Omega}. \tag{9.75}$$

We next consider

$$\mathscr{M} = \{(u,v) \in C(\overline{\Omega}) \times C(\overline{\Omega}) : C_1 < u, v < C_2 \text{ in } \overline{\Omega}\},$$

and define

$$\mathscr{H} : \mathscr{M} \times [0,1] \to C(\overline{\Omega}) \times C(\overline{\Omega}),$$

by

$$\mathscr{H}(\mathbf{w}, t) = (\mathbf{I} - \Delta)^{-1} \begin{pmatrix} u + \left(\frac{1-t}{d_1} + \frac{t}{D}\right)(a - u + u^p v) \\ \\ v + \frac{1}{d_1}(b - u^p v) \end{pmatrix}.$$

Remark that $\mathscr{H}$ is a compact perturbation of the identity and by (9.75) we have $\mathscr{H}(\mathbf{w}) \neq \mathbf{0}$ on $\partial \mathscr{M}$. Therefore, the Leray–Schauder degree $\deg(\mathscr{H}, \mathscr{M}, \mathbf{0})$ is well defined.

Further, it is easy to see that solving (9.2) is equivalent to finding a fixed point of $\mathscr{H}(\cdot, 1)$ in $\mathscr{M}$. According to the above choice of δ we have that $\mathbf{w}_0$ is the only fixed point of $\mathscr{H}(\cdot, 0)$. Furthermore by (9.74) we have

$$\deg(\mathbf{I} - \mathscr{H}(\cdot, 0), \mathscr{M}, 0) = \operatorname{index}(\mathbf{I} - \mathscr{H}(\cdot, 0), \mathbf{w}_0) = 1. \tag{9.76}$$

Using (9.68) we have $\mathbf{I} - \mathscr{H}(\cdot, 1) = \mathscr{F}$. Thus, if (9.2) has no other solutions except the constant one $\mathbf{w}_0$, then by (9.69)–(9.71) we have

$$\deg(\mathbf{I} - \mathscr{H}(\cdot, 1), \mathscr{M}, 0) = \operatorname{index}(\mathscr{F}, \mathbf{w}_0) = (-1)^{\sum_{k=j+1}^{i} e(\mu_k)} = -1. \tag{9.77}$$

On the other hand, from (9.76), (9.77) and the invariance of the Leray–Schauder degree to the homotopy we deduce

$$1 = \deg(\mathbf{I} - \mathscr{H}(\cdot, 0), \mathscr{M}, 0) = \deg(\mathbf{I} - \mathscr{H}(\cdot, 1), \mathscr{M}, 0) = -1,$$

a contradiction. Therefore, there exists a nonconstant solution of (9.2). This finishes the proof. □

Corollary 9.27 *Let $a, b, d_2 > 0$ be fixed. Assume that*

$$b(p-1) > a \tag{9.78}$$

and all the eigenvalues μ_i have odd multiplicity. Then, there exists a sequence of intervals $\{(k_n, K_n)\}_n$ with $0 < k_n < K_n < k_{n-1} \to 0$ (as $n \to \infty$) such that system (9.2) has at least one nonconstant solution for all $d_1 \in \bigcup_{n\geq 1}(k_n, K_n)$.

Proof. In view of (9.78), condition (9.73) holds for small values of $d_1 > 0$. Also for $a, b, d_2 > 0$ fixed we have

$$\mu^+(a,b,d_1,d_2) \to \infty \qquad \text{as } d_1 \to 0,$$

and

$$\mu^-(a,b,d_1,d_2) \to \frac{(a+b)^{p+1}}{d_2(b(p-1)-a)} \qquad \text{as } d_1 \to 0.$$

Therefore we can find a sequence of intervals $\{(k_n, K_n)\}_n$ with $0 < k_n < K_n < k_{n-1} \to 0$ as $n \to \infty$ and such that

$$\sum_{\substack{i\geq 0,\\ \mu^- < \mu_i < \mu^+}} e(\mu_i) \quad \text{is odd} \tag{9.79}$$

for all $d_1 \in \bigcup_{n\geq 1}(k_n, K_n)$. Now, conditions (i)–(ii) in Theorem 9.26 are fulfilled, whence the conclusion of the corollary. □

Corollary 9.28 *Let $a, b, d_1 > 0$ be fixed. Assume that (9.78) holds and*

$$\sum_{\substack{i\geq 0,\\ 0 < \mu_i < \frac{b(p-1)-a}{d_1(a+b)}}} e(\mu_i) \quad \textit{is odd.} \tag{9.80}$$

Then there exists $D > 0$ such that the steady-state system (9.2) has at least one nonconstant solution for any $d_2 > D$.

Proof. By virtue of (9.78), for any $d_2 > 0$ large enough condition (9.73) holds. Also for any a, b, d_1 fixed we have

$$0 < \mu^-(a,b,d_1,d_2) < \mu^+(a,b,d_1,d_2) < \frac{b(p-1)-a}{d_1(a+b)}$$

and

$$\mu^-(a,b,d_1,d_2) \to 0\,,\ \ \mu^+(a,b,d_1,d_2) \to \frac{b(p-1)-a}{d_1(a+b)} \qquad \text{as } d_2 \to \infty.$$

Therefore, for $d_2 > 0$ large, condition (9.80) implies (i)–(ii) in Theorem 9.26. This concludes the proof. □

Corollary 9.29 *Let $a,b,d_1,d_2>0$ be fixed such that $0<a+b\leq 1$ and all the eigenvalues μ_i have odd multiplicity. Then, there exists a sequence of intervals $\{(p_n,P_n)\}_n$ with $0<p_n<P_n<p_{n+1}\to\infty$ (as $n\to\infty$) such that system (9.2) has at least one nonconstant solution for all $p\in\bigcup_{n\geq 1}(p_n,P_n)$.*

Proof. The proof is similar to that of Corollary 9.27. It relies on the fact that for large values of $p>1$ condition (9.73) holds and for $a,b,d_2>0$ fixed we have

$$\mu^+(a,b,d_1,d_2)\to\infty \qquad \text{as } p\to\infty,$$

and

$$\mu^-(a,b,d_1,d_2)\to 0 \qquad \text{as } p\to\infty.$$

□

9.4 Lengyel–Epstein Model

The Lengyel–Epstein model accounts for the chlorite–iodide–malonic acid and starch reaction (CIMA) in an open unstirred gel reactor. There are five reactants involved in the CIMA reaction, three of them having a slowly oscillating concentration. In a first approach, one can assume that three of the concentrations remain constant over the reaction process. Using this assumption, Lengyel and Epstein deduce the mathematical model of the CIMA reaction which, after a proper rescaling, reads as

$$\begin{cases} u_t-\Delta u=a-u-\dfrac{4uv}{1+u^2} & \text{in } \Omega\times(0,\infty),\\ v_t-d\Delta v=b\left(u-\dfrac{uv}{1+u^2}\right) & \text{in } \Omega\times(0,\infty),\\ u(x,0)=u_0(x),v(x,0)=v_0(x) & \text{on } \Omega,\\ \dfrac{\partial u}{\partial \nu}(x,t)=\dfrac{\partial u}{\partial \nu}(x,t)=0 & \text{on } \partial\Omega\times(0,\infty), \end{cases} \tag{9.81}$$

where $\Omega\subset\mathbb{R}^N$ $(N\geq 1)$, $a,b,d>0$. Here u and v represent the concentrations of iodide and chlorite ions respectively. The parameters a, $b>0$ are related to the feed concentrations and to experimentally determined rate constants. In our case, a shift towards higher values of a represents an increase in the supply of malonic acid relative to the supply of the chloride dioxide, and increasing b corresponds to a higher supply of iodine.

9.4.1 Global Solutions in Time

We are concerned in this section with the parabolic system (9.81). The main result is the following.

Theorem 9.30 *Assume that $u_0, v_0 > 0$ in $\overline{\Omega}$. Then, system (9.81) has a unique solution (u,v) which is global in time.*

Furthermore we have

(i) There exist two constants $C_1, C_2 > 0$ depending only on a, u_0 and v_0 such that

$$C_1 < u, v < C_2 \quad \text{in } \overline{\Omega} \times (0,\infty). \tag{9.82}$$

(ii) $\limsup_{t\to\infty} \|u(\cdot,t)\|_{L^\infty(\Omega)} < a$ *and* $\limsup_{t\to\infty} \|v(\cdot,t)\|_{L^\infty(\Omega)} < a^2+1$.

Proof. (i) The existence and uniqueness part follows once we exhibit an invariant region for our system. To this aim, let

$$M_1 = \max\{a, \max_{\overline{\Omega}} u_0\}, \quad M_2 = \max\{2+M_1^2, \max_{\overline{\Omega}} v_0\}$$

$$m_1 = \min\left\{\frac{a}{1+4M_2}, \min_{\overline{\Omega}} u_0\right\}, \quad m_2 = \min\left\{\frac{1}{2}, \min_{\overline{\Omega}} v_0\right\}.$$

It is easily seen that the rectangle $(m_1, M_1) \times (m_2, M_2)$ is an invariant region with respect to the vector field generated by (9.81). Therefore, there exists a unique solution (u,v) of (9.81) which is global in time. The two constants in (9.82) can now be chosen as

$$C_1 = \min\{m_1, m_2\}, \quad C_2 = \max\{M_1, M_2\}.$$

(ii) In view of (9.82), there exists $\varepsilon > 0$ such that

$$\varepsilon < 4uv/(1+u^2) \quad \text{in } \overline{\Omega} \times (0,\infty)$$

and let U be the unique global solution of

$$\begin{cases} U_t = a - \varepsilon - U & \text{in } (0,\infty), \\ U(0) = 2\|u_0\|_{L^\infty(\Omega)}, \end{cases} \tag{9.83}$$

that is,

$$U(t) = a - \varepsilon + e^{-t}\Big(2\|u_0\|_{L^\infty(\Omega)} - a + \varepsilon\Big), \quad t \geq 0.$$

By the maximum principle we find $u \leq U$ in $\overline{\Omega} \times (0,\infty)$ so

$$\limsup_{t\to\infty} \|u(\cdot,t)\|_{L^\infty(\Omega)} \leq \lim_{t\to\infty} U(t) = a - \varepsilon < a.$$

The proof of the second part in (ii) is more involved. To this aim, let us fix $m > 0$ such that $(a-\varepsilon)^2 + m < a^2$ and define

$$H(U,V) = \sup_{C_1<\xi<U} \xi\left(1 - \frac{V-m}{1+\xi^2}\right).$$

Let now V be the unique global solution of

$$\begin{cases} V_t = bH(U,V) & \text{in } (0,\infty), \\ V(0) = 2\|v_0\|_{L^\infty(\Omega)}, \end{cases} \tag{9.84}$$

Let $w = v - V$. Then w satisfies

$$\begin{cases} -w_t + d\Delta w = b\left[H(U,V) - \left(u - \dfrac{uv}{1+u^2}\right)\right] & \text{in } \Omega \times (0,\infty), \\ \dfrac{\partial w}{\partial \nu} = 0 & \text{on } \partial\Omega \times (0,\infty). \end{cases} \tag{9.85}$$

We claim that $w < 0$ in $\overline{\Omega} \times (0,\infty)$. This is clearly true for $t = 0$. Assume that $w < 0$ is not true in $\Omega \times (0,\infty)$. Then, using the fact that $w(x,0) < 0$ in $\overline{\Omega}$ it follows that there exists $(x_0,t_0) \in \Omega \times (0,\infty)$ such that

$$w(x,t) < 0 \quad \text{in } \Omega \times (0,t_0) \quad \text{and } w(x_0,t_0) = 0.$$

This yields

$$w_t(x_0,t_0) \geq 0 \quad \text{and} \quad \max_{\overline{\Omega}} w(\cdot,t_0) = 0. \tag{9.86}$$

Further, from the fact that $v(x_0,t_0) = V(t_0)$ we have

$$
\begin{aligned}
H(U(x_0,t_0),V(x_0,t_0)) &= \sup_{C_1<\xi<U(x_0,t_0)} \xi\left(1-\frac{V(t_0)-m}{1+\xi^2}\right)\\
&> \sup_{C_1<\xi<U(x_0,t_0)} \xi\left(1-\frac{v(x_0,t_0)}{1+\xi^2}\right)\\
&> \sup_{C_1<\xi<u(x_0,t_0)} \xi\left(1-\frac{v(x_0,t_0)}{1+\xi^2}\right)\\
&\geq u(x_0,t_0)\left(1-\frac{v(x_0,t_0)}{1+u^2(x_0,t_0)}\right).
\end{aligned}
$$

Combining this last estimate with (9.85) we find

$$
-w_t(x_0,t_0+d\Delta w(x_0,t_0)>0. \tag{9.87}
$$

Our analysis splits up into two distinct cases.

Case 1. $x_0 \in \Omega$. From (9.86) we find $\Delta w(x_0,t_0) \leq 0$ so

$$
-w_t(x_0,t_0)+d\Delta w(x_0,t_0) \leq 0
$$

which contradicts (9.87).

Case 2. $x_0 \in \partial\Omega$. Then, from (9.87) it follows that $\Delta w(x_0,t_0) > 0$ so there exists a smooth open set $\omega \subset \Omega$ such that $x_0 \in \partial\omega$ and w satisfies

$$
\begin{cases}
\Delta w(x,t_0) \geq 0, w<0 & \text{in } \omega,\\
w(x_0)=0. &
\end{cases}
$$

By Hopf's boundary point lemma it follows now that

$$
\frac{\partial w}{\partial \nu}(x_0,t_0)>0
$$

which contradicts the fact that the outer normal derivative $\partial w/\partial \nu$ vanishes on $\partial\Omega \times (0,\infty)$.

Hence, in both the above cases we raise a contradiction, which means that $v<V$ in $\overline{\Omega}\times(0,\infty)$.

Let us now remark that

$$
(U_0,V_0)=(a-\varepsilon, 1+(a-\varepsilon)^2+m)
$$

is the unique equilibrium point of the ODE system (9.83)–(9.85). Also, the above equilibrium point is asymptotically stable and so

$$\lim_{t\to\infty} V(t) = 1+(a-\varepsilon)^2+m < 1+a^2.$$

The last estimates in (ii) follows now from the above relation and the fact that $v < V$ in $\overline{\Omega}\times(0,\infty)$. This finishes our proof. □

9.4.2 Turing Instabilities

Remark that

$$(u,v) = (\alpha, 1+\alpha^2), \alpha = a/5 \tag{9.88}$$

is the only equilibrium state of (9.81). In this subsection we discuss the Turing instability of (9.88). this occurs when (9.88) is unstable for (9.81) but stable for the associated ODE system, that is

$$\begin{cases} \dfrac{du}{dt} = a - u - \dfrac{4uv}{1+u^2} \\ \dfrac{dv}{dt} = b\left(u - \dfrac{uv}{1+u^2}\right). \end{cases} \tag{9.89}$$

As in the previous section, we first rewrite (9.81) in vectorial form as follows:

$$\mathbf{w}_t = \begin{pmatrix} 1 & 0 \\ 0 & d \end{pmatrix} \Delta \mathbf{w} + T(\mathbf{w}) \quad \text{in } \overline{\Omega}\times(0,\infty), \tag{9.90}$$

where

$$\mathbf{w} = \begin{pmatrix} u \\ v \end{pmatrix} \quad \text{and} \quad T(\mathbf{w}) = \begin{bmatrix} a-u-\frac{4uv}{1+u^2} \\ bu - \frac{buv}{1+u^2} \end{bmatrix}.$$

Denote by

$$0 = \mu_0 < \mu_1 \le \mu_2 \le \cdots \le \mu_n < \cdots$$

the eigenvalues of $-\Delta$ with homogeneous Neumann boundary condition and by $m(\mu_k)$ the multiplicity of μ_k, $k \ge 0$. Consider also the space $\mathbf{X}$ and its decomposition as in (9.65) and (9.66).

The linearization of (9.90) at the uniform steady state $\mathbf{w}_0 = (\alpha, 1+\alpha^2)^T$ is

$$\mathscr{L} : \mathbf{X} \to C(\overline{\Omega})\times C(\overline{\Omega}), \quad \mathscr{L} = \begin{pmatrix} 1 & 0 \\ 0 & d \end{pmatrix} \Delta + T(\mathbf{w}_0).$$

Then $\mathbf{X}_k$ is invariant for $\mathscr{L}$ and ξ_k is an eigenvalue of $\mathscr{L}$ on $\mathbf{X}_k$ if and only if ξ is an eigenvalue of the matrix

$$A_k = \begin{pmatrix} -\mu_k + \dfrac{3\alpha^2-5}{1+\alpha^2} & -\dfrac{4\alpha}{1+\alpha^2} \\ \dfrac{2b\alpha^2}{1+\alpha^2} & -d\mu_k - \dfrac{b\alpha}{1+\alpha^2} \end{pmatrix}.$$

Further,

$$\det(A_k) = d\mu_k\left(\mu_k - \frac{3\alpha^2-5}{1+\alpha^2}\right) + \frac{b\alpha}{1+\alpha^2}(\mu_k+5),$$

$$\mathrm{Tr}(A_k) = -(1+d)\mu_k - \frac{(b+4)\alpha}{1+\alpha^2} < 0.$$

If $\mu_1 \geq (3\alpha^2-5)/(1+\alpha^2)$ then $\det(A_k) > 0$ for all $k \geq 0$.

If $\mu_1 < (3\alpha^2-5)/(1+\alpha^2)$, then let $p \geq 1$ be the largest positive integer such that

$$0 = \mu_0 < \mu_1 \leq \mu_2 \leq \cdots \leq \mu_p < \frac{3\alpha^2-5}{1+\alpha^2} \leq \mu_{p+1} \leq \mu_{p+2} \leq \cdots.$$

and let

$$D = \frac{b\alpha}{1+\alpha^2} \min_{1\leq k\leq p} \frac{\mu_k+5}{\frac{3\alpha^2-5}{1+\alpha^2} - \mu_k}. \tag{9.91}$$

Following a similar approach to that in Theorems 9.3 and 9.5 we find:

Theorem 9.31 *Assume $b\alpha > 3\alpha^2 - 5$.*

(i) If one of the following holds:

(i1) $\mu_1 \geq (3\alpha^2-5)/(1+\alpha^2)$.
(i2) $\mu_1 < (3\alpha^2-5)/(1+\alpha^2)$ *and* $d < D$.
then the uniform steady state (9.88) is asymptotically stable.

(ii) If $\mu_1 < (3\alpha^2-5)/(1+\alpha^2)$ and $d > D$ then (9.88) is stable for the ODE system (9.89) but unstable for (9.81), so Turing instabilities occur.

Proof. Remark first that condition $b\alpha > 3\alpha^2 - 5$ is equivalent to the stability of (9.88) for the ODE system (9.89).

(i) both conditions (i1) and (i2) imply $\mathrm{tr}(A_k) < 0 < \det(A_k)$ for all $k \geq 0$ so the spectrum of the linearized operator $\mathscr{L}$ lies in the region $\{z \in \mathbb{C} : \mathrm{Re}(z) < -c\}$, where $c > 0$ is a positive real number. By standard methods (see e.g. Theorem 5.1.1. in [110, Theorem 5.11]) it follows that (9.88) is asymptotically stable for (9.81).

(ii) Let $k \geq 1$ be the index that achieves the minimum in (9.91). Then $\det(A_k) < 0$ which by Corollary 5.1 in [110] implies that (9.88) is unstable for (9.81). This concludes the proof. □

9.4.3 A Priori Estimates for Stationary Solutions

In this section we shall be concerned with the steady state system corresponding to (9.81), that is,

$$\begin{cases} -\Delta u = a - u - \dfrac{4uv}{1+u^2} & \text{in } \Omega, \\ -d\Delta v = b\left(u - \dfrac{uv}{1+u^2}\right) & \text{in } \Omega, \\ \dfrac{\partial u}{\partial \nu}(x) = \dfrac{\partial u}{\partial \nu}(x) = 0 & \text{on } \partial\Omega. \end{cases} \tag{9.92}$$

Using Theorem 1.1 we have:

Lemma 9.32 *Any solution (u,v) of (9.46) satisfies*

$$\frac{a}{5+4a^2} < u < a \qquad in\ \Omega, \tag{9.93}$$

and

$$1+\left(\frac{a}{5+4a^2}\right)^2 < v < 1+a^2 \qquad in\ \Omega. \tag{9.94}$$

For any solution (u,v) of (9.46) we denote by $\bar{u}$ and $\bar{v}$ the average over Ω of u respectively v. Then, by integrating the two equations in (9.92) we find $\bar{u} = a$.

Let now

$$\phi = u - \bar{u}, \quad \psi = v - \bar{v}.$$

Our next result provides energy estimates for ϕ and ψ.

Proposition 9.33 *Let (u,v) be a nonconstant solution of (9.46). Then:*

(i) $\int \phi\psi > 0$ *and* $\int \nabla\phi \cdot \nabla\psi > 0$;

(ii) $\dfrac{3d^2\mu_1^2}{b^2(3\mu_1^2+15\mu_1+20)} \le \dfrac{\|\nabla\phi\|_2^2}{\|\nabla\psi\|_2^2} \le \dfrac{16d^2}{b^2}$;

(iii) $\dfrac{\mu_1^3}{(\mu_1+1)(2\mu_1^2+2\mu_1+1)}\left(\dfrac{d_2}{d_1}\right)^2 \le \dfrac{\|\nabla\phi\|_2^2+\|\phi\|_2^2}{\|\nabla\psi\|_2^2+\|\psi\|_2^2} \le \dfrac{\mu_1+1}{\mu_1}\left(\dfrac{d_2}{d_1}\right)^2.$

Proof. (i) From (9.92) we have

$$-\Delta(bu-4dv) = b(a-5u) = -5b\phi \quad \text{in } \Omega. \tag{9.95}$$

We next multiply this equality by $bu-4dv$ and integrate over Ω. We obtain

$$\int |\nabla(bu-4dv)|^2 = -5b\int \phi(bu-4dv) = -5b^2\int \phi^2 + 20bd\int \phi\psi.$$

This yields

$$\int \phi\psi \geq \frac{b}{4d}\int \phi^2 > 0. \tag{9.96}$$

We now multiply (9.95) by ϕ and integrate over Ω. We have

$$-5b\int \phi^2 = \int \nabla\phi\nabla(bu-4dv) = b\int |\nabla\phi|^2 - 4d\int \nabla\phi\nabla\psi,$$

so

$$4d\int \nabla\phi\nabla\psi = 5b\int \phi^2 + b\int |\nabla\phi|^2 > 0. \tag{9.97}$$

(ii) We multiply (9.95) by $bu-4dv$ and integrate over Ω. Using (9.97) we find

$$\begin{aligned}16d^2\int |\nabla\psi|^2 &= b^2\int |\nabla\phi|^2 + 5b^2\int \phi^2 + 20bd\int \phi\psi \\ &\leq b^2\int |\nabla\phi|^2 + 5b^2\int \phi^2 + 20bd\left(\frac{b}{3d\mu_1}\int \phi^2 + \frac{3d\mu_1}{4d}\int \psi^2\right) \\ &= b^2\int |\nabla\phi|^2 + \left(5b^2 + \frac{4b^2}{3\mu_1}\right)\int \phi^2 + 15d^2\mu_1\int \psi^2.\end{aligned}$$

Next, by Poincaré's inequality (9.60) we derive

$$16d^2\int |\nabla\psi|^2 \leq \left[b^2 + \frac{1}{\mu_1}\left(5b^2 + \frac{4b^2}{3\mu_1}\right)\right]\int |\nabla\phi|^2 + 15d^2\int |\nabla\psi|^2,$$

so

$$\int |\nabla\psi|^2 \leq \frac{b^2(3\mu_1^2+15\mu_1+20)}{3\mu_1^2}\int |\nabla\phi|^2,$$

which implies the first inequality in (ii). For the second inequality in (ii) we start from

$$0 \leq \int |\nabla(bu-4dv)|^2 = 16d^2\int |\nabla\psi|^2 - 10b^2\int \phi^2 - b^2\int |\nabla\phi|^2,$$

which yields

$$16d^2\int |\nabla\psi|^2 - b^2\int |\nabla\phi|^2 \geq 0,$$

and so

$$\frac{\|\nabla\phi\|_2^2}{\|\nabla\psi\|_2^2} \leq \frac{16d^2}{b^2}.$$

(iii) By Poincaré's inequality we find

$$\frac{\mu_1}{\mu_1+1}\frac{\|\nabla\phi\|_2^2}{\|\nabla\psi\|_2^2} \leq \frac{\|\nabla\phi\|_2^2+\|\phi\|_2^2}{\|\nabla\psi\|_2^2+\|\psi\|_2^2} \leq \frac{\mu_1+1}{\mu_1}\frac{\|\nabla\phi\|_2^2}{\|\nabla\psi\|_2^2}.$$

It only remains to use (ii) and the above estimates to deduce (iii). This finishes the proof. □

9.4.4 Nonexistence Results

Theorem 9.34 *(i) Let $a,b>0$ be fixed. Then, there exists $D=D(a,b,\mu_1)>0$ such that if $0<d<D$ then (9.92) has no nonconstant solutions.*

(ii) Let $a,d>0$, then there exists $B=B(a,d,\mu_1)>0$ such that (9.92) has no nonconstant solutions if $b>B$.

Proof. Assume that (u,v) is a nonconstant solution of (9.92). Let us first multiply by ψ in the second equation of (9.92). Integrating over Ω we find

$$\begin{aligned}
d\int|\nabla\psi|^2 &= \int\phi\psi-\int\frac{uv}{1+u^2}\psi\\
&=\int\phi\psi-\int\left(\frac{uv}{1+u^2}-\frac{u\bar{v}}{1+u^2}\right)\psi-\int\left(\frac{u\bar{v}}{1+u^2}-\frac{\bar{u}\bar{v}}{1+\bar{u}^2}\right)\psi\\
&=\int\phi\psi-\int\frac{u}{1+u^2}\psi^2+\int\frac{(u\bar{u}-1)\bar{v}}{(1+u^2)(1+\bar{u}^2)}\psi.
\end{aligned}$$

From Theorem 9.32, there exist $C_1,C_2>0$ depending on a such that

$$d\int|\nabla\psi|^2a\le C_1\int|\phi||\psi|-C_2\int\psi^2.$$

By Poincaré's inequality and Theorem 9.33(ii) we now derive

$$\begin{aligned}
d\int|\nabla\psi|^2 &\le C_1\left(\frac{C_2}{C_1}\int\psi^2+\frac{C_1}{4C_2}\int\phi^2\right)-C_2\int|\psi|^2\\
&=\frac{C_1^2}{4C_2}\int\phi^2\le\frac{C_1^2}{4C_2\mu_1}\int|\nabla\phi|^2\\
&\le\frac{C}{\mu_1}\frac{16d^2}{b^2}\int|\nabla\psi|^2,
\end{aligned}$$

where $C>0$ depends only on a. This last inequality yields

$$d\int|\nabla\psi|^2\le\frac{C}{\mu_1}\frac{16d}{b^2}\int|\nabla\psi|^2.$$

Now, it is clear that if d is small or $b>0$ is large enough the above inequality implies ψ =const so u and v are both constant solutions, which contradicts our assumption. □

Theorem 9.35 *Assume that* $a \le 5\sqrt{3}$ *and*

$$\frac{b}{d} > \frac{8a}{5} - \frac{25}{a}.$$

Then, (9.92) has no nonconstant solutions.

Proof. Let (u,v) be a nonconstant solution of (9.92). We multiply the first equation in (9.92) by $(1+u^2)\phi$ and integrate over Ω. We obtain

$$\int \nabla\phi\nabla(\phi+u^2\phi) = \int (a-u)(1+u^2)\phi - 4\int uv\phi,$$

and hence

$$\begin{aligned}\int (1+2u\phi+u^2)|\nabla\phi|^2 &= \int (a-u)\phi + \int (a-u)u^2\phi - 4\int uv\phi \\ &= -\int \phi^2 + \int (a-u)u^2\phi - 4\int uv\phi.\end{aligned} \tag{9.98}$$

Remark that

$$\begin{aligned} 1+2u\phi+u^2 &= 3u^2 - \frac{2a}{5}u + 1 \\ \int (a-u)u^2\phi &= \int \left(-u^2 + \frac{4a}{5}u + \frac{4a^2}{25}\right)\phi^2 \\ \int uv\phi &= \int \left(u - \frac{a}{5}\right)v\phi + \frac{a}{5}\int v\phi = \int v\phi^2 + \frac{a}{5}\phi\psi. \end{aligned}$$

Thus, from (9.98) we find

$$\int \left(3u^2 - \frac{2a}{5}u + 1\right)|\nabla\phi|^2 = \int \left(-u^2 + \frac{4a}{5}u + \frac{4a^2}{25} - 1\right)\phi^2 - 4\int v\phi^2 + \frac{4a}{5}\int \phi\psi.$$

From Theorem 9.33 we have $v \ge 1$ in Ω so

$$\int \left(3u^2 - \frac{2a}{5}u + 1\right)|\nabla\phi|^2 = \int \left(-u^2 + \frac{4a}{5}u + \frac{4a^2}{25} - 5\right)\phi^2 + \frac{4a}{5}\int \phi\psi. \tag{9.99}$$

We next use the fact that

$$3u^2 - \frac{2a}{5}u + 1 \ge 1 - \frac{a^2}{75}, \quad -u^2 + \frac{4a}{5}u + \frac{4a^2}{25} \le \frac{8a^2}{25}.$$

Also, from (9.96) we find

$$\frac{4a}{5}\int \phi\psi \ge \frac{ab}{5d}\int \phi^2.$$

Using all these estimates in (9.99) we now obtain

$$\left(1-\frac{a^2}{75}\right)\int|\nabla\phi|^2 \le \left(\frac{8a^2}{25}-5-\frac{ab}{5d}\right)\int\phi^2.$$

It is now clear that under the condition on a, b and d stated in theorem 9.35 the left-hand side of the above inequality is positive while the right-hand side is negative, which is a contradiction. □

9.4.5 Existence

We first formulate the steady state system (9.92) in a framework in which the Leray–Schauder degree theory can be easily applied. Remark first that (9.92) is equivalent to

$$\mathscr{F}(\mathbf{w}) = \mathbf{w} - (\mathbf{I}-\Delta)^{-1}(\mathscr{G}(\mathbf{w})+\mathbf{w}), \qquad \mathbf{w}\in\mathbf{X}^+,$$

where

$$\mathscr{G}(\mathbf{w}) = \begin{bmatrix} a-u-\frac{4uv}{1+u^2} \\ \\ \frac{b}{d}\left(u-\frac{uv}{1+u^2}\right) \end{bmatrix}.$$

Further, if $\mathbf{w}_0$ is the uniform steady state defined in (9.88), then

$$\nabla\mathscr{F}(\mathbf{w}_0) = \mathbf{I} - (\mathbf{I}-\Delta)^{-1}(\mathbf{I}+A),$$

where

$$A := \nabla\mathscr{G}(\mathbf{w}_0) = \frac{1}{1+\alpha^2}\begin{pmatrix} 3\alpha^2-5 & -4\alpha \\ \\ -\frac{2b\alpha^2}{d} & -\frac{b\alpha}{d} \end{pmatrix}.$$

If $\nabla\mathscr{F}(\mathbf{w}_0)$ is invertible, then the index of $\mathscr{F}$ at $\mathbf{w}_0$ is given by

$$\text{index}(\mathscr{F},\mathbf{w}_0) = (-1)^{\gamma}, \tag{9.100}$$

where γ denotes the number of negative eigenvalues of $\nabla\mathscr{F}(\mathbf{w}_0)$.

To better quantify the index of $\mathscr{F}$ at $\mathbf{w}_0$, let us introduce

$$P(a,b,d,\mu) = \det(\mu\mathbf{I}-A).$$

Then, it is easy to see that $\nabla\mathscr{F}(\mathbf{w}_0)$ is invertible if and only if $\mu_k I - A$ is invertible for all $k \ge 0$, that is

$$P(a,b,d,\mu_k) \neq 0 \quad \text{for all } k \geq 0.$$

Remark now that

$$P(a,b,d,\mu) = \mu^2 - \Big(\frac{3\alpha^2-5}{1+\alpha^2} - \frac{b\alpha}{d(1+\alpha^2)}\Big)\mu + \frac{5b\alpha^3}{d(1+\alpha^2)}.$$

Therefore, if

$$3\alpha^2 - 5 > \frac{b\alpha}{d} + 2\sqrt{\frac{5b\alpha}{d}(1+\alpha^2)}, \tag{9.101}$$

then the equation $P(a,b,d,\mu) = 0$ has exactly two positive solutions $\mu^{\pm}(a,b,d)$ given by

$$\mu^{\pm}(a,b,d) = \frac{1}{2}\left[\sigma^2 \pm \sqrt{\sigma^2 - \frac{20b\alpha}{d(1+\alpha^2)}}\right],$$

where

$$\sigma = \frac{3\alpha^2-5}{1+\alpha^2} - \frac{b\alpha}{d(1+\alpha^2)}.$$

Our main existence result is the following.

Theorem 9.36 *Assume that condition (9.101) holds and there exist $i > j \geq 0$ such that*

(i) $\mu_i < \mu^+(a,b,d) < \mu_{i+1}$ and $\mu_j < \mu^-(a,b,d) < \mu_{j+1}$.
(ii) $\sum_{k=j+1}^{i} m(\mu_k)$ is odd.

Then (9.92) has at least one nonconstant solution.

The next results that follow from Theorem 9.36 show that the Turing patterns may occur in the Lengyel–Epstein model (9.81) provided $a, 1/b$ and d are large.

Corollary 9.37 *Let $b,d > 0$ be fixed. Assume that all the eigenvalues μ_k have odd multiplicity. Then, there exists a sequence of intervals $\{(a_n,A_n)\}_n$ with $0 < a_n < A_n < a_{n+1} \to \infty$ (as $n \to \infty$) such that system (9.92) has at least one nonconstant solution for all $a \in \bigcup_{n\geq 1}(a_n,A_n)$.*

Proof. Note first that for large values of $a > 0$ condition (9.101) is satisfied. Further, since

$$\mu^+(a,b,d) \to \infty \quad \text{as } a \to \infty,$$

and

$$\mu^-(a,b,d) \to \infty \quad \text{as } a \to \infty,$$

one can find a sequence of intervals $\{(a_n,A_n)\}_n$ as in the statement of Corollary 9.38 such that

$$\sum_{\substack{k\geq 0,\\ \mu^-<\mu_k<\mu^+}} m(\mu_k) \quad \text{is odd} \tag{9.102}$$

for all $a \in \bigcup_{n\geq 1}(a_n,A_n)$. The conclusion follows now from Theorem 9.36. □

Corollary 9.38 *Let $d>0$, $a>\frac{5\sqrt{15}}{3}$ be fixed and assume that*

$$\sum_{\substack{k\geq 0,\\ 0<\mu_k<\frac{3\alpha^2-5}{1+\alpha^2}}} e(\mu_i) \quad \textit{is odd.} \tag{9.103}$$

Then, there exists $B>0$ such that the steady-state system (9.92) has at least one nonconstant solution for any $0<b<B$.

Proof. It is easy to see that (9.101) holds for small values of $b>0$. Also

$$0<\mu^-(a,b,d)<\mu^+(a,b,d)<\frac{3\alpha^2-5}{1+\alpha^2}$$

and

$$\mu^-(a,b,d)\to 0\,,\ \mu^+(a,b,d)\to\frac{3\alpha^2-5}{1+\alpha^2} \qquad \text{as } b\to\infty.$$

Therefore, for $b>0$ small enough, conditions (i) and (ii) in Theorem 9.36 are satisfied. □

Corollary 9.39 *Let $a>\frac{5\sqrt{15}}{3}$, $b>0$ be fixed and assume that (9.103) holds. Then there exists $D>0$ such that the steady-state system (9.92) has at least one nonconstant solution for any $d>D$.*

Proof. First, let us remark that for large $d>0$ condition (9.101) holds. Also for any a,b fixed we have

$$0<\mu^-(a,b,d)<\mu^+(a,b,d)<\frac{3\alpha^2-5}{1+\alpha^2}$$

and

$$\mu^-(a,b,d)\to 0\,,\ \mu^+(a,b,d)\to\frac{3\alpha^2-5}{1+\alpha^2} \qquad \text{as } d\to\infty.$$

Therefore, for $d>0$ large, conditions (9.103) and (i)–(ii) in Theorem 9.36 are fulfilled. This implied the existence of a nonconstant solution for large $d>0$. □

Combining Corollary 9.39 with Theorem 9.35 we now obtain

Corollary 9.40 *Let $b > 0$ be fixed.*

(i) If $0 < a \leq \frac{5\sqrt{15}}{3}$ and $0 < \frac{d}{b} < \frac{\sqrt{15}}{25}$ then (9.92) has no nonconstant solutions.
(ii) If $a > \frac{5\sqrt{15}}{3}$ and (9.103) holds, then there exists $D > 0$ such that (9.92) has at least one nonconstant solution for all $d > D$.

Chapter 10
Pattern Formation and the Gierer–Meinhardt Model in Molecular Biology

> Mathematics is a part of physics. Physics is an experimental science, a part of natural sciences. Mathematics is the part of physics where experiments are cheap.
>
> Vladimir Arnold (1937– 2010)

10.1 Introduction

In 1972 Gierer and Meinhardt [98] proposed a mathematical model for pattern formation of spatial tissue structures in morphogenesis, a biological phenomenon discovered by Trembley [196] in 1744. The mechanism behind the Gierer–Meinhardt's model is based on the existence of two chemical substances: a slowly diffusing activator and a rapidly diffusing inhibitor. The ratio of their diffusion rates is assumed to be small.

The model introduced by Gierer and Meinhardt reads as

$$\begin{cases} u_t = d_1 \Delta u - \alpha u + c\rho \dfrac{u^p}{v^q} + \rho_0 \rho & \text{in } \Omega \times (0,T), \\ v_t = d_2 \Delta v - \beta v + c' \rho' \dfrac{u^r}{v^s} & \text{in } \Omega \times (0,T), \end{cases} \tag{10.1}$$

subject to Neumann boundary conditions in a smooth bounded domain Ω. Here the unknowns u and v stand for the concentration of activator and inhibitor with the source distributions ρ and ρ' respectively. In system (10.1), d_1, d_2 are the diffusion coefficients and $\alpha, \beta, c, c', \rho_0$ are positive constants. The exponents $p, q, r, s > 0$

M. Ghergu and V. Rădulescu, *Nonlinear PDEs*, Springer Monographs in Mathematics, DOI 10.1007/978-3-642-22664-9_10, © Springer-Verlag Berlin Heidelberg 2012 verify the relation

$$qr > (p-1)(s+1) > 0.$$

The model introduced by Gierer and Meinhardt has been used with satisfactory quantitative results for modelling the head regeneration process of hydra, an animal of a few millimeters in length, consisting of 100,000 cells of about 15 different types and having a polar structure.

The Gierer–Meinhardt system originates in the Turing one [197] introduced in 1952 as a mathematical model for the development of complex organisms from a single cell. It has been emphasized that localized peaks in concentration of chemical substances, known as inducers or morphogenesis, could be responsible for a group of cells developing differently from the surrounding cells. Turing discovered through linear analysis that a large difference in relative size of diffusivities for activating and inhibiting substances carries instability of the homogeneous, constant steady state, thus leading to the presence of nontrivial, possibly stable stationary configurations.

A global existence result for a more general system than (10.1) is given in the recent paper of Jiang [112]. It has also been shown that the dynamics of the system (10.1) exhibit various interesting behaviors such as periodic solutions, unbounded oscillating global solutions, and finite time blow-up solutions. We refer the reader to Ni, Suzuki, and Takagi [150] for a description of the dynamics concerning the system (10.1).

Many works have been devoted to the study of the steady-state solutions of (10.1), that is, solutions of the stationary system

$$\begin{cases} d_1\Delta u - \alpha u + c\rho\dfrac{u^p}{v^q} + \rho_0\rho = 0 & \text{in } \Omega, \\ d_2\Delta v - \beta v + c'\rho'\dfrac{u^r}{v^s} = 0 & \text{in } \Omega, \end{cases} \tag{10.2}$$

subject to Neumann boundary conditions. The main difficulty in the treatment of (10.2) is the lack of variational structure. Another direction of research is to consider the *shadow system* associated to (10.2), an idea due to Keener [114]. This system is obtained by dividing by d_2 in the second equation and then letting $d_2 \to \infty$. It has been shown that nonconstant solutions of the shadow system associated to (10.2) exhibit interior or boundary concentrating points. Among the large number of works in this direction we refer the interested reader to [151–153], [204], [205] as well as to the survey papers of Ni [148], [149].

In the following, new features of Gierer–Meinhardt type systems are emphasized. More exactly, we shall be concerned with systems of the type

$$\begin{cases} \Delta u - \alpha u + \dfrac{u^p}{v^q} + \rho(x) = 0,\ u > 0 & \text{in } \Omega, \\ \Delta v - \beta v + \dfrac{u^r}{v^s} = 0,\ v > 0 & \text{in } \Omega, \\ u = 0,\ v = 0 & \text{on } \partial\Omega, \end{cases} \tag{10.3}$$

in a smooth bounded domain $\Omega \subset \mathbb{R}^N$ $(N \geq 1)$. Here u and v represent the concentration of the activator and inhibitor and $\rho \in C^{0,\gamma}(\overline{\Omega})$ $(0 < \gamma < 1)$ represents the source distribution of the activator. We assume that $\rho \geq 0$ in Ω, $\rho \not\equiv 0$ and α, β are nonnegative real numbers. The case $\rho \equiv 0$ is more delicate and involves a more careful analysis of the Gierer–Meinhardt system. This situation has been analyzed in the recent works [38], [39], [150], [153], [204], [205].

We are mainly interested in the case where the activator and inhibitor have different source terms, that is, $(p,q) \neq (r,s)$.

Let us notice that the homogeneous Dirichlet boundary condition in (10.3) (instead of Neumann's one as in (10.2)) turns the system singular in the sense that the nonlinearities $\frac{u^p}{v^q}$ and $\frac{u^r}{v^s}$ become unbounded around the boundary.

The existent results in the literature for (10.3) concern the case of common sources of the concentrations, that is, $(p,q) = (r,s)$. If $p = q = r = s = 1$ and $\rho \equiv 0$, the system (10.3) was studied in Choi and McKenna [38]. In Kim [116], [117] it is studied the system (10.3) with $p = r$ and $q = s$. In the case of common sources, a *decouplization* of the system is suitable in order to provide *a priori* estimates for the unknowns u and v. More precisely, if $p = r$ and $q = s$ then, subtracting the two equations in (10.3) and letting $w = u - v$ we get the equivalent form

$$\begin{cases} \Delta w - \alpha w + (\beta - \alpha) w v + \rho(x) = 0 & \text{in } \Omega, \\ \Delta v - \beta v + \dfrac{(v+w)^p}{v^q} = 0 & \text{in } \Omega, \\ v = w = 0 & \text{on } \partial\Omega. \end{cases} \tag{10.4}$$

Thus, the study of system (10.3) amounts to the study of (10.4) in which the first equation is linear. This is more suitable to derive upper and lower barriers for u and v (see [38], [116], [117]). For more applications of the decouplization method in the context of elliptic systems we refer the reader to [118]. We also mention here the paper of Choi and McKenna [39] where the existence of radially symmetric solutions in the case $p = r > 1$, $q = 1$, $s = 0$ and $\Omega = B_1 \subset \mathbb{R}^2$ is discussed. In [39], *a priori* bounds for concentrations u and v are obtained through sharp estimates for the associated Green's function.

In our case, such a decouplization is not possible due to the fact that $(p,q) \neq (r,s)$. In order to overcome this, we shall exploit the boundary behavior of solutions of single singular equations associated to system (10.3). In turn, this approach requires uniqueness or suitable comparison principles for single singular equations that come from our system. These features are usually associated with nonlinearities having a sublinear growth and that is why we restrict our attention to the case $p < 1$. Our results extend those presented in [94], [93] and give precise answers to some questions raised in Choi and McKenna [38], [39] and Kim [116], [117]. Also the approach we give here enables us to deal with various type of exponents. For instance, we shall consider the case $p < 0$ (see Theorems 10.15 and 10.16) which means that the nonlinearity in the first equation of (10.3) is singular in both its variables u and v. Furthermore, these results can be successfully applied to treat the case $-1 < s \leq 0$ (see Remark 26).

We are interested in the following range of exponents

$$-\infty < p < 1,$$

$$q, r, s > 0 \text{ and } s \geq r - 1. \tag{10.5}$$

In our approach we do not require any order relation between the nonnegative numbers α and β. Also we do not impose any growth condition on the source distribution $\rho(x)$ of the activator. A major role in our analysis will be played by the number

$$\sigma = \min\left\{1, \frac{2+r}{1+s}\right\}. \tag{10.6}$$

10.2 Some Preliminaries

Throughout this section $\|\cdot\|_\infty$ denotes the $L^\infty(\Omega)$ norm. Also we denote by λ_1 and φ_1 the first eigenvalue and the first normalized eigenfunction of $-\Delta$ in $H_0^1(\Omega)$ with $\|\varphi_1\|_\infty = 1$. As is well known, $\varphi_1 \in C^2(\overline{\Omega})$, $\varphi_1 > 0$ in Ω, and there exists $C > 0$ such that

$$Cd(x) \le \varphi_1 \le \frac{1}{C} d(x) \quad \text{in } \Omega. \tag{10.7}$$

We also recall the following useful result which is due to Lazer and McKenna.

Lemma 10.1 (Lazer and McKenna [130]) $\int_\Omega \varphi_1^\tau dx < \infty$ *if and only if* $\tau > -1$.

Proposition 10.2 *Let* $0 \le p < 1$, $0 < q < p+1$ *and* $a \in C^{0,\gamma}(\Omega)$ $(0 < \gamma < 1)$ *be such that*

$$a_1 \varphi_1^{-q}(x) \le a(x) \le a_2 \varphi_1^{-q}(x) \quad \text{in } \Omega, \tag{10.8}$$

for some $a_1, a_2 > 0$. *Then, the problem*

$$\begin{cases} \Delta u - \alpha u + a(x) u^p + \rho(x) = 0 & \text{in } \Omega, \\ u > 0 & \text{in } \Omega, \\ u = 0 & \text{on } \partial\Omega, \end{cases} \tag{10.9}$$

has a unique solution $u \in C^2(\Omega) \cap C(\overline{\Omega})$. *Moreover, there exist* $m_1, m_2 > 0$ *such that*

$$m_1 \varphi_1 \le u \le m_2 \varphi_1 \quad \text{in } \Omega. \tag{10.10}$$

Proof. Let w be the unique solution of

$$\begin{cases} \Delta w - \alpha w + \rho(x) = 0 & \text{in } \Omega, \\ w > 0 & \text{in } \Omega, \\ w = 0 & \text{on } \partial\Omega. \end{cases} \tag{10.11}$$

By standard elliptic arguments and the maximum principle we have $w \in C^2(\overline{\Omega})$. Obviously $\underline{u} := w$ is a subsolution of (10.9). Furthermore, by virtue of (10.7) we can find $c_1, c_2 > 0$ such that

$$c_1 \varphi_1 \le w \le c_2 \varphi_1 \quad \text{in } \Omega. \tag{10.12}$$

Since $q < p+1$, by a result in Wei [206], there exists $h \in C^2(0,1) \cap C^1[0,1]$ such that

$$\begin{cases} -h''(t) = t^{-q} h^p(t), & \text{for all } 0 < t < 1, \\ h > 0 & \text{in } (0,1), \\ h(0) = h(1) = 0. \end{cases}$$

Using the fact that $h'(0) > 0$ we have

$$c_3 t \le h(t) \le c_4 t, \tag{10.13}$$

for $t > 0$ small enough and for some $c_3, c_4 > 0$. Furthermore, we may find $c > 0$ such that $h'(c\varphi_1) > 0$ in $\overline{\Omega}$.

We are looking for a supersolution of (10.9) in the form $\overline{u} := Mh(c\varphi_1) + w$, for $M > 1$ large enough. For this purpose we have to check that the inequality $-\Delta\overline{u} + \alpha\overline{u} \geq a(x)\overline{u}^p + \rho(x)$ holds in Ω provided that $M > 1$ is sufficiently large.

We have

$$\begin{aligned} -\Delta\overline{u} + \alpha\overline{u} &\geq -\Delta(Mh(c\varphi_1)) + \rho(x) \\ &= Mc^{2-q}\varphi_1^{-q}h^p(c\varphi_1)|\nabla\varphi_1|^2 + M\lambda_1 c\varphi_1 h'(c\varphi_1) + \rho(x) \quad \text{in } \Omega. \end{aligned} \tag{10.14}$$

By (10.13) we may write

$$-\Delta\overline{u} + \alpha\overline{u} \geq Mc^{2+p-q}c_3^p\varphi_1^{p-q}|\nabla\varphi_1|^2 + M\lambda_1 c\varphi_1 h'(c\varphi_1) + \rho(x) \quad \text{in } \Omega. \tag{10.15}$$

On the other hand, by (10.8), (10.12) and (10.13) we have

$$a(x)\overline{u}^p \leq a_2\varphi_1^{-q}(Mh(c\varphi_1) + w)^p \leq a_2\varphi_1^{p-q}(Mcc_4 + c_2)^p \quad \text{in } \Omega. \tag{10.16}$$

Using Hopf's maximum principle, there exist $\omega \subset\subset \Omega$ and $\delta > 0$ such that

$$|\nabla\varphi_1| > \delta \text{ in } \Omega\setminus\omega \quad \text{and } \varphi_1 > \delta \text{ in } \omega. \tag{10.17}$$

Since $0 \leq p < 1$, we may choose $M > 1$ such that

$$Mc^{2+p-q}c_3^p\delta^2 > a_2(Mcc_4 + c_2)^p, \tag{10.18}$$

$$M\lambda_1 c \min_{\overline{\omega}} \varphi_1 h'(c\varphi_1) \geq a_2(Mcc_4 + c_2)^p \max_{\overline{\omega}} \varphi_1^{p-q}. \tag{10.19}$$

Combining (10.15), (10.16) and (10.18) we obtain

$$-\Delta\overline{u} + \alpha\overline{u} \geq Mc^{2+p-q}c_3^p\varphi_1^{p-q}|\nabla\varphi_1|^2 + \rho(x) \geq a(x)\overline{u}^p + \rho(x) \quad \text{in } \Omega\setminus\omega. \tag{10.20}$$

Furthermore, by (10.15), (10.16) and (10.19) we deduce

$$-\Delta\overline{u} + \alpha\overline{u} \geq M\lambda_1 c\varphi_1 h'(c\varphi_1) + \rho(x) \geq a(x)\overline{u}^p + \rho(x) \quad \text{in } \omega. \tag{10.21}$$

Now the claim follows by (10.20) and (10.21). Thus, the problem (10.9) has a solution $u \in C^2(\Omega) \cap C(\overline{\Omega})$ such that $\underline{u} \leq u \leq \overline{u}$ in Ω. By (10.12) and (10.13) we obtain the estimate (10.10). This also implies that

$$C_1\varphi_1^{p-q} \leq a(x)u^p \leq C_2\varphi_1^{p-q} \quad \text{in } \Omega,$$

for some $C_1, C_2 > 0$. Since $p-q > -1$, by Lemma 10.1 we get $a(x)u^p \in L^1(\Omega)$ which finally yields $\Delta u \in L^1(\Omega)$. Now the uniqueness follows by Theorem 1.2. This concludes the proof of Proposition 10.2. □

We next consider the problem

$$\begin{cases} \Delta v - \beta v + a(x)v^{-s} + b(x) = 0 & \text{in } \Omega, \\ v > 0 & \text{in } \Omega, \\ v = 0 & \text{on } \partial\Omega, \end{cases} \tag{10.22}$$

where $a \in C^{0,\gamma}(\Omega)$ $(0 < \gamma < 1)$ satisfies

$$a_1 \varphi_1^r(x) \le a(x) \le a_2 \varphi_1^r(x) \quad \text{in } \Omega, \tag{10.23}$$

for some $a_1, a_2 > 0$ and $r \in \mathbb{R}$. We also assume that $b \in C^{0,\gamma}(\overline{\Omega})$, $\beta \ge 0$ and $s > 0$.

For convenience, let us introduce $\Gamma_{s,r} : (0,\infty) \to (0,\infty)$ defined by

$$\Gamma_{s,r}(t) = \begin{cases} t & , \quad \text{if } s < r+1, \\ t(1+|\log t|)^{1/(1+s)} & , \quad \text{if } s = r+1, \\ t^{(2+r)/(1+s)} & , \quad \text{if } s > r+1, \end{cases} \tag{10.24}$$

for all $r > -2$ and $s > 0$. It is easy to see that

$$\Gamma_{s,r}(t) \ge t^{\sigma} \quad \text{for all } t > 0, \tag{10.25}$$

where σ is defined in (10.6). Moreover, for all $m > 0$ there exists $m_1, m_2 > 0$ such that

$$m_1 \Gamma_{s,r}(t) \le \Gamma_{s,r}(mt) \le m_2 \Gamma_{s,r}(t) \quad \text{for all } t > 0. \tag{10.26}$$

Proposition 10.3 (i) *If $r \le -2$ then the problem (10.22) has no classical solutions.*

(ii) *If $r > -2$ then the problem (10.22) has a unique solution $u \in C^2(\Omega) \cap C(\overline{\Omega})$. Moreover, there exist $c_1, c_2 > 0$ such that*

$$c_1 \Gamma_{r,s}(\varphi_1) \le v \le c_2 \Gamma_{r,s}(\varphi_1) \quad \text{in } \Omega. \tag{10.27}$$

A general nonexistence result for singular elliptic equations with unbounded potentials can be found in [66]. Also a nonexistence result in the case $b \equiv 0$, $\beta = 0$ and $r \le -2$ is presented in [214]*Theorem 1.2. Concerning the existence part in Proposition 10.3, a similar result can be found in [107] in the case $b \equiv 0$, $\beta = 0$ and $r \ge 0$. We shall give here a different proof which relies on a direct construction of a sub and supersolution. This will provide the estimate (10.27).

Proof. (i) Assume that there exist $r \geq -2$ and $v \in C^2(\Omega) \cap C(\overline{\Omega})$ a classical solution of (10.22). For $0 < \varepsilon < 1$ consider the problem

$$\begin{cases} \Delta z - \beta z + a_1(\varphi_1 + \varepsilon)^r (z+\varepsilon)^{-s} = 0 & \text{in } \Omega, \\ z > 0 & \text{in } \Omega, \\ z = 0 & \text{on } \partial\Omega. \end{cases} \tag{10.28}$$

Obviously, $\underline{z} = 0$ is a subsolution and $\overline{z} = v$ is a supersolution of (10.28). Hence, for all $0 < \varepsilon < 1$ there exists $z_\varepsilon \in C^2(\overline{\Omega})$ a solution of (10.28) such that $0 < z_\varepsilon \leq v$ in Ω. Multiplying by φ_1 in (10.28) and then integrating over Ω we get

$$(\beta + \lambda_1) \int_\Omega z_\varepsilon \varphi_1 dx = a_1 \int_\Omega \varphi_1 (\varphi_1 + \varepsilon)^r (z_\varepsilon + \varepsilon)^{-s} dx.$$

Since $z_\varepsilon \leq v$ in Ω, the above equality yields

$$(\beta + \lambda_1) \int_\Omega v \varphi_1 dx \geq a_1 (1 + \|v\|_\infty)^{-s} \int_\Omega \varphi_1 (\varphi_1 + \varepsilon)^r dx.$$

This implies

$$\int_\omega \varphi_1 (\varphi_1 + \varepsilon)^r dx < M \qquad \text{for all } \omega \subset\subset \Omega,$$

where $M > 0$ does not depend on ε. Passing to the limit with $\varepsilon \to 0$ in the above inequality we find $\int_\omega \varphi_1^{1+r} dx < M$, for all $\omega \subset\subset \Omega$, that is, $\int_\Omega \varphi_1^{1+r} dx < \infty$. Since $r \geq -2$, the last inequality contradicts Lemma 10.1. Therefore, the problem (10.22) has no classical solutions if $r \geq -2$.

(ii) Let $r > -2$ and $s \geq r - 1$. According to [189]*Theorem 1, there exists $H \in C^2(0,1) \cap C[0,1]$ such that

$$\begin{cases} -H''(t) = t^r H^{-s}(t), & \text{for all } 0 < t < 1, \\ H > 0 & \text{in } (0,1), \\ H(0) = H(1) = 0. \end{cases} \tag{10.29}$$

Since H is concave, there exists $H'(0+) > 0$. Hence, taking $0 < \eta < 1$ sufficiently small, we can assume that $H' > 0$ in $(0, \eta)$. From [189]*p. 904 (see also Theorem 3.5 in [66]), there exist $c_1, c_2 > 0$ such that

$$c_1 \Gamma_{s,r}(t) \leq H(t) \leq c_2 \Gamma_{s,r}(t) \quad \text{in } (0, \eta). \tag{10.30}$$

As a consequence of (10.30) and the fact that $s \geq r - 1$ we derive

$$H(t)^{s+1} \leq c_3 t^r \quad \text{in } (0, \eta), \tag{10.31}$$

for some positive constant $c_3 > 0$. Let $c > 0$ be such that $c\varphi_1 < \eta$ in Ω. We claim that we can find $0 < m < 1$ small enough such that $\underline{v} := mH(c\varphi_1)$ satisfies

$$2\beta \underline{v}^{1+s} \leq a(x) \quad \text{in } \Omega \quad \text{and} \quad -\Delta \underline{v} \leq \frac{1}{2} a(x) \underline{v}^{-s} \quad \text{in } \Omega. \tag{10.32}$$

Then, from (10.32) we deduce

$$\Delta \underline{v} - \beta \underline{v} + a(x) \underline{v}^{-s} \geq 0 \quad \text{in } \Omega, \tag{10.33}$$

that is, $\underline{v}$ is a subsolution of (10.22).

By virtue of (10.23) and (10.31) we have

$$2\beta \underline{v}^{1+s} = 2\beta m^{1+s} H^{1+s}(c\varphi_1) \leq 2\beta m^{1+s} c_3 (c\varphi_1)^r \leq \frac{2\beta m^{1+s} c^r c_3}{a_1} a(x) \quad \text{in } \Omega.$$

Let us chose now $m > 0$ such that $2\beta m^{1+s} c^r c_3 < a_1$. This concludes the first inequality in (10.32).

In order to establish the second inequality in (10.32), a straightforward computation yields

$$\begin{aligned} -\Delta \underline{v} &= -mc^2 |\nabla \varphi_1|^2 H''(c\varphi_1) + m\lambda_1 c\varphi_1 H'(c\varphi_1) \\ &= mc^{2+r} \varphi_1^r |\nabla \varphi_1|^2 H^{-s}(c\varphi_1) + m\lambda_1 c\varphi_1 H'(c\varphi_1) \\ &= m^{1+s} c^{2+r} \varphi_1^r |\nabla \varphi_1|^2 \underline{v}^{-s} + m\lambda_1 c\varphi_1 H'(c\varphi_1) \quad \text{in } \Omega. \end{aligned} \tag{10.34}$$

Since H' is decreasing on $(0,\eta)$, it follows that $tH'(t) \leq H(t)$ for all $t \in (0,\eta)$. Furthermore, from (10.31) we deduce

$$c\varphi_1 H'(c\varphi_1) \leq H(c\varphi_1) \leq c_3 (c\varphi_1)^r H^{-s}(c\varphi_1) \quad \text{in } \Omega. \tag{10.35}$$

Combining (10.34) and (10.35), for $0 < m < 1$ we obtain

$$\begin{aligned} -\Delta \underline{v} &\leq m^{1+s} c^{2+r} \varphi_1^r |\nabla \varphi_1|^2 \underline{v}^{-s} + m\lambda_1 c^r c_3 \varphi_1^r H^{-s}(c\varphi_1) \\ &\leq mc^{2+r} \varphi_1^r |\nabla \varphi_1|^2 \underline{v}^{-s} + m\lambda_1 c^r c_3 \varphi_1^r \underline{v}^{-s} \\ &= mc^r \varphi_1^r \underline{v}^{-s} (c^2 |\nabla \varphi_1|^2 + c_3 \lambda_1) \\ &\leq \frac{mc^r}{a_1} (c^2 \|\nabla \varphi_1\|_\infty^2 + c_3 \lambda_1) a(x) \underline{v}^{-s} \quad \text{in } \Omega. \end{aligned}$$

Now, it suffices to choose $0 < m < 1$ such that $\frac{mc^r}{a_1}(c^2\|\nabla\varphi_1\|_\infty^2 + c_3\lambda_1) < \frac{1}{2}$. This establishes the second inequality in (10.32) and the fact that $\underline{v}$ is a subsolution of (10.22).

Next we provide a supersolution $\overline{v}$ of (10.22) such that $\underline{v} \le \overline{v}$ in Ω. To this aim we first claim that there exists $M > 1$ large enough such that $z := MH(c\varphi_1)$ satisfies

$$\Delta z + a(x)z^{-s} \le 0 \quad \text{in } \Omega. \tag{10.36}$$

As before we have

$$\Delta z = -M^{1+s}c^{2+r}\varphi_1^r|\nabla\varphi_1|^2 z^{-s} - M\lambda_1 c\varphi_1 H'(c\varphi_1) \quad \text{in } \Omega. \tag{10.37}$$

Let $\omega \subset\subset \Omega$ and $\delta > 0$ be such that (10.17) holds and let us consider $M > 1$ such that

$$M^{1+s}c^{2+r}\delta^2 > a_2, \tag{10.38}$$

$$M^{1+s}c\lambda_1 \min_{\overline{\omega}} \varphi_1 H'(c\varphi_1) \ge a_2 \max_{\overline{\omega}} \varphi_1^r H^{-s}(c\varphi_1). \tag{10.39}$$

Then, as in the proof of Proposition 10.2, by (10.17) and (10.38)–(10.39) we get

$$\begin{aligned}
\Delta z + a(x)z^{-s} &\le -M^{1+s}c^{2+r}\varphi_1^r|\nabla\varphi_1|^2 z^{-s} + a(x)z^{-s} \\
&\le -M^{1+s}c^{2+r}\delta^2\varphi_1^r z^{-s} + a_2\varphi_1^r z^{-s} \\
&= -\left(M^{1+s}c^{2+r}\delta^2 - a_2\right)\varphi_1^r z^{-s} \le 0 \quad \text{in } \Omega \setminus \omega,
\end{aligned}$$

and

$$\begin{aligned}
\Delta z + a(x)z^{-s} &\le -M\lambda_1 c\varphi_1 H'(c\varphi_1) + a(x)z^{-s} \\
&\le -M\lambda_1 c\varphi_1 H'(c\varphi_1) + a_2\varphi_1^r z^{-s} \\
&= -\frac{1}{M^s}\left(M^{1+s}c\lambda_1\varphi_1 H'(c\varphi_1) - a_2\varphi_1^r H^{-s}(c\varphi_1)\right) \\
&\le 0 \quad \text{in } \omega.
\end{aligned}$$

Hence, we have obtained the inequality in (10.36).

Let $\tilde{w} \in C^2(\overline{\Omega})$ be the unique solution of

$$\begin{cases}
\Delta\tilde{w} - \beta\tilde{w} + b(x) = 0 & \text{in } \Omega, \\
\tilde{w} > 0 & \text{in } \Omega, \\
\tilde{w} = 0 & \text{on } \partial\Omega.
\end{cases}$$

Then $\overline{v} := z + \tilde{w}$ satisfies $\overline{v} > 0$ in Ω, $\overline{v} = 0$ on $\partial\Omega$ and by (10.36) we have

$$\Delta\overline{v} - \beta\overline{v} + a(x)\overline{v}^{-s} + b(x) \le \Delta z - \beta z + a(x)z^{-s} \le 0 \quad \text{in } \Omega.$$

Hence, $\overline{v}$ is a supersolution of (10.22) and clearly we have $\underline{v} \le \overline{v}$ in Ω. It follows that problem (10.22) has a classical solution $v \in C^2(\Omega) \cap C(\overline{\Omega})$ such that $\underline{v} \le v \le \overline{v}$ in Ω.

On the other hand, since $\tilde{w} \in C^2(\overline{\Omega})$, we deduce that there exists $\tilde{c}_1 > 0$ such that $\tilde{w} \le \tilde{c}_1\varphi_1$ in Ω. This implies $\tilde{w} \le \tilde{c}_2\Gamma_{s,r}(\varphi_1)$ in Ω, for some $\tilde{c}_2 > 0$. Finally, using the last inequality, the definition of $\underline{v}, \overline{v}$ and (10.30), we get the estimate (10.27). The uniqueness of the solution follows by Corollary 1.3. This finishes the proof of Proposition 10.3. □

10.3 Case $0 \le p < 1$

10.3.1 Existence

The main existence result in this case is the following:

Theorem 10.4 *Assume that $0 \le p < 1$, $q\sigma < p+1$ and q,r,s satisfy (10.5). Then the system (10.3) has at least one classical solution and there exist $c_1, c_2 > 0$ such that any solution (u,v) of (10.3) satisfies the following estimates in Ω:*

$$c_1 d(x) \le u \le c_2 d(x) \quad \textit{in } \Omega,$$

and

$$c_1 d(x) \le v \le c_2 d(x) \quad \textit{if } s < r+1,$$

$$c_1 d(x)\,(1+|\ln d(x)|)^{1/(1+s)} \le v \le c_2 d(x)\,(1+|\ln d(x)|)^{1/(1+s)} \quad \textit{if } s = r+1,$$

$$c_1 d(x)^{(2+r)/(1+s)} \le v \le c_2 d(x)^{(2+r)/(1+s)} \quad \textit{if } s > r+1,$$

where $d(x) = \operatorname{dist}(x, \partial\Omega)$.

Proof. Let $0 < \varepsilon_0 < 1$ and set

$$\Omega_\varepsilon := \{x \in \Omega : d(x) > \varepsilon\}, \quad \text{for all } 0 < \varepsilon < \varepsilon_0. \tag{10.40}$$

For ε_0 small enough, Ω_ε remains a smooth domain. The existence of a solution to (10.3) will be proved by considering the approximated system

$$\begin{cases} \Delta u - \alpha u + \dfrac{u^p}{v^q} + \rho(x) = 0,\ u > 0 & \text{in } \Omega_\varepsilon, \\ \Delta v - \beta v + \dfrac{u^r}{v^s} = 0,\ v > 0 & \text{in } \Omega_\varepsilon, \\ u = \varepsilon,\ \ v = \Gamma_{s,r}(\varepsilon) & \text{on } \partial\Omega_\varepsilon. \end{cases} \tag{10.41}$$

The existence of a classical solution to (10.41) is obtained by using the Schauder's fixed point theorem. For $0 < \varepsilon < \varepsilon_0$ and $m_1, m_2 < 1 < M_1, M_2$ consider

$$\mathscr{A} = \left\{ (u,v) \in C(\overline{\Omega}_\varepsilon) \times C(\overline{\Omega}_\varepsilon) : \begin{array}{ll} m_1\varphi_1 \le u \le M_1\varphi_1 & \text{in } \Omega_\varepsilon, \\ m_2\Gamma_{s,r}(\varphi_1) \le v \le M_2\Gamma_{s,r}(\varphi_1) & \text{in } \Omega_\varepsilon, \\ u = \varepsilon,\ v = \Gamma_{s,r}(\varepsilon) & \text{on } \partial\Omega_\varepsilon \end{array} \right\}.$$

Next, we define the mapping $\mathscr{T} : \mathscr{A} \to C(\overline{\Omega}_\varepsilon) \times C(\overline{\Omega}_\varepsilon)$ as follows. For $(u,v) \in \mathscr{A}$ we set

$$\mathscr{T}(u,v) = (Tu, Tv), \tag{10.42}$$

where Tu and Tv satisfy

$$\begin{cases} \Delta(Tu) - \alpha(Tu) + \dfrac{(Tu)^p}{v^q} + \rho(x) = 0,\ Tu > 0 & \text{in } \Omega_\varepsilon, \\ \Delta(Tv) - \beta(Tv) + \dfrac{u^r}{(Tv)^s} = 0,\ Tv > 0 & \text{in } \Omega_\varepsilon, \\ Tu = \varepsilon,\ \ Tv = \Gamma_{s,r}(\varepsilon) & \text{on } \partial\Omega_\varepsilon. \end{cases} \tag{10.43}$$

Using the definition of $\mathscr{A}$, by the sub and supersolution method combined with Theorem 1.2 and Corollary 1.3, the above system has a unique solution (Tu, Tv) with $Tu, Tv \in C^2(\overline{\Omega}_\varepsilon)$. Basic to our approach are the following two results which allows us to apply Schauder's fixed point theorem.

Lemma 10.5 *There exist $m_1 < 1 < M_1$ and $m_2 < 1 < M_2$ which are independent of ε such that $\mathscr{T}(\mathscr{A}) \subseteq \mathscr{A}$, for all $0 < \varepsilon < \varepsilon_0$.*

Proof. Let $w \in C^2(\overline{\Omega})$ be the unique solution of problem (10.11). In view of (10.7) and (10.12) we have

$$w(x) \le c_2\varphi_1 \le \frac{c_2}{C} d(x) = \frac{c_2}{C}\varepsilon \quad \text{on } \partial\Omega_\varepsilon.$$

Hence, if $\delta_1 = \min\{1, \frac{C}{c_2}\}$ then $\delta_1 w \le \varepsilon$ on $\partial\Omega_\varepsilon$. Furthermore,

$$\Delta(Tu) - \alpha(Tu) + \rho(x) \le 0 \le \Delta(\delta_1 w) - \alpha(\delta_1 w) + \rho(x) \quad \text{in } \Omega_\varepsilon,$$

$$Tu = \varepsilon \ge \delta_1 w \quad \text{on } \partial\Omega_\varepsilon.$$

By the maximum principle, we obtain $Tu \ge \delta_1 w$ in Ω_ε. In view of (10.12), let us choose $m_1 = \delta_1 c_1$ in the definition of $\mathscr{A}$ (where c_1 is the constant in (10.12)). Then, (10.12) combined with the last estimates yields

$$Tu \ge m_1 \varphi_1 \quad \text{in } \Omega_\varepsilon. \tag{10.44}$$

From the second equation in (10.43) and the fact that $u \ge m_1\varphi_1$ in Ω_ε we have

$$\Delta(Tv) - \beta(Tv) + \frac{m_1^r \varphi_1^r}{(Tv)^s} \le 0 \quad \text{in } \Omega_\varepsilon. \tag{10.45}$$

Let us consider the problem

$$\begin{cases} \Delta\xi - \beta\xi + \varphi_1^r \xi^{-s} = 0 & \text{in } \Omega, \\ \xi > 0 & \text{in } \Omega, \\ \xi = 0 & \text{on } \partial\Omega. \end{cases} \tag{10.46}$$

Using Proposition 10.3 (ii), there exists $\xi \in C^2(\Omega) \cap C(\overline{\Omega})$ a unique solution of (10.46) with the additional property

$$c_3 \Gamma_{s,r}(\varphi_1) \le \xi \le c_4 \Gamma_{s,r}(\varphi_1) \quad \text{in } \Omega, \tag{10.47}$$

for some $c_3, c_4 > 0$. Moreover, by (10.47), (10.7) and the property (10.26) of $\Gamma_{s,r}$ we can find $c_5, c_6 > 0$ such that

$$c_5 \Gamma_{s,r}(d(x)) \le \xi \le c_6 \Gamma_{s,r}(d(x)) \quad \text{in } \Omega. \tag{10.48}$$

Let $\delta_2 = \min\{1, m_1^{r/(1+s)}, \frac{1}{c_6}\}$. Then

$$\Delta(\delta_2\xi) - \beta(\delta_2\xi) + m_1^r \varphi_1^r (\delta_2\xi)^{-s} \ge \delta_2\left(\Delta\xi - \beta\xi + \varphi_1^r \xi^{-s}\right) = 0 \quad \text{in } \Omega, \tag{10.49}$$

and by (10.48) we have

$$\delta_2\xi \le \delta_2 c_6 \Gamma_{s,r}(d(x)) \le \Gamma_{s,r}(\varepsilon) \quad \text{on } \partial\Omega_\varepsilon. \tag{10.50}$$

Therefore, from (10.45), (10.49) and (10.50) we have obtained

$$\Delta(Tv) - \beta(Tv) + m_1^r \varphi_1^r (Tv)^{-s} \le 0 \le \Delta(\delta_2\xi) - \beta(\delta_2\xi) + m_1^r \varphi_1^r (\delta_2\xi)^{-s} \quad \text{in } \Omega,$$

$$Tv = \Gamma_{s,r}(\varepsilon) \ge \delta_2\xi \quad \text{on } \partial\Omega_\varepsilon.$$

By Corollary 1.3 it follows that $Tv \geq \delta_2 \xi$ in Ω_ε. In view of (10.47), the last inequality leads us to $Tv \geq \delta_2 c_3 \Gamma_{s,r}(\varphi_1)$ in Ω_ε. Thus, we consider

$$m_2 = \min\{1, \delta_2 c_3\} > 0$$

in the definition of the set $\mathscr{A}$. Note that m_2 is independent of ε and $Tv \geq m_2 \Gamma_{s,r}(\varphi_1)$ in Ω_ε.

The definition of $\mathscr{A}$ and (10.25) yield

$$v \geq m_2 \Gamma_{s,r}(\varphi_1) \geq m_2 \varphi_1^\sigma \quad \text{in } \Omega_\varepsilon.$$

Using the estimate $v \geq m_2 \varphi_1^\sigma$ in the first equation of (10.43) we get

$$\Delta(Tu) - \alpha(Tu) + m_2^{-q} \varphi_1^{-q\sigma} (Tu)^p + \rho(x) \geq 0 \quad \text{in } \Omega_\varepsilon. \tag{10.51}$$

As above, we next consider the problem

$$\begin{cases} \Delta \zeta - \alpha \zeta + m_2^{-q} \varphi_1^{-q\sigma} \zeta^p + \rho(x) = 0 & \text{in } \Omega, \\ \zeta > 0 & \text{in } \Omega, \\ \zeta = 0 & \text{on } \partial\Omega. \end{cases} \tag{10.52}$$

Since $q\sigma < p+1$, by Proposition 10.2 there exists $\zeta \in C^2(\Omega) \cap C(\overline{\Omega})$ a unique solution of (10.52) such that

$$c_7 \varphi_1 \leq \zeta \leq c_8 \varphi_1 \quad \text{in } \Omega, \tag{10.53}$$

for some $c_7, c_8 > 0$. Note that $q\sigma < p+1$, (10.53) and Lemma 10.1 imply $\Delta \zeta \in L^1(\Omega)$. Let $A_1 = \max\{1, \frac{1}{Cc_7}\}$. Then

$$\Delta(A_1 \zeta) - \alpha(A_1 \zeta) + m_2^{-q} \varphi_1^{-q\sigma} (A_1 \zeta)^p + \rho(x) \leq 0 \quad \text{in } \Omega_\varepsilon.$$

Also by (10.7) and (10.53) we have

$$A_1 \zeta \geq A_1 c_7 \varphi_1 \geq A_1 C c_7 d(x) \geq \varepsilon \quad \text{on } \partial\Omega_\varepsilon.$$

Define

$$\Psi(x,t) = -\alpha t + m_2^{-q} \varphi_1^{-q\sigma}(x) (A_1 t)^p + \rho(x), \quad (x,t) \in \Omega_\varepsilon \times (0, \infty).$$

Then Ψ satisfies the hypotheses in Theorem 1.2 and

$$\Delta(A_1\zeta)+\Psi(x,A_1\zeta)\le 0\le \Delta(Tu)+\Psi(x,Tu)\quad \text{in } \Omega_\varepsilon,$$

$$Tu, A_1\zeta>0 \text{ in } \Omega_\varepsilon,\ \ Tu=\varepsilon\le A_1\zeta \text{ on } \Omega_\varepsilon,$$

$$\Delta(A_1\zeta)\in L^1(\Omega_\varepsilon).$$

By Theorem 1.2 it follows that $Tu\le A_1\zeta$ in Ω_ε. In view of (10.53), let us take $M_1:=\max\{1,A_1c_8\}$ in the definition of the set $\mathscr{A}$. Then M_1 does not depend on ε and by (10.53) we have

$$Tu\le M_1\varphi_1 \quad \text{in } \Omega_\varepsilon.$$

The definition of $\mathscr{A}$ yields $u\le M_1\varphi_1$ in Ω_ε. Then, the second equation of system (10.43) produces

$$\Delta(Tv)-\beta(Tv)+\frac{M_1^r\varphi_1^r}{(Tv)^s}\ge 0 \quad \text{in } \Omega_\varepsilon. \tag{10.54}$$

Let $A_2=\max\{1,M_1^r,\frac{1}{c_5}\}$. If ξ is the unique solution of (10.46), then

$$\Delta(A_2\xi)-\beta(A_2\xi)+M_1^r\varphi_1^r(A_2\xi)^{-s}\le 0 \quad \text{in } \Omega_\varepsilon,$$

and, by (10.48) we also have

$$A_2\xi\ge A_2c_5\Gamma_{s,r}(d(x))\ge \Gamma_{s,r}(\varepsilon) \quad \text{on } \partial\Omega_\varepsilon.$$

Therefore, by Corollary 1.3 it follows that $Tv\le A_2\xi$ in Ω_ε. Now, we take $M_2:=\max\{1,A_2c_4\}$ in the definition of the set $\mathscr{A}$. It follows that M_2 is independent of ε and, by virtue of (10.47), we obtain $Tv\le M_2\Gamma_{s,r}(\varphi_1)$ in Ω_ε. This finishes the proof of our Lemma 10.5. □

Lemma 10.6 *The mapping $\mathscr{T}:\mathscr{A}\to\mathscr{A}$ defined in (10.42)–(10.43) is compact and continuous.*

Proof. Let us fix $(u,v)\in\mathscr{A}$. Then u,v,Tu and Tv are bounded away from zero in $\overline{\Omega_\varepsilon}$ which yields

$$\left\|\frac{(Tu)^p}{v^q}\right\|_{L^\infty(\Omega_\varepsilon)},\left\|\frac{u^r}{(Tv)^s}\right\|_{L^\infty(\Omega_\varepsilon)}\le c_\varepsilon=c(\varepsilon,m_1,m_2,M_1.M_2,p,q,r,s).$$

Hence, by Hölder estimates, for all $\tau>N$ we obtain

$$\|Tu\|_{W^{2,\tau}(\Omega_\varepsilon)},\|Tv\|_{W^{2,\tau}(\Omega_\varepsilon)}\le c_{1,\varepsilon},$$

for some $c_{1,\varepsilon} > 0$ independent of u and v. Since the embedding $W^{2,\tau}(\Omega_\varepsilon) \hookrightarrow C^{1,\gamma}(\overline{\Omega}_\varepsilon)$, $0 < \gamma < 1 - N/\tau$ is compact, we derive that the mapping $\mathscr{T} : \mathscr{A} \to \mathscr{A} \subset C(\overline{\Omega}_\varepsilon) \times C(\overline{\Omega}_\varepsilon)$ is also compact.

It remains to prove that $\mathscr{T}$ is continuous. To this aim, let $\{(u_n, v_n)\}_{n\geq 1} \subset \mathscr{A}$ be such that $u_n \to u$ and $v_n \to v$ in $C(\overline{\Omega}_\varepsilon)$ as $n \to \infty$. Since $\mathscr{T}$ is compact, there exists $(U,V) \in \mathscr{A}$ such that up to a subsequence we have

$$\mathscr{T}(u_n, v_n) \to (U,V) \quad \text{in } \mathscr{A} \text{ as } n \to \infty.$$

Using the $L^\infty(\Omega_\varepsilon)$ bounds of $\left(\frac{(Tu_n)^p}{v_n^q}\right)_{n\geq 1}$ and $\left(\frac{u_n^r}{(Tv_n)^s}\right)_{n\geq 1}$, it follows that $(Tu_n)_{n\geq 1}$ and $(Tv_n)_{n\geq 1}$ are bounded in $W^{2,\tau}(\Omega_\varepsilon)$ for all $\tau > N$. As before, this implies that $(Tu_n)_{n\geq 1}$ and $(Tv_n)_{n\geq 1}$ are bounded in $C^{1,\gamma}(\overline{\Omega}_\varepsilon)$ $(0 < \gamma < 1 - N/\tau)$. Next, by Schauder estimates, it follows that $(Tu_n)_{n\geq 1}$ and $(Tv_n)_{n\geq 1}$ are bounded in $C^{2,\gamma}(\overline{\Omega}_\varepsilon)$. Since $C^{2,\gamma}(\overline{\Omega}_\varepsilon)$ is compactly embedded in $C^2(\overline{\Omega}_\varepsilon)$, we deduce that up to a subsequence, we have that

$$Tu_n \to U \quad \text{and } Tv_n \to V \quad \text{in } C^2(\overline{\Omega}_\varepsilon) \text{ as } n \to \infty.$$

Passing to the limit in (10.43) we get that (U,V) satisfies

$$\begin{cases} \Delta U - \alpha U + \dfrac{U^p}{v^q} + \rho(x) = 0,\ U > 0 & \text{in } \Omega_\varepsilon, \\ \Delta V - \beta V + \dfrac{u^r}{V^s} = 0,\ V > 0 & \text{in } \Omega_\varepsilon, \\ U = \varepsilon,\ V = \Gamma_{s,r}(\varepsilon) & \text{on } \partial\Omega_\varepsilon. \end{cases}$$

Using the uniqueness of (10.43), it follows that $Tu = U$ and $Tv = V$. Thus, we have obtained that any subsequence of $\{\mathscr{T}(u_n, v_n)\}_{n\geq 1}$ has a subsequence converging to $\mathscr{T}(u,v)$ in $\mathscr{A}$. But this implies that the entire sequence $\{\mathscr{T}(u_n, v_n)\}_{n\geq 1}$ converges to $\mathscr{T}(u,v)$ in $\mathscr{A}$, whence the continuity of $\mathscr{T}$. The proof of Lemma 10.6 is now complete. □

We now come back to the proof of Theorem 10.4. According to Lemmas 10.5 and 10.6 we are now in position to apply Schauder's fixed point theorem. Thus, for all $0 < \varepsilon < \varepsilon_0$, there exists $(u_\varepsilon, v_\varepsilon) \in \mathscr{A}$ such that $\mathscr{T}(u_\varepsilon, v_\varepsilon) = (u_\varepsilon, v_\varepsilon)$. By standard elliptic regularity arguments, we deduce $u_\varepsilon, v_\varepsilon \in C^2(\overline{\Omega}_\varepsilon)$. Therefore, for all $0 < \varepsilon < \varepsilon_0$ we have proved the existence of a solution $(u_\varepsilon, v_\varepsilon) \in C^2(\overline{\Omega}_\varepsilon) \times C^2(\overline{\Omega}_\varepsilon)$ of system (10.41). Next, we extend $u_\varepsilon = \varepsilon$, $v_\varepsilon = \Gamma_{s,r}(\varepsilon)$ in $\Omega \setminus \overline{\Omega}_\varepsilon$. Furthermore, by the definition of $\mathscr{A}$ we have

$$m_1\varphi_1 \le u_\varepsilon \le M_1\varphi_1 + \varepsilon \le M_1\varphi_1 + \varepsilon_0 \quad \text{in } \Omega, \tag{10.55}$$

$$m_2\Gamma_{s,r}(\varphi_1) \le v_\varepsilon \le M_2\Gamma_{s,r}(\varphi_1) + \Gamma_{s,r}(\varepsilon) \le M_1\varphi_1 + c_{\varepsilon_0} \quad \text{in } \Omega. \tag{10.56}$$

As above, L^∞ bounds together with Hölder estimates yield $(u_\varepsilon)_{0<\varepsilon<\varepsilon_0}, (v_\varepsilon)_{0<\varepsilon<\varepsilon_0}$ are bounded in $W^{2,\tau}_{\text{loc}}(\Omega)$, for all $\tau > N$. With similar arguments, there exist $u, v \in C^2(\Omega)$ such that for all $\omega \subset\subset \Omega$, $(u_\varepsilon)_{0<\varepsilon<\varepsilon_0}$ and $(v_\varepsilon)_{0<\varepsilon<\varepsilon_0}$ converge up to a subsequence to u and v respectively in $C^2(\overline{\omega})$ as $\varepsilon \to 0$. Passing to the limit with $\varepsilon \to 0$ in (10.41) and (10.55)–(10.56) we get

$$\begin{cases} \Delta u - \alpha u + \dfrac{u^p}{v^q} + \rho(x) = 0 & \text{in } \Omega, \\ \Delta v - \beta v + \dfrac{u^r}{v^s} = 0 & \text{in } \Omega, \end{cases}$$

and

$$m_1\varphi_1 \le u \le M_1\varphi_1 \quad \text{in } \Omega, \tag{10.57}$$

$$m_2\Gamma_{s,r}(\varphi_1) \le v \le M_2\Gamma_{s,r}(\varphi_1) \quad \text{in } \Omega. \tag{10.58}$$

Now, we extend $u = v = 0$ on $\partial\Omega$. From (10.57) and (10.58) we deduce that $u, v \in C(\overline{\Omega})$. Hence, the system (10.3) has a classical solution (u, v).

It remains to establish the boundary estimates of the solution to (10.3). This follows essentially by using the same arguments as above. Let (u, v) be an arbitrary solution of (10.3). Then $\Delta u - \alpha u + \rho(x) \le 0$ in Ω which implies that $u \ge w$ in Ω, where w is the unique solution of (10.11). By (10.12) it follows that $u \ge c_1\varphi_1$ in Ω. Using this inequality in the second equation of (10.3) we deduce $\Delta v - \beta v + c_2\varphi_1^r v^{-s} \le 0$ in Ω for some $c_2 > 0$ (we actually have $c_2 = c_1^r > 0$). Next, let ξ be the unique solution of (10.46). A similar argument to that used in before yields $v \ge c_3\xi$ in Ω. In view of estimate (10.27) in Proposition 10.3 we derive that $v \ge c_4\Gamma_{s,r}(\varphi_1)$ in Ω for some $c_4 > 0$. According to (10.25) it follows that $v \ge c_5\varphi_1^\sigma$ in Ω. This inequality combined with the first equation in system (10.3) produces $\Delta u - \alpha u + c_6\varphi_1^{-q\sigma}u^p + \rho(x) \ge 0$ in Ω.

Consider the problem

$$\begin{cases} \Delta z - \alpha z + c_6\varphi_1^{-q\sigma} z^p + \rho(x) = 0 & \text{in } \Omega, \\ z > 0 & \text{in } \Omega, \\ z = 0 & \text{on } \partial\Omega. \end{cases} \tag{10.59}$$

Since $q\sigma < p+1$, by Proposition 10.2 there exists a unique solution of (10.59) such that $z \le c_7\varphi_1$ in Ω. Thus, by Theorem 1.2 we get $u \le z \le c_7\varphi_1$ in Ω. Using this last inequality in the second equation of (10.3) we finally obtain $\Delta v - \beta v + c_8\varphi_1^r v^{-s} \ge 0$ in Ω for some $c_8 > 0$. By virtue of Proposition 10.3 we have $v \le c_9\Gamma_{s,r}(\varphi_1)$ in Ω. Thus, we have obtained

$$m_1\varphi_1 \le u \le m_2\varphi_1 \quad \text{in } \Omega,$$

$$m_1\Gamma_{s,r}(\varphi_1) \le u \le m_2\Gamma_{s,r}(\varphi_1) \quad \text{in } \Omega,$$

for some fixed constants $m_1, m_2 > 0$. Now, the boundary estimates in Theorem 10.4 follows from the above inequalities combined with (10.7). This concludes the proof. □

10.3.2 Further Results on Regularity

Further regularity of the solution to (10.3) can be obtained using the same arguments as in Gui and Lin [107]. More precisely, it is proved in [107] that if $u \in C^2(\Omega) \cap C(\overline{\Omega})$ satisfies $-\Delta u = u^{-\nu}$ in a smooth bounded domain Ω and $u = 0$ on $\partial\Omega$, then $u \in C^{1,1-\nu}(\overline{\Omega})$. Using the conclusion in Theorem 10.4 we have

Corollary 10.7 *Assume that $0 \le p < 1$ and q, r, s satisfy (10.5).*

(i) If $q \le p$ and $s \le r$, then the system (10.3) has at least one classical solution. Moreover, any solution of (10.3) belongs to $C^2(\overline{\Omega}) \times C^2(\overline{\Omega})$.

(ii) If $-1 < p-q < 0$ and $-1 < r-s < 0$, then the system (10.3) has at least one classical solution. Moreover, any solution (u,v) of (10.3) satisfies $u \in C^2(\Omega) \cap C^{1,1+p-q}(\overline{\Omega})$ and $v \in C^2(\Omega) \cap C^{1,1+r-s}(\overline{\Omega})$.

Proof. Let (u,v) be a classical solution of (10.3). We rewrite the system (10.3) in the form

$$\Delta u = f_1(x) \text{ in } \Omega, \quad \Delta v = f_2(x) \text{ in } \Omega, \quad u = v = 0 \text{ on } \partial\Omega,$$

where

$$f_1(x) = \alpha u(x) - \frac{u^p(x)}{v^q(x)} - \rho(x),\ f_2(x) = \beta v(x) - \frac{u^r(x)}{v^s(x)}, \quad \text{for all } x \in \Omega.$$

Note that in our settings we have $\sigma = 1$ in (10.6) and by virtue of Theorem 10.4 there exist $c_1, c_2 > 0$ such that $c_1 d(x) \le u, v \le c_2 d(x)$ in Ω. Hence,

$$|f_1(x)| \le m_1 d^{p-q}(x) + \rho(x) \quad \text{in } \Omega \quad \text{and } |f_2(x)| \le m_2 d^{r-s}(x) \quad \text{in } \Omega. \quad (10.60)$$

(i) Since $0 \le p-q$ and $0 \le r-s$, by (10.60) we get $f_1, f_2 \in L^\infty(\Omega)$. Next, standard elliptic arguments lead us to $u, v \in C^2(\overline{\Omega})$.

(ii) Let us assume that $-1 < p-q < 0$ and $-1 < r-s < 0$. From (10.60) we derive

$$|f_1(x)| \le c d^{p-q}(x) \quad \text{in } \Omega \quad \text{and } |f_2(x)| \le c d^{r-s}(x) \quad \text{in } \Omega,$$

for some positive constant $c > 0$.

If $N = 1$ then, for all $x_1, x_2 \in \Omega$ we have

$$|u'(x_1) - u'(x_2)| \le \left| \int_{x_1}^{x_2} |f_1(t)| dt \right| \le c \left| \int_{x_1}^{x_2} d^{p-q}(t) dt \right| \le \tilde{c} |x_1 - x_2|^{1+p-q},$$

where $\tilde{c} > 0$ does not depend on x_1, x_2. This yields $u \in C^{1,1+p-q}(\overline{\Omega})$ and similarly $v \in C^{1,1+r-s}(\overline{\Omega})$.

If $N \ge 2$, the conclusion follows exactly in the same way as in [107]. More precisely, let $\mathscr{G}$ denote the Green's function for the Laplace operator. Then for all $x \in \Omega$ we have

$$u(x) = \int_\Omega \mathscr{G}(x,y) f_1(y) dy, \qquad v(x) = \int_\Omega \mathscr{G}(x,y) f_2(y) dy,$$

and

$$\nabla u(x) = \int_\Omega \mathscr{G}_x(x,y) f_1(y) dy, \qquad \nabla v(x) = \int_\Omega \mathscr{G}_x(x,y) f_2(y) dy.$$

Then, for all $x_1, x_2 \in \Omega$, $x_1 \ne x_2$ we have

$$\begin{aligned} |\nabla u(x_1) - \nabla u(x_2)| &\le \int_\Omega |\mathscr{G}_x(x_1,y) - \mathscr{G}_x(x_2,y)| |f_1(y)| dy \\ &\le c \int_\Omega |\mathscr{G}_x(x_1,y) - \mathscr{G}_x(x_2,y)| d^{p-q}(y) dy, \end{aligned}$$

and similarly

$$|\nabla v(x_1) - \nabla v(x_2)| \le c \int_\Omega |\mathscr{G}_x(x_1,y) - \mathscr{G}_x(x_2,y)| d^{r-s}(y) dy.$$

From now on, we need only to employ the sharp estimates given in [107, Theorem 1.1] in order to obtain $u \in C^{1,1+p-q}(\overline{\Omega})$ and $v \in C^{1,1+r-s}(\overline{\Omega})$. This finishes the proof of Corollary 10.7. □

10.3.3 Uniqueness of a Solution

The issue of uniqueness is a delicate matter even in one dimension. In this case the system (10.3) reads

$$\begin{cases} u'' - \alpha u + \dfrac{u^p}{v^q} + \rho(x) = 0 & \text{in } (0,1), \\ v'' - \beta v + \dfrac{u^r}{v^s} = 0 & \text{in } (0,1), \\ u(0) = u(1) = 0,\ v(0) = v(1) = 0. \end{cases} \tag{10.61}$$

In [38] it is proved that the system (10.61) has a unique solution provided that $p = q = r = s = 1$. The main idea is to write (10.61) as a linear system with smooth coefficients and then to use the $C^2[0,1] \times C^2[0,1]$ regularity of the solution. This approach has been used in [94] (see also [95] or [93, Theorem 2.7]) in the case $\beta \leq \alpha$, $0 < q \leq p \leq 1$ and $r - p = s - q \geq 0$.

We are able to show that the uniqueness of the solution to (10.61) still holds provided that

$$-1 < p - q < 1, \quad -1 < r - s < 1. \tag{10.62}$$

Note that for the above range of exponents, the solutions of (10.61) do not necessarily belong to $C^2[0,1] \times C^2[0,1]$. We prove that a $C^{1+\delta}$-regularity up to the boundary of the solution suffices in order to have uniqueness. Therefore, we prove

Theorem 10.8 *Let* $\Omega = (0,1)$, $0 \leq p < 1$ *and* $q, r, s > 0$ *verify (10.62). Then the system (10.61) has a unique classical solution.*

Unlike the Neumann boundary condition, in which large multiplicities of solutions are observed, the uniqueness in the above result seems to be a particular feature of the Dirichlet boundary condition together with the sublinear character of the first equation in the system (10.61).

Proof. Let (u, v) be a classical solution of (10.61). Then, by virtue of Corollary 10.7, we have

$$u, v \in C^2[0,1] \times C^2[0,1] \quad \text{if } 0 \le p-q,\ 0 \le r-s,$$

and

$$u \in C^2(0,1) \cap C^{1,1+p-q}[0,1],\ v \in C^2(0,1) \cap C^{1,1+r-s}[0,1],$$

if $-1 < p-q < 0$, $-1 < r-s < 0$. Furthermore, by Hopf's maximum principle we also have that $u'(0) > 0$, $v'(0) > 0$, $u'(1) < 0$ and $v'(1) < 0$.

Assume that there exist (u_1, v_1) and (u_2, v_2) two different solutions of (10.61).

First we claim that we cannot have $u_2 \ge u_1$ or $v_2 \ge v_1$ in [0,1]. Assume by contradiction that $u_2 \ge u_1$ in [0,1]. Then

$$v_2'' - \beta v_2 + \frac{u_2^r}{v_2^s} = 0 = v_1'' - \beta v_1 + \frac{u_1^r}{v_1^s} \quad \text{in } (0,1),$$

and by Corollary 1.3 we get $v_2 \ge v_1$ in $[0,1]$. This implies that

$$u_1'' - \alpha u_1 + \frac{u_1^p}{v_2^q} + \rho(x) \le 0 = u_2'' - \alpha u_2 + \frac{u_2^p}{v_2^q} + \rho(x) \quad \text{in } (0,1). \tag{10.63}$$

On the other hand, the mapping $\Psi(x,t) = -\alpha t + \frac{t^p}{v_2^q(x)} + \rho(x)$, $(x,t) \in (0,1) \times (0,\infty)$ satisfies the hypotheses in Theorem 1.2. Hence $u_2 \le u_1$ in $[0,1]$, that is $u_1 \equiv u_2$. This also implies $v_1 \equiv v_2$, which is a contradiction. Replacing u_1 by u_2 and v_1 by v_2, we also get that the situation $u_1 \ge u_2$ or $v_1 \ge v_2$ in [0,1] is not possible.

Set $U = u_2 - u_1$ and $V = v_2 - v_1$. From the above arguments, both U and V change sign in $(0,1)$. The key result in our approach is the following.

Proposition 10.9 *U and V vanish only at finitely many points in the interval* $[0,1]$.

Proof. Subtracting the corresponding equations for (u_1, v_1) and (u_2, v_2) we obtain the following linear problem

$$\begin{cases} \mathbf{W}''(x) + A(x)\mathbf{W}(x) = \mathbf{0} \ \text{ in } (0,1), \\ \mathbf{W}(0) = \mathbf{W}(1) = \mathbf{0}, \end{cases} \tag{10.64}$$

where $\mathbf{W} = (U,V)^T$ and $A(x) = (A_{ij}(x))_{1 \le i,j \le 2}$ is a 2×2 matrix defined as

$$A_{11}(x) = -\alpha + \begin{cases} \dfrac{1}{v_2^q(x)} \cdot \dfrac{u_2^p(x) - u_1^p(x)}{u_2(x) - u_1(x)}, & u_1(x) \ne u_2(x) \\ p\dfrac{u_1^{p-1}(x)}{v_1^q(x)}, & u_1(x) = u_2(x) \end{cases}$$

$$A_{12}(x)=\begin{cases}-\dfrac{u_1^p(x)}{v_1^q(x)v_2^q(x)}\cdot\dfrac{v_2^q(x)-v_1^q(x)}{v_2(x)-v_1(x)}, & v_1(x)\neq v_2(x)\\ -q\dfrac{u_1^p(x)}{v_1^{q+1}(x)}, & v_1(x)=v_2(x)\end{cases}$$

$$A_{21}(x)=\begin{cases}\dfrac{1}{v_2^s(x)}\cdot\dfrac{u_2^r(x)-u_1^r(x)}{u_2(x)-u_1(x)}, & u_1(x)\neq u_2(x)\\ r\dfrac{u_1^{r-1}(x)}{v_1^s(x)}, & u_1(x)=u_2(x)\end{cases}$$

$$A_{22}(x)=-\beta-\begin{cases}\dfrac{u_1^r(x)}{v_1^s(x)v_2^s(x)}\cdot\dfrac{v_2^s(x)-v_1^s(x)}{v_2(x)-v_1(x)}, & v_1(x)\neq v_2(x)\\ s\dfrac{u_1^r(x)}{v_1^{s+1}(x)}, & v_1(x)=v_2(x).\end{cases}$$

Lemma 10.10 *We have*

(i) $A_{ij}\in C(0,1)$, *for all* $1\le i,j\le 2$.
(ii) $A_{12}(x)\neq 0$ *and* $A_{21}(x)\neq 0$ *for all* $x\in(0,1)$.
(iii) $d^{1-(p-q)}(x)A_{1j}\in L^\infty(0,1)$ *and* $d^{1-(r-s)}(x)A_{2j}\in L^\infty(0,1)$, *for* $j=1,2$.

Proof. The claims in (i) and (ii) are easy to verify. We prove only the statement in (iii). To this aim, let us notice first that by the regularity of solutions, there exist $c_1,c_2>0$ such that

$$c_1d(x)\le u_i,v_i\le c_2d(x)\quad\text{in }(0,1),\ 1\le i\le 2. \tag{10.65}$$

By (10.65) and the fact that

$$|a^q-b^q|\le q|a-b|\max\{a^{q-1},b^{q-1}\}\quad\text{for all } a,b>0,$$

we have

$$\begin{aligned}d(x)|A_{12}(x)| &\le qd(x)\frac{u_1^p(x)}{v_1^q(x)v_2^q(x)}\max\{v_1^{q-1}(x),v_2^{q-1}(x)\}\\ &\le qd^{p-q}(x)\left(\frac{u_1(x)}{d(x)}\right)^p\max\left\{\left(\frac{d(x)}{v_1(x)}\right)^{q+1},\left(\frac{d(x)}{v_2(x)}\right)^{q+1}\right\}\\ &\le cd^{p-q}(x)\quad\text{for all } 0<x<1.\end{aligned}$$

Hence $d^{1-(p-q)}(x)A_{12}\in L^\infty(0,1)$. We obtain similar estimates for $d^{1-(p-q)}(x)A_{11}$ and $d^{1-(r-s)}(x)A_{2j}$, $j=1,2$. This concludes the proof. □

Next, Lemma 10.10 (i)–(ii) allows us to employ the following result which is proved in [38, Lemma 7].

Lemma 10.11 (see [38]) *Let $0 < a < b < 1$ and $A = (A_{ij})_{1\le i,j\le 2}$ be such that*

(i) $A_{ij} \in C[a,b]$, for all $1 \le i,j \le 2$.
(ii) $A_{12}(x) \ne 0$ and $A_{21}(x) \ne 0$ for all $x \in [a,b]$.

Assume that there exists $\mathbf{W} = (U,V)^T \in C^2[a,b] \times C^2[a,b]$ such that $\mathbf{W} \not\equiv 0$ and $\mathbf{W}''(x) + A(x)\mathbf{W}(x) = \mathbf{0}$ in $[a,b]$. Then, neither U nor V can have infinitely many zeros in $[a,b]$.

As a consequence, we deduce that if $\mathbf{W}$ is a solution of (10.64) and $\mathbf{W}$ vanishes for infinitely many times in an interval $[a,b] \subset (0,1)$ then, applying Lemma 10.11 in $[\varepsilon, 1-\varepsilon]$ for all $\varepsilon > 0$ sufficiently small, we get $\mathbf{W} \equiv \mathbf{0}$.

It remains to show that U and V cannot vanish infinitely many times in the neighborhood of $x = 0$ and $x = 1$. We shall consider only the case $x = 0$; the situation where U or V have infinitely many zeros near $x = 1$ can be handled in the same manner.

Without losing the generality, we may assume that V has infinitely many zeros in a neighborhood of $x = 0$. By the continuity of V it follows that $V(0) = 0$. Furthermore, since $V \in C^2(0,1) \cap C^1[0,1]$, by Rolle's theorem we get that both V' and V'' have infinitely many zeros near $x = 0$. Therefore, $V'(0) = 0$, that is, $v_1'(0) = v_2'(0)$.

If $U'(0) = 0$, then $\mathbf{W}(0) = \mathbf{W}'(0) = \mathbf{0}$. Let $\gamma = \min\{0, p-q, r-s\}$. Then $-1 < \gamma \le 0$ and by Lemma 10.10 (iii) it follows that $x^{1-\gamma}A_{ij} \in L^\infty(0,1/2)$. Thus, we can use Proposition 10.12 in order to get that $\mathbf{W} \equiv \mathbf{0}$ in $[0,1/2]$. Then, by Lemma 10.11 we obtain $\mathbf{W} \equiv 0$ in $[0,1]$, which is a contradiction. Hence $U'(0) \ne 0$. Subtracting the second equation corresponding to v_1 and v_2 in the system (10.61) we have

$$\begin{aligned} V''(x) &= \beta V(x) + \frac{u_1^r(x)}{v_1^s(x)} - \frac{u_2^r(x)}{v_2^s(x)} \\ &= x^{r-s}\left\{\beta\frac{V(x)}{x^{r-s}} + \left(\frac{u_1(x)}{x}\right)^r\left(\frac{x}{v_1(x)}\right)^s - \left(\frac{u_2(x)}{x}\right)^r\left(\frac{x}{v_2(x)}\right)^s\right\}. \end{aligned} \tag{10.66}$$

Since $r - s < 1$, $v_1'(0) = v_2'(0) > 0$ and $u_1'(0) \ne u_2'(0)$ we get

$$\begin{aligned} &\lim_{x\to 0^+}\left\{\beta\frac{V(x)}{x^{r-s}} + \left(\frac{u_1(x)}{x}\right)^r\left(\frac{x}{v_1(x)}\right)^s - \left(\frac{u_2(x)}{x}\right)^r\left(\frac{x}{v_2(x)}\right)^s\right\} \\ &= \frac{u_1'^r(0) - u_2'^r(0)}{v_1'^s(0)} \ne 0. \end{aligned} \tag{10.67}$$

From (10.66) and (10.67) we derive that V'' has constant sign in a small neighborhood of $x = 0$ which contradicts the fact that V vanishes infinitely many times in the neighborhood of $x = 0$. This finishes the proof of Proposition 10.9. □

Proposition 10.12 *Let* $0 < a < 1$, $-1 < \gamma \le 0$ *and* $A = (A_{ij})_{1 \le i,j \le 2}$ *be a* 2×2 *matrix such that for all* $1 \le i, j \le 2$ *we have*

$$A_{ij} \in C(0,a] \quad \textit{and} \quad x^{1-\gamma} A_{ij} \in L^\infty(0,a).$$

Assume that there exists $\mathbf{W} = (W_1, W_2)^T \in (C^2(0,a] \cap C^1[0,a])^2$ *a solution of*

$$\begin{cases} \mathbf{W}''(x) + A(x)\mathbf{W}(x) = \mathbf{0} \text{ in } (0,a], \\ \mathbf{W}(0) = \mathbf{W}'(0) = \mathbf{0}. \end{cases}$$

Then $\mathbf{W} \equiv \mathbf{0}$ *in* $[0,a]$.

Proof. First we need the following result whose proof is a simple exercise of calculus.

Lemma 10.13 *Let* $f \in C(0,a] \cap L^{1+\delta}(0,a)$ *for some* $a, \delta > 0$ *and* $u \in C^2(0,a] \cap C^1[0,a]$ *be such that* $u(0) = u'(0) = 0$ *and* $u'' = f$ *in* $(0,a)$. *Then*

$$u(x) = \int_0^x (x-t) f(t) dt, \quad \textit{for all } 0 \le x \le a.$$

Since $\mathbf{W} \in C^1[0,1] \times C^1[0,1]$ we have $A\mathbf{W} \in C(0,a] \cap L^{1+\delta}(0,a)$ provided that $0 < \delta < -1 - \gamma^{-1}$. Therefore, by Lemma 10.13 we get

$$\mathbf{W}(x) = -\int_0^x (x-t) A(t) \mathbf{W}(t) dt \quad \text{for all } 0 \le x \le a. \tag{10.68}$$

Define $B = (B_{ij})_{1 \le i,j \le 2}$ by $B_{ij}(x) = x^{1-\gamma} A_{ij}(x)$, $0 < x \le a$, $1 \le i, j \le 2$. Then $B_{ij} \in C(0,a] \cap L^\infty(0,a)$. Set

$$M = \max_{1 \le i,j \le 2} \|B_{ij}\|_\infty, \quad k = \max \left\{ \frac{|\mathbf{W}(x)|}{x}; 0 < x \le a \right\},$$

where $|\mathbf{W}(x)| = \max\{ |W_1(x)|, |W_2(x)| \}$. Notice that both M and k are finite, since $\mathbf{W} \in C^1[0,a]$. From (10.68) we have

$$\mathbf{W}(x) = -\int_0^x (x-t) B(t) \frac{\mathbf{W}(t)}{t} t^\gamma dt \quad \text{for all } 0 \le x \le a,$$

which yields

$$|\mathbf{W}(x)| \le M \int_0^x (x-t)\frac{|\mathbf{W}(t)|}{t} t^\gamma dt \quad \text{for all } 0 \le x \le a. \tag{10.69}$$

It follows that

$$|\mathbf{W}(x)| \le Mk \int_0^x (x-t)t^\gamma dt = \frac{Mk}{(1+\gamma)(2+\gamma)} x^{2+\gamma} \quad \text{for all } 0 \le x \le a. \tag{10.70}$$

Using (10.70) in (10.69) we obtain

$$\begin{aligned}
|\mathbf{W}(x)| &\le \frac{M^2 k}{(1+\gamma)(2+\gamma)} \int_0^x (x-t)t^{1+2\gamma} dt \\
&= \frac{M^2 k}{(1+\gamma)(2+\gamma)(2+2\gamma)(3+2\gamma)} x^{3+2\gamma} \\
&\le \frac{M^2 k}{2(1+\gamma)^2} x^{3+2\gamma} \quad \text{for all } 0 \le x \le a.
\end{aligned}$$

By induction, we deduce that for all $n \ge 2$ we have

$$|\mathbf{W}(x)| \le \frac{M^n k}{n!(1+\gamma)^n} x^{n+1+n\gamma} \quad \text{for all } 0 \le x \le a.$$

Since $-1 < \gamma \le 0$, we can pass to the limit in the last inequality in order to get $\mathbf{W} \equiv \mathbf{0}$. This completes the proof. □

Proof of Theorem 10.8 continued. Let us define

$$\mathscr{I}^+ = \{x \in [0,1] : U(x) \ge 0\}, \quad \mathscr{I}^- = \{x \in [0,1] : U(x) \le 0\},$$

$$\mathscr{J}^+ = \{x \in [0,1] : V(x) \ge 0\}, \quad \mathscr{J}^- = \{x \in [0,1] : V(x) \le 0\}.$$

Since both U and V have a finite number of zeros, it follows that the above sets consist of finitely many disjoint closed intervals. Therefore, $\mathscr{I}^+ = \cup_{i=1}^m I_i^+$. For our convenience, let I^+ denote any interval I_i^+ and similar notations will be used for I^-, J^+ and J^-. We have

Lemma 10.14 *For all intervals I^+, I^-, J^+ and J^- defined above, the following situations can not occur:*

(i) $I^+ \subset J^+$; (ii) $I^- \subset J^-$; (iii) $J^+ \subset I^-$; (iv) $J^- \subset I^+$.

Proof. (i) Assume by contradiction that $I^+ \subset J^+$. This yields $u_2 \ge u_1$ and $v_2 \ge v_1$ in I^+. Furthermore we have

$$u_1'' - \alpha u_1 + \frac{u_1^p}{v_2^q} + \rho(x) \le 0 = u_2'' - \alpha u_2 + \frac{u_2^p}{v_2^q} + \rho(x) \quad \text{in } I^+,$$

$$u_1, u_2 > 0 \quad \text{in } I^+, \ u_1 = u_2 = 0 \quad \text{on } \partial I^+, \ u_1'' \in L^1(0,1).$$

Thus, by Theorem 1.2 with $\Psi(x,t) = -\alpha t + \frac{t^p}{v_2^q(x)} + \rho(x)$, $(x,t) \in I^+ \times (0,\infty)$, it follows that $u_2 \le u_1$ in I^+. Since $u_2 \ge u_1$ in I^+, we deduce $u_1 = u_2$ in I^+, that is, $U \equiv 0$ which contradicts Proposition 10.9. Replacing u_1, v_1 with u_2, v_2 in the above arguments we deduce the statement (ii).

(iii) Suppose that $J^+ \subset I^-$. Then $v_2 \ge v_1$ and $u_1 \ge u_2$ in J^+ which yield

$$\frac{u_1^r}{v_1^s} \ge \frac{u_2^r}{v_2^s} \quad \text{in } J^+.$$

Hence $V = v_2 - v_1$ satisfies

$$\begin{cases} V'' - \beta V = \dfrac{u_1^r}{v_1^s} - \dfrac{u_2^r}{v_2^s} \ge 0 & \text{in } J^+, \\ V = 0 & \text{on } \partial J^+. \end{cases}$$

Therefore, by the maximum principle, we have $V \le 0$ in J^+. Since $V \ge 0$ in J^+, it follows that $V \equiv 0$ in J^+ which again contradicts Proposition 10.9. The proof of (iv) follows in a similar way. □

From now on, the proof of Theorem 10.8 follows in the same manner as in [38, Theorem 6]. □

10.4 Case $p < 0$

10.4.1 A Nonexistence Result

Theorem 10.15 *Suppose* $-\infty < p < 0$, $q, r, s > 0$ *and one of the following hold*

(i) $q \ge 2$ *and* $s < 1$.

(ii) $q > 2$ *and* $s = 1$.

(iii) $q > s + 1$ *and* $s > 1$.

Then the system (10.3) has no classical solutions.

Proof. Assume that the system (10.3) has a classical solution (u,v) and set $M = \|u\|_\infty$. Then v satisfies

$$\Delta v - \beta v + c_1 v^{-s} \geq 0 \quad \text{in } \Omega,$$

where $c_1 = M^r > 0$. By Corollary 1.3 we get $v \leq z$, where z is the unique solution of the problem

$$\begin{cases} \Delta z - \beta z + c_1 z^{-s} = 0 & \text{in } \Omega, \\ z > 0 & \text{in } \Omega, \\ z = 0 & \text{on } \partial\Omega. \end{cases}$$

Furthermore, from the estimate (10.27) in Proposition 10.3 (with $r = 0$) there exists $c_2 > 0$ such that $v \leq z \leq c_2 \Gamma_{s,0}(\varphi_1)$ in Ω, which yields

$$\begin{cases} v \leq c_2 \varphi_1 \text{ in } \Omega & \text{if } s < 1, \\ v \leq c_2 \varphi_1 (1 + |\log \varphi_1|)^{1/(1+s)} \text{ in } \Omega & \text{if } s = 1, \\ v \leq c_2 \varphi_1^{2/(1+s)} \text{ in } \Omega & \text{if } s > 1. \end{cases} \tag{10.71}$$

If $s = 1$ and $q > 2$, we fix $0 < \theta < 1$ such that $q\theta \geq 2$. Let us set

$$k = \begin{cases} 1, & \text{if } s < 1, \\ \theta, & \text{if } s = 1, \\ 2/(s+1), & \text{if } s > 1. \end{cases}$$

Then $qk \geq 2$ and by (10.71) we get $v \leq c_3 \varphi_1^k$ in Ω, for some $c_3 > 0$. Using this inequality in the first equation of (10.3) we deduce $\Delta u - \alpha u + c \varphi_1^{-qk} u^p + \rho(x) \leq 0$ in Ω, where $c = c_3^{-q}$. This means that u is a supersolution of the problem

$$\begin{cases} \Delta z - \alpha z + c \varphi_1^{-qk} z^p + \rho(x) = 0 & \text{in } \Omega, \\ z > 0 & \text{in } \Omega, \\ z = 0 & \text{on } \partial\Omega. \end{cases} \tag{10.72}$$

Note that $w \in C^2(\overline{\Omega})$ defined as the unique solution of (10.11) is a subsolution of (10.72). By the standard maximum principle it is easy to get $u \geq w$ in Ω. Hence, the problem (10.72) has classical solutions, but this contradicts Proposition 10.3 (i), since $qk \geq 2$. Therefore, the system (10.3) has no solutions. The proof of Theorem 10.15 is now complete. □

10.4.2 Existence

Theorem 10.16 *Assume that $-\infty < p < 0$, q,r,s satisfy (10.5) and $q\sigma < 2$. Then, the system (10.3) has classical solutions. Moreover, if $q < p+1$ and $s < r+1$ any classical solution (u,v) of (10.3) satisfies $u,v \in C^2(\Omega)\cap C^1(\overline{\Omega})$.*

Proof. For all $0 < \varepsilon < \varepsilon_0$ let Ω_ε be defined as in (10.40). For $m_1 < 1 < M_1$ and $m_2 < 1 < M_2$ we set

$$\mathscr{B}_\varepsilon = \left\{ (u,v) \in C(\overline{\Omega}_\varepsilon)\times C(\overline{\Omega}_\varepsilon) : \begin{array}{ll} m_1\varphi_1 \le u \le M_1\varphi_1^\nu & \text{in } \Omega_\varepsilon, \\ m_2\Gamma_{s,r}(\varphi_1) \le v \le M_2\varphi_1^\tau & \text{in } \Omega_\varepsilon, \\ u=\varepsilon,\ v=\Gamma_{s,r}(\varepsilon) & \text{on } \partial\Omega_\varepsilon, \end{array} \right\},$$

where

$$\nu = \begin{cases} 1, & \text{if } q\sigma < 1+p \\ 1/2, & \text{if } q\sigma = 1+p \\ \dfrac{2-q\sigma}{1-p}, & \text{if } q\sigma > 1+p \end{cases} \quad \text{and } \tau = \begin{cases} 1, & \text{if } s < 1+r\nu \\ 1/2, & \text{if } s = 1+r\nu \\ \dfrac{2+r\nu}{1+s}, & \text{if } s > 1+r\nu \end{cases}. \tag{10.73}$$

In order to prove that $\mathscr{B}_\varepsilon$ is not empty, we first remark that $\nu \le 1$. Therefore, we only need to check that

$$\Gamma_{s,r}(t) \le c_0 t^\tau \quad \text{for all } 0 < t \le 1, \tag{10.74}$$

for some fixed $c_0 > 0$. To this aim we analyze the cases $s < 1+r$, $s = 1+r$ and $s > 1+r$.

If $s < 1+r$, since $\tau \le 1$ we have $\Gamma_{s,r}(t) = t \le t^\tau$ for all $0 < t \le 1$.

If $s > 1+r$, from $\nu \le 1$ we have $s > 1+r\nu$ which implies $\tau = \frac{2+r\nu}{1+s} \le \frac{2+r}{1+s}$. Hence

$$\Gamma_{s,r}(t) = t^{(2+r)/(1+s)} \le t^{(2+r\nu)/(1+s)} = t^\tau \quad \text{for all } 0 < t \le 1.$$

Finally, if $s = 1+r$ then $s \ge 1+r\nu$ which implies $\tau = 1/2$ or $\tau = \frac{2+r\nu}{1+s}$. In both cases we have $\tau < 1$. Then

$$\Gamma_{s,r}(t) = t(1+|\ln t|)^{1/(1+s)} \le c_0 t^\tau \quad \text{for all } 0 < t \le 1,$$

and for some fixed $c_0 > 0$.

Remark 25 *Since $t^{\theta}(1+|\log t|)^{1/(s+1)} \to 0$ as $t \to 0$, for all $\theta > 0$, we could replace the value $1/2$ in the definition of ν and τ by any number $\theta \in (0,1)$ in the case $q\sigma = p+1$ and $s = 1+r\nu$ respectively.*

Therefore, for small $0 < m_1, m_2 < 1$ and for large values of $M_1, M_2 > 1$ the set $\mathscr{B}_\varepsilon$ is not empty.

As in the previous section, for all $(u,v) \in \mathscr{B}_\varepsilon$ let us denote by (Tu, Tv) the unique solution of

$$\begin{cases} \Delta(Tu) - \alpha(Tu) + \dfrac{(Tu)^p}{v^q} + \rho(x) = 0,\ Tu > 0 & \text{in } \Omega_\varepsilon, \\ \Delta(Tv) - \beta(Tv) + \dfrac{u^r}{(Tv)^s} = 0,\ Tv > 0 & \text{in } \Omega_\varepsilon, \\ Tu = \varepsilon,\ Tv = \Gamma_{s,r}(\varepsilon) & \text{on } \partial\Omega_\varepsilon. \end{cases} \tag{10.75}$$

In this way we have defined a mapping

$$\mathscr{T} : \mathscr{B}_\varepsilon \to C(\overline{\Omega}_\varepsilon) \times C(\overline{\Omega}_\varepsilon),\ \mathscr{T}(u,v) = (Tu, Tv).$$

Now, we proceed as in the proof of Theorem 10.4. The main point is to show that there exist $0 < m_1, m_2 < 1$ and $M_1, M_2 > 1$ which are independent of ε such that $\mathscr{T}(\mathscr{B}_\varepsilon) \subseteq \mathscr{B}_\varepsilon$. This allows us to employ the Schauder's fixed point theorem.

Following the proof of Lemma 10.5 we get the existence of $m_1, m_2 \in (0,1)$ which are independent of ε and such that

$$Tu \geq m_1\varphi_1,\ Tv \geq m_2\Gamma_{s,r}(\varphi_1) \quad \text{in } \Omega_\varepsilon.$$

Since $\Gamma_{s,r}(t) \geq t^\sigma$ for all $0 < t \leq 1$, the definition of $\mathscr{B}_\varepsilon$ yields $v \geq m_2\varphi_1^\sigma$ in Ω_ε. Furthermore, the first equation in (10.75) produces

$$\Delta(Tu) - \alpha(Tu) + m_2^{-q}\varphi_1^{-q\sigma}(Tu)^p + \rho(x) \geq 0 \quad \text{in } \Omega_\varepsilon.$$

Let $\zeta \in C^2(\Omega) \cap C(\overline{\Omega})$ be the unique solution of (10.52). Since $p < 0$ and $q\sigma < 2$, we shall make use of Proposition 10.3 (ii) instead of Proposition 10.2 as we did in the proof of Theorem 10.4. Therefore, there exist $c_1, c_2 > 0$ such that

$$c_1\Gamma_{-p,-q\sigma}(\varphi_1) \leq \zeta \leq c_2\Gamma_{-p,-q\sigma}(\varphi_1) \quad \text{in } \Omega. \tag{10.76}$$

Note that $\Gamma_{-p,-q\sigma}(t) \geq t$ for all $0 < t \leq 1$. Hence, by (10.7) and (10.76) we get

$$\zeta \geq c_1\varphi_1 \geq Cc_1 d(x) \quad \text{in } \Omega.$$

Let us fix $A > 1$ such that $ACc_1 > 1$. Since $p < 0$ we find

$$\Delta(A\zeta) - \alpha(A\zeta) + m_2^{-q}\varphi_1^{-q\sigma}(A\zeta)^p + \rho(x) \leq 0 \quad \text{in } \Omega_\varepsilon,$$

$$A\zeta \geq \varepsilon = Tu \quad \text{on } \partial\Omega_\varepsilon.$$

In view of Corollary 1.3 we derive $A\zeta \geq Tu$ in Ω_ε and by (10.76) it follows that

$$Tu \leq Ac_2\Gamma_{-p,-q\sigma}(\varphi_1) \quad \text{in } \Omega_\varepsilon.$$

Note that $\Gamma_{-p,-q\sigma}(t) \leq \tilde{c}t^\nu$ for all $0 < t \leq 1$ and for some fixed constant $\tilde{c} > 0$. Therefore, we can find $M_1 > 1$ sufficiently large such that $Tu \leq M_1\varphi_1^\nu$ in Ω_ε.

Using the estimate $u \leq M_1\varphi_1^\nu$ in Ω_ε, from the second equation in (10.75) we deduce

$$\Delta(Tv) - \beta(Tv) + M_1^r\varphi_1^{r\nu}(Tv)^{-s} \geq 0 \quad \text{in } \Omega_\varepsilon.$$

Since $\nu \leq 1$, we can easily prove that $\Gamma_{s,r}(t) \leq c_0\Gamma_{s,r\nu}(t)$, for all $0 < t \leq 1$ and for some positive constant c_0. This implies that

$$Tv = \Gamma_{s,r}(\varepsilon) \leq c_0\Gamma_{s,r\nu}(\varepsilon) \quad \text{on } \partial\Omega_\varepsilon.$$

Next, similar arguments to those in the proof of Lemma 10.5 yield $Tv \leq c\Gamma_{s,r\nu}(\varphi_1)$ in Ω_ε. It remains to notice that $\Gamma_{s,r\nu}(t) \leq \bar{c}t^\tau$ for all $0 < t \leq 1$ and for some $\bar{c} > 0$. Hence, $Tv \leq M_2\varphi_1^\tau$ in Ω_ε for some $M_2 > 1$ independent of ε. Therefore $\mathscr{T}(\mathscr{B}_\varepsilon) \subseteq \mathscr{B}_\varepsilon$. From now on, we proceed exactly in the same way as in the proof of Theorem 10.4.

Assume next that $q < p+1$ and $s < r+1$. Then, by (10.6) and (10.73) we get $\sigma = \nu = \tau = 1$. With the same arguments as in the proof of Theorem 10.4 we get $m_1 d(x) \leq u, v \leq m_2 d(x)$ in Ω, for some $m_1, m_2 > 0$ and for all solutions (u,v) of (10.3). Then we use the same approach as in Corollary 10.7 in order to get that $u, v \in C^2(\Omega) \cap C^{1,\gamma}(\overline{\Omega})$, for some $0 < \gamma < 1$. This finishes the proof of Theorem 10.16. □

Remark 26 *The above approach can be employed to extend the study of system (10.3) to the following class of exponents:*

$$0 \leq p < 1,\ 0 < q < p+1,\ r > 0,\ -1 < s \leq 0.$$

In this sense, we need the smooth variant of Proposition 10.3 concerning the sublinear case $-1 < s \leq 0$. Taking into account the fact that $r > 0$, if $-1 < s \leq 0$ then the problem (10.22) has a unique solution $v \in C^2(\overline{\Omega})$. One can show that system (10.3) has classical solutions and any solution (u,v) of (10.3) satisfies

$$c_1 d(x) \leq u, v \leq c_2 d(x) \quad \text{in } \Omega,$$

for some $c_1, c_2 > 0$. Furthermore, with the same idea as in the proof of Corollary 10.7 we get

(i) if $p \geq q$ then $u, v \in C^2(\overline{\Omega})$;
(ii) if $-1 < p - q < 0$ then $u \in C^2(\Omega) \cap C^{1,1+p-q}(\overline{\Omega})$ and $v \in C^2(\overline{\Omega})$.

Appendix A
Caffarelli–Kohn–Nirenberg Inequality

Inequality is the cause of all local movements.

Leonardo da Vinci (1452–1519)

The Hardy–Sobolev inequality states that for any given domain $\Omega \subset \mathbb{R}^N$, $N \geq 3$ and any $u \in C_0^\infty(\Omega)$,

$$K^2 \int_\Omega \frac{u^2}{|x|^2} dx \leq \int_\Omega |\nabla u|^2 dx, \tag{A.1}$$

where $K = (N-2)/2$. Though the constant K^2 is optimal, in the sense that

$$K^2 = \inf_{u \in C_0^\infty(\Omega)\setminus\{0\}} \frac{\int_\Omega |\nabla u|^2 dx}{\int_\Omega u^2/|x|^2 dx},$$

inequality in relation (A.1) is never achieved. This fact has led to the improvement of the Hardy–Sobolev inequality in various ways. For instance, Brezis and Vázquez [31] showed that if Ω is bounded then for some $\gamma > 0$,

$$\gamma \left(\int_\Omega |u|^p dx \right)^{2/p} + K^2 \int_\Omega \frac{u^2}{|x|^2} dx \leq \int_\Omega |\nabla u|^2 dx, \tag{A.2}$$

with $1 \leq p < 2N/(N-2)$. One of the consequences of (A.2) is that the operator $-\Delta - \mu/|x|^2$ is *coercive*, in the sense that

$$\inf_{\|u\|_{L^2(\Omega)}=1} \int_\Omega \left(|\nabla u|^2 - \mu \frac{u^2}{|x|^2} \right) > 0,$$

whenever $\mu \leq K^2$. For other improvements on the Hardy–Sobolev inequality we refer to Adimurthi, Chaudhuri and Ramaswamy [1], Brezis and Marcus [28], Davies

M. Ghergu and V. Rădulescu, *Nonlinear PDEs*, Springer Monographs in Mathematics,
DOI 10.1007/978-3-642-22664-9, © Springer-Verlag Berlin Heidelberg 2012

[58], Dávila and Dupaigne [61], Filippas, Maz'ya and Tertikas [74], Rădulescu, Smets and Willem [170].

The Caffarelli–Kohn–Nirenberg inequality generalizes the Hardy–Sobolev inequality and was first obtained in 1984 (see [35]). We provide here a direct proof of this basic inequality.

Theorem A.1 *Let $N \geq 1$ and a, b and p be such that*

(i) if $N \geq 3: \ -\infty < a < \dfrac{N-2}{2},\ a \leq b \leq a+1$ *and* $p = \dfrac{2N}{N-2+2(b-a)}$,

(ii) if $N = 2: \ -\infty < a < 0,\ a < b \leq a+1$ *and* $p = \dfrac{2}{b-a}$,

(iii) if $N = 1: \ -\infty < a < -\dfrac{1}{2},\ a+\dfrac{1}{2} < b \leq a+1$ *and* $p = \dfrac{2}{-1+2(b-a)}$.

Then, there exists a positive constant $C_{a,b} = C(a,b) > 0$ *such that*

$$\left(\int_{\mathbb{R}^N} \frac{|u|^p}{|x|^{bp}} dx\right)^{1/p} \leq C_{a,b} \left(\int_{\mathbb{R}^N} \frac{|\nabla u|^2}{|x|^{2a}} dx\right)^{1/2}, \tag{A.3}$$

for any $u \in C_0^\infty(\mathbb{R}^N)$.

Let us remark that inequality (A.3) contains as particular cases the classical Sobolev inequality (if $a = b = 0$) and the Hardy inequality (if $a = 0$ and $b = 1$); we refer to [28, 58, 170] for further details.

Proof. We shall consider here the case $N \geq 3$; the cases $N = 2$ and $N = 1$ are similar. We divide the proof into three steps.

Step 1: $b = a+1$. Let

$$\mathbf{F}(x) = \frac{1}{2b-N} \frac{x}{|x|^{-2b}}.$$

By the divergence theorem we have

$$\begin{aligned}
\int_{\mathbb{R}^N} \frac{|u|^2}{|x|^{-2b}} dx &= -\int_{\mathbb{R}^N} |u|^2 \mathrm{div}\mathbf{F} dx \\
&= 2\int_{\mathbb{R}^N} |u| \mathbf{F} \cdot \nabla |u| dx \\
&\leq \alpha^2 \int_{\mathbb{R}^N} |u|^2 \frac{\|\mathbf{F}\|^2}{|x|^{2a}} dx + \alpha^{-2} \int_{\mathbb{R}^N} \frac{|\nabla u|^2}{|x|^{2a}} dx.
\end{aligned} \tag{A.4}$$

On the other hand,

$$\frac{\|\mathbf{F}\|^2}{|x|^{2a}} = \frac{1}{(n-2-2a)^2} |x|^{-2b},$$

so from (A.4) we obtain

$$\left(1-\frac{\alpha^2}{(n-2-2a)^2}\right)\int_{\mathbb{R}^N}\frac{|u|^2}{|x|^{2b}}dx \le \alpha^{-2}\int_{\mathbb{R}^N}\frac{|\nabla u|^2}{|x|^{2a}}dx.$$

This implies

$$\begin{aligned}\int_{\mathbb{R}^N}\frac{|u|^2}{|x|^{2b}}dx &\le \frac{(N-2-2a)^2}{\alpha^2(N-2-2a)^2-\alpha^4}\int_{\mathbb{R}^N}\frac{|\nabla u|^2}{|x|^{2a}}dx\\ &\le \frac{4}{\alpha^2(N-2-2a)^2}\int_{\mathbb{R}^N}\frac{|\nabla u|^2}{|x|^{2a}}dx.\end{aligned}$$

Step 2: $b=a$. By the Sobolev inequality we have

$$\left(\int_{\mathbb{R}^N}\frac{|u|^p}{|x|^{ap}}dx\right)^{1/p} \le C_{N,p}\left(\int_{\mathbb{R}^N}\nabla\left(\frac{u}{|x|^a}\right)dx\right)^{1/2}. \tag{A.5}$$

Also a direct computation and Step 1 yields

$$\begin{aligned}\int_{\mathbb{R}^N}\nabla\left(\frac{u}{|x|^a}\right)dx &= \int_{\mathbb{R}^N}\frac{|\nabla u|^2}{|x|^{2a}}dx + a^2\int_{\mathbb{R}^N}\frac{|u|^2}{|x|^{2a+2}} - 2a\int_{\mathbb{R}^N}\frac{u}{|x|^{2a+2}}x\cdot\nabla u dx\\ &\le C_a\int_{\mathbb{R}^N}\frac{|\nabla u|^2}{|x|^{2a}} - 2a\int_{\mathbb{R}^N}\frac{u}{|x|^{2a+2}}x\cdot\nabla u dx.\end{aligned} \tag{A.6}$$

We next estimate the second integral in the right-hand side of (A.6). We have

$$\begin{aligned}-2a\int_{\mathbb{R}^N}\frac{u}{|x|^{2a+2}}x\cdot\nabla u dx &\le 2|a|\int_{\mathbb{R}^N}\frac{u}{|x|^{2a+2}}x\cdot\nabla u dx\\ &\le |a|\left(\int_{\mathbb{R}^N}\frac{u^2}{|x|^{2a+2}}dx + \int_{\mathbb{R}^N}\frac{|\nabla u|^2}{|x|^{2a}}dx\right)\\ &\le C_a\int_{\mathbb{R}^N}\frac{|\nabla u|^2}{|x|^{2a}}dx.\end{aligned} \tag{A.7}$$

Now, the inequality follows from (A.5) to (A.7).

Step 3: $a<b<a+1$. We obtain the Caffarelli–Kohn–Nirenberg inequality by interpolation. Let $p=2(1-\theta)+2^*\theta$, where $2^*=2N/(N-2)$ which implies

$$b=a+1-\frac{N\theta}{N-2+2\theta}.$$

From the Hölder inequality and the two previous steps we obtain

$$\begin{aligned}
\int_{\mathbb{R}^N} \frac{|u|^p}{|x|^{bp}} dx &= \int_{\mathbb{R}^N} \frac{|u|^{2(1-\theta)+2^*\theta}}{|x|^{2(1-\theta)(a+1)+2^*\theta a}} dx \\
&\leq \left(\int_{\mathbb{R}^N} \frac{|u|^2}{|x|^{2a+2}} dx \right)^{1-\theta} \left(\int_{\mathbb{R}^N} \frac{|u|^{2^*}}{|x|^{2^a}} dx \right)^{\theta} \\
&\leq C_{a,b} \int_{\mathbb{R}^N} \frac{|\nabla u|^2}{|x|^{2a}} dx.
\end{aligned}$$

This finishes the proof. □

The best embedding constant appearing in (A.3) is given by

$$S(a,b) = \inf_{u \in D \setminus \{0\}} \frac{\left(\int_{\mathbb{R}^N} \frac{|\nabla w|^2}{|x|^{-2a}} dx \right)^{1/2}}{\left(\int_{\mathbb{R}^N} \frac{|w|^p}{|x|^{-bp}} dx \right)^{1/p}}, \tag{A.8}$$

where D is the completion of $C_0^\infty(\mathbb{R}^N)$ with respect to the inner product

$$(u,v) = \int_{\mathbb{R}^N} |x|^{-2a} \nabla u \cdot \nabla v dx.$$

Remark A.1. (see [36]). If one of the following conditions holds

(i) $N \geq 2$ and $a < b < a+1$.
(ii) $N = 1$ and $a + \frac{1}{2} < b < a+1$, then the infimum in (A.8) is always achieved.

The extremal functions for (6.3) are ground state solutions of the singular Euler equation

$$-\text{div}\left(|x|^{-2a} \nabla u\right) = |x|^{-bp} |u|^{p-2} u \qquad \text{in } \mathbb{R}^N.$$

This equation has been recently studied (see [36,201]) in connection with a complete understanding of the best constants, the qualitative properties of extremal functions, the existence (or nonexistence) of minimizers and their symmetry.

Appendix B
Estimates for the Green Function Associated to the Biharmonic Operator

> We are what we repeatedly do. Excellence then, is not an act, but a habit.
>
> Aristotle (384 BC–322 BC)

Higher order differential operators have been studied starting with the contributions of Jacob Bernoulli [17], who tried to study the nodal line patterns of vibrating plates. He modeled this phenomenon by means of the fourth order operator $\frac{\partial^4}{\partial x^4} + \frac{\partial^4}{\partial y^4}$. However, Bernoulli's model was not accepted for reasons of lack of rotationally symmetry. It seems that the first use of the biharmonic operator Δ^2 is due to Lagrange about 1811, who corrected a manuscript by Sophie Germain on the modeling of elastic plates. We refer to the recent book by Gazzola, Grunau and Sweers [83] for an excellent overview of results in the theory of higher order elliptic equations.

Let $\Omega \subset \mathbb{R}^N$ $(N \geq 2)$ be a bounded domain whose boundary $\partial\Omega$ is of class C^{16} if $N = 2$ and of class C^{12} if $N \geq 3$. Let also $\delta(x) = \mathrm{dist}(x, \partial\Omega)$.

We denote by $G(\cdot,\cdot)$ the Green function associated with the biharmonic operator Δ^2 subject to Dirichlet boundary conditions, that is, for all $y \in \Omega$, $G(\cdot,y)$ satisfies in the distributional sense:

$$\begin{cases} \Delta^2 G(\cdot,y) = \delta_y(\cdot) & \text{in } \Omega, \\ G(\cdot,y) = \partial_\nu G(\cdot,y) = 0 & \text{on } \partial\Omega. \end{cases}$$

The study of the Green function for the biharmonic equation goes back to Boggio [20] in 1901. He proved that the Green function is positive in any ball of $\mathbb{R}^n$. Boggio [21] and Hadamard [108] conjectured that this fact should be true at least in any

M. Ghergu and V. Rădulescu, *Nonlinear PDEs*, Springer Monographs in Mathematics,
DOI 10.1007/978-3-642-22664-9,

smooth *convex* domain of $\mathbb{R}^n$. In fact, Hadamard [109] already knew in 1908 that this conjecture fails in annuli with small inner radius.

Starting in the late 1940s, various counterexamples have been constructed that disprove the Boggio–Hadamard conjecture. For instance, if a domain in $\mathbb{R}^2$ has a right-angle, then the associated Green function fails to be everywhere positive (see Coffman and Duffin [52]). A similar result holds for thin ellipses: Garabedian [81] found that in an ellipse in $\mathbb{R}^2$ with the ratio of the half axes $\simeq 2$, the Green function for the biharmonic operator changes sign (for an elementary proof, see also Shapiro and Tegmark [179]). In turn, if the ellipse is close to a ball in the plane, Grunau and Sweers [103] proved that the Green function is positive. Recently, Grunau and Sweers [104–106] and Grunau and Robert [101] provided interesting characterizations of the regions where the Green function is negative. They also obtained that if a domain is sufficiently close to a unit ball in a suitable $C^{4,\gamma}$-sense, then the biharmonic Green function under Dirichlet boundary condition is positive.

Recall first that the biharmonic Green function

$$G : \Omega \times \Omega \setminus \{(z,z) : z \in \Omega\} \to (0,\infty)$$

is continuous. Also, by the estimates in [102, Theorem 1] we have

$$\lim_{(x,y)\to(z,z)} G(x,y) = +\infty, \qquad \text{for all } z \in \Omega.$$

Hence $G : \Omega \times \Omega \to (0,\infty]$ is continuous (in the extended sense).

We recall here some useful estimates regarding the biharmonic Green function presented in Dall'Acqua and Sweers [55] (see also Krasovskiĭ [122]).

Proposition B.1. (see [55]) *Let k be an N−dimensional multi-index. Then, there exists a positive constant c depending on Ω and k such that for any $x,y \in \Omega$ we have*

(i) For $|k| \geq 2$:

(i1) if $N > 4 - |k|$ then

$$|D_x^k G(x,y)| \leq c|x-y|^{4-N-|k|} \min\left\{1, \frac{\delta(y)}{|x-y|}\right\}^2,$$

(i2) if $N = 4 - |k|$ then

$$|D_x^k G(x,y)| \le c \log\left(2 + \frac{\delta(y)}{|x-y|}\right) \min\left\{1, \frac{\delta(y)}{|x-y|}\right\}^2,$$

(i3) if $N < 4 - |k|$ *then*

$$|D_x^k G(x,y)| \le c\delta(y)^{4-N-|k|} \min\left\{1, \frac{\delta(y)}{|x-y|}\right\}^{N+|k|-2}.$$

(ii) For $|k| < 2$*:*

(ii1) if $N > 4 - |k|$ *then*

$$|D_x^k G(x,y)| \le c|x-y|^{4-N-|k|} \min\left\{1, \frac{\delta(x)}{|x-y|}\right\}^{2-|k|} \min\left\{1, \frac{\delta(y)}{|x-y|}\right\}^2,$$

(ii2) if $N = 4 - |k|$ *then*

$$|D_x^k G(x,y)| \le c \log\left(2 + \frac{\delta(y)}{|x-y|}\right) \min\left\{1, \frac{\delta(x)}{|x-y|}\right\}^{2-|k|} \min\left\{1, \frac{\delta(y)}{|x-y|}\right\}^2,$$

(ii3) if $2(2 - |k|) \le N < 4 - |k|$ *then*

$$|D_x^k G(x,y)| \le c\delta(y)^{4-N-|k|} \min\left\{1, \frac{\delta(x)}{|x-y|}\right\}^{2-|k|} \min\left\{1, \frac{\delta(y)}{|x-y|}\right\}^{N+|k|-2},$$

(ii4) if $N < 2(2 - |k|)$ *then*

$$|D_x^k G(x,y)| \le c\delta^{2-|k|-N/2}(x)\delta^{2-N/2}(y) \min\left\{1, \frac{\delta(x)}{|x-y|}\right\}^{N/2} \min\left\{1, \frac{\delta(y)}{|x-y|}\right\}^{N/2}.$$

References

1. Adimurthi, N. Chaudhuri, and M. Ramaswamy, An improved Hardy–Sobolev inequality and its application, *Proc. Amer. Math. Soc.* **130** (2002), 489–505.
2. A. Aftalion and W. Reichel, Existence of two boundary blow-up solutions for semilinear elliptic equations, *J. Differential Equations* **141** (1997), 400–421.
3. R. Agarwal and D. O'Regan, Existence theory for single and multiple solutions to singular positone boundary value problems, *J. Differential Equations* **175** (2001), 393–414.
4. S. Alama and G. Tarantello, On the solvability of a semilinear elliptic equation via an associated eigenvalue problem, *Math. Z.* **221** (1996), 467–493.
5. S. Alama and G. Tarantello, Elliptic problems with nonlinearities indefinite in sign, *J. Funct. Anal.* **141** (1996), 159–215.
6. A. Ambrosetti, H. Brezis, and G. Cerami, Combined effects of concave and convex nonlinearities in some elliptic problems, *J. Funct. Anal.* **122** (1994), 519–543.
7. A. Ambrosetti and P. Rabinowitz, Dual variational methods in critical point theory and applications, *J. Funct. Anal.* **7** (1973), 349–381.
8. V.I. Arnold, *Geometrical Methods in the Theory of Ordinary Differential Equations*, Grundlehren Math. Wiss., vol. 250, Springer-Verlag, Berlin, 1983.
9. P. Ashwin and Z. Mei, Normal form for Hopf bifurcation of partial differential equations on the square, *Nonlinearity* **8** (1995), 715–734.
10. A. Bahri and P.-L. Lions, Solutions of superlinear elliptic equations and their Morse indices, *Commun. Pure Appl. Math.* **45** (1992), 1205–1215.
11. C. Bandle and E. Giarrusso, Boundary blow-up for semilinear elliptic equations with nonlinear gradient terms, *Advances in Differential Equations* **1** (1996), 133–150.
12. C. Bandle and M. Marcus, 'Large' solutions of semilinear elliptic equations: Existence, uniqueness, and asymptotic behaviour, *J. Anal. Math.* **58** (1992), 9–24.
13. C. Bandle and M. Marcus, Dependence of blowup rate of large solutions of semilinear elliptic equations on the curvature of the boundary, *Complex Variables, Theory Appl.* **49** (2004), 555–570.
14. P. Bénilan, H. Brezis, and M. Crandall, A semilinear equation in $L^1(\mathbb{R}^N)$, *Ann. Scuola Norm. Sup. Pisa* **4** (1975), 523–555.
15. D.L. Benson, J.A. Sherratt, and P.K. Maini, Diffusion driven instability in an inhomogeneous domain, *Bull. Math. Biol.* **55** (1993), 365–384.
16. H. Berestycki and L. Nirenberg, On the method of moving planes and the sliding method, *Bol. Soc. Brasil Math.* **22** (1991), 1–37.
17. J. Bernoulli, Essay théoretique sur les vibrations des plaques élastiques, rectangulaires et libres, *Nova Acta Acad. Sci. Imperialis Petropolitanae* **5** (1789), 197–219.
18. L. Bieberbach, $\Delta u = e^u$ und die automorphen Funktionen, *Math. Ann.* **77** (1916), 173–212.
19. N.H. Bingham, C.M. Goldie, and J.L. Teugels, *Regular Variation*, Cambridge University Press, Cambridge, 1987.

M. Ghergu and V. Rădulescu, *Nonlinear PDEs*, Springer Monographs in Mathematics, DOI 10.1007/978-3-642-22664-9,

20. T. Boggio, Sull'equilibrio delle piastre elastiche incastrate, *Rend. Acc. Lincei* **10** (1901), 197–205.
21. T. Boggio, Sulle funzioni di Green d'ordine m, *Rend. Circ. Mat. Palermo* **20** (1905), 97–135.
22. P. Bolle, On the Bolza problem, *J. Differential Equations* **152** (1999), 274–288.
23. H. Brezis, *Analyse Fonctionnelle. Théorie et Applications*, Masson, Paris, 1983.
24. H. Brezis, Degree theory: old and new, *Topological Nonlinear Analysis, II* (Frascati, 1995), 87–108, Progr. Nonlinear Differential Equations Appl., 27, Birkhäuser, Boston, 1997.
25. H. Brezis, Symmetry in Nonlinear PDE's, *Proceedings of Symposia in Pure Mathematics*, Amer. Math. Soc., 1999.
26. H. Brezis and S. Kamin, Sublinear elliptic equations in $\mathbb{R}^N$, *Manuscripta Math.* **74** (1992), 87–106.
27. H. Brezis and E.H. Lieb, A relation between pointwise convergence of functions and convergence of functionals, *Proc. Amer. Math. Soc* **88** (1983), 486–490.
28. H. Brezis and M. Marcus, Hardy's inequalities revisited, *Ann. Scuola Norm. Sup. Pisa Cl. Sci.* **25** (1997), 217–237.
29. H. Brezis and L. Nirenberg, Positive solutions of nonlinear elliptic equations involving critical Sobolev exponent, *Comm. Pure Appl. Math.* **36** (1983), 486–490.
30. H. Brezis and L. Oswald, Remarks on sublinear elliptic equations, *Nonlinear Anal., T.M.A.* **10** (1986), 55–64.
31. H. Brezis and J.L. Vázquez, Blow-up solutions of some nonlinear elliptic problems, *Rev. Mat. Univ. Complut. Madrid* **10** (1997), 443–469.
32. K.J. Brown and F.A. Davidson, Global bifurcation in the Brusselator system, *Nonlinear Anal.* **24** (1995), 1713–1725.
33. J. Busca and R. Manásevich, A Liouville-type theorem for Lane–Emden system, *Indiana Univ. Math. J.* **51** (2002), 37–51.
34. J. Byeon and Z.-Q. Wang, Standing waves with a critical frequency for nonlinear Schrödinger equations, *Calc. Var. Partial Differential Equations* **18** (2003), 207–219.
35. L. Caffarelli, R. Kohn, and L. Nirenberg, First order interpolation inequalities with weights, *Compositio Math.* **53** (1984), 259–275.
36. F. Catrina and Z.-Q. Wang, On the Caffarelli–Kohn–Nirenberg inequalities: sharp constants, existence (and nonexistence) and symmetry of extremal functions, *Comm. Pure Appl. Math.* **54** (2001), 229–258.
37. S. Chandrasekhar, *An Introduction to the Study of Stellar Structure*, Dover Publications, New York, 1967.
38. Y.-S. Choi and J.P. McKenna, A singular Gierer–Meinhardt system of elliptic equations, *Ann. Inst. H. Poincaré, Anal. Non Linéaire* **17** (2000), 503–522.
39. Y.-S. Choi and J.P. McKenna, A singular Gierer–Meinhardt system of elliptic equations: the classical case, *Nonlinear Anal.* **55** (2003), 521–541.
40. Ph. G. Ciarlet, *Introduction à l'Analyse Numérique Matricielle et à l'Optimisation*, Masson, Paris, 1988.
41. G.R. Cirmi and M.M. Porzio, L^1-solutions for some nonlinear degenerate elliptic and parabolic equations, *Ann. Mat. Pura Appl.* **(4) 169** (1995), 67–86.
42. F. Cîrstea and V. Rădulescu, Blow-up solutions for semilinear elliptic problems, *Nonlinear Analysis, T.M.A.* **48** (2002), 541–554.
43. F. Cîrstea and V. Rădulescu, Uniqueness of the blow-up boundary solution of logistic equations with absorption, *C. R. Acad. Sci. Paris, Ser. I* **335** (2002), 447–452.
44. F. Cîrstea and V. Rădulescu, Existence and uniqueness of blow-up solutions for a class of logistic equations, *Commun. Contemp. Math.* **4** (2002), 559–586.
45. F. Cîrstea and V. Rădulescu, Asymptotics for the blow-up boundary solution of the logistic equation with absorption, *C. R. Acad. Sci. Paris, Ser. I* **336** (2003), 231–236.
46. F. Cîrstea and V. Rădulescu, Extremal singular solutions for degenerate logistic-type equations in anisotropic media, *C. R. Acad. Sci. Paris, Ser. I* **339** (2004), 119–124.
47. F. Cîrstea and V. Rădulescu, Nonlinear problems with boundary blow-up: a Karamata regular variation theory approach, *Asymptotic Analysis* **46** (2006), 275–298.

48. F. Cîrstea and V. Rădulescu, Boundary blow-up in nonlinear elliptic equations of Bieberbach–Rademacher type, *Transactions Amer. Math. Soc.* **359** (2007), 3275–3286.
49. Ph. Clément, J. Fleckinger, E. Mitidieri, and F. de Thélin, Existence of positive solutions for a nonvariational quasilinear elliptic system, *J. Differential Equations* **166** (2000), 455–477.
50. Ph. Clément and G. Sweers, Existence and multiplicity results for a semilinear elliptic eigenvalue problem, *Ann. della Scola Norm. Sup. di Pisa* **14** (1987), 97–121.
51. Ph. Clément and G. Sweers, Getting a solution between sub and supersolutions without monotone iteration, *Rend. Istit. Mat. Univ. Trieste* **19** (1987), 189–194.
52. C.V. Coffman and R.J. Duffin, On the structure of biharmonic functions satisfying the clamped condition on a right angle, *Adv. Appl. Math.* **1** (1950), 373–389.
53. M. Coclite and G. Palmieri, On a singular nonlinear Dirichlet problem, *Commun. Partial Diff. Equations* **14** (1989), 1315–1327.
54. M.G. Crandall, P.H. Rabinowitz, and L. Tartar, On a Dirichlet problem with a singular nonlinearity, *Commun. Partial Diff. Equations* **2** (1977), 193–222.
55. A. Dall'Acqua and G. Sweers, Estimates for Green function and Poisson kernels of higher order Dirichlet boundary value problems, *J. Differential Equations* **205** (2004), 466–487.
56. R. Dalmasso, Solutions d'équations elliptiques semi-linéaires singulières, *Ann. Mat. Pura Appl.* **153** (1989), 191–201.
57. R. Dautray and J.-L. Lions, *Mathematical Analysis and Numerical Methods for Science and Technology, Vol. 1: Physical Origins and Classical Methods*, Springer-Verlag, Berlin Heidelberg, 1985.
58. E.B. Davies, A review of Hardy inequalities, in *The Maz′ya Anniversary Collection, Vol. 2 (Rostock, 1998)*, pp. 55–67, Oper. Theory Adv. Appl., 110, Birkhäuser, Basel, 1999.
59. J. Dávila, M. del Pino, and M. Musso, The supercritical Lane–Emden–Fowler equation in exterior domains, *Commun. Partial Differential Equations* **32** (2007), 1225–1243.
60. J. Dávila, M. del Pino, M. Musso, and J. Wei, Fast and slow decay solutions for supercritical elliptic problems in exterior domains, *Calc. Var. Partial Differential Equations* **32** (2008), 453–480.
61. J. Dávila and L. Dupaigne, Hardy-type inequalities, *J. Eur. Math. Soc.* **6** (2004), 335–365.
62. P.G. de Gennes, Wetting: statics and dynamics, *Review of Modern Physics* **57** (1985), 827–863.
63. D.G. de Figueiredo and B. Ruf, Elliptic systems with nonlinearities and arbitrary growth, *Mediterr. J. Math.* **1** (2004), 417–431.
64. Y. Du and Q. Huang, Blow-up solutions for a class of semilinear elliptic and parabolic equations, *SIAM J. Math. Anal.* **31** (1999), 1–18.
65. S. Dumont, L. Dupaigne, O. Goubet, and V. Rădulescu, Back to the Keller–Osserman condition for boundary blow-up solutions, *Advanced Nonlinear Studies* **7** (2007), 271–298.
66. L. Dupaigne, M. Ghergu, and V. Rădulescu, Lane–Emden–Fowler equations with convection and singular potential, *J. Math. Pures Appliquées* **87** (2007), 563–581.
67. I. Ekeland, Nonconvex minimization problems, *Bull. Amer. Math. Soc.* **1** (1979), 443–473.
68. V.R. Emden, *Gaskugeln, Anwendungen der mechanischen Warmetheorie auf kosmologische und meteorologische Probleme*, Teubner-Verlag, Leipzig, 1907.
69. T. Erneux and E. Reiss, Brusselator isolas, *SIAM J. Appl. Math.* **43** (1983), 1240–1246.
70. A. Farina, Some Liouville-type theorems for elliptic equations and their consequences, *Boll. Soc. Esp. Mat. Appl.* **45** (2008), 9–34.
71. E. Fabes, C. Kenig, and R. Serapioni, The local regularity of solutions of degenerate elliptic operators, *Comm. Partial Diff. Eqns.* **7** (1982), 77–116.
72. P. Felmer and D.G. de Figueiredo, A Liouville-type theorem for elliptic systems, *Ann. Sc. Norm. Super. Pisa* **21** (1994), 387–397.
73. R.J. Field and R.M. Noyes, Oscillations in Chemical Systems IV. Limit cycle behavior in a model of a real chemical reaction, *J. Chem. Phys.* **60** (1974), 1877–1884.
74. S. Filippas, V. Maz'ya, and A. Tertikas, Critical Hardy–Sobolev inequalities, *J. Math. Pures Appl.* **87** (2007), 37–56.
75. D.G. de Figueiredo and B. Sirakov, Liouville type theorems, monotonicity results and a priori bounds for positive solutions of elliptic systems, *Math. Ann.* **333** (2005), 231–260.

76. R. Filippucci, P. Pucci, and V. Rădulescu, Existence and nonexistence results for quasilinear elliptic exterior problems with nonlinear boundary conditions, *Comm. in Partial Differential Equations* **33** (2008), 706–717.
77. R.H. Fowler, Further studies of Emden's and similar differential equations, *Q. J. Math. (Oxford Series)* **2** (1931), 259–288.
78. B. Franchi, R. Serapioni, and F. Serra Cassano, Approximation and imbedding theorems for weighted Sobolev spaces associated to Lipschitz continuous vector fields, *Boll. Un. Mat. Ital. B* **11** (1977), 83–117.
79. A. Friedman and B. McLeod, Blow-up solutions of nonlinear parabolic equations, *Arch. Rat. Mech. Anal.* **96** (1986), 55–80.
80. V. Galaktionov and J.L. Vázquez, The problem of blow-up in nonlinear parabolic equations, *Discrete Contin. Dynam. Systems, Ser. A* **8** (2002), 399–433.
81. P.R. Garabedian, A partial differential equation arising in conformal mapping, *Pacific J. Math.* **1** (1951), 485–524.
82. J. García-Melián, R. Letelier-Albornoz, and J. Sabina de Lis, Uniqueness and asymptotic behaviour for solutions of semilinear problems with boundary blow-up, *Proc. Amer. Math. Soc.* **129** (2001), 3593–3602.
83. F. Gazzola, H-Ch. Grunau, and G. Sweers, *Polyharmonic Boundary Value Problems, A monograph on positivity preserving and nonlinear higher order elliptic equations in bounded domains*, Springer Lecture Notes in Mathematics, vol. 1991, 2010.
84. M. Ghergu, Nonconstant steady-state solutions for Brusselator type systems, *Nonlinearity* **21** (2008), 2331–2345.
85. M. Ghergu, Steady-state solutions for Gierer–Meinhardt type systems with Dirichlet boundary condition, *Trans. Amer. Math. Soc.* **361** (2009), 3953–3976.
86. M. Ghergu, Lane–Emden systems with negative exponents, *J. Funct. Anal.* **258** (2010), 3295–3318.
87. M. Ghergu, C. Niculescu, and V. Rădulescu, Explosive solutions of elliptic equations with absorption and nonlinear gradient term, *Proc. Indian Acad. Sci. (Math. Sci.)* **112** (2002), 441–451.
88. M. Ghergu and V. Rădulescu, Bifurcation and asymptotics for the Lane–Emden–Fowler equation, *C. R. Acad. Sci. Paris, Ser. I* **337** (2003), 259–264.
89. M. Ghergu and V. Rădulescu, Sublinear singular elliptic problems with two parameters, *J. Differential Equations* **195** (2003), 520–536.
90. M. Ghergu and V. Rădulescu, Existence and nonexistence of entire solutions to the logistic differential equation, *Abstract and Applied Analysis* **17** (2003), 995–1003.
91. M. Ghergu and V. Rădulescu, Nonradial blow-up solutions of sublinear elliptic equations with gradient term, *Commun. Pure Appl. Anal.* **3** (2004), 465–474.
92. M. Ghergu and V. Rădulescu, Bifurcation for a class of singular elliptic problems with quadratic convection term, *C. R. Acad. Sci. Paris, Ser. I* **338** (2004), 831–836.
93. M. Ghergu and V. Rădulescu, On a class of singular Gierer–Meinhardt systems arising in morphogenesis, *C. R. Acad. Sci. Paris, Ser. I* **344** (2007), 163–168.
94. M. Ghergu and V. Rădulescu, A singular Gierer–Meinhardt system with different source terms, *Proceedings of the Royal Society of Edinburgh: Section A (Mathematics)* **138A** (2008), 1215–1234.
95. M. Ghergu and V. Rădulescu, *Singular Elliptic Problems. Bifurcation and Asymptotic Analysis*, Oxford Lecture Series in Mathematics and Its Applications, vol. 37, Oxford University Press, New York, 2008.
96. M. Ghergu and V. Rădulescu, Turing patterns in general reaction-diffusion systems of Brusselator type, *Communications in Contemporary Mathematics* **12** (2010), 661–679.
97. B. Gidas, W-M. Ni, and L. Nirenberg, Symmetry and related properties via the maximum principle, *Comm. Math. Phys.* **68** (1979), 209–243.
98. A. Gierer and H. Meinhardt, A theory of biological pattern formation, *Kybernetik* **12** (1972), 30–39.
99. D. Gilbarg and N.S. Trudinger, *Elliptic Partial Differential Equations of Second Order*, 2nd edn, Springer Verlag, Berlin, 1983.

100. P. Gray and S.K. Scott, Autocatalytic reactions in the isothermal, continuous stirred tank reactor: isolas and other forms of multistability, *Chem. Eng. Sci.* **38** (1983), 29–43.
101. H.-Ch. Grunau and F. Robert, Positivity and almost positivity of biharmonic Green's functions under Dirichlet boundary conditions, *Arch. Rational Mech. Anal.* **195** (2010), 865–898.
102. H.-Ch. Grunau, F. Robert, and G. Sweers, Optimal estimates from below for biharmonic Green functions, *Proc. Amer. Math. Soc.* **139** (2011), 2151–2161.
103. H.-Ch. Grunau and G. Sweers, Positivity for perturbations of polyharmonic operators with Dirichlet boundary conditions in two dimensions, *Math. Nachr.* **179** (1996), 89–102.
104. H.-Ch. Grunau and G. Sweers, Positivity for equations involving polyharmonic operators with Dirichlet boundary conditions, *Math. Ann.* **307** (1997), 589–626.
105. H.-Ch. Grunau and G. Sweers, Sign change for the Green function and for the first eigenfunction of equations of clamped plate type, *Arch. Rational Mech. Anal.* **150** (1999), 179–190.
106. H.-Ch. Grunau and G. Sweers, Regions of positivity for polyharmonic Green functions in arbitrary domains, *Proc. Amer. Math. Soc.* **135** (2007), 3537–3546.
107. C. Gui and F. H. Lin, Regularity of an elliptic problem with a singular nonlinearity, *Proc. Royal Soc. Edinburgh Sect. A* **123** (1993), 1021–1029.
108. J. Hadamard, *Mémoire sur le problème d'analyse relatif à l'équilibre des plaques élastiques encastrées*, in: Œuvres de Jacques Hadamard, Tome II, 515–641, CNRS Paris (1968).
109. J. Hadamard, *Sur certains cas intéressants du problème biharmonique*, in: Œuvres de Jacques Hadamard, Tome III, pp. 1297–1299. CNRS Paris (1968). Reprint of: Atti IV Congr. Intern. Mat. Rome 12–14, 1908.
110. D. Henry, *Geometric Theory of Semilinear Parabolic Equations*, Lectures Notes in Mathematics No. 840, Springer-Verlag, 1993.
111. L. Hörmander, *The Analysis of Linear Partial Differential Operators I*, Springer Verlag, Berlin, 1983.
112. H. Jiang, Global existence of solutions of an activator-inhibitor system, *Discrete Contin. Dyn. Syst.* **14** (2006), 737–751.
113. H. Kang, Dynamics of local map of a discrete Brusselator model: eventually trapping regions and strange attractos, *Discrete Contin. Dyn. Syst.* **20** (2008), 939–959.
114. J.P. Keener, Activators and inhibitors in pattern formation, *Stud. Appl. Math.* **59** (1978), 1–23.
115. J.B. Keller, On solutions of $\Delta u = f(u)$, *Comm. Pure Appl. Math.* **10** (1957), 503–510.
116. E.H. Kim, A class of singular Gierer–Meinhardt systems of elliptic boundary value problems, *Nonlinear Anal.* **59** (2004), 305–318.
117. E.H. Kim, Singular Gierer–Meinhardt systems of elliptic boundary value problems, *J. Math. Anal. Appl.* **308** (2005), 1–10.
118. D. Kinderlehrer and G. Stampacchia, *An Introduction to Variational Inequalities and Their Applications*, Academic Press, New York, 1980.
119. S. Kishenassamy, Recent progress in boundary blow up, in *Elliptic and Parabolic Problems: A Special Tribute to the Work of Haim Brezis* (C. Bandle et al., eds.), Progress in Nonlinear Differential Equations and Their Applications, vol. 63, Birkhäuser, 2005, pp. 289–301.
120. V. Kondratiev, V. Liskevich and Z. Sobol, Second-order semilinear elliptic inequalities in exterior domains, *J. Differential Equations* **187** (2003), 429–455.
121. V. Kondratiev, V. Liskevich and Z. Sobol, Positive solutions to semi-linear and quasi-linear second-order elliptic equations on unbounded domains, *Handbook of Differential Equations*, vol. VI (2008), 177–267.
122. J.P. Krasovskiĭ, Isolation of singularities of the Green function (Russian), *Izv. Akad. Nauk SSSR Ser. Mat.* **31** (1967), 977–1010, English translation in: Math. USSR, Izv. **1** (1967), 935–966.
123. A. Kristály, V. Rădulescu, and Cs. Varga, *Variational Principles in Mathmatical Physics, Geometry, and Economics: Qualitative Analysis of Nonlinear Equations and Unilateral Problems*, Encyclopedia of Mathematics and its Applications, No. 136, Cambridge University Press, Cambridge, 2010.
124. T. Kolokolnikov, T. Erneux, and J. Wei, Mesa-type patterns in the one-dimensional Brusselator and their stability, *Phys. D* **214** (2006), 63–77.

125. T. Kusano and C.A. Swanson, Entire positive solutions of singular elliptic equations, *Japan J. Math.* **11** (1985), 145–155.
126. A.V. Lair and A.W. Wood, Large solutions of semilinear elliptic equations with nonlinear gradient terms, *Internat. J. Math. Math. Sci.* **22** (1999), 869–883.
127. S. Lancelotti, A. Musesti, and M. Squassina, Infinitely many solutions for polyharmonic elliptic problems with broken symmetries, *Math. Nachr.* **253** (2003), 35–44.
128. J.H. Lane, On the theoretical temperature of the sun under hypothesis of a gaseous mass maintaining its volume by its internal heat and depending on the laws of gases known to terrestrial experiment, *Amer. J. Sci.* **50** (1869), 57–74.
129. J.M. Lasry and P.-L. Lions, Nonlinear elliptic equations with singular boundary conditions and stochastic control with state constraints; the model problem, *Math. Ann.* **283** (1989), 583–630.
130. A.C. Lazer and P.J. McKenna, On a singular nonlinear elliptic boundary value problem, *Proc. Amer. Math. Soc.* **3** (1991), 720–730.
131. A.C. Lazer and P.J. McKenna, On a problem of Bieberbach and Rademacher, *Nonlinear Anal., T.M.A.* **21** (1993), 327–335.
132. J.-F. Le Gall, A path-valued Markov process and its connections with partial differential equations, in *First European Congress of Mathematics*, Vol. II (Paris, 1992), 185–212, Progr. Math., 120, Birkhäuser Verlag, Basel, 1994.
133. I. Lengyel and I.R. Epstein, Modeling of Turing structures in the chlorite–iodide–malonic acid-starch reaction system, *Science* **251** (1991), 650–652.
134. C. Li, Local asymptotic symmetry of singular solutions to nonlinear elliptic equations, *Invent. Math.* **123** (1996), 221–231.
135. J.-L. Lions and E. Magenes, *Non-homogenoeus boundary value problems and applications*, Vol. I and II, Springer-Verlag, Berlin, 1972.
136. V. Liskevich, S. Lyakhova, and V. Moroz, Positive solutions to nonlinear p-Laplace equations with Hardy potential in exterior domains, *J. Differential Equations* **232** (2007), 212–252.
137. J. Karamata, Sur un mode de croissance régulière de fonctions. Théorèmes fondamentaux, *Bull. Soc. Math. France* **61** (1933), 55–62.
138. C. Loewner and L. Nirenberg, Partial differential equations invariant under conformal or projective transformations, in *Contribution to Analysis*, Academic Press, New York, 1974, 245–272.
139. Y. Lou and W.-M. Ni, Diffusion, self-diffusion and cross-diffusion, *J. Differential Equations* **131** (1996), 79–131.
140. Y. Lou and W.-M. Ni, Diffusion vs. cross-diffusion: an elliptic approach, *J. Differential Equations* **154** (1999), 157–190.
141. M. Marcus, On solutions with blow-up at the boundary for a class of semilinear elliptic equations, in *Developments in Partial Differential Equations and Applications to Mathmatical Physics* (G. Buttazzo *et al.*, Eds.), Plenum Press, New York (1992), 65–77.
142. M. Marcus and L. Véron, Uniqueness and asymptotic behavior of solutions with boundary blow-up for a class of nonlinear elliptic equations, *Ann. Inst. H. Poincaré, Anal. Non Linéaire* **14** (1997), 237–274.
143. M. Marcus and L. Véron, Existence and uniqueness results for large solutions of general nonlinear elliptic equations, *J. Evol. Equations* **3** (2003), 637–652.
144. P. Mironescu and V. Rădulescu, The study of a bifurcation problem associated to an asymptotically linear function, *Nonlinear Anal., T.M.A.* **26** (1996), 857–875.
145. M.K.V. Murthy and G. Stampacchia, Boundary value problems for some degenerate elliptic operators, *Ann. Mat. Pura Appl.* **80** (1968), 1–122.
146. M. Naito and H. Usami, Existence of nonoscillatory solutions to second-order elliptic systems of Emden-Fowler type, *Indiana Univ. Math. J.* **55** (2006), 317–340.
147. A.A. Neville, P.C. Matthews, and H.M. Byrne, Interactions between pattern formation and domain growth, *Bull. Math. Biol.* **68** (2006), 1975–2003.
148. W.-M. Ni, Diffusion, cross-diffusion, and their spike-layer steady states, *Notices of Amer. Math. Soc.* **45** (1998), 9–18.

149. W.-M. Ni, Diffusion and cross-diffusion in pattern formation, *Atti Accad. Naz. Lincei Cl. Sci. Fis. Mat. Natur. Rend. Lincei Mat. Appl.* **15** (2004), 197–214.
150. W.-M. Ni, K. Suzuki, and I. Takagi, The dynamics of a kinetics activator-inhibitor system, *J. Differential Equations* **229** (2006), 426–465.
151. W.-M. Ni and I. Takagi, On the shape of least-energy solutions to a semilinear Neumann problem, *Comm. Pure Appl. Math.* **44** (1991), 819–851.
152. W.-M. Ni and I. Takagi, Locating the peaks of least-energy solutions to a semilinear Neumann problem, *Duke Math. J.* **70** (1993), 247–281.
153. W.-M. Ni and J. Wei, On positive solutions concentrating on spheres for the Gierer–Meinhardt system, *J. Differential Equations* **221** (2006), 158–189.
154. L. Nirenberg, *Topics in Nonlinear Functional Analysis*, 2nd edn, Courant Institute, 2001.
155. R. Osserman, On the inequality $\Delta u \geq f(u)$, *Pacific J. Math.* **7** (1957), 1641–1647.
156. D. Passaseo, Some concentration phenomena in degenerate semilinear elliptic problems, *Nonlinear Anal.* **24** (1995), 1011–1025.
157. B. Pena and C. Perez-Garcia, Stability of Turing patterns in the Brusselator model, *Phys. Rev. E* **64** (2001), 056213, 9 pp.
158. R. Peng, J. Shi, and M. Wang, On stationary patterns of a reaction-diffusion model with autocatalysis and saturation law, *Nonlinearity* **21** (2008), 1471–1488.
159. R. Peng and M. Wang, Pattern formation in the Brusselator system, *J. Math. Anal. Appl.* **309** (2005), 151–166.
160. A. Poretta and L. Véron, Symmetry of large solutions of nonlinear elliptic equations in a ball, *J. Funct. Anal.* **236** (2006), 581–591.
161. I. Prigogine and R. Lefever, Symmetry breaking instabilities in dissipative systems II, *J. Chem. Phys.* **48** (1968), 1665–1700.
162. M.H. Protter and H.F. Weinberger, *Maximum Principles in Differential Equations*, Prentice Hall, Englewood Cliffs NJ, 1967.
163. P. Pucci and J. Serrin, Remarks on the first eigenspace for polyharmonic operators, *Atti Sem. Mat. Fis. Univ. Modena* **36** (1988), 107–117; and Proc. 1986–87 *Focused Research Program on Spectral Theory and Boundary Value Problems*, Argonne National Laboratory, Report ANL-87-26, **3** (1989), 135–145.
164. P. Pucci and J. Serrin, A note on the strong maximum principle for elliptic differential inequalities, *J. Math. Pures Appl.* **79** (2000), 57–71.
165. P. Pucci, J. Serrin, and H. Zou, A strong maximum principle and a compact support principle for singular elliptic inequalities, *J. Math. Pures Appl.* **78** (1999), 769–789.
166. P. Quittner and Ph. Souplet, A priori estimates and existence for elliptic systems via bootstrap in weighted Lebesgue spaces, *Arch. Ration. Mech. Anal.* **174** (2004), 49–81.
167. P. Quittner, Blow-up for semilinear parabolic equations with a gradient term, *Math. Meth. Appl. Sci.* **14** (1991), 413–417.
168. P. Rabinowitz, Variational methods for nonlinear elliptic eigenvalue problems, *Indiana Univ. Math. J.* **23** (1974), 729–745.
169. H. Rademacher, Einige besondere Probleme der partiellen Differentialgleichungen, in *Die Differential und Integralgleichungen der Mechanik und Physik I*, 2nd edn, (P. Frank und R. von Mises, eds.), Rosenberg, New York, 1943, pp. 838–845.
170. V. Rădulescu, D. Smets, and M. Willem, Hardy–Sobolev inequalities with remainder terms, *Topol. Meth. Nonlinear Anal.* **20** (2002), 145–149.
171. W. Reichel and H. Zou, Non-existence results for semilinear cooperative elliptic systems via moving spheres, *J. Differential Equations* **161** (2000), 219–243.
172. F. Rothe, *Global Solutions in Reaction-Diffusion Systems*, Lecture Notes in Mathematics, Springer-Verlag, 1983.
173. G.V. Rozenbljum, Distribution of the discrete spectrum of singular differential operators, *Dokl. Acad. Nauk SSSR* **202** (1972), 1012–1015; *Soviet Math. Dokl.* **13** (1972), 245–249.
174. J. Schnakenberg, Simple chemical reaction system with limit cycle behaviour, *J. Theor. Biol.* **81** (1979), 389–400.
175. E.E. Sel'kov, Self-oscillations in glycolysis, *Eur. J. Biochem.* **4** (1968), 79–86.

176. E. Seneta, *Regularly Varying Functions*, Lecture Notes in Mathematics 508, Springer Verlag, Berlin Heidelberg, 1976.
177. J. Serrin and H. Zou, Existence of positive solutions of the Lane–Emden system, *Atti del Sem. Mat. Univ. Modena* XLVI (1998), 369–380.
178. J. Serrin and H. Zou, Non-existence of positive solutions of Lane–Emden systems, *Differential Integral Equations* **9** (1996), 635–653.
179. H.S. Shapiro and M. Tegmark, An elementary proof that the biharmonic Green function of an eccentric ellipse changes sign, *SIAM Rev.* **36** (1994), 99–101.
180. J. Shi and M. Yao, On a singular nonlinear semilinear elliptic problem, *Proc. Royal Soc. Edinburgh, Sect. A* **128** (1998), 1389–1401.
181. B. Sirakov, Standing wave solutions of the nonlinear Schrödinger equation in $\mathbb{R}^N$, *Ann. Mat. Pura Appl.* **(4) 181** (2002), 73–83.
182. J. Smoller, *Shock-Waves and Reaction-Diffusion Equations*, 2nd edn, Springer-Verlag, 1994.
183. Ph. Souplet, The proof of the Lane–Emden conjecture in four space dimensions, *Adv. Math.* **221** (2009), 1409–1427.
184. L.E. Stephenson and D.J. Wollkind, Weakly nonlinear stability analyses of one-dimensional Turing pattern formation in activator-inhibitor/immobilizer model systems, *J. Math. Biol.* **33** (1995), 771–815.
185. W. Strauss, Existence of solitary waves in higher dimensions, *Comm. Math. Phys.* **55** (1977), 149–162.
186. E. W. Stredulinsky, *Weighted Inequalities and Degenerate Elliptic Partial Differential Equations*, Springer-Verlag, Berlin, New York, 1984.
187. M. Struwe, Infinitely many critical points for functionals which are not even and applications to superlinear boundary value problems, *Manuscripta Math.* **32** (1980), 335–364.
188. M. Struwe, *Variational Methods, Applications to Nonlinear Partial Differential Equations and Hamiltonian Systems*, 4th edn, Ergebnisse der Mathematik und ihrer Grenzgebiete, Springer Verlag, Berlin, 2008.
189. S. Taliaferro, A nonlinear singular boundary value problem, *Nonlinear Anal., T.M.A.* **3** (1979), 897–904.
190. S. Taliaferro, On the growth of superharmonic functions near an isolated singularity I, *J. Differential Equations* **158** (1999), 28–47
191. S. Taliaferro, On the growth of superharmonic functions near an isolated singularity II, *Commun. Partial Differential Equations* **26** (2001), 1003–1026.
192. S. Taliaferro, Isolated singularities of nonlinear elliptic inequalities, *Indiana Univ. Math. J.* **50** (2001), 1885–1897.
193. S. Taliaferro, Local behavior and global existence of positive solutions of $au^\lambda \le -\Delta u \le u^\lambda$, *Ann. Inst. H. Poincaré Anal. Non Linéaire* **19** (2002), 889–901.
194. S. Taliaferro, Isolated singularities of nonlinear elliptic inequalities II. Asymptotic behavior of solutions, *Indiana Univ. Math. J.* **55** (2006), 1791–1812.
195. K. Tanaka, Morse indices and critical points related to the symmetric mountain pass theorem and applications, *Commun. Partial Diff. Equations* **14** (1989), 99–128.
196. A. Trembley, Mémoires pour servir à l'histoire d'un genre de polypes d'eau douce, à bras en forme de cornes. Leiden, Netherlands, 1744.
197. A. Turing, The chemical basis of morphogenesis, *Philos. Trans. Roy. Soc. B* **237** (1952), 37–72.
198. E.H. Twizell, A.B. Gumel, and Q. Cao, A second-order scheme for the Brusselator reaction-diffusion system, *J. Math. Chem.* **26** (1999), 297–316.
199. L. Véron, *Singularities of Solutions of Second Order Quasilinear Equations*, Pitman Res. Notes Math. Ser., Vol. 353, Longman, Harlow, 1996.
200. X. Wang, On concentration of positive bound states of nonlinear Schrödinger equations, *Comm. Math. Phys.* **153** (1993), 229–244.
201. Z.-Q. Wang and M. Willem, Singular minimization problems, *J. Differential Equations* **161** (2000), 317–320.
202. M.J. Ward and J. Wei, The existence and stability of asymmetric spike patterns for the Schnakenberg model, *Stud. Appl. Math.* **109** (2002), 229–264.

203. M.J. Ward and J. Wei, Stationary multiple spots for reaction-diffusion systems, *J. Math. Biol.* **57** (2008), 53–89.
204. J. Wei and M. Winter, Spikes for the Gierer–Meinhardt system in two dimensions: the strong coupling case, *J. Differential Equations* **178** (2002), 478–518.
205. J. Wei and M. Winter, Existence and stability analysis of asymmetric patterns for the Gierer–Meinhardt system, *J. Math. Pures Appl.* **83** (2004), 433-476.
206. Z. Wei, Positive solutions of singular sublinear second order boundary value problems, *Systems Sci. Math. Sci.* **11** (1998), 82–88.
207. M. Wiegner, A degenerate diffusion equation with a nonlinear source term, *Nonlinear Anal. T.M.A.*, 12 (1997), 1977–1995.
208. M. Winkler, On the Cauchy problem for a degenerate parabolic equation, *Z. Anal. Anwendungen*, 20 (2001), 677–690.
209. M. Winkler, A critical exponent in a degenerate parabolic equation, *Math. Methods Appl. Sci.*, 25 (2002), 911–925.
210. M. Winkler, Blow-up of solutions to a degenerate parabolic equation not in divergence form, *J. Differential Equations*, 192 (2003), 445–474.
211. M. Winkler, Blow-up in a degenerate parabolic equation, *Indiana Univ. Math. J.*, 53 (2004), 1414–1442.
212. M. Winkler, Nontrivial ordered ω-limit sets in a linear degenerate parabolic equation, *Discrete Continuous Dynam. Systems*, 17 (2007), 739–750.
213. J.S.W. Wong, On the generalized Emden-Fowler equation, *SIAM Rev.* **17** (1975), 339–360.
214. Z. Zhang and J. Cheng, Existence and optimal estimates of solutions for singular nonlinear Dirichlet problems, *Nonlinear Anal.* **57** (2004), 473–484.
215. N.S. Zheng, Weakly invariant regions for reaction-diffusion systems and applications, *Proc. Royal Soc. Edinburgh Sect. A (Mathematics)* **130** (2000), 1165–1180.
216. H. Zou, A priori estimates for a semilinear elliptic system without variational structure and their applications, *Math. Ann.* **323** (2002), 713–735.
217. Y. You, Global dynamics of the Brusselator equations, *Dyn. Partial Diff. Eqns.* **4** (2007), 167–196.

Index

M. Ghergu and V. Rădulescu, *Nonlinear PDEs*, Springer Monographs in Mathematics,
DOI 10.1007/978-3-642-22664-9, © Springer-Verlag Berlin Heidelberg 2012